The Apoplast of higher plants: Compartment of Storage, Transport and Reactions

The Apoplast of Higher Plants: Compartment of Storage, Transport and Reactions

The significance of the apoplast for the mineral nutrition of higher plants

edited by

Burkhard Sattelmacher† and Walter J. Horst
University of Hannover, Germany

Springer

A C.I.P. Catalogue record for this book is available from the Library of Congress.

ISBN 978-1-4020-5842-4 (HB)
ISBN 978-1-4020-5843-1 (e-book)

Published by Springer,
P.O. Box 17, 3300 AA Dordrecht, The Netherlands.

www.springer.com

Printed on acid-free paper

IN MEMORY OF BURKHARD SATTELMACHER

Burkhard Sattelmacher was an internationally highly estimated scientist in the area of plant mineral nutrition. He contributed substantially to the scientific excellence of Plant Nutrition especially through his engagement within the German Research foundation particularly through the initiation and contribution to coordinated research programmes and as a member of the International Council for Plant Nutrition. He was a stimulating teacher, mentor, and colleague. He found research in plant nutrition fascinating, and was able to transmit that fascination to those around him. He died in November 2005 at the age of 58 after many months of courageous fighting against his disease.

Born in Kiel he studied Botany at the Technical University of Berlin. He got his PhD in Plant Nutrition at the same University under the guidance of Horst Marschner. Deeply concerned about poverty alleviation through plant-production research he continued his work on the physiology of potato for 4 years as a post doc at the International Potato Center (CIP), Lima, Peru. This and follow-up research in Hohenheim represented the basis for his habilitation at the University of Hohenheim in 1986. In 1985 he accepted the call as professor for Plant Nutrition in Kiel. Since 1992 he was head and chairholder of Plant Nutrition at the Institute of Plant Nutrition and Soil Science, Faculty of Agricultural and Nutritional Sciences, University of Kiel.

In the centre of the scientific interest of Burkhard Sattelmacher was the physiology of crops. He was convinced that its basic understanding is a prerequisite for solving practical problems related to crop management. In the early nineties Burkhard Sattelmacher developed a research area on nutrient fluxes in agricultural land-use systems comparing conventional and "biological" plant-production systems. Over 9 years he participated in a German Research Foundation (DFG)-funded Special Research Project with research projects on root turn-over, N uptake particularly from manure, ammonia and dinitrogen-oxide emission in a winter rape-seed winter-barley rotation. He extended his interest to the nutrient budgets of natural ecosystems in the ecosystem research programme Bornhöveder Seenkette.

Among the research projects he initiated during the last years were particularly two to which he devoted his full force until the last days of his life: the DFG Special Research Programme "The apoplast of higher plants: compartment of storage, transport, and reactions" and the DFG Research Group "Matter fluxes in grasslands of Inner Mongolia as influenced by stocking rate".

Burkhard Sattelmacher always maintained an interest in new developments in agronomy, botany, and soil science. He especially enjoyed discussing ideas with colleagues and students. He was highly estimated as a referee for scientific journals, as well as for funding agencies not only because of his wide knowledge and experience, but because his interest was in the progress of science, without personal bias.

We have lost in Burkhard Sattelmacher an extraordinary person, teacher, scientist, and colleague. We will miss his stimulating contributions to scientific progress. The co-editor and the authors dedicate this book summarising the main achievement of the special research programme "Apoplast" which he initiated and of which he was the speaker. Unfortunately, he did not have the pleasure to finish this book himself.

Walter. Horst

CONTENTS

PREFACE

It was the botanist Ernst Münch, who separated the plant into two principal compartments, the "dead" apoplast and the living symplast. While Münch thought, that water and solute transport were the sole functions of this new plant compartment, we know today that apoplastic functions are much more diverse. It has been suggested to consider "the apoplast as the internal physiological environment of plant bodies", that essentially maintains homeostasis. The response to phytohormones such as auxins or pathogen attack may illustrate that in many cases environmental stimuli are not received directly by the cell, but perceived via changes within the plant's internal environment.

It is not before the last 20 years, that the cell wall attracted the interest of a broader group of plant scientists. It soon became evident, that the term cell wall may be misleading, since it is not appropriate for a highly complex and flexible matrix consisting of cellulose, hemicellulose, pectins and proteins interacting with metabolites. By now we know that the chemical and physical properties of cell walls are not static but depend on a number of parameters including ontogeny and environmental parameters such as temperature, light, nutrient supply, and biotic and abiotic stresses. This is why it was suggested to replace the term "cell wall" by the more precise term "extracellular matrix". The more we learned about the extracellular matrix, the more it became apparent that only few processes during growth and development of a plant do not involve cell walls.

From the viewpoint of plant mineral nutrition, the apoplast is of interest in many respects: nutrients do not simply pass through the apoplast before being taken up into the symplast, but they may also be adsorbed to cell-wall components, complexed, and oxidized/reduced which may be of significance for nutrient acquisition, nutrient function and tolerance of deficiency and toxicity stresses. Also the regulation of long-distance ion transport in the apoplast, the xylem, is not understood. However, this process is of great significance for the understanding of deficiency and toxicity symptoms.

The book summarizes the experimental work conducted during a trans-disciplinary research programme funded for 6 years by the German Research Foundation (DFG) within the Priority Research Project SPP 717. This financial support is highly acknowledged.

In their contributions, the authors from different disciplines not only report original research but also review the state of knowledge in their particular research fields: nutrient acquisition, short and long distance (xylem) transport, tolerance of nutrient deficiencies and mineral toxicities, and the role of micro-organisms colonizing the apoplast. Introductory remarks are written to each of the sections by internationally highly recognized scientists in their research areas.

We hope that this book will contribute to stimulate further research leading to a better understanding of the role of the apoplast in plant mineral nutrition.

Burkhard Sattelmacher† Walter J. Horst

Foreword

The Plant–Leaf Apoplast

THE PLANT–LEAF APOPLAST

D.T. CLARKSON
Biological Sciences, University of Exeter, UK, clarkson@gallini10.freeserve.co.uk

Abstract. The general properties of the extracellular matrix (the cell wall) in roots and leaves are compared and it is suggested that there may be several components of the apoplast in either tissue system where barriers such as the endodermis and bundle sheath are present. There can be very large amounts of xylem fluid delivered to leaves, all of which must pass through the apoplast at some point. The apoplast is a relatively small volume and processes such as ion uptake into the cells, phloem export and sequestration in the matrix are important in preventing a rapid build-up of extracellular solutes. The techniques developed to study such matters are briefly described. The proposition that the apoplast can serve as a store for nutrients is examined. Invasion of the apoplast by endophytic organisms and by environmental pollutants and toxic metals provokes an intriguing cascade of responses, some of which increase the resistance of the plant to biotic and abiotic stress.

1. INTRODUCTION

The extra-cellular matrix of the walls around most living cells is porous, pores being waterfilled in all but very exceptional circumstances. Between the cells of the root cortex leaf mesophyll there are much larger spaces. In leaves, carbon dioxide, oxygen and water vapour diffuse through these spaces, the former gases dissolving in the water film around the cells while the latter evaporates from the surfaces. It is possible to give a misleading emphasis to the differences between the root and leaf apoplasts due to these prominent spaces. In actuality, the water layers associated with cell surfaces and the porosity, chemical composition and electrostatic properties of the walls are of a similar order in both tissue systems. The cardinal difference is that water in the apoplast of the root cortex is usually continuous with the bulk water in the external medium while in leaves it is not. Continuity of water allows diffusion of solutes into and out from a root, even when the root has a barrier resistance created by an exodermis (Hose et al., 2001).

B. Sattelmacher† and W.J. Horst (eds.), The Apoplast of Higher Plants: Compartment of Storage, Transport and Reactions, 3–12.

Thus, materials that accumulate, for one reason or another, in the root apoplast can be unloaded directly into the external medium. This cannot occur in leaves unless a substance is volatile. Discharging other sorts of materials from leaves can occur only if they can reach, and be loaded into the phloem. For some very important inorganic ions, e.g. Ca^{2+} and H^{+}, this does not occur to any significant extent.

In roots there are at least two apoplast components that are separated by the endodermis. Thus there is a cortical apoplast and a stelar apoplast; the xylem contents may be regarded as part of the latter system. There is an analogous structure, the bundle sheath, separating the extracellular space within the leaf veins and the apoplast of the mesophyll. This structure is present in many species and is by no means restricted to the C4 species in which it has a special significance is separating stages of carbon fixation in photosynthesis. In leaves, therefore, we may find that there are two quite separate apoplast compartments. This has relevance in the discussion of methodology of apoplast compositional analysis.

The above, fairly obvious, points bear on some of the principal difficulties that readers will encounter in the sections that follow in this book. In pre-empting that discussion, we might ask ourselves how the conditions can be regulated in such a highly ramified, but small space can be measured reliably, how those conditions can be regulated and how the apoplast might serve as a storage compartment?

2. SOME PHYSICAL FACTS

Using several averaging, extractive techniques (see below) estimates of the volume of the apoplast in leaves are in the range of 4–11% of the leaf fresh weight. Thus, in a leaf weighing 1 g FW the expected apoplast volume might be 40–110 μl. Much of the water in this volume is present in pores present in the interstices of interwoven cellulose micro fibrils of the wall, or more properly, the extracellular matrix. These pores have been estimated to be in the order of 3–5 nm diameter (Gogarten, 1988; Sattelmacher, 2001). The porosity of the wall can be imagined in both the vertical plane, in which case the pores will usually be no more than 1–3 μm in length or in the horizontal plane in which case they can be enormously longer, and highly tortuous (Clarkson, 1991). Diffusion of solutes in this horizontal plane will be greatly hindered because of chemical interactions with the pectins of the middle lamella and by the great length of the pores. These considerations suggest that it may be possible for considerable gradients of solute concentration to build up at tissue sites where there may be rapid efflux of materials from the cells.

Close to the surface of the plasma membrane there are water layers whose exact physical state is hard to measure. Water molecules closest to the plasma membrane will be bound to the surface and have a significantly greater viscosity than bulk water, resembling liquid ice. Water molecules can be exchanged readily between this layer and the "free" water beyond it. As long as the rate constant for this exchange is faster than that for water movement across the plasma membrane, the bound water layer will be kinetically invisible. For solutes, however, diffusion across the bound water layer(s) involves movement through a layer of greater viscosity than in "free" water and solute transfer across the layer will be slowed down.

Rates of xylem delivery of water and solutes to leaves during the day can be large relative to the volume of the apoplast. Consider that it would not be uncommon for the mass of water lost by evapo-transpiration during the day to be five times greater than the fresh weight of a leaf. At some point, all of this water must move through the leaf apoplast. In a leaf with FW 1g, the apoplast volume might be as little 40 μl. In an hour during the photoperiod, as much as 500 μl may pass through this compartment. This is an equivalent to a complete volume change every 4.8 min. Since the xylem sap may contain major solutes, such as K^+, Ca^{2+} and NO_3^- in millimolar concentrations, they must either enter the cells quickly or be translocated out of the leaf in the phloem to prevent a massive build-up of electrolytes in the apoplast. In the special case of leaves receiving a heavy input of salt because they are in a saline medium, this may occur so that the water potential between the apoplast and the cells will be decreased (Flowers et al., 1991).

3. MEASUREMENTS OF APOPLAST COMPOSITION

The foregoing description of some of the physical features of the apoplast indicates two things. First that a method that produces a general composition of the fluid that can be extracted from the apoplast may obscure important gradients of composition between sites in the leaf where metabolism and transport activity are more or less intense (see Canny, 1990). Second, that there is a risk that physical intervention into such thin, tortuous layers may report the composition where the apoplast has been disrupted by a probe or electrode. In the sections that follow we will see that either of the above approaches can give consistent results that can be useful in interpreting the relationship between the extra- and intra-cellular environment in leaves. For obvious reasons though, they do not give the complete picture and, at worst, may lead us to heresy in which part of the truth is mistaken for the whole truth.

The earliest attempts to analyse apoplast fluids made use of centrifugation or displacement (e.g. Bernstein, 1971). Another approach to estimate the free space composition of leaves made use of tracer wash-out techniques that were the exact counterparts of those employed in transport kinetics of root segments. Leaves, preloaded with radioactive tracers, were cut into fine strips and suspended in a non-radioactive bathing medium that was sampled and changed at intervals (Pitman et al., 1974). This allowed the kinetics of the tracer exchange to be measured and the ionic content of various pools, or phases to be estimated. These techniques showed that the general exchange properties of leaf cells, in these conditions, were quite similar to those of roots.

In the sections that follow, the reader will find several examples of methods for estimating the composition of the solution in the apoplast. Centrifugation/displacement techniques are evaluated in the paper by Lohaus (this volume, pp. 323–336) and in exemplary studies published earlier (Dannel et al., 1995; Lohaus et al., 2001). In addition there is direct intervention of micro-electrode probes to measure ion activities (Felle and Hanstein, this volume, pp. 295–306), and ion-specific fluorescent probes that can be targeted, in transgenic plants, to the apoplast (Plieth et al., this volume, pp. 373–392).

Evidently, when a detached leaf is centrifuged, with its petiole parallel to the centrifugal force, a number of fractions of different composition may be gathered (e.g. Dannel et al., 1995). The first will comprise the contents of the xylem vessels, followed by the solution obtained from the apoplast around the leaf veins (or within the bundle sheath, where this is present). Later fractions will contain the solution migrating from the walls. In the paper of Lohaus (this volume, pp. 323–336) this fractionation is described and certain markers enzymes, that indicate actual damage to the cells, are assayed (Dannel et al., 1995). Clearly, evidence of damage invalidates the whole approach.

Ion selective micro electrodes, with several barrels that can measure two or three ion activities simultaneously, can be introduced into leaves via the open stomata (Felle et al., 2000). This allows the electrode tip to make contact with mesophyll cells with the minimum of cellular disruption so that is less risk that ion leakage from damaged cells can contaminate the apoplast fluids. In the paper by Felle and Hanstein (this volume, pp. 295–306), it is shown that reproducible data can be readily obtained from a number of species. The electrodes can be left in position while conditions affecting the leaf are changed. Thus continuous recording of changes over several hours can be obtained. The results are, however, from a very small, specific site in a widely dispersed compartment. Unless a great deal of sampling is done, it is hard to see how significant gradients of composition in the apoplast can be detected. The physiological significance of this approach would be greatly increased if leaves if intact plants could be attached to measuring cuvettes.

Plieth et al. (this volume, pp. 373–392) developed the use of ion-specific fluorescent indicators in a very promising direction. Various indicator dyes have been available for some time and assessments have been made of leaf apoplast pH (Hoffmann and Kosegarten, 1995), K^+ (Mühling and Sattelmacher, 1997), Ca^{2+} (Mühling et al., 1997) and Na^+ (Mühling and Läuchli, 2002). The technique can reveal micro-variation in the activity of an ion where it can be combined with *in vivo* ratio-imaging techniques. The problem about the exact points of origin of the fluorescent signal can now be tackled by combining a newer generation of fluorescent proteins, e.g. aequorins and green fluorescent protein, with targetting sequences and expressing the genes that encode them in transgenic plants. The potential of this exciting approach is demonstrated by Plieth et al. (this volume, pp. 373–392). The approach should be able to reveal some of the subtlety of the dynamics of transfers between the apoplast and the cell interiors. It depends, however, on the availability of transformable species and, ideally, the optical sectioning of intact leaves. For the time being, these two constraints limit the widespread application of the technique.

4. STORAGE IN THE APOPLAST?

In the sections that follow, the significance of accumulations of silicon (Wiese et al., this volume, pp. 33–48), boron (Wimmer and Goldbach, this volume, pp. 19–32), manganese (Fecht-Christoffers et al., this volume, pp. 307–322), iron (Nikolic and Römheld, this volume, pp. 353–372) and calcium (Felle and Hanstein, this volume, pp. 295–306, Plieth et al., this volume, pp. 373–392) in the apoplast is discussed. At the risk of seeming over-concerned with semantics, it might be said that there is more to a "store" than simply the presence of a given material. In the case of B, for example, the rapidity, with which B-deficiency can occur in growing tissues of plants previously well supplied with B, suggests that the large proportion of plant B harboured in the apoplast is not a very effective store for B, at least, as far as developing cells are concerned. The accumulation of motor vehicles at some bottleneck on the autobahn cannot be properly described as a storage area, whereas an adjacent car park in a rest area might be. The car park, like any kind of store, has a finite capacity and organised entrances and exits. A functional biological store would be a place where material can be deposited or withdrawn in a regulated manner. Central to such a notion is that there should be some control over the process of accumulation. The cell vacuole fits well with these general ideas because fluxes of materials across the tonoplast are regulated by transporters and channels. Similar regulation of transport occurs at the plasma membrane (PM). Perhaps the salient test of the idea that the apoplast can function as a store is that materials accumulating in it can be mobilized when required and moved either into surrounding cells, or exported for use in another part of the plant. The reader might like to keep this in mind.

It should be remembered that the water-filled spaces in the cell wall comprise a small volume. This places an obvious osmotic constraint on the amount of a solute that can accumulate. Processes such as adsorption, precipitation and polymerization can all reduce the osmotic activity of a material. With a simple salt, such as NaCl, none of the above processes can be expected to play much part in lowering the osmotic strength of the solution in the apoplast, especially when delivery to the leaf is faster than processes such as salt excretion, vacuolar sequestration or export in the phloem. If NaCl accumulation takes place in the apoplast, the cells might respond by the synthesis of compatible solutes, to keep the water potential of the cytosol lower than that of the apoplast, but failing this, one might envisage cells losing turgor. This idea was advanced some time ago by Oertli (1968) and other authors have found evidence to suggesting that cells in the leaf can be damaged by extracellular salt in this way (e.g. Flowers et al., 1991). In the paper by Lohaus (this volume, pp. 323–336) this idea is challenged. This matter provides a suitable project for more refined imaging of Na and probing with Na-selective micro-electrodes.

The xylem stream delivers divalent Mn^{2+} ions to the leaf apoplast. Mn^{2+} is probably the ion species that crosses the plasma membrane of leaf cells via divalent ion channels (Clarkson, 1988). Where the Mn supply is excessive, there appears to be stimulation of peroxidase activity in the apoplast that, in turn oxidizes Mn^{II} to Mn^{III} (Fecht-Christoffers et al., this volume, pp. 307–322). Among other effects, this leads to the accumulation of manganese oxide. In leaves, autoradiography shows that there can be sizeable, discrete deposits of this form of Mn (Horst and Marschner, 1978). There is also a poorly-understood reaction between apoplastic silica and Mn which is probably due to adsorption. This seems to ameliorate the effects of Mn over-supply and toxicity. Both the precipitation and adsorption processes can be viewed as defence mechanisms that may counteract excessive Mn (Rogalla and Römheld, 2002; Wiese et al., this volume, pp. 33–48). Whether or not these accumulations should be regarded as a store is much less certain. Mn in solution will be in a chemically determined equilibrium with Mn adsorbed onto silica and it may be that organic acids can solubilize precipitated Mn oxides, but neither process satisfies the criteria for a useful store.

The situation for Fe is more difficult to interpret. In the paper by Nikolic and Römheld (this volume, pp. 353–372) there is evidence that Fe^{III} iron delivered to the apoplast predominantly as ferric citrate complexes, becomes adsorbed onto polymeric structural components of the mesophyll extracellular matrix. The concentration of free iron in the apoplast fluid appears to be always very low. The authors suggest that the availability of free Fe^{III} for subsequent reduction (via PM redox system) and transport to the cytoplasm depends on release from the extracellular pool. This equilibrium, affected as it is by factors such as apoplast fluid pH, makes the extracellular iron a buffer; or so it would seem. Nikolic and Römheld (this volume, pp. 353–372), however, found little difference between the quantity of Fe in the extracellular matrix of green and chlorotic leaves of sunflower. This makes it seem that access to the "stored" apoplastic iron may not respond to the biochemical "need" for iron within the cell. Apoplastic iron is not, perhaps, so effective a store as stromal phytoferritin that is rapidly mobilized in greening leaves (Marschner, 1986).

Calcium ions play an essential role in cross-linking elements of the extracellular matrix. They are always present but remain exchangeable with other divalent cations, depending on their activity and the pH of the apoplast solution. There will always be free Ca^{2+} in equilibrium with this exchangeable Ca^{2+} in the matrix Since there is a very large free energy gradient directed towards the cell interior, an increase in the rate of Ca^{2+} entry via channels in the PM will draw in the free Ca^{2+} from the apoplast solution but this will be replaced rapidly by the Ca^{2+} buffer in the extracellular matrix. Thus, changes in the Ca^{2+} fluxes across the PM might be expected not to cause major shifts in the apoplasmic free Ca^{2+}.

In general, the results by Felle and Hanstein (this volume, pp. 295–306) and Plieth et al. (this volume, pp. 373–392) support this notion. There is, however, a very large internal stockage of vacuolar Ca^{2+}; this would seem to be the more obvious store for cytoplasmic homeostasis. The apoplast Ca^{2+} might be viewed as a buffer for Ca^{2+} at the outer surface of the PM and would, therefore, play an important role in maintaining membrane integrity.

5. THE APOPLAST AS THE FIRST LINE OF DEFENCE IN STRESS RESPONSES

Stresses in leaves can have a biotic or an abiotic origin. A leaf exists in air and in water that is rich in the spores of fungi and bacteria. Some of these are pathogenic and may have specialised means of invading the leaf by penetrating the epidermal layers. These are beyond the scope of the present work. The pores of open stomata are sufficiently large, however, to permit entry of airborne microbial propagules into the leaf. Systematic inspection of a number of plant species leads to the conclusion that microbes are of common occurrence in leaf tissues (Hallmann et al., 1977). Autotrophic nitrifiers are present in significant numbers in the needles of spruce trees (Teubner et al., this volume, pp. 405–426). Bacterial endophytes are found in intercellular spaces because they are too large to enter the pores of the wall matrix. The apoplast solution provides C and N sources for bacterial growth, but the endophytic bacterial numbers are clearly kept in check either by substrate supply of by other means. In exchange for this substrate supply the plant may gain certain advantages. Some species of endophytic bacteria are dinitrogen fixers; much has been made of this but the consensus seems to be that N fixed in this way does not contribute more than 5% of the N requirements of species such as sugar cane (Dong et al., 1994).

Of more general significance is the suggestion that commensural microbes may increase resistance of the host to biotic stress (Hallmann et al., 1997), thus being analogous to a simple immune system. Plants can activate differentially distinct defence pathways depending on the type of challenge presented to them (Van Loon, 1997). This is an intriguing and complex subject but, in summary it can be said that the increased resistance can be of two kinds. Traumatic injury to cells in one part of the plant may trigger a defence pathway in which the accumulation of salicylic acid leads to the induction of pathogenesis-related, PR, proteins; the pathway is known as the systemic acquired resistance, SAR (Van Loon, 1997). This expression leads to an enhancement of the hypersensitive reactions in parts of the plant remote from the primary injury; these contain subsequent pathogenic attacks and limit their systemic development. A second kind of induced systemic resistance, ISR, has been extensively studied by van Loon and colleagues (Van Loon, 1997; van Wees et al., 2000). The non-pathogenic challenge presented by endophytic microbes, such as the rhizosphere bacterium. *Psuedomonas fluorescens*, activates an ISR pathway. In this, increased jasmonate level is a crucial signal eliciting the expression of PR genes and the eventual release of PR proteins. Most published work has been concerned with the exposure of roots to the bacteria, but ISR also occurs when they are applied to leaves (van Loon, personal communication).

Leaves can be subjected to intense biochemical oxidations and reductions during metabolism as well as experiencing atmospheric pollutants, such as ozone. These processes can lead to oxidative stress that needs to be dealt with if serious damage to cells is to be avoided. This has become a major research activity in recent decades (Dietz, 1997). Although the context is specifically related to excessive Mn supply via xylem delivery, in the paper of Fecht-Christoffers et al., (this volume, pp. 307–322), the response in the apoplast to the presence of a strong oxidant is an example of a more generalized set of responses. Substances such as ascorbic acid and dehydroascorbic acid, NADH-oxidase activity and peroxidases all of which are involved in the transformations of Mn in the apoplast. There is also the release of a raft of PR proteins with glucanase, chitinase activities into the apoplast of cowpea leaves; similar observations have been made in sunflower (Jung et al., 1995). The "defence" proteins are released in response to cell trauma (see above). The appearance of PR proteins in the apoplast fluid suggests that high Mn levels may disrupt the integrity of cell membranes. Earlier work by Wissemeier and Horst (1987) showed that callose is deposited on walls of cowpea (*Vigna unguiculata*) subjected to excessive Mn. Callose is often an indication of disruption of PM integrity and of the normal controls over cellular Ca^{2+} homeostasis.

Many sorts of cell injury or stress lead to H_2O_2 production and events downstream of this seem, at present, to be of a rather general kind. The history of research into biochemical and physiological "cascades" shows that initially, a puzzling variety of responses follow from a single event. It is then questioned as to how a simple signal can generate such a diversity of biochemical and physiological consequences.

A classic example would be the consequences of elevated cytosolic calcium ion activities. Many different kinds of perturbation raise cell Ca^{2+} activity, some of which are described by Plieth et al. (this volume, pp. 373–392). Refinement of techniques revealed the subtleties of this elevation (Trewavas and Mahlo, 1998). We now know, for instance, that both the amplitude and frequency of transient elevations of cell calcium are important in what happens subsequently. A large, sustained spike of cell Ca^{2+} has a different effect to frequent smaller spikes. It seems likely that the cell is able to interpret the nature of the signal from information of this kind (Knight, 2000). Other responses to environmental signals and challenges have yet to be investigated as intensively as those that elevate cell calcium. When this has been done, it is highly probable that the responses in the apoplast will be seen to be more subtle and diverse than they appear at present.

ACKNOWLEDGEMENTS

The author is indebted to Professor LC van Loon of Utrecht University, NL, for his most helpful discussion of systemic and acquired resistance in plants, and to Dilys Parry, of Bristol University for bibliographic assistance.

REFERENCES

Bernstein L (1971) Methods for determining solutes in the cell walls of leaves. *Plant Physiol.*, 47, 361–365.

Canny MJ (1990) Rates of apoplast diffusion in wheat leaves. *New Phytol.*, 116, 263–268.

Clarkson DT (1988) The uptake and translocation of manganese by plant roots. In: *Manganese in Soils and Plants*. Graham RD, Hannam RJ and Uren NC (eds). Kluwer, Dordrecht., pp.101–111.

Clarkson DT (1991) Roots and the delivery of solutes to the xylem. *Philos. Trans. R. Soc. London*, B, 341, 5–17.

Dannel F, Pfeffer H and Marschner H (1995) Isolation of apoplasmic fluid from sunflower leaves and its use for studies on the influence of nitrogen supply on apoplasmic pH. *J. Plant Physiol.*, 146, 273–278.

Dietz KJ (1997) Functions and response of the leaf apoplast under stress. *Prog. Bot.*, 58, 221–254.
Dong ZM, Canny MJ, McCully ME, Roboredo MR, Cabadilla CF, Ortega E and Rodes R (1994) A nitrogen-fixing endophyte of sugar cane stems. A new role for the apoplast. *Plant Physiol.*, 105, 1139–1147.
Felle HH, Hanstein S, Steinmeyer R, and Hedrick R (2000) Dynamics of ion activities in the apoplast of the sub-stomatal cavity of intact *Vicia faba* leaves during stomatal closure evoked by ABA and darkness. *Plant J.*, 24, 297–304.
Flowers TJ, Hajibagheri MA and Yeo AR (1991) Ion accumulation in cell walls of rice plants growing under saline conditions; evidence for the Oertli hypothesis. *Plant Cell Environ.*, 14, 319–325.
Gogarten JP (1988) Physical properties of the cell wall of photoatutotrophic suspension cells from *Chenopodium rubrumn* L. *Planta*, 174, 333–339.
Hallmann J, Quadt-Hallmann A, Mahaffee WF and Kloepper JW (1997) Bacterial endophytes in agricultural crops. *Can. J. Microbiol.*, 43, 895–914.
Hoffmann B and Kosegarten H (1995) FITC-dextran for measuring apoplast pH and apparent pH gradients between various cell types in sunflower leaves. *Physiol. Plant.*, 95, 327–335.
Horst WJ and Marschner H (1978) Effect of silicon on manganese tolerance of bean plants (*Phaseolus vulgaris* L.). *Plant Soil*, 50, 287–304.
Hose E, Clarkson DT, Steudle E, Schreiber L and Hartung W (2001) The exodermis; a variable apoplastic barrier. *J. Exp. Bot.*, 52, 2245–2264.
Jung JL, Maurel S, Fritig B and Günther H (1995) Different pathogenesis-related proteins are expressed in sunflower (*Helianthus annuus* L.) in response to physical, chemical and stress factors. *J. Plant Physiol.*, 145, 153–160.
Knight H (2000) Calcium signalling during abiotic stress in plants. *Int. Rev. Cytol.*, 195, 269–324.
Lohaus G, Pennewiss K, Sattelmacher B, Hussmann m and Mühling KH (2001) Is the infiltration-centrifugation technique appropriate for the isolation of apoplastic fluid? A critical evaluation with different plant species. *Physiol. Plant.*, 111, 457–465.
Marschner H (1986) *Mineral Nutrition of Higher Plants*. Academic Press, London. 674 pp.
Mühling KH and Sattelmacher B (1997) Determination of apoplastic K^+ in intact leaves by ratio imaging of PBFI fluorescence. *J. Exp. Bot.*, 48, 337.
Mühling KH and Läuchli A (2002) Determination of apoplastic Na^+ in intact leaves of cotton by *in vivo* ratio imaging. *Functional Plant Biol.*, 29, 1491–149.
Mühling KH, Wimmer M and Goldbach H (1997) Apoplastic and membrane associated Ca^{2+} in leaves and roots as affected by boron deficiency. *Physiol. Plant.*, 102, 179–184.
Oertli JJ (1968) Extracellular salt accumulation, a possible mechanism of salt injury in plants. *Agrochimica*, 12, 461–469.
Pitman MG, Lüttge U, Läuchli A and Ball E (1974) Free space characteristics of barley leaf slices. *Austr. J. Plant Physiol.*, 1, 65–75.
Rogalla H and Römheld V (2002) Role of leaf apoplast in silicon-mediated manganese tolerance of *Cucumis sativus* L. *Plant Cell Environ.*, 25, 549–555.
Sattelmacher B (2001) The apoplast and its significance in plant mineral nutrition. *New Phytol.*, 149, 167–192.
Trewavas AJ and Mahlo R (1998) Ca^{2+} signalling in plant cells: the big network! *Curr. Opin. Plant Biol.*, 1, 428–433.
Van Loon LC (1997) Induced resistance in plants and the role of pathogenesis-related proteins. *Eur. J. Plant Pathol.*, 103, 753–765.
Van Wees SCM, de Swart EAM, van Pelt JA, van Loon, LC and Pieterse, CMJ (2000) Enhancement of induced disease risistance by simultaneous activation of slicylate- and jasmonate-dependent defence pathways in *Arabidopsis thaliana*. Proceedings of the National Academy of Sciences of the USA, 97, 871–8716.
Wissemeier AH and Horst WJ (1987) Callose deposition in leaves of cowpea (*Vigna unguiculata* L.) as a sensitive response to high Mn supply. *Plant Soil*, 102, 283–286.

Section 1

Cell Wall–Ion Interactions: Significance for Nutrition of Plants and their Stress Tolerance

CELL WALL–ION INTERACTIONS

Significance for the nutrition of plants and their stress tolerance

N. CARPITA

Department of Botany and Plant Pathology, Purdue University, U.S.A, carpita@purdue.edu

Key words: aluminium, boron, cell wall, ion interactions, pectin matrix, silica

In addition to its central role in plant growth and development, the cell wall is a plant's interface with the environment. This dynamic composite of cross-linking glycans embedded in a gel of pectin provides plant cells with a rich variety of shapes and sizes (McCann and Roberts, 1991). Different types of walls are made by flowering plants, a "Type I" pectin-rich, xyloglucan-cellulosic wall is made by most dicotyledonous and non-gramineous monocotyledonous plants, and a "Type II" pectin-poor, arabinoxylan-cellulosic wall is made by commelinoid monocots, including grasses and cereals (Carpita and Gibeaut, 1993). Unique to the walls of Poales species (grasses and cereals), is a mixed-linkage $(1\rightarrow3),(1\rightarrow4)$- β-D-glucan that appears transiently during cell expansion phases of growth (Carpita, 1996; Smith and Harris, 1999). Type I walls are further distinguished from Type II walls in the way their structures are reinforced when differentiation begins. The Type I is characterized by the appearance of several cross-linking structural proteins, such as the extensins and other kinds of hydroxyproline-rich glycoproteins, glycine-rich proteins, and proline-rich proteins (Cassab and Varner, 1988). Except in rare instances of tough cells, such as those of the maize periderm (Hood et al., 1988), the Type II wall has very little structural protein. Instead, the walls are cross-linked mostly by phenylpropanoids, such as esterified and etherified hydroxycinnamic acids and other aromatic lignin-like substances (Scalbert et al., 1985). This property imparts a strong autofluorescence in the non-lignified cells of the Type II, permitting a facile means to classify them (Rudall and Caddick, 1994).

B. Sattelmacher† and W.J. Horst (eds.), The Apoplast of Higher Plants: Compartment of Storage, Transport and Reactions, 15–18.

The pectin-rich Type I wall provides a complex, dynamic matrix that is the major determinant of wall pH, ionic balance, porosity, and electrical status (Carpita and Gibeaut, 1993). Pectins comprise two principal uronic acid-rich polymers, homogalacturonan (HG) and rhamnogalacturonan I (RG I). The HGs have a repeating backbone of α-D-(1→4)-linked galacturonosyl residues and are secreted to the wall in largely esterified form, with 70% or more of the residues bearing methyl esters. They may be de-esterified in muro, in random and block-wise fashion, at precise stages of cell development (Willats et al., 2001). Calcium ions play a key structural role once the methyl esters are removed. If long runs of unesterified uronosyl residues exist, then anti-parallel chains of HGs in are tightly cross-linked into "junction zones". Calcium can also cross-link parallel chains of partly de-esterified HGs (Carpita and Gibeaut, 1993).

HGs may also be substituted with neutral sugars, such as xylose to form xylogalacturonan enriched in flowering and fruit tissues, and/or acetylated (Schols et al., 1995). The HGs also form the backbone of one of the most complex polysaccharides found in Nature, called rhamnogalacturonan II (RG II). RG IIs contain many rare sugars and saccharinic acids, including apiose, aceric acid, Kdo and Dha, in four distinct side chains (O'Neill et al., 1990). One of the two apiose residues forms a boron-di-diester, coupling to RG II molecules, and forming structures essential for growth and development (O'Neill et al., 2001), porosity (Fleischer et al., 1999), and wall elasticity (Findeklee and Goldbach, 1996) and tensile strength (Ryden et al., 2003). The RG I is a polymer of the repeating disaccharide, →2)- α-L-rhamnosyl-(1→4)- α-D-galactosyl-(1→, which possesses neutral side-chains of highly branched (1→5)- α-L-arabinans, (1→4)- β-D-galactans, and type I arabino-(1→4)- β-D-galactans (Willats et al., 2001).

Given the enormous complexities and dynamics of the negatively charge pectin matrix, it is not surprising that it plays a central role in the mineral nutrition and ion status of plants, both at the sites of absorption in the root and utilization at the shoot meristems. This section contains four articles that describe unique features of mineral nutrition and the impacts of environmental stress. Two ions, boron and silica, are considered to impact structural elements of the wall directly. Wimmer and Goldbach (this volume, pp. 19-32) review boron in the type I cell wall and provide new data on its interactions with calcium ions in the formation of boron complexes. Because the only known function for boron in the apoplast is in the dimerization of RG II molecules (Ishii and Mansunaga, 1996; O'Neill et al., 1996), and given the function of these dimers in wall architecture (Findeklee and Goldbach, 1996; Fleischer et al., 1999; Ryden et al., 2003), divalent cations and other ions have important functions in regulating monomer to dimer ratios. In fact, Ca^{2+} has been shown to stabilize boron-RG II dimers (Kobayashi et al.,

1999). Wimmer and Goldbach (this volume, pp. 19–32) document such an interaction with free calcium. Within minutes of boron deficiency, stabilizing boron and calcium complexes form, resulting in retention of the remaining boron. Sodium ions interfere with these complexes, indicating that a consequence of saline stress could be the loss of boron from the apoplast. In addition to their studies of calcium-boron interactions, the authors show that boronic acid coupled to fluorescent dyes are new tools to locate the boron binding sites in muro.

An often overlooked element is silicon, which is accumulated to considerable amounts in several species. As Wiese et al. (this volume, pp. 33–48) describe, silicon is not known as an essential element, but has many direct and indirect beneficial effects through growth and development. The near impossible task of generating silicon-free plants precludes determination of its necessity. Silicon is absorbed as $Si(OH)_4$ and accumulates mainly in the apoplast. Wiese et al. (this volume, pp. 33–48) explain that plants range from "excluders" to "accumulators", the latter of which include many grass species with Type II walls, with cucumber being an exception among Type I-walled plants in accumulating silicon. Their article focuses on three silicon-mineral interactions: the impact of silicon on phosphate-induced zinc deficiency, on exchange capacity and binding forms of manganese, and on interactions with boron.

The key enzyme in creating a charged pectin matrix that carries cationic exchange capacity is pectin methyl esterase (PME). Thus, controls of this activity in the apoplast are essential determinants of the whole of mineral nutrition. Because they are cationic at physiological pH, polyamines could interact directly with de-methylated uronosyl residues of pectic substances. Gerendás (this volume, pp. 67–84) provides an extensive review of PME activity and some of the downstream consequences with respect to charge density, metal ions, calcium ions, and matrix pH. He focuses on the possible role of polyamines in regulation of PME activity. Levels of polyamines in the apoplast are greatly influenced by nutritional status, but very little if any direct effect on PME activity could be demonstrated.

While several authors point out that too much of an essential element can be as deleterious as too little, one of the principal agronomic problems today is aluminium (Al^{3+}) toxicity in acidic soils. Horst et al. (this volume, pp. 49–66) examine further the mechanism of Al^{3+} toxicity in root systems. They point out that the actual mechanism of toxicity remains to be established, although the cell wall is thought to play a role. One of the principal effects of Al^{3+} toxicity is a rapid inhibition of root elongation with a switch from cellulose synthesis to callose synthesis. Horst et al. (this volume, pp. 49–66) show a tight quantitative correlation between the levels of Al^{3+} and callose synthesis. They also discuss other factors involved in tolerance and resistance.

The pectin matrix has been implicated in contributing to resistance, presumably by binding to the Al^{3+} ion (Blamey et al., 1990), and Horst et al. (this volume, pp. 49–66) correlate Al^{3+} and pectin content. However, a role for pectins in resistance is tempered by the fact that Al^{3+}-tolerant grass species possess a pectin-poor Type II wall. Roles for organic acids, such as citric acid, and Ca^{2+}, silicates and phosphorous in mitigating Al^{3+} toxicity through sequestration is discussed.

REFERENCES

Blamey, F. P. C., Edmeades, D. C. and Wheeler, D. M. (1990). Role of root cation-exchange capacity in differential aluminum tolerance of Lotus species. *Journal of Plant Nutrition*, 13, 729–744.

Carpita, N. C. (1996). Structure and biogenesis of the cell walls of grasses. *Annual Review of Plant Physiology and Plant Moecular Biology*, 47, 445–476.

Carpita, N. C. and Gibeaut, D. M. (1993). Structural models of primary-cell walls in flowering plants: consistency of molecular structure with the physiological properties of the cell wall during the growth. *Plant Journal*, 3, 1–30.

Cassab, G. I. and Varner, J. E. (1988). Cell wall proteins. *Annual Review of Plant Physiology and Plant Molecular Biology*, 39, 321–353.

Findeklee, P. and Goldbach, H. E. (1996). Rapid effects of boron deficiency on cell wall elasticity modulus in *Cucurbita pepo* roots. *Botanica Acta*, 109, 463–465.

Fleischer, A., O'Neill, M. A. and Ehwald, R. (1999). The pore size of non-graminaceous plant cell walls is rapidly decreased by borate ester cross-linking of the pectic polysaccharide rhamnogalacturonan II. *Plant Physiology*, 121, 829–838.

Hood, E. E., Qing, X. S. and Varner, J. E. (1988). A developmentally regulated hydroxyproline-rich glycoprotein in maize pericarp cell walls. *Plant Physiology*, 87, 138–142.

Ishii, T. and Mansunaga, T. (1996). Isolation and characterization of boron-rhamnogalacturonan II complex from cell walls of sugar beet pulp. *Carbohydrate Research*, 284, 1–9.

Kobayashi, M., Nakagawa, H., Asaka, T. and Matoh, T. (1999). Borate-rhamnogalacturon II bonding reinforced by Ca^{2+} retains pectic polysaccharides in higher plant cell walls. *Plant Physiology.*, 119, 199–203.

McCann, M. C. and Roberts, K. (1991). Architecture of the primary cell wall. In C. W. Lloyd (ed.) The Cytoskeletal Basis of Plant Growth and Form, Academic Press, London, pp. 109–129

O'Neill, M. A., Albersheim, P. and Darvill, A. G. (1990). The pectic polysaccharides of primary cell walls. In P. M. Dey and J. B. Harbourne (eds) Methods in Plant Biochemistry, Vol 2: Carbohydrates, Academic Press, London, pp. 415–441.

O'Neill, M. A., Eberhard, S., Albersheim, P. and Darvill, A. G. (2001). Requirement of borate cross-linking of cell wall rhamnogalacturonan II for *Arabidopsis* growth. *Science*, 294, 846–849.

O'Neill, M. A., Warrenfeltz, D., Kates, K., Pellerin, P., Doco, T., Darvill, A. G. and Albersheim, P. (1996). Rhamnogalacturonan-II, a pectic polysaccharide in the walls of growing plant cell, forms a dimer that is covalently cross-linked by a borate ester – in vitro conditions for the formation and hydrolysis of the dimer. *The Journal of Biological Chemistry*, 271, 22923–22930.

Rudall, P. J. and Caddick, L. R. (1994). Investigation of the presence of phenolic compounds in monocotyledonous cell walls, using UV fluorescence microscopy. *Annals of Botany*, 74, 483–491.

Ryden, P., Sugimoto-Shirasu, K., Smith, A.C., Findlay, K., Reiter, W-D. and McCann, M. C. (2003). Tensile properties of *Arabidopsis* cell walls depend on both a xyloglucan cross-linked microfibrillar network and rhamnogalacturonan II-borate complexes. *Plant Physiology*, 132, 1033–1040.

Scalbert, A., Monties, B., Lallemand, J-Y., Guittet, E. and Rolando, C. (1985). Ether linkage between phenolic acids and lignin fractions from wheat straw. *Phytochemistry*, 24, 1359–1362.

Schols, H. A., Bakx, E. J., Schipper, D. and Voragen, A. G. J. (1995). A xylogalacturonan subunit present in the modified hairy regions of apple pectin. *Carbohydrate Research*, 279, 265–279.

Smith, B.G. and Harris, P.J. (1999). The polysaccharide composition of Poales cell walls: Poaceae cell walls are not unique. Biochemical Systematics and Ecology 27(1), pp. 33–53.

Willats, W. G. T., McCartney, L., Mackie, W. and Knox, J. P. (2001). Pectin: cell biology and prospects for functional analysis. *Plant Molecular Biology*, 47, 9–27.

BORON IN THE APOPLAST OF HIGHER PLANTS

Relevance for rapid deficiency reactions, interaction with calcium activity, and characterization of soluble boron complexes

M.A. WIMMER and H.E. GOLDBACH
Institute of Crop Science and Resource Conservation – Plant Nutrition, University of Bonn, Germany, e-mail: m.wimmer@uni-bonn.de and h.goldbach@uni-bonn.de

Abstract. The relevance of different apoplastic B fractions for rapid B deficiency reactions and their interaction with apoplastic Ca activity was investigated. Soluble apoplastic B was reduced quickly under B deficiency in both, roots and leaves, and could be used as a tool to detect early or latent B deficiency. The critical level of soluble B was determined to be below 0.5 μM in *Vicia faba* roots. Negatively charged B complexes are potential reaction sites for polyvalent cations. In *V. faba* root tips, short term and long term interactions between B and Ca were observed. Within minutes of B deficiency, an immediate and mutual stabilizing interaction occurred between B and free Ca. A reduction of membrane-bound Ca was seen after 4 h of B-deficient conditions. In B-sufficient roots, 5% of total B was present in the form of a Ca-stabilized soluble B complex, which disappeared within 2 h of B deficiency. Besides soluble B, this B complex is likely involved in early B deficiency reactions. Results indicate that this complex could represent cell-wall precursors. An interaction between Na and Ca is likely involved in a reduction of B binding in the root apoplast under saline conditions and could be one of several factors influencing B tolerance. Boronic acid-coupled fluorescent dyes are shown to be useful tools to visualize B binding sites *in vivo*, and a novel technique using immunogold-labelled antibodies might allow the sub-cellular localization of B binding sites in the future.

Key words: boron deficiency, toxicity, soluble boron, calcium, boron complex, FITC

B. Sattelmacher† and W.J. Horst (eds.), The Apoplast of Higher Plants: Compartment of Storage, Transport and Reactions, 19–32.

1. INTRODUCTION

The apoplast harbours 50–98% of plant B, depending on species, plant organ and B supply. Symptoms of B deficiency are related to an impaired structural integrity of the cell wall (Goldbach, 1997; Loomis and Durst, 1992). Only recently the essentiality of B-RG II cross-links for Arabidopsis growth was shown (O'Neill et al., 2001) and remains the only demonstrated function of B in plants. Boron cross-links have been shown to affect the cell-wall pore-size (Fleischer et al., 1998, 1999) as well as the cell-wall tensile-strength (Ryden et al., 2003). Other possible functional sites include cell membranes and the cell-wall cytoskeleton continuum (Bassil et al., 2004; Goldbach et al., 2001; Yu et al., 2001, 2002, 2003), which may be indirectly affected by altered cell-wall physics (Findeklee and Goldbach, 1996; Fleischer et al., 1999) or a hence unknown mechanism with other ligands than RGII (Ralston and Hunt, 2001).

It was the objective of this project to improve our understanding of primary functions of B in higher plants. We aimed at determining soluble B concentrations and their relevance for early B deficiency reactions, as well as investigating the interaction between B and Ca in the apoplast. Based on obtained results, we then focused on the characterization of Ca-stabilized B fractions and their relevance for B tolerance. In a third part B binding sites were visualized using fluorescent dyes and the suitability of the method assessed.

Experiments were conducted using *Vicia faba* L. cv. Troy plants grown at 10 μM (+B) or without B (–B). An enlarged apoplastic fluid was obtained from roots by equilibration with different solutions. In leaves, apoplastic washing fluid was collected using a modified infiltration/centrifugation technique (Mühling and Sattelmacher, 1995). A spectrophotometric curcumin method was miniaturized in order to allow determination of B in small sample volumes (50–150 μl) with high sensitivity (detection limit 0.003 mg B l^{-1}) (Wimmer and Goldbach, 1999b). Reproducibility is good with a relative standard deviation of 1–5% at B concentrations between 0.05 and 0.40 mg B l^{-1}. The method has successfully been applied to determine B concentrations in waters, nutrient solutions, plant parts, phloem sap, xylem sap and apoplastic washing fluids of several plant species. Results are comparable to those determined with ICP-MS.

2. EARLY BORON DEFICIENCY REACTIONS

2.1 Chemical form of boron in the apoplast

Critical for determining B functions in the apoplast is an understanding of the physical and chemical properties of B and its complexes (reviews in

Brown et al., 2002; Goldbach, 1997; Loomis and Durst, 1992). Boric acid and borate can readily form esters with a variety of compounds possessing vicinal cis-diols or proximal hydroxyls in the correct orientation. Reactions are spontaneous, pH-dependent and rapid. A range of sugars, polyols, phenolics, amino acids and several other biomolecules have been predicted or shown to react with boric acid, although the stability of B complexes with these compounds under physiological conditions is likely low (Loomis and Durst, 1992; Ralston and Hunt, 2000, 2001). The functional significance of these putative B-containing complexes has not yet been determined.

Within the last few years, the chemical form of cell wall-bound B has been identified. A B polysaccharide complex was isolated from different plants and identified as a dimeric B-Rhamnogalacturonan II (dB-RGII) complex, where two chains of monomeric RGII (mRGII) are cross-linked by a 1:2 borate ester with two of the four apiosyl redidues of RGII side chains (Ishii and Matsunaga, 1996; Kaneko et al., 1997; Kobayashi et al., 1996; Matoh et al., 1993; O'Neill et al., 1996; Pellerin et al., 1996). Although mRGII is present in cells adapted to low B levels (Pellerin et al., 1996), B seems to be associated solely with the dimeric form (Ishii et al., 1999). Even though B-RGII accounts for 40–80 % of apoplastic B, the remaining 20–60 % have neither been identified, nor do we know their functions.

2.2 Relevance of apoplastic B concentrations for early deficiency reactions

Because of the rapidity and wide variety of B deficiency symptoms, it is a great challenge to determine primary functions of B in plants (Blevins and Lukaszewski, 1998). In growing tissues, primary effects of B deficiency occur within few minutes, but separating primary from secondary reactions remains difficult (Findeklee and Goldbach, 1996; Goldbach et al., 2001).

In *V. faba* roots, total B of older root parts decreased more slowly under B deficiency than that of growing root tips (Table 1), confirming the high stability of mature B-RGII complexes reported earlier (O'Neill et al., 1996). We therefore suggest that changes in soluble apoplastic B rather than changes in B-RGII complexes are responsible for early deficiency reactions and that a "critical level" of soluble B needs to be maintained at all times.

Table 1. Total B content (μg B/g DW) in V. faba, means of 6 replicates ± S.D.

	+B	-B 10 h	-B 24 h
Shoots	15.4 (±1.4)	n.d.	15.5 (±1.9)
Roots, older parts	17.5 (±3.4)	19.8 (±6.7)	13.5 (±3.2)
Roots, tips	14.3 (±3.3)	11.6 (±4.7)	4.6 (±2.8)*

* $p<0.001$

Knowing that ferricyanide-inducible reductase activity (RA) is influenced by B, we tested its suitability to determine a "critical level" of soluble B. At low B concentrations, RA correlated with apoplastic soluble B concentrations. However, even though RA rapidly reacted to changes in B supply, it was also highly variable depending on other factors (e.g. plant age) making it unsuitable for the determination of either B deficiency or a critical level of soluble B (Goldbach et al., 2000; Wimmer, 2000).

Using an equilibration assay, we determined soluble apoplastic B concentrations of control *V. faba* root tips to range from 0.4–0.7 μM. After 24 h of –B treatment this B concentration fell below detectable limits. Irrespective of B supply, only 4–9% of the total B was present in a soluble form, indicating that B supply determines the "total B pool", but does not affect the binding strength of B. In the apoplastic washing fluid of *V. faba* leaves, the soluble B concentration was slightly higher (1.0–1.5 μM). After only 2 days of –B treatment it decreased to 0.7 μM, whereas at the same time no change in total B and no deficiency symptoms could be observed.

The results suggest that for growing *V. faba* roots, a soluble B concentration of 0.5–1.0 μM is sufficient for growth or, in other words, the critical level lies below 0.5 μM. Higher soluble B concentrations in leaves might explain the lack of rapid deficiency reactions as compared to roots. The results are in line with the hypothesis that soluble B is involved in early B deficiency reactions.

3. BORON – CALCIUM INTERACTIONS

An interaction of B and Ca has been repeatedly reported, although no clear relationship was established. In some cases, high supply of Ca reduced B deficiency symptoms, possibly by enhancing B mobilization from root to shoot (Carpena et al., 2000). B supply affected Ca translocation to the shoot and fruit (Ganmore-Neumann and Davidov, 1993; Yamauchi et al., 1986). The plant membrane is a possible site of interaction between the two elements. Recently, a complementary role of B and Ca in establishing a symbiosis and in nodule formation has been demonstrated (Redondo-Nieto et al., 2001). In rape cells, increasing B supply reduced the amount of NaCl-extractable Ca, whereas the effect on total Ca in leaves was dependent on the B efficiency of the cultivar used (Wang et al., 2003).

3.1 Effect of B supply on free and membrane-bound calcium

In cell walls, the dB-RGII complex bears a negative charge, indicating a possible site for interaction with Ca^{2+}. Additionally, pectic chains contain a large number of carboxyl groups that are negatively charged at physiological pH (pK_a of pectic acid is 3.5–4.0: Jarvis et al., 1984). Van Duin et al. (1987) demonstrated the possibility of a direct B/Ca interaction in B-Ca-polycarboxylate complexes.

We hypothesize that, under B-deficient conditions, a reduction in cell wall negative charge should result in an increase in "free" or easily exchangeable Ca. In cooperation with the group Mühling/Sattelmacher, Kiel, we used a combination of fluorescein-isothiocyanate (FITC), coumarin-benzothiazol (BTC) and chloro-tetracycline (CTC) fluorescence to monitor the effect of B supply on free and membrane-bound Ca^{2+} (Mühling et al., 1998). Adding B to B-deficient intact seedling roots immediately decreased the concentration of free Ca^{2+} (Fig. 1, left). In excised roots, a 7–14% increase in "free" Ca^{2+} was seen after 15 min of -B (not shown). Similarly, Ca deficiency increased the amount of soluble B, irrespective of the pH (Table 2). B deficiency of more than 4 h decreased membrane-bound Ca^{2+} (Fig. 1, right).

In leaves, water-extractable apoplastic Ca and K increased significantly on the 4th day of B deficiency, clearly preceded by a decline in soluble B on the 2nd day, but still before visible B deficiency symptoms were observed (Mühling et al., 1998).

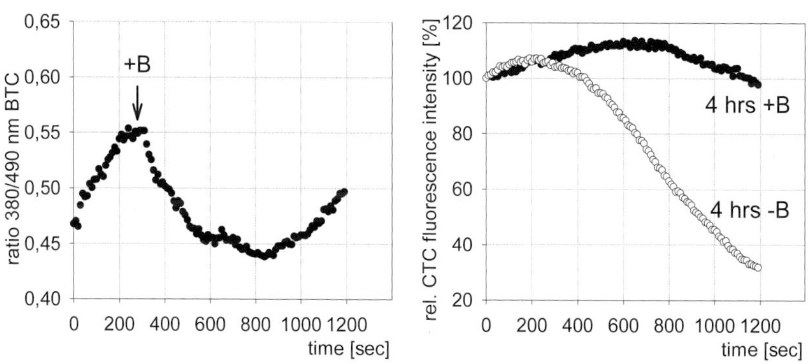

Fig. 1 Effect of B supply on free apoplastic Ca^{2+}(left: plants were grown 24 h –B before addition of 10 mmol m^{-3} B) and on membrane-bound Ca^{2+} (right: plants were grown 4 h with 10 μM B (+B) or without B (–B)).

Table 2. Soluble B content in V. faba root tips incubated at different pH; means of 4 replicates; different letters within rows indicate significant differences at $p<0.05$.

Ca concentration [mM]	0	0.5	1.0	1.5	2.0
Soluble B [μg/g DW] pH 5.8	0.54^{a}	0.34^{ab}	0.32^{ab}	0.26^{b}	0.25^{b}
pH 3.6	0.42^{a}	0.36^{a}	0.24^{a}	n.d.	0.20^{a}

The results show that there are short term and long term interactions between apoplastic B and Ca. Within minutes of B deficiency, the ratio of free/bound Ca is altered independently of the pH. However, we do not know whether this change is due to the formation of B-Ca-ligand complexes or to steric factors affecting the pectic network. It is concluded that low Ca supply could aggravate B deficiency effects in cell walls, and vice versa.

3.2 Stabilization of different B fractions by Ca

Addition of Ca^{2+} in vitro enhanced the formation and delayed the decomposition of dB-RGII (Ishii et al., 1999; Kobayashi et al., 1999; O'Neill et al., 1996). In several plant species, isolated dB-RGII contained Ca (Kobayashi et al., 1999; O'Neill et al., 1996; Pellerin et al., 1996). The latter could not be removed by treatment with cation-exchange resin, indicating that coordinate bonds were likely involved in Ca binding (Kobayashi et al., 1999). These results are consistent with earlier observations that B deficiency induces a decrease in pectin-associated Ca (Yamauchi et al., 1986). It is still unclear, whether Ca interacts directly with the borate anion or bridges RGII at another site, as indicated by ^{11}B-NMR studies (O'Neill et al., 1996). Stabilization of the dB-RGII complex by Ca^{2+} could be one reason, why dB-RGII is surprisingly stable at pH 4.8 (Ishii and Matsunaga, 1996) and why dB-RGII formation is highest at pH values between 3 and 4 (O'Neill et al., 1996).

In *V. faba* root tips, more than 90% of total B was present in a water-insoluble form, irrespective of B supply. This fraction was not measurably altered by up to 6 h of B deprivation (Fig. 2). After removal of exchangeably bound Ca (using $BaCl_2$), insoluble B was significantly lowered to 35% of the initial value after 6 h or more of B depletion. This effect was neither observed in +B plants nor without Ca removal. This is in line with observations, where EDTA treatment led to a break-down of B-RGII complexes (Fleischer et al., 1999). Since Ca removal did not equally affect insoluble B in older root parts, we suggest that in growing root tips, B is bound to other ligands (possibly RGII precursors) before being incorporated into stable B-RGII complexes.

Using different buffer treatments, it was shown that part of the Ca-stabilized B fraction (approx. 5% of total B) was present in the form of a B complex. Without Ca, the complex became soluble, but did not fall apart, indicating that Ca was necessary to hold the complex in its place (Wimmer and Goldbach, 1999a). The Ca-stabilized B complex was not detectable after only 2 h of B deficiency.

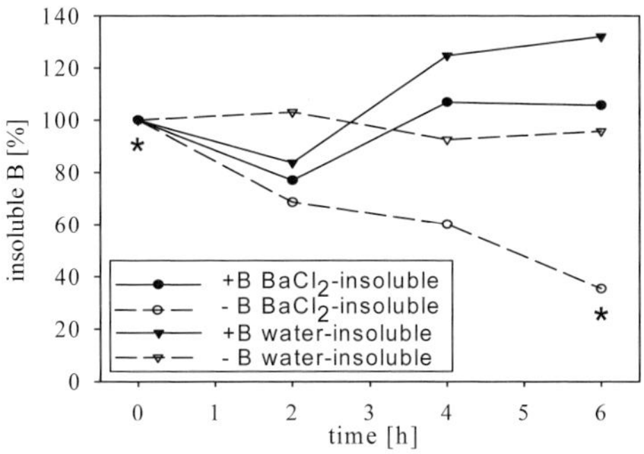

*Fig. 2 Insoluble B in root tips after equilibration in water or $BaCl_2$ solution, relative to the initial B content; mean of 5 replicates; *significant difference at $p<0.05$.*

The results indicate that a relatively large B fraction in root tips is stabilized by Ca under *in vivo* conditions and becomes soluble only under B deficiency and after Ca removal. Our results suggest that a breakdown of weak B complexes attached to Ca in root tips might be involved in early deficiency reactions.

3.3 Characterization of Ca-stabilized B complexes

To test the hypothesis that Ca-stabilized soluble B complexes represent RGII precursors, *V. faba* root segments were analyzed for soluble uronic acids (as marker for cell-wall precursors) and Ca-stabilized soluble B (Fig. 3). Both components are predominantly released from the elongation zone behind the root tip where we also observed the highest concentration of soluble B complexes (not shown).

A direct identification of soluble B complexes using MALDI-FTMS was not possible due to the high amount of unknown components released from roots. We tried to purify the extracts using column chromatography, but most resins interacted with B making them unsuitable for purification.

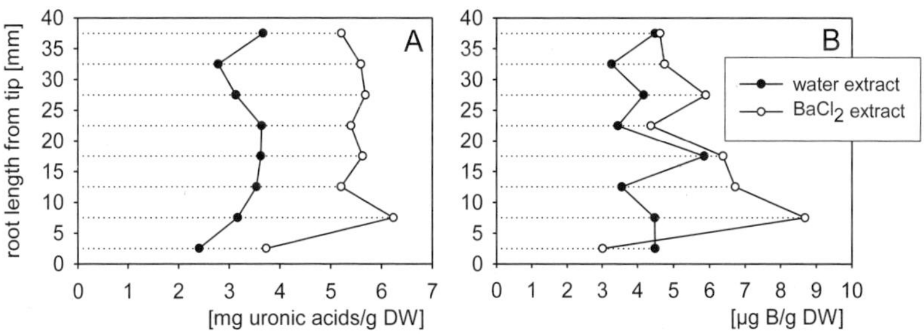

Fig. 3 Profiles of uronic acids (A) and boron (B) released along the root tips of V. faba.

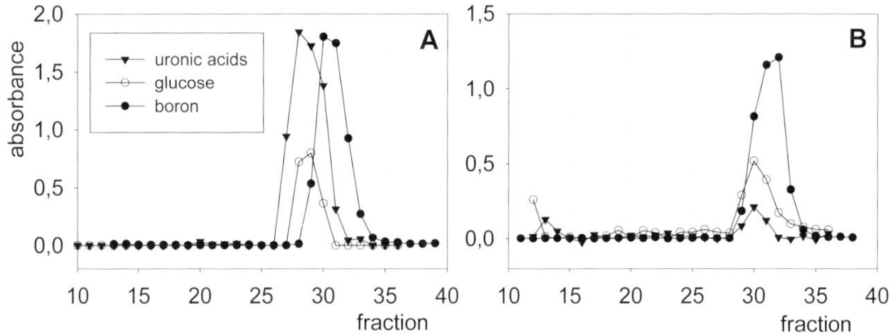

Fig. 4. Column chromatography on Superdex 75 column, standards (A) and V. faba root extract, pre-concentrated by dialysis (B).

The best resin was Superdex 75, where interaction with boric acid was minimal (Fig. 4A). Chromatography of root extracts required pre-concentrations steps (i.e. dialysis), and did not indicate the presence of detectable amounts of B complexes of higher molecular weights (no B-containing peak in early fractions, Fig. 4B). Since soluble B complexes are likely not very stable, they could have been destroyed either during the concentration step or through interactions with the column resin.

Although soluble B complexes could not be positively identified, our results (together with evidence from the literature) are in line with the assumption that Ca-bound soluble B complexes in roots represent cell-wall precursors, and that B might be involved their ordered deposition into the cell wall.

3.4 Relevance of soluble B complexes for B tolerance and influence of salinity

High external B concentrations result in B toxicity. The mechanism of B toxicity is still a matter of speculation, but it is likely that soluble B is involved in the occurrence of B toxicity (Loomis and Durst, 1992). High external B supply could lead to a passive influx of boric acid into the cell, where it is partially converted into borate due to the higher internal pH and might form complexes with a variety of putative ligands in the symplasm (Wimmer et al., 2002). The formation of Ca-bound apoplastic soluble B complexes could represent a tolerance mechanism against toxic B concentrations. Since Na interacts with membrane-bound Ca (Läuchli and Schubert, 1989; Lynch et al., 1987), we wanted to determine the effect of salinity on B tolerance. Wheat was used, because differently B-tolerant cultivars were available.

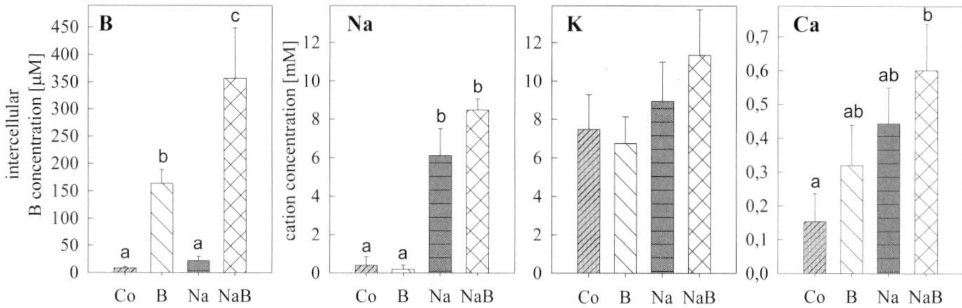

Fig. 5 Soluble intercellular B, Na, K and Ca ion concentrations in wheat leaves; Plants were grown in Control (Co): 5 µM B, B; 200 µM B, Na; 5 µM B, 75 mM NaCl; NaB; 200 µM B and 75 mM NaCl; means of 4 replicates; bars: standard deviations; different letters indicate significant differences between treatments, p<0.05.

In roots, salinity significantly reduced total Ca and K and slightly reduced total B contents (not shown). Without saline conditions, intercellular soluble B concentrations were 8 µM in controls (supply 5 µM) and 163 µM in high B treatments (supply 200 µM), respectively. These values closely reflect the external B supply. Soluble B concentrations were significantly increased by salinity in the intercellular fluids, resulting in an overall enhanced B toxicity (Fig. 5). Intercellular Na^+ concentrations were also increased, but reached a maximum value of only about 8 mM at an external supply of 75 mM. Although Ca^{2+} is mainly localized in the apoplastic space, its soluble intercellular concentration was well below 1 mM. Similar to earlier results with *V. faba*, high B supply increased apoplastic soluble Ca concentrations, an effect that was further enhanced by additional salt stress (Fig. 5).

The results indicate that an interaction between Na and Ca affects B binding in the root apoplast under saline conditions. In combination with observed differences in B uptake rates of the wheat cultivars, this could be one factor influencing B tolerance. An increase in leaf apoplastic soluble B, Ca and protein concentrations and a change in protein pattern could be due to structural alterations of the cell wall (Wimmer et al., 2001, 2002, 2003).

4. VISUALIZATION OF B BINDING SITES USING FLUORESCENT DYES

Metabolic disruptions under B-deficient conditions are highly localized and not all plant tissues are equally responsive to B depletion (Blevins and Lukaszewski, 1998; Lovatt, 1985). Since B function is likely related to the binding of B to an acceptor molecule (Brown et al., 2002), visualization and identification of the respective B binding sites would greatly enhance our understanding of primary B functions.

Boronic acid-coupled fluorescent dyes (B-FITC, B-rhodamine) have been developed (Glüsenkamp et al., 1997). Based on the chemical behavior of boronic acid it was expected that they bind to diols in the same way as boric acid, and can thus be used to visualize B binding sites *in vivo*. When plants were grown in dye-containing solution and root cross-sections incubated in buffers at pH 1–6, binding of both dyes was stable between pH 3–6, as indicated by low release of fluorescence (Fig. 6).

The greatest stability was obtained at pH 3–4, which is the pH maximum of dB-RGII formation (Fleischer et al., 1999). Additionally the dyes were at least partially replaced by increasing concentrations of boric acid (not shown), which further indicates that the dyes and B are competing for the same binding places in plant roots.

Generally, staining with both dyes was most intense in the extension zone of root tips and in tips of root hairs. Within the elongation zone, binding was not uniform, possibly indicating pulsating growth or uneven secretion of B-binding molecules. In root cross-sections, labeling with both dyes was found predominantly in the walls of xylem vessels, the endodermis and epidermal cells (Fig. 7A, B) indicating a high B binding capacity of these cells. Staining was especially intense in B-deficient plants, where the cell shapes appeared irregular (Fig. 7C, D).

The main disadvantage of the method is that the bright fluorescence of the bound dye prohibits the identification of B binding sites at a sub-cellular level. In order to visualize the sub-cellular distribution of B binding sites, a novel technique was developed using immunogold-labelled anti-FITC antibodies as a probe for B-FITC (Edelmann et al., 2000). The immunogold label was mainly concentrated in the cell-wall region and near the

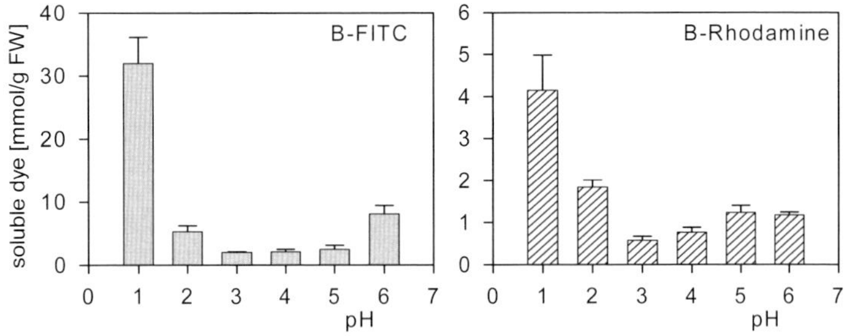

Fig. 6. B-FITC and B-rhodamine released from root cross-sections after incubation in buffers of different pH.

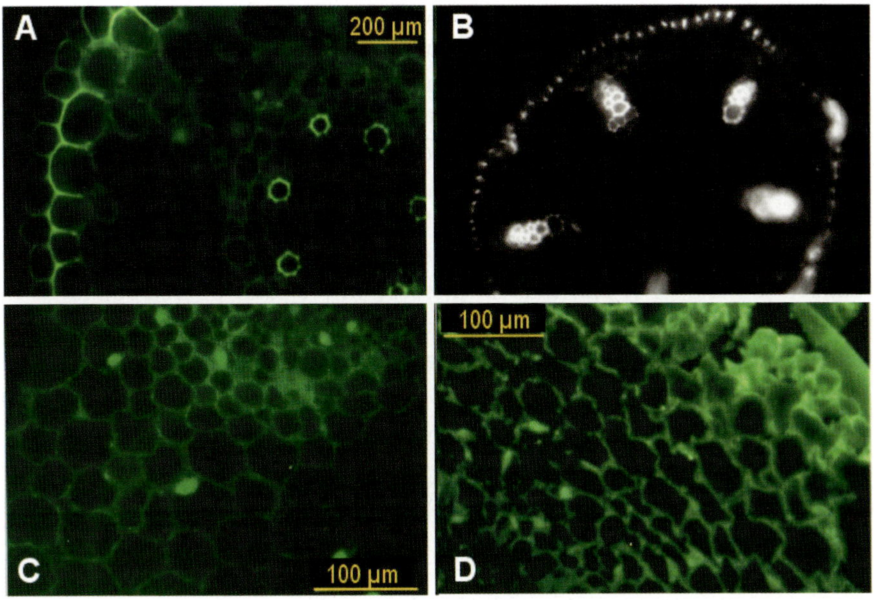

Fig. 7. Visualization of B binding sites in root cross-sections: B-Rhodamine in wheat (A); B-FITC in sunflower (B); B-FITC in +B (C) and –B (D) wheat roots.

plasmalemma within the cytoplasm. High density of label was also observed in a zone likely corresponding to the Casparian strip. Although the method clearly needs optimization, results indicate its suitability for localizing B binding sites with high resolution.

The results show that the distribution of B binding sites in roots corresponds to the reported sites of primary B deficiency reactions. This sustains the assumption that binding of B is a prerequisite for B functions. Boronic acid-coupled fluorescent dyes are useful to visualize B binding sites in plant tissues. However, further development of electron microscopic methods is considered indispensable in order to elucidate the sub-cellular distribution of B binding sites.

5. CONCLUSION

The earliest effects of B deficiency occur within minutes of B deprivation and are unlikely caused by changes in the highly stable B-RGII complex. Our results indicate that soluble apoplastic B and a Ca-stabilized soluble B complex, possibly representing cell-wall precursors, are involved in primary B deficiency reactions. We suggest that a critical level of soluble apoplastic B in the low µM range has to be maintained at all times, similar to that determined for *Vicia faba* roots (0.5 μM) and leaves (approx. 1 μM). The higher critical level in leaves might explain why leaves react more slowly to B deficiency than roots. Results also confirm that low Ca supply might aggravate B deficiency effects in cell walls, and vice versa. The interactions between Ca and B are multifaceted and comprise an immediate alteration in the ratio of free/bound Ca upon a change in B supply, and a much slower reaction of the membrane-bound Ca fraction. Under saline conditions, Na displaces Ca from exchange sites of the roots, which is likely the cause for a reduced binding of B and thus an increase in the ratio of free/bound B. In saline and high B soils, this interaction is suggested to be one of several factors affecting B tolerance. Visualization of B binding sites in roots using boronic acid-coupled fluorescent dyes confirms that B preferably binds to sites known to be especially reactive to B deficiency. This is further evidence that the reaction of B with an acceptor molecule is required for any given function of B in plant tissues. Overall, our results suggest that B plays roles in the apoplast of higher plants beyond that of cross-linking the cell wall, and that a certain level of soluble B fractions, albeit low, has to be maintained in order to avoid B deficiency reactions.

REFERENCES

Bassil, E., Hu, H. and Brown, P. H. (2004). Use of phenylboronic acids to investigate boron function in plants. Possible role of boron in transvacuolar cytoplasmic strands and cell-to-wall adhesion. *Plant Physiol.*, 136, 3383–3395.

Blevins, D. G. and Lukaszewski, K. M. (1998). Boron in plant structure and function. *Annu. Rev. Plant Physiol. Plant Mol. Biol.*, 49, 481–500.

Brown, P. H., Bellaloui, N., Wimmer, M. A., Bassil, E. S., Ruiz, J., Hu, H., Pfeffer, H., Dannel, F. and Römheld, V. (2002). Boron in plant biology. *Plant Biol.*, 4, 211–229.

Carpena, R. O., Esteban, E., Sarro, M. J., Penalosa, J., Garate, A., Lucena, J. J. and Zornoza, P. (2000). Boron and calcium distribution in nitrogen-fixing pea plants. *Plant Science*, 151, 163–170.

Edelmann, H. G., Wimmer, M. A. and Goldbach, H. E. (2000). Immunogold labelling of boric acid binding sites by immunoreaction to the FITC-moiety of FITC-boronic acid: first results. *J. Trace and Microprobe Tech.*, 18, 451–459.

Findeklee, P. and Goldbach, H. E. (1996). Rapid effects of boron deficiency on cell wall elasticity modulus in Cucurbita pepo roots. *Bot. Acta*, 109, 463–465.

Fleischer, A., O'Neill, M. A. and Ehwald, R. (1999). The pore size of non-graminaceaous plant cell walls is rapidly decreased by borate ester cross-linking of the pectic polysaccharide rhamnogalacturonan II. *Plant Physiol.*, 121, 829–838.

Fleischer, A., Titel, C. and Ehwald, R. (1998). The boron requirement and cell wall properties of growing and stationary suspension-cultured *Chenopodium album* L. cells. *Plant Physiol.*, 117, 1401–1410.

Ganmore-Neumann, R. and Davidov, S. (1993). Uptake and distribution of calcium in rose plantlets as affected by calcium and boron concentration in culture solution. *Plant Soil*, 155/156, 151–154.

Glüsenkamp, K.-H., Kosegarten, H., Mengel, K., Grolig, F., Esch, A. and Goldbach, H. E. (1997). A fluorescein boronic acid conjugate as a marker for borate binding sites in the apoplast of growing roots of *Zea mays* L. and *Helianthus annuus* L. In R. W. Bell and B. Rerkasem (eds), *Boron in Soils and Plants*. Kluwer Academic Publishers, Dordrecht, pp. 229–235.

Goldbach, H. E. (1997). A critical review on current hypotheses concerning the role of boron in higher plants: suggestions for further research and methodological requirements. *J. Trace and Microprobe Tech.*, 15(1), 51–91.

Goldbach, H. E., Wimmer, M. A. and Findeklee, P. (2000). Discussion paper: Boron – how can the critical level be defined? *Z. Pflanzenern. Bodenk.*, 163, 115–121.

Goldbach, H. E., Yu, Q., Wingender, R., Schulz, M., Wimmer, M., Findeklee, P. and Baluska, F. (2001). Rapid response reactions of roots to boron deprivation. *J. Plant Nutr. Soil Sci.*, 164, 173–181.

Ishii, T. and Matsunaga, T. (1996). Isolation and characterization of a boron-rhamnogalacturonan-II complex from cell walls of sugar beet pulp. *Carbohydrate Research*, 284, 1–9.

Ishii, T., Matsunaga, T., Pellerin, P., O'Neill, M. A., Darvill, A. and Albersheim, P. (1999). The plant cell wall polysaccharide Rhamnogalacturonan II self-assembles into a covalently cross-linked dimer. *J. Biol. Chem.*, 274, 13098–13104.

Jarvis, B. C., Yasmin, S., Ali, A. H. and Hunt, R. (1984). The interaction between auxin and boron in adventitious root development. *New Phytol.*, 97, 197–204.

Kaneko, S., Ishii, T. and Matsunaga, T. (1997). A boron-rhamnogalacturonan-II complex from bamboo shoot cell walls. *Phytochemistry*, 44, 243–248.

Kobayashi, M., Matoh, T. and Azuma, J.-I. (1996). Two chains of rhamnogalacturonan II are cross-linked by borate-diol ester bonds in higher plant cell walls. *Plant Physiol.*, 110, 1017–1020.

Kobayashi, M., Nakagawa, H., Asaka, T. and Matoh, T. (1999). Borate-rhamnogalacturonan II bonding reinforced by Ca^{2+} retains pectic polysaccharides in higher-plant cell walls. *Plant Physiol.*, 119, 199–203.

Läuchli, A. and Schubert, S. (1989). The role of calcium in the regulation of membrane and cellular growth processes under salt stress. In J. H. Cherry (ed.), *Environmental Stress in Plants*, Vol. G19, Springer-Verlag, Berlin Heidelberg, pp. 131–138.

Loomis, W. D. and Durst, R. W. (1992). Chemistry and biology of boron. *BioFactors*, 3(4), 229–239.

Lovatt, C. J. (1985). Evolution of xylem resulted in a requirement for boron in the apical meristems of vascular plants. *New Phytol.*, 99, 509–522.

Lynch, J., Cramer, G. R. and Läuchli, A. (1987). Salinity reduces membrane-associated calcium in corn root protoplasts. *Plant Physiol.*, 83, 290–294.

Matoh, T., Ishigaki, K.-I., Ohno, K. and Azuma, J.-I. (1993). Isolation and characterization of a boron-polysaccharide complex from radish roots. *Plant Cell Physiol.*, 34, 639–643.

Mühling, K. H. and Sattelmacher, B. (1995). Apoplastic ion concentration of intact leaves of field bean *(Vicia faba)* as influenced by ammonium and nitrate nutrition. *J. Plant Physiol.*, 147, 81–86.

Mühling, K. H., Wimmer, M. and Goldbach, H. E. (1998). Apoplastic and membrane-associated Ca^{2+} in leaves and roots as affected by boron deficiency. *Physiologia Plantarum*, 102, 179–184.

O'Neill, M., Eberhard, S., Albersheim, P. and Darvill, A. (2001). Requirement of borate cross-linking of cell wall rhamnogalacturonan II for *Arabidopsis* growth. *Science*, 294, 846–849.

O'Neill, M. A., Warrenfeltz, D., Kates, K., Pellerin, P., Doco, T., Darvill, A. G. and Albersheim, P. (1996). Rhamnogalacturonan-II, a pectic polysaccharide in the walls of growing plant cell, forms a dimer that is covalently cross-linked by a borate ester. *J. Biol. Chem.*, 271, 22923–22930.

Pellerin, P., Doco, T., Vidal, S., Williams, P., Brillouet, J.-M. and O'Neill, M. A. (1996). Structural characterization of red wine rhamnogalacturonan II. *Carbohydrate Research*, 290, 183–197.

Ralston, N. V. C. and Hunt, C. D. (2000). Biological boron interactions: Charge and structure characteristics required for boroester formation with biomolecules. *FASEB Journal*, 14, A538.

Ralston, N. V. C. and Hunt, C. D. (2001). Diadenosine phosphates and S-adenosylmethionine: novel boron binding biomolecules detected by capillary electrophoresis. *Biochimica et Biophysica Acta*, 1527, 20–30.

Redondo-Nieto, M., Rivilla, R., El-Hamdaoui, A., Bonilla, I. and Bolanos, L. (2001). Boron deficiency affects early infection events in the pea-Rhizobium symbiotic interaction. *Aust. J. Plant Physiol.*, 28, 819–823.

Ryden, P., Sugimoto-Shirasu, K., Smith, A. C., Findlay, K., Reiter, W. D. and McCann, M. C. (2003). Tensile properties of *Arabidopsis* cell walls depend on both a xyloglucan cross-linked microfibrillar network and rhamnogalacturonan II-borate complexes. *Plant Physiol.*, 132, 1033–1040.

van Duin, M., Peters, J. A., Kieboom, A. P. G. and VanBekkum, H. (1987). Synergic coordination of calcium in borate-polyhydroxycarboxylate systems. *Carbohydrate Research*, 162, 65–78.

Wang, H. Y., Wang, Y. H., Du, C. W., Xu, F. S. and Yang, Y. H. (2003). Effects of boron and calcium supply on calcium fractionation in plants and suspension cells of rape cultivars with different boron efficiency. *J. Plant Nutr.*, 26(4), 789–806.

Wimmer, M. (2000). Untersuchungen zur Funktion von Bor im Apoplasten der Ackerbohne (*Vicia faba* L.). Bonner Agrikulturchemische Reihe, Band 3. Bonn, Germany.

Wimmer, M. A. and Goldbach, H. E. (1999a). Influence of Ca^{2+} and pH on the stability of different boron fractions in intact roots of *Vicia faba* L. *Plant Biol.*, 1, 632–637.

Wimmer, M. A. and Goldbach, H. E. (1999b). A miniaturized curcumin method for the determination of boron in solutions and biological samples. *Zeitschrift für Pflanzenernährung und Bodenkunde*, 162, 15–18.
Wimmer, M. A., Mühling, K. H., Läuchli, A., Brown, P. H. and Goldbach, H. E. (2001). Interaction of salinity and boron toxicity in wheat (*Triticum aestivum* L.). In W. J. Horst et al. (eds), *Plant Nutrition: Food security and sustainability of agro-ecosystems through basic and applied research*. Kluwer Academic Publishers, Dordrecht.
Wimmer, M. A., Mühling, K. H., Läuchli, A., Brown, P. H. and Goldbach, H. E. (2002). Boron toxicity: the importance of soluble boron. In H. E. Goldbach et al. (eds.), Boron in Plant and Animal Nutrition. Kluwer Academic Publishers, New York.
Wimmer, M. A., Mühling, K. H., Läuchli, A., Brown, P. H. and Goldbach, H. E. (2003). The interaction between salinity and boron toxicity affects the subcellular distribution of ions and proteins in wheat leaves. *Plant, Cell and Environ.*, 26, 1267–1274.
Yamauchi, T., Hara, T. and Sonoda, Y. (1986). Distribution of calcium and boron in the pectin fraction of tomato leaf cell wall. *Plant Cell Physiol.*, 27(4), 729–732.
Yu, Q., Baluska, F., Jasper, F., Menzel, D. and Goldbach, H. E. (2003). Short-term boron deprivation enhances levels of cytoskeletal proteins in maize, but not zucchini, root apices. *Physiol. Plant.*, 117(2), 270–278.
Yu, Q., Hlavacka, A., Matoh, T., Volkmann, D., Menzel, D., Goldbach, H. E. and Baluska, F. (2002). Short-term boron deprivation inhibits endocytosis of cell wall pectins in meristematic cells of maize and wheat root apices. *Plant Physiol.*, 130(1), 415–421.
Yu, Q., Wingender, R., Schulz, M., Baluska, F. and Goldbach, H. E. (2001). Short-term boron deprivation induces increased levels of cytoskeletal proteins in *Arabidopsis* roots. *Plant Biol.*, 3(4), 335–340.

SILICON IN PLANT NUTRITION

Effects on zinc, manganese and boron leaf concentrations and compartmentation

H. WIESE[1], M. NIKOLIC[2] and V. RÖMHELD[3]

[1]*Institut für Pflanzenernährung, Justus-Liebig-Universität Gießen, Germany, heike.wiese@ernaehrung.uni-giessen.de*

[2]*Centre for Multidisciplinary Studies of the Belgrade University, Serbia*

[3]*Institut für Pflanzenernährung (330), Universität Hohenheim, Germany*

Abstract. Silicon (Si), taken up as $Si(OH)_4$ by plants, is transported and deposited mainly in the apoplast since Si transport and distribution follows that of water. This makes it rather likely that it influences the physical and chemical properties of the apoplast. In order to investigate the effect of Si on the properties of the leaf apoplast, mineral concentrations and binding forms of ions in the cell walls and intercellular washing fluid were determined. Three mineral element/silicon interactions were the focus of our study: a) the influence of Si on phosphate-induced zinc deficiency, b) effects of Si on exchange capacity and binding forms of manganese in the leaf apoplast and c) silicon/boron interactions. Silicon was shown to influence in particular the compartmentation of zinc, boron, and manganese.

Key words: binding forms, cell wall, compartmentation, silicon–boron interaction, silicon–manganese interaction, silicon–zinc interaction

1. INTRODUCTION

Although silicon (Si) is not known to be an essential mineral element for most higher plants, it has many direct and indirect beneficial effects on their growth and development (Epstein, 1994, 1999). The supply of Si can increase resistance to fungal diseases (Samuels et al., 1994; Blaich and Grundhofer, 1998; Liang et al., 2005; Wiese et al., 2005), improve mechanical stability of stems and leaf blades (Idris et al., 1975; Adatia and Besford, 1986; Rafi et al., 1997), raise tolerance to excess supply of manganese (Williams and Vlamis, 1957; Horst and Marschner 1978a,

B. Sattelmacher† and W.J. Horst (eds.), The Apoplast of Higher Plants: Compartment of Storage, Transport and Reactions, 33–47.

1978b; Rogalla and Römheld 2001, 2002a) and aluminum (Galvez et al., 1987; Barceló et al., 1993; Corrales et al., 1997; Cocker et al., 1998a, b, c; Wang et al., 2004), as well as to salt stress (Ahmad et al., 1992; Liang et al., 1996).There has also been some discussion that Si can increase the physiological availability of zinc in leaf tissue (Marschner et al., 1990).

There is great variation in uptake of Si between plant species. Takahashi et al. (1990) divided plants into four classes depending on leaf Si concentration. Excluders ("–") contain less than 0.5% Si (expressed in relation to the dry weight), and the three groups of accumulators contain Si over a range from "+" with 0.5–2% Si; "++" with 2–4% Si to "+++" with more than 4% Si. Rice (*Oryza sativa* L.), is an example of the last group ("+++"). Most grasses have Si concentrations between 1–2%, whereas most dicots are excluders. An exception of this is cucumber (*Cucumis sativus* L.) in which Si concentration in the leaves is comparable to those of graminaceous species (Table 1).

In this context the high concentration of Si as compared with other minerals should be emphasized. Plants treated with Si result in Si concentrations comparable to those of the macronutrients K and Ca. Even in plants cultivated on a standard nutrient solution prepared from salts puriss. pro-analysis grade using double distilled water and without any additional Si supply, reach concentrations comparable to those of the micronutrients Fe, Zn, Mn and Cu (Table 1). This observation illustrates the difficulties in obtaining plants free from Si to prove the essentiality of this element (Epstein, 1994).

It has been supposed that plants which accumulate high amounts of Si, such as rice, do so by an active uptake mechanism (Barber and Shone, 1966) whereas Si excluders actively exclude Si. However, the Si concentration of the plant may not be a reliable indicator of uptake mechanisms. Therefore, Raven (1983) characterized plant species depending on their Si concentration

Table 1. Si concentration of old leaves (in relation to the dry weight) in a variety of plant species treated with 0.9 mM or 1.8 mM Si (according to Rogalla, 2001).

		Si concentration					
		0 mM Si supply		0.9 mM Si supply		1.8 mM Si supply	
Plant species	D	(mmol g^{-1} DW)	(%)	(mmol g^{-1} DW)	(%)	(mmol g^{-1} DW)	(%)
Zea mays	+	n.d.	–	0.28	0.8	0.60	1.7
Hordeum vulgare	++	0.004	0.01	0.71	2.0	1.00	2.8
Ricinus communis	–	n.d.	–	0.04	0.1	0.04	0.1
Vicia faba	–	n.d.	–	0.04	0.1	0.04	0.1
Solanum tuberosum	–	n.d.	–	0.04	0.1	0.04	0.1
Cucumis sativus	+/++	0.053	0.15	0.64	1.8	0.86	2.4

*n.d. not determined; D: degree of accumulation according to Takahashi et al. (1990).

in the xylem sap as compared with that of the soil solution. This could be less (e.g. legumes), equal (e.g. oat, *Avena sativa* L.) or higher (e.g. rice and *Equisetum* sp.) corresponding to an active Si exclusion, a passive, or an active Si uptake mechanism, respectively. Nevertheless, this classification is also not clear–cut since both the findings of Vorm (1980) and our own results (Rogalla, 2001) showed that uptake mechanisms can shift from exclusion to passive uptake or from passive to active uptake when Si supply in the growth medium becomes rather low. In addition, barley (*Hordeum vulgare* L.) takes up Si actively when transpiration is low e.g. under conditions of high humidity (Barber and Shone, 1966). These authors also showed that Si uptake in barley depended on temperature and could be inhibited by metabolic inhibitors. These results are not in agreement with a solely passive Si uptake mechanism. Recently, active uptake in rice has been further characterized (Ma et al., 2002, 2004; Mitani and Ma, 2005) and also proved for cucumber (Liang et al., 2005; M. Nikolic, Y. Liang and V. Römheld, unpublished data). Preliminary results indicate that, xylem loading of Si is the primary active process in rice and cucumber and similar to boron uptake by *Arabidopsis thaliana* L. and tomato (*Lycopersicon esculentum* Mill.) (Dannel et al., 2002), whereas uptake into the root symplast is of lesser importance (Y. Liang and V. Römheld, unpublished data).

Since the uptake of undissociated silicic acid can be passive (Raven, 1983) and Si root-to-shoot transport takes place in the xylem, it was concluded that Si distribution follows the transpiration stream (Jones and Handreck, 1965; Handreck and Jones, 1968). When the water is transpired into the atmosphere, monosilicic acid concentrates in the remaining solution and the concentrated silicic acid solution polymerizes in the cell walls of epidermis and mesophyll cells (Handreck and Jones, 1968). Therefore, with the exception of Si excluders, Si is deposited mainly at sites of high transpiration. Accordingly, the Si content of a plant organ reflects its transpiration. This applies e.g. to graminaceous species such as oat, which has a passive uptake mechanism (Jones and Handreck, 1965). In accordance with this, the distribution of Si within cucumber and barley plants can be explained by the transport of Si along the transpiration stream (Rogalla, 2001). Thus, Si concentrations of shoots were higher than of roots, Si concentrations of old leaves were higher than of young leaves, and comparing the basal parts of graminaceous leaves with leaf tips, the Si concentration in old parts of the leaves were higher than in younger parts (Table 2).

The significance of the apoplast for Si deposition can be shown by the Si compartmentation in leaves of cucumber and barley (Table 3). More than

Table 2. Si concentration and distribution in 46 days barley and 42 days old cucumber plants grown on a nutrient solution containing 1.8 mM Si (Rogalla, 2001).

Barley	Root	Stalk	Basis of old leaves	Tips of old leaves	Young leaves
(mmol g^{-1} DW)	0.17	0.21	0.33	1.63	0.47
(%)	8.0	6.3	9.7	51.3	24.7
Cucumber	Root	Stalk	Petiole	Old leaves	Young leaves
(mmol g^{-1} DW)	0.27	0.23	0.08	0.86	0.41
(%)	21.2	2.0	0.5	71.2	5.2

Table 3. Si concentration and distribution in leaf compartments of barley and cucumber plants grown on a nutrient solution containing 1.8 mM Si (Rogalla, 2001).

Leaf compartment	Barley (mmol g^{-1} DW)	(%)	Cucumber (mmol g^{-1} DW)	(%)
Total	0.990	100.0	0.860	100.0
Cell wall	0.900	92.0	0.780	91.0
Cell sap	0.007	0.7	0.004	0.5
Apoplastic fluid	0.001	0.1	0.001	0.1
Loss by cell wall isolation		7.2		8.4

90% of total Si content was localized in cell walls. Only 0.1% occurred in the apoplastic fluid and less than 1% in the cell sap, thus in the symplast. This is in accordance with the results of Heine et al. (2005).

Kaufman et al. (1985) found only negligible amounts of intracellular deposited Si in barley, contrary to other grasses, which store Si in certain types of epidermis cells. The difference between the sum of Si concentrations in particular fractions and the total Si content of leaves was interpreted as a loss of Si binding sites during cell wall isolation.

In contrast to rice (Islam and Saha, 1969; Ma and Takahashi, 1990, 1991, 1993), Si supply did not change mineral contents of cucumber or barley (Rogalla, 2001). Hammond et al. (1995), however, observed a decrease in Ca content by increased Si supply in barley, but only at a much higher Si supply (2.8 mM) than used by Rogalla (2001).

In spite of equal total mineral contents, changes in mineral concentration in the intercellular washing fluid of cucumber leaves by Si could be observed (Rogalla, 2001). Thus, manganese (Mn) and boron (B) concentrations were significantly lower and zinc (Zn) concentration tended to be higher in Si-treated plants compared with non Si-treated plants. These differences in compartmentation due to Si supply are discussed separately in detail with respect to zinc, manganese and boron nutrition (see below). In barley no effect of Si supply on mineral concentration in the different fractions of apoplastic fluid could be shown (Rogalla, 2001).

2. THE INFLUENCE OF SILICON ON PHOSPHATE-INDUCED ZINC DEFICENCY

Miyake and Takahashi (1978, 1983) reported that tomato and cucumber growing in nutrient solution without additional Si supply, showed symptoms such as leaf chlorosis and a reduced growth. The authors concluded that these symptoms of disorder were due to Si deficiency. However, the nutrient solutions of these experiments contained high concentrations of P and low concentrations of Zn. It is well known that there are strong interactions between Zn and P in plants. Thus, a high P supply might lead to Zn deficiency and subsequently to a Zn deficiency-induced high accumulation of P in the shoot (Cakmak and Marschner, 1986; Marschner and Cakmak, 1986). Marschner et al. (1990) were able to show that a reduction of P supply as well as an increase in Zn supply prevented this disorder described by Miyake and Takahashi (1978, 1983). Therefore, the so-called Si deficiency symptoms were caused by a P-induced Zn deficiency. Since Si supply also prevented the symptoms of that disorder, Si influenced adverse P/Zn interactions in plants. Marschner et al. (1990) were able to demonstrate an increase in the proportion of water-soluble to total Zn, and an avoidance of toxic P accumulation as a consequence of an elevated Si supply. Consequently, symptoms of P-induced Zn deficiency did not appear if plants were treated with Si. They postulated that Si supply might increase the Zn availability within the leaves and in particular in the leaf appoplast. Cakmak and Marschner (1987) suggested that the relative portion of water-soluble Zn is a better marker for the Zn nutritional status of a plant than the concentration of total Zn particularly under condition of high P supply. A higher portion of water-soluble Zn would lead to a more balanced Zn/P-relationship within the plant (Cakmak and Marschner, 1986; Marschner and Cakmak, 1986). However, this effect of Si supply seems to appear only under very special conditions. Contrary to the investigations of Marschner et al. (1990) no influence of Si on P-induced Zn deficiency could be observed in different nutrient solution experiments by Rogalla (2001). As a consequence of the high P and low Zn supply in the nutrient solution, Zn concentration in the leaves of cucumber were below the critical range for Zn deficiency and P concentrations far exceeded the critical range for P toxicity according to Bergmann (1988) (Table 4).

The plants showed typical symptoms of P-induced Zn deficiency like leaf chlorosis. However, plant growth was hardly affected. Both Si concentration as well as Zn and P concentration in plants of these different experiments were comparable to that of Marschner et al. (1990). Therefore, differences in Si, P or Zn nutritional status could not account for the different results.

Table 4. Plant dry weight, P concentration and content, Zn concentration, and water-soluble Zn in leaves of cucumber plants growing at different Zn supplies (0.05, 0.1 or 1.0 μM) with Si (+Si; 1.8 mM) or without (–Si) Si supply (Rogalla, 2001).

Si supply	–Si	+Si	–Si	+Si	–Si	+Si
Zn supply (μM)	0.05	0.05	0.1	0.1	1.0	1.0
Dry weight per plant	4.96	5.09	5.49	4.69	5.79	6.00
(g)	± 1.4	± 0.9	± 0.6	± 1.5	± 1.8	± 2.3
Zn concentration	0.31	0.36	0.37	0.45	2.12	1.87
(μmol g^{-1} DW)	± 0.05	± 0.03	± 0.05	± 0.14	± 0.33	± 0.28
P concentration	1.56	1.10	1.35	1.11	0.66	0.78
(mmol g^{-1} DW)	± 0.09	± 0.10	± 0.20	± 0.56	± 0.23	± 0.27
P content	2.16	2.19	2.04	2.14	0.98	1.54
(mmol $leaf^{-1}$)	± 0.90	± 0.73	± 0.92	± 1.82	± 0.65	± 1.10
Water-soluble Zn	37.3	40.5	49.5	40.6	29.9	28.6
(% of total Zn content)	± 10.9	± 8.5	± 11.4	± 13.6	± 16.3	± 8.2

P supply was 1 mM; critical value for Zn deficiency 0.3 μmol g^{-1} DW, and for P toxicity 0.65 mmol g^{-1} DW (according to Bergmann, 1988); data are means of 3 replicates ± SD.

However, an increase in the ratio of water-soluble to total Zn content by Si, as reported by Marschner et al. (1990), could not been shown. The fact that in the experiments of Rogalla (2001) Si supply did not result in the expected avoidance of an accumulation of toxic P concentrations in the shoots could probably be attributed to the fact that the ratio of water-soluble Zn to the total Zn content was not increased by Si supply because of so far unknown growth factors (e.g. growth conditions, developmental stage of the plants). This suggestion leaves open the possibility that an increase in the ratio of water-soluble to total Zn plays an important role in the avoidance of P-induced Zn deficiency. In order to avoid toxic levels of P accumulation, slight changes in this relationship may be critical. Furthermore, the high variability of Zn concentration in the apoplastic fluid (Rogalla, 2001) is an indication that this parameter is very variable. In conclusion, a Si-mediated increase in internal Zn availability, seems to be achieved only under very specific conditions. This corresponds to the fact that Miyake and Takahashi (1978) also reported that the "Si deficiency symptoms" only appeared under distinct conditions.

3. EFFECTS OF SILICON ON EXCHANGE CAPACITY AND BINDING FORMS OF MANGANESE IN THE LEAF APOPLAST

Manganese (Mn) is an essential element for plant growth. However, it can be toxic when applied in excess to the growth medium. Mn toxicity symptoms differ very much between species (summarized in El-Jaoual and

Cox, 1998). Under conditions of high light intensities the formation of reactive oxygen species, especially hydroxyl radicals, can be an important factor for the appearance of Mn toxicity symptoms (González et al., 1998; J. Dragisic Maksimovic, pers. comm.). Often the development of brown spots on older leaves, primarily in the cell walls of epidermic cells, and the browning of roots are characteristic of the toxicity (Williams and Vlamis, 1957; Horst and Marschner, 1978a, 1978b). These brown spots are locations of high concentrations of oxidised Mn (Horiguchi, 1987) and oxidised phenolic compounds (Wissemeier and Horst, 1992; Fecht-Christoffers et al., this volume, pp. 307–322).

There are great differences in Mn tolerances between plant species and also between genotypes. Resistance against high Mn concentration in the growth medium can be achieved by decreased Mn uptake, a decreased Mn transport to the shoots, or by an increased tissue tolerance (El-Jaoual and Cox, 1998). Furthermore many factors can affect Mn tolerance, e.g. leaf age and form of N nutrition (Horst et al., 1999). Recent reports emphasise the importance of the leaf epidermis in Mn tolerance. Thus, González and Lynch (1999) found higher Mn concentrations in the epidermis of Mn-tolerant than in Mn-sensitive bean (*Phaseolus vulgaris* L.) cultivars. It would be of interest to know whether Si can further enhance Mn detoxification in the epidermis of the Mn-tolerant cultivars, since it is well known, that Si can increase Mn tolerance in many plant species and Si is mainly located in cell walls of epidermis cells (Hodson and Sangster, 1989a; Bode et al., 1994).

Understanding the mechanism of Si-mediated increased Mn tolerance has been the object of many investigations. In rice it is assumed that Si enhances aerenchym formation and stability (Islam and Saha, 1969; Ma and Takahashi, 1990, 1991). Therefore, oxygen transport to the root and further

Table 5. Effect of increasing Mn supply on fresh weight and Mn toxicity symptoms of cucumber plants growing in nutrient solution with (+Si; 1.8 mM) or without Si (–Si) (according to Rogalla and Römheld, 2002a).

Si supply	Manganese concentration (μM)				
	0.5	5	50	500	1000
	Appearence of manganese toxicity symptoms				
–Si	–	–	+	++	+++
+Si	–	–	–	+	++
	Fresh weight per plant (g)				
Shoot, –Si	17.43 ± 6.96	14.98 ± 4.99	8.32 ± 0.35	6.13 ± 5.81	4.20 ± 0.99
Root, –Si	22.90 ± 9.14	18.01 ± 6.00	6.64 ± 0.28	6.19 ± 6.01	5.52 ± 1.30
Shoot, +Si	16.27 ± 0.42	15.70 ± 3.82	16.77 ± 0.21	15.83 ± 0.51	8.24 ± 5.10
Root, +Si	20.76 ± 0.54	22.40 ± 5.45	20.19 ± 0.26	24.30 ± 0.78	7.42 ± 4.59

*– none, + marginal, ++ strong, +++ severe Mn toxicity symptoms; the data are means of 2 replicates ± SD.

to the rhizosphere is enhanced and soluble Mn^{2+} is oxidised and precipitated as MnO_2 at the root surface. This results in a decreased uptake of Mn in Si-treated rice plants. Also in plant species, like Sudan grass [*Sorghum sudanense* (Piper) Stapf.] (Bowen, 1972) and sorghum [*Sorghum bicolor* (L.) Moench] (Galvez et al., 1989), Si supply decreased uptake of Mn. In other plant species, however, such as barley and bean Si does not affect Mn uptake in spite of increased Mn tolerance of leaves. For both plant species, the distribution of Mn obviously is more homogeneous in the leaves of Si-treated plants than in those not treated with Si (Williams and Vlamis, 1957; Horst and Marschner, 1978c). Also in cucumber Si clearly decreased symptoms of Mn toxicity (Table 5).

The development of brown spots on leaves, a typical symptom of Mn toxicity (Williams and Vlamis, 1957; Horst and Marschner, 1978b), is reduced by Si application. Despite equal total Mn concentrations in the leaves of plants treated with high Mn, plants not treated with Si have higher Mn concentrations in the intercellular washing fluid compared to plants treated with Si, especially in the $BaCl_2$- and DTPA-exchangeable fraction of the apoplast (Rogalla, 2001; Rogalla and Römheld, 2001, 2002a). In Si-treated plants less Mn was located in the symplast (<10%) and more Mn

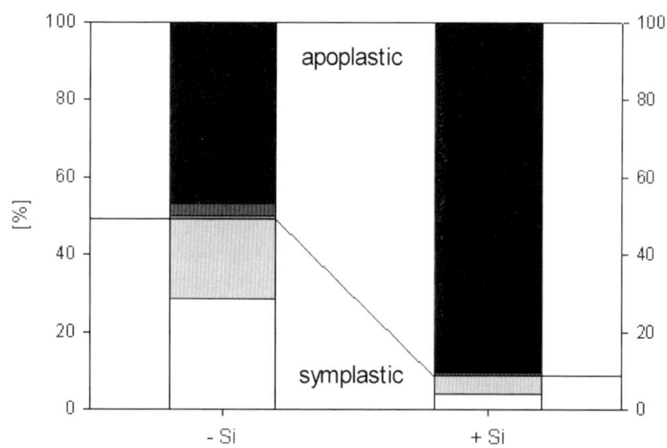

Fig. 1. Relative Mn distribution in leaves calculated on leaf fresh weight basis (according to Rogalla and Römheld, 2002a). Cell walls were isolated according to the method of Goldberg (1985) from leaves of cucumber plants treated with high Mn (100 µM) without (–Si) or with 1.8 mM Si supply (+Si). The isolated cell walls were washed subsequently with $BaCl_2$, DTPA, and HCl. The different Mn fractions are plotted as follows: symplastic – water-extractable, white; Triton-X-extractable, light grey; apoplastic – apoplastic washin fluid, grey; cell wall (weakly bound), dark grey; cell wall (strongly bound), black.

was bound to the cell wall (>90%) compared with non Si-treated plants (about 50% in each compartment). Therefore, it is concluded, that Si causes a stronger binding of Mn to the cell wall (Rogalla and Römheld, 2002a; Fig. 1). Thus, the Mn-concentration is reduced both in the intercellular washing fluid and in the press sap. Therefore, Mn present in Si-treated plants is less available and thus less toxic than in Si-untreated plants. Even low Si concentrations are able to increase Mn tolerance (Rogalla and Römheld, 2001). Furthermore, Si already bound to the cell wall prior to the introduction of a high supply of Mn can cause enhanced Mn tolerance which implies that there is no need for a simultaneous Si and Mn supply. This is in contrast to the Si-induced amelioration of Al toxicity as a consequence of the formation of less phytotoxic hydroxyaluminium-silicates in the apoplast of the root apex (Wang et al., 2004). The Mn concentration of the intercellular washing fluid correlated positively with the severity of Mn-toxicity symptoms and negatively with Si supply. This strong correlation is evidence that Si-mediated binding of Mn to the cell wall is the main mechanism of increased Mn tolerance in cucumber (Rogalla and Römheld, 2001). Furthermore, it has recently been demonstrated that Si-induced binding of Mn in the cell walls is followed by diminishing an EPR signal for hydroxyl radicals (directly generated from free Mn via Fenton reaction) in the apoplastic fluid of cucumber leaves (J. Dragisic Maksimovic and M. Nikolic, unpublished data).

Silicon-mediated binding of Mn to the cell wall as a means of detoxification is not restricted to Mn, because binding of Zn (Neumann et al., 1997) and Al (Hodson and Evans, 1995; Cocker et al., 1998b) may serve a similar process. In the case of Zn, Neumann et al. (1997) could show that this metal is detoxified in heavy-metal-tolerant *Minuartia verna* (L.) Hiern, as zinc silicate in the leaf epidermal cell walls. However, the methods employed in the study of Rogalla and Römheld (2001, 2002a) do not allow any conclusion concerning the nature of Mn binding to the cell wall. Other methods, such as NMR spectroscopy would have to be used in future to clarify the nature of Mn binding to the cell wall in Si-treated plants of cucumber.

This mechanism may also explain Si-mediated enhanced tissue tolerance of high Mn in other plant species than cucumber. For instance, in pumpkin (*Cucurbita moschata* Duch.) it appears that Si can alleviate Mn toxicity through a localised accumulation of Mn together with Si in a metabolically inactive form around the base of trichomes on the leaf surface (Iwasaki and Matsumura, 1999). The mechanism of this accumulation and detoxification of Mn in the epidermis is not yet clear but one may speculate that Mn binding to the cell wall of the epidermal cells may be involved as in the case of cucumber. It is probable, however, that a mechanism of Si-mediated binding of Mn to the cell wall is not the only mechanism by which Si is able to affect tissue tolerance to Mn. Detoxification of Mn by storage in the vacuoles may play an important role in Si-mediated Mn tolerance in bean (Horst and Marschner, 1978a, c). Since in cowpea (*Vigna unguiculata* (L.) Walp.) higher Mn concentrations were found in the apoplastic washing fluid of $BaCl_2$-infiltrated leaves of non Si-treated plants than in Si-treated plants (Horst et al., 1999), Mn binding to the cell wall similarly to that in cucumber is to be expected. However, in cowpea unlike in cucumber this appeared to play only a minor role in Si-mediated Mn tolerance (Iwasaki et al., 2002a,b).

Williams and Vlamis (1957) reported an increased tissue tolerance of barley due to Si treatment. However, in the experiments of Rogalla (2001) no differences in the Mn concentrations in the apoplastic fluid of Si-treated and non Si-treated barley were detected. A reason for this could be the low Mn supply (0.05 μM Mn) for barley which did not lead to substantial binding of Mn to the cell wall. The above shown effects in cucumber could only be demonstrated at Mn supplies >0.5 μM (Rogalla, 2001). In addition, Si-treated cucumber plants were not more sensitive to Mn deficiency than plants without additional Si supply (Rogalla, 2001) as it was described for soybean (*Glycine max* (L.) *Merrill.*, Kluthcouski, 1980). This could have been expected assuming a stronger binding of Mn to the cell wall causing a decreased physiological Mn availability.

4. SILICON/BORON INTERACTIONS

Silicon and boron (B) are elements with many similar chemical properties. Both are weak, undissociated acids in aqueous solution and can complex readily with polyhydroxy compounds like sorbitol or mannitol (Brown et al., 1999; Kinrade et al., 1999). Plants take up Si as monosilicic acid and B as boric acid. The distribution of both elements within the plant depends strongly on the transpiration stream. Dicotyledonous and graminaceous species differ not only in their capacity for Si uptake but also in their B and Ca requirement, which is inversely related (Loomis and Durst, 1992). Due to these similarities, interactions between Si and B are probable, but there is little information available regarding this aspect. Nable et al. (1990) reported positive correlations between Si and B uptake within different genotypes of barley. Thus, genotypes with a lower Si uptake also showed a lower B uptake and were, therefore, more tolerant of high B supply. However, there are no reports on a direct Si/B interaction within these genotypes of barley as shown for oilseed rape (*Brassica napus* L.) (Liang and Shen, 1994). These authors found that under conditions of sufficient or elevated B supply Si supply led to a decrease in B uptake, whereas under conditions of B deficiency Si supply enhanced B uptake in oilseed rape. Also in *Lilium longiflorum* L., Si/B interactions were observed (Polster and Schwenk, 1992). In this plant species Si supply increased the range between critical deficiency and toxicity concentrations for B. In cucumber Si supply had no effect on total B concentration of leaves, but interacted with B compartmentation especially at high B supply (Fig. 2; Rogalla 2001). Thus, more B was bound to the cell wall and less B was in the press sap of leaves of Si-treated plants, which was further correlated with a lower B concentration of the intercellular washing fluid.

It is well known that B plays an important role in phenol metabolism, ascorbic acid synthesis, and peroxidase binding (Goldbach et al., 1991). However, in spite of the above-mentioned effect of Si on B availability, no significant changes in phenol metabolism and redox status of the leaf apoplast could be found (Rogalla and Römheld, 2002b). Furthermore, Si/B interactions seem to have a negligible effect on the physiology of plants, e.g. tolerance to B deficiency or B toxicity.

B uptake was long considered as a purely passive mechanism (Brown and Hu, 1994). Independent of the Si supply cucumber showed a concentration mechanism for B in the xylem exudate at a low external B supply (1 and 10 μM, respectively) (Rogalla, 2001). Therefore, like sunflower (*Helianthus annuus* L.) (Dannel et al., 1998) cucumber also seems to have an active B uptake mechanism when B supply is low.

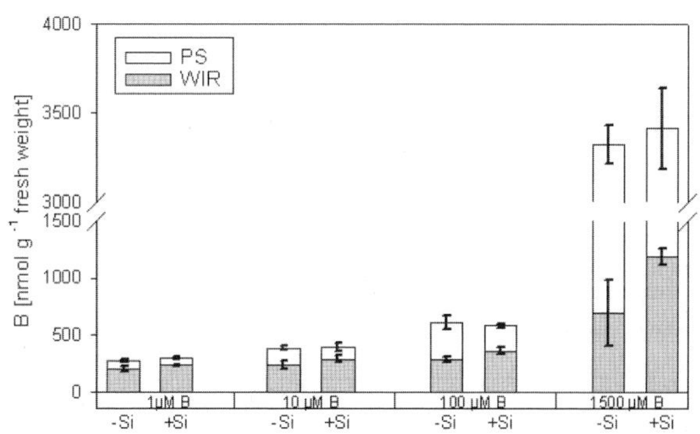

Fig. 2. B concentration of press sap (PS) and water-insoluble residue (WIR) of leaves of 32 days-old cucumber plants at varied Si and B supplies (Rogalla and Römheld, 2002b). The data are means of 5 replicates ± SD.

5. CONCLUSIONS

Even though the proof of Si as an essential mineral element for plant nutrition is still lacking due to the difficulty of cultivating Si-free plants, it can be clearly shown that Si nutrition leads to changes in the mineral nutrition. Above all, Si mediated changes in mineral element compartmentation seem to play an important role, as in the case of Mn and B in cucumber plants. These changes become distinct mainly under conditions of imbalanced nutrient supply when Si is able to directly affect plant physiology, e.g. in Si-enhanced Mn tolerance, Al resistance and the Zn-deficiency/P toxicity interaction.

ACKNOWLEDGEMENT

The authors dedicate this work to the late Prof. Dr. Drs. h.c. Horst Marschner who initiated this project. The authors are thankful to Ernest Kirkby for critical comments to the manuscript. M.N. thanks the Serbian Ministry of Science and Environmental Protection (grant no. 143020B).

REFERENCES

Adatia, M.H. and Besford, R.T. (1986). The effects of silicon on cucumber plants grown in recirculating nutrient solution. *Ann. Bot.*, 58, 343–351.

Ahmad, R., Zaheer, S.H. and Ismail, S. (1992). Role in silicon in salt tolerance of wheat (*Triticum aestivum* L.). *Plant Sci.*, 85, 43–50.

Barber, D.A. and Shone, M.G.T. (1966). The absorption of silica from aqueous solutions by plants. *J. Exp. Bot.*, 17, 569–578.

Barceló, J., Guevara, P. and Poschenrieder, C. (1993). Silicon amelioration of aluminium toxicity in teosinte (*Zea mays* L. ssp. mexicana). *Plant Soil*, 154, 249–255.

Bergmann, W. (1988). Ernährungsstörungen bei Kulturpflanzen. Jena: Gustav Fischer Verlag.

Blaich, R. and Grundhofer, H. (1998). Silicate incrusts induced by powdery mildew in cell walls of different plant species. *Z. Pflanzenkr. Pflanzenschutz*, 105, 114–120.

Bode, E., Kozik, S., Kunz, U. and Lehmann, H. (1994). Vergleichende elektronenmikroskopische Untersuchungen zur Lokalisation von Silizium in Blättern zweier verschiedener Gräserarten. *Dtsch. Tierärztl. Wschr.*, 101, 367–372.

Bowen, J.E. (1972). Manganese-silicon interaction and its effect on growth of Sudan grass. *Plant Soil*, 37, 577–588.

Brown, P.H. and Hu, H. (1994). Boron uptake by sunflower, squash and cultured tobacco cells. *Physiol. Plant.*, 91, 435–441.

Brown, P.H., Bellaloui, N., Hu, H. and Dandekar, A. (1999). Transgenically enhanced sorbitol synthesis facilitates phloem boron transport and increases tolerance of tobacco to boron deficiency. *Plant Physiol.*, 119, 17–20.

Cakmak, I. and Marschner, H. (1986). Mechanism of phosphorus-induced zinc deficiency in cotton. I. Zinc deficiency-enhanced uptake rate of phosphorus. *Physiol. Plant.*, 68, 483–490.

Cakmak, I. and Marschner, H. (1987). Mechanism of phosphorus-induced zinc deficiency in cotton III. Changes in physiological availability of zinc in plants. *Physiol. Plant.*, 70, 13–20.

Cocker, K.M., Evans, D.E. and Hodson, M.J. (1998a). The amelioration of aluminium toxicity by silicon in higher plants: Solution chemistry or an in planta mechanism? *Physiol. Plant.*, 104, 608–614.
Cocker, K.M., Evans, D.E. and Hodson, M.J. (1998b). The amelioration of aluminium toxicity by silicon in wheat (*Triticum aestivum* L.) – malate exudation as evidence for an in planta mechanism. *Planta*, 204, 318–323.
Cocker, K.M., Hodson, M.J., Evans, D.E. and Sangster, A.G. (1998c). Interaction between silicon and aluminium in *Triticum aestivum* L. (cv. Celtic). *Israel J. Plant. Sci.*, 45, 285–292.
Corrales, I., Poschenrieder, C. and Barceló, J. (1997). Influence of silicon pretreatment on aluminium toxicity in maize roots. *Plant Soil*, 190, 203–209.
Dannel, F., Pfeffer, H. and Römheld, V. (1998) Compartmentation of boron in roots and leaves of sunflower as affected by boron supply. *J. Plant Physiol.*, 153, 615–622.
Dannel, F., Pfeffer, H. and Römheld, V. (2002) Update on boron in higher plants–Uptake, primary translocation and compartmentation. *Plant Biol.*, 4, 193–204.
El-Jaoual, T. and Cox, D.A. (1998). Manganese toxicity in plants. *J. Plant Nutr.*, 21, 353–386.
Epstein, E. (1994). The anomaly of silicon in plant biology. *Proc. Natl. Acad. Sci. USA*, 91, 11–17.
Epstein, E. (1999). Silicon. *Annu. Rev. Plant Physiol. Plant Mol. Biol.*, 50, 641–664.
Galvez, L., Clark, R.B., Gourley, L.M. and Maranville, J.W. (1987). Silicon interactions with manganese and aluminium toxicity in sorghum. *J. Plant Nutr.*, 10, 1139–1147.
Galvez, L,, Clark, R.B., Gourley, L.M. and Maranville, J.W. (1989). Effects of silicon on mineral composition of sorghum grown with excess manganese. *J. Plant Nutr.*, 12, 547–561.
Goldbach, H.E., Blaser-Grill, J., Lindemann, N., Porzelt, M., Hörrmann, C., Lupp, B. and Gessner, B. (1991). Influence of boron on net proton release and its relation to other metabolic processes. *Curr. Top. Plant Biochem. Physiol.*, 10, 195–220.
Goldberg, R. (1985). Cell-wall isolation, general growth aspects. In: Cell Components, Modern Methods of Plant Analysis. Springer Verlag Berlin.
González, A. and Lynch, J.P. (1999). Subcellular and tissue Mn compartmentation in bean leaves under Mn toxicity stress. *Aust. J. Plant Physiol.*, 26, 811–822.
González, A., Steffen, K.L. and Lynch, J.P. (1998). Light and excess manganese. *Plant Physiol.*, 118, 493–504.
Hammond, K.E., Evans, D.E. and Hodson, M.J. (1995). Aluminium silicon interactions in barley (*Hordeum vulgare* L.) seedlings. *Plant Soil*, 173, 89–95.
Handreck, K.A. and Jones, L.H.P. (1968). Studies of silica in the oat plant. IV. Silica content of plant parts in relation to stage of growth, supply of silica, and transpiration. *Plant Soil*, 24, 449–458.
Heine, G., Tikum, G. and Horst, W.J. (2005). Silicon nutrition of tomato and bitter gourd with special emphasis on silicon distribution in root fractions. *J. Plant Nutr. Soil Sci.*, 168, 600–606.
Hodson, M.J. and Evans, D.E. (1995). Aluminium silicon interactions in higher plants. *J. Exp. Bot.*, 46 (283), 161–171.
Hodson, M.J. and Sangster, A.G. (1989a). Silica deposition in the inflorescence bracts of wheat (*Triticum aestivum*). II. X-ray microanalysis and backscattered electron imaging. *Can. J. Bot.*, 67, 281–287.
Hodson, M.J. and Sangster, A.G. (1993). The interaction between silicon and aluminium in *Sorghum bicolor* (L.) Moench: growth analysis and X-ray microanalysis. *Ann. Bot.*, 72, 389–400.
Horiguchi, T. (1987). Mechanism of manganese toxicity and tolerance of plants, II. Deposition of oxidized manganese in plant tissues. *Soil Sci. Plant Nutr.*, 33, 595–606.
Horst, W.J. and Marschner, H. (1978a) Symptome von Mangan-Überschuss bei Bohnen (*Phaseolus vulgaris*). *Z. Pflanzenernaehr. Bodenkd.*, 141, 129–142.
Horst, W.J. and Marschner H (1978b) Einfluß von Silizium auf den Bindungszustand von Mangan im Blattgewebe von Bohnen (*Phaseolus vulgaris*). *Z. Pflanzenernaehr. Bodenkd.*, 141, 487–497.
Horst, W.J. and Marschner, H. (1978c). Effect of silicon on manganese tolerance of bean plants (*Phaseolus vulgaris* L.). *Plant Soil*, 50, 287–303.
Horst, W.J., Fecht, M., Naumann, A., Wissemeier, A.H. and Maier, P. (1999). Physiology of manganese toxicity and tolerance in *Vigna unguiculata* (L.) Walp. *J. Plant Nutr. Soil Sci.*, 162, 263–274.
Idris, M., Hossain, M. and Choudhury, F.A. (1975). The effect of silicon on lodging of rice in presence of added nitrogen. *Plant Soil*, 43, 691–695.
Islam, A. and Saha, R.C. (1969). Effects of silicon on the chemical composition of rice plants. *Plant Soil*, 30, 446–457.
Iwasaki, K. and Matsumura, A. (1999). Effect of Silicon on Alleviation of Manganese,Toxicity in Pumpkin (*Cucurbita moschata* Duch. cv. Shintosa). *Soil Sci. Plant Nutr.*, 45, 909–920.

Iwasaki, K., Maier, P., Fecht, M. and Horst, W.J. (2002a). Effects of silicon supply on apoplastic manganese concentrations in leaves and their relation to manganese tolerance in cowpea (*Vigna unguiculata* (L.) Walp.). *Plant Soil*, 238, 281–288.

Iwasaki, K., Maier, P., Fecht, M. and Horst, W.J. (2002b). Leaf apoplastic silicon enhances manganese tolerance of cowpea (*Vigna unguiculata* (L.) Walp.). *J. Plant Physiol.*, 159, 167–173.

Jones, L.H.P. and Handreck, K.A. (1965). Studies of silica in the oat plant. III. Uptake of silica from soils by the plant. *Plant Soil*, 23, 79–96.

Kaufman, P.B., Dayanandan, P. and Franklin, C.A. (1985). Structure and function of silica bodies in the epidermal system of grass shoots. *Ann. Bot.*, 55, 487–507.

Kinrade, S.D., Del Nin, J.W., Schach, A.S., Sloan, T.A., Wilson, K.L. and Knight, C.T.G. (1999). Stable five- and six-coordinated silicate anions in aqueous solution. *Science*, 285, 1542–1545.

Kluthcouski, J. (1980). The effect of silicon on the manganese nutrition of soybeans (*Glycine max* (L.) Merrill). *Plant Soil*, 56, 157–160.

Liang, Y. and Shen, Z. (1994). Interaction of silicon and boron in oilseed rape plants. *J. Plant Nutr.*, 17, 415–425.

Liang, Y., Shen, Q., Shen, Z. and Ma, T. (1996). Effects of silicon on salinity tolerance of two barley cultivars. *J. Plant Nutr.*, 19, 173–183.

Liang, Y., Si, J. and Römheld, V. (2005).,Silicon uptake and transport is an active process in *Cucumis sativus* L. *New Phytol.*, 167, 797–804.

Liang, Y.C., Sun, W.C., Si, J. and Romheld, V. (2005). Effects of foliar- and root-applied silicon on the enhancement of induced resistance to powdery mildew in *Cucumis sativus*. *Plant Pathol.*, 54, 678–685.

Loomis, W.D. and Durst, R.W. (1992). Chemistry and biology of boron. *BioFactors*, 3, 229–239.

Ma, J. and Takahashi, E. (1990). The effect of silicic acid on rice in a P-deficient soil. *Plant Soil*, 126, 121–125.

Ma. J. and Takahashi, E. (1991). Effect of silicate on phosphate availability for rice in a P- deficient soil. *Plant Soil*, 133, 151–155.

Ma, J. and Takahashi, E. (1993). Interaction between calcium and silicon in water-cultured rice plants. *Plant Soil*, 148, 107–113.

Ma, J.F., Tamai, K., Ichii, M. and Wu, G.F. (2002). A rice mutant defective in Si uptake. *Plant Physiol.*, 130, 2111–2117.

Ma, J.F., Mitani, N., Nagoa, S., Konishi, S., Tamai, K., Iwashita, T. and Yano, M. (2004). Characterization of the silicon uptake system and molecular mapping of the silicon transporter gene in rice. *Plant Physiol.*, 136, 3248–3289.

Marschner, H., Oberle, H., Cakmak, I. and Römheld, V. (1990). Growth enhancement by silicon in cucumber (*Cucumis sativus*) plants depends on imbalance in phosphorus and zinc supply. In M.L. van Beusichem (ed.), *Plant nutrition – Physiology and applications* (pp. 241–249). Dordrecht: Kluwer Academic Publishers.

Marschner, H. and Cakmak, I. (1986). Mechanism of phosphorus-induced zinc deficiency in cotton. II. Evidence for impaired shoot control of phosphorus uptake and translocation under zinc deficiency. *Physiol. Plant.*, 68, 491–496.

Mitani, N. and Ma, J.F. (2005). Uptake system of silicon in different plant species. *J. Exp. Bot.*, 414, 1255–1261.

Miyake, Y. and Takahashi, E. (1978). Silicon deficiency of tomato plant. *Soil Sci. Plant Nutr.*, 24, 175–189.

Miyake, Y. and Takahashi, E. (1983). Effect of silicon on the growth of solution-cultured cucumber plant. *Soil Sci. Plant Nutr.*, 29, 71–83.

Nable, R.O., Lance, R.C.M. and Cartwright, B. (1990). Uptake of boron and silicon by barley genotypes with differing susceptibilities to boron toxicity. *Ann. Bot.*, 66, 83–90.

Neumann, D., Zurnieden, U., Schwieger, W., Leopold, I. and Lichtenberger, O. (1997). Heavy metal tolerance of *Minuartia verna*. *J. Plant Physiol.*, 151, 101–108.

Polster J and Schwenk M (1992) The role of boron, silicon and nucleic bases on pollen tube growth of *Lilium longiflorum* (L.). *Z. Naturforsch.*, 47, 102–108.

Rafi, M.M., Epstein, E. and Falk, R.H. (1997). Silicon deprivation causes physical abnormalities in wheat (*Triticum aestivum* L.). *J. Plant Physiol.*, 151, 497–501.

Raven, J.A. (1983). The transport and function of silicon in plants. *Biol. Rev. Camb. Philos. Soc.*, 58, 179–207.

Rogalla, H. (2001). Einfluss von Silizium auf Austauschereigenschaften des Apoplasten und indungszustand von Nährstoffen in Blättern. Dissertation. Universität Hohenheim, Institut für Botanik und Botanischer Garten und Institut für Pflanzenernährung. Stuttgart: Verlag Grauer (ISBN 3-86186-363-4).

Rogalla, H. and Römheld, V. (2001). Mechanism of silicon-mediated manganese tolerance of *Cucumis sativus* L.: Effect of silicon nutrition on manganese concentration in the intercellular washing fluid. In W.J. Horst et al. (eds), Plant nutrition – Food security and sustainability of agro-ecosystems (pp. 258–259). Dordrecht: Kluwer Academic Publishers.

Rogalla, H. and Römheld, V. (2002a). Role of leaf apoplast in silicon-mediated manganese tolerance of *Cucumis sativus* L. *Plant Cell Environ.*, 25, 549–555.

Rogalla, H. and Römheld, V. (2002b). Effects of silicon on the availability of boron: possible effects on the phenol pathway and on the redox status in *Cucumis sativus* L. In H. Goldbach et al. (eds), Boron nutrition in animals and plants (pp. 205–213). London: Kluwer Plenum Academic Publishers.

Samuels, A.L., Glass, A.D.M., Menzies, J.G. and Ehret, D.L. (1994). Silicon in cell walls and papillae of *Cucumis sativus* during infection by *Sphaerotheca fuliginea*. *Physiol. Mol. Plant Pathol.*, 44, 237–242.

Takahashi, E., Ma, J.F. and Miyake, Y. (1990). The possibility of silicon as an essential element for higher plants. *Comm. Agric. Food Chem.*, 2, 99–122.

Vorm, P.D.J., v.d. (1980). Uptake of Si by five plant species, as influenced by variations in Si-supply. *Plant Soil*, 56, 153–156.

Wang, Y., Stass, A and Horst, W.J. (2004). Apoplastic binding of aluminum is involved in silicon-induced amlioration of aluminum toxicity in maize. *Plant Physiol.*, 136, 3762–3770.

Wiese, J., Wiese, H., Schwartz, J. and Schubert, S. (2005). Osmotic stress and silicon act additively in enhancing pathogen resistance in barley against barley powdery mildew. *J. Plant Nutr. Soil Sci.*, 168, 1–6.

Williams, E.D. and Vlamis, J. (1957). The effect of silicon on yield and manganese-54 uptake and distribution in the leaves of barley plants grown in culture solutions. *Plant Physiol.*, 32, 404–409.

Wissemeier, A.H. and Horst, W.J. (1992). Effect of light intensity on manganese toxicity symptoms and callose formation in cowpea (*Vigna unguiculata* (L.) Walp). *Plant Soil*, 143, 299–309.

SIGNIFICANCE OF THE ROOT APOPLAST FOR ALUMINIUM TOXICITY AND RESISTANCE OF MAIZE

W.J. HORST[1], M. KOLLMEIER[1], N. SCHMOHL[1], M. SIVAGURU[2], Y. WANG[3], H.H. FELLE[4], R. HEDRICH[5], W. SCHRÖDER[6], and A. STAß[1]

[1]Institute of Plant Nutrition, Leibniz University of Hannover, Germany, horst@pflern.uni-hannover.de; [2]Division of Biological Sciences and Molecular Cytology Core, Biomolecular Research Facilities, USA; [3]College of Agriculture, Yangzhou University, China; [4]Botanisches Institut I, Justus-Liebig Universität Gießen, Germany; [5]Julius-von-Sachs-Institut für Biowissenschaften, Universität Würzburg, Germany; [6] Institute for Chemistry and Dynamics of the Geosphere: ICG-III Phytosphere, Forschungszentrum Jülich GmbH, Germany.

Abstract. The mechanism of aluminium-induced inhibition of root elongation is still not well understood. It is a matter of debate whether the primary lesions of Al toxicity are apoplastic or symplastic. The present paper summarises evidence from own experimental work and the literature which could contribute to the understanding of Al toxicity and resistance in maize focussing on the role of the apoplast.

Short-term application of Al to specific zones of the root apex revealed that the 1–2 mm distal transition zone is the most Al-sensitive apical root zone. The inhibition of basipetal transport of auxin by Al indicates that auxin is part of the Al-signal transduction from this to the 3–6 mm central elongation zone. Al induced callose formation, membrane depolarization, and prominent alterations in the cytoskeleton especially in this zone. Accumulation of Al in the root apex is modulated by the pectin content of the cell walls and the degree of methylation of the pectin. After short-term Al supply, Al accumulates particularly in the cell walls of the outer cortical cells modifying the apoplastic transport of higher molecular weight solutes. Apoplastic Al also affects plasma-membrane properties as expressed by changes in the trans- membrane potential, ion fluxes and enhanced callose synthesis.

Higher Al resistance of root apices is sensitively reflected by a lower Al-induced callose formation. Si-enhanced Al resistance was not related to Al exclusion but to modification of the Al-binding properties of the cell walls. Genotypic Al resistance is related to less accumulation of Al due to lower negativity of the cell walls and the plasma membrane, and particularly to an enhanced release of organic acid anions via an Al-induced citrate and malate-permeable large conductance anion channel in the most Al-sensitive root zone. In an Al-resistant cultivar the channel open probability was greater than in the Al-sensitive cultivar and a higher level of citrate in the root could be maintained.

B. Sattelmacher† and W.J. Horst (eds.), The Apoplast of Higher Plants: Compartment of Storage, Transport and Reactions, 49–66.

It is concluded that the apoplast of the most Al-sensitive root zone plays an important role in Al toxicity and resistance in maize. Although symplastic lesions of Al toxicity cannot be excluded, the protection of the apoplast appears to be a prerequisite for Al resistance. An in-depth molecular characterisation of Al-induced apoplastic reaction in the most sensitive root zone is urgently required.

Key words: aluminum, callose, cell wall, pectin, organic acids, silicon

1. INTRODUCTION

On acid mineral soils aluminium (Al) toxicity is the most important soil factor determining the composition of the natural vegetation and limiting crop yields. It has been estimated that worldwide on 30–40% of the arable land surface crop productivity is limited by soil acidity (Uexküll and Mutert, 1995). Aluminium may be present in the soil solution in mononuclear and polynuclear forms. Although the potential specific role of the Al_{13} hydroxy polymer in Al toxicity is not yet fully understood, it is generally agreed that mononuclear Al species, particularly Al^{3+} rather than $Al(OH)^{2+}$ and $Al\,(OH)_2^{+}$, are the most phytotoxic Al species (Kinraide, 1991). Organic and inorganic Al complexes are considered not or less phytotoxic (Kerven et al., 1989).

One of the most rapid primary lesions of Al toxicity is an inhibition of root elongation which can be measured within less than one hour after the exposure of the roots to excess Al supply (Llugany et al., 1995). An equally or even more sensitive indicator of Al effects on roots is the induction of callose synthesis (Wissemeier et al., 1987), particularly in the root apex (Wissemeier and Horst 1995). Aluminium-induced callose formation is an indicator of Al-sensitivity and a reliable parameter for the classification of genotypes of different plant species for Al resistance (Wissenmeier et al., 1992; Schreiner et al., 1994; Horst et al., 1997). Collet and Horst, (2001) developed a rapid non-destructive screening procedure for maize cultivars for adaptation to acid soils with high Al supply, and Eticha et al. (2005a) successfully used Al-induced callose formation to study inheritance of Al resistance using a 13X13 diallel of maize cultivars of largely different origin. Inhibition of shoot growth by Al treatment may be regarded as a secondary effect due to Al-induced deficiencies particularly of Mg, Ca, and P, phytohormone imbalance, and drought stress as a consequence of impaired root growth and root activity.

Although much progress has been made during recent years, the mechanisms of Al-induced inhibition of root elongation and Al resistance are still not well understood. There are a number of excellent reviews in recent years summarising the state of knowledge and addressing knowledge gaps (Taylor, 1991; Kochian, 1995; Delhaize and Ryan, 1995; Matsumoto

2000; Kochian et al., 2004). Particularly the relative importance of symplastic versus apoplastic lesions of Al toxicity remains a matter of debate. Rengel (1996) and especially Horst (1995) focused the attention on the role of the apoplast in Al toxicity regarding short-term inhibition of root elongation by Al. These reviews represented the baseline for our studies on Al toxicity and resistance conducted in the scope of this research project.

2. ALUMINIUM TOXICITY

2.1 Root spatial sensitivity

Ryan et al. (1993) were the first who unequivocally demonstrated the role of the root apex in the perception of Al toxicity in maize. They also clearly refuted the hypothesis put forward by Bennet et al. (1985) that the root cap plays a decisive role in the expression of Al-induced inhibition of root elongation. In a more refined methodological approach we presented evidence that in the Al-sensitive maize cultivar "Lixis" the distal part of the transition zone (DTZ, 1–2 mm) is the most Al-sensitive apical root zone in maize (Sivaguru and Horst, 1998). This agreed with the fact that in this zone, Al-induced callose formation and Al accumulation was most intense. Application of Al only to the DTZ reduced cell elongation in the elongation zone (EZ) to the same extent as application to the entire 10 mm root apex.

Application of Al only to the EZ did not inhibit root elongation (Fig. 1, Kollmeier et al., 2000). This result demands for a signal transduction

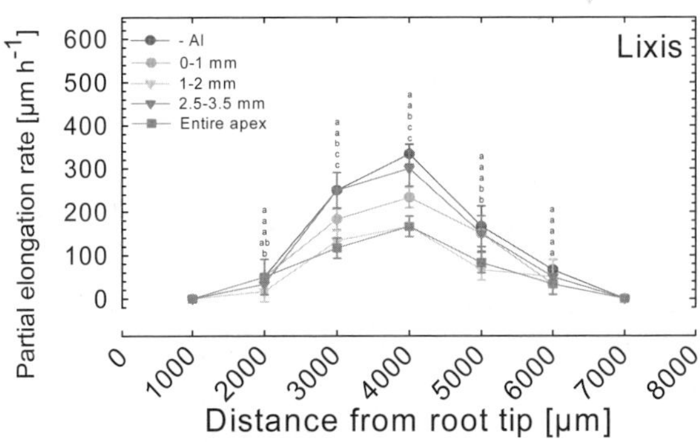

Fig. 1. Effect of Al supply (90 µM, 1 h) to the entire root apex or to specific 1-mm root zones on partial elongation rates of 1-mm root segments of the primary roots of the maize cultivar Lixis. Values are means of five independent measurements SD. Different letters indicate significant differences at P<0.05 (Tuckey's test). From Kollmeier et al., 2000.

mediating the Al signal between DTZ and EZ. Following up on the results of Hasenstein and Evans (1988) who had demonstrated inhibition of basipetal transport of IAA in roots, Kollmeier et al. (2000) provided evidence that auxin is part of this signal transduction or even the signal because basipetal auxin transport from the root apex to the EZ was inhibited by Al (Table 1), and externally applied auxin to the EZ could partly restore the Al-induced root-growth inhibition.

Table 1. Effect of Al on [^{3}H]IAA distribution relative to total IAA uptake into the apex of primary roots of maize. Al, NPA and TIBA were applied in nutrient solution, pH 4.3, in 0.6% (w/v) agarose to the DTZ for 30 min. [^{3}H]IAA (0.1 µM in 1.2% agarose in nutrient solution) was applied to the MZ for 30 min. Values are means of five independent replicates SD. Different letters indicate significant differences at $P<0.05$ (Tukey test). From Kollmeier et al., 2000.

		Control	90 µM Al	10 µM NPA	10 µM TIBA
Root segment	Root zone	0.1 µM [^{3}H] IAA	0.1 µM [^{3}H] IAA	0.1 µM [^{3}H] IAA	0.1 µM [^{3}H] IAA
0–1 mm	MZ	56.6	68.5	70.4	67.6
1–2 mm	DTZ	13.4b	17.6ab	19.4ab	21.4a
2–3 mm	EZ	10.7a	4.4b	3.4b	5.3b
3–4 mm	EZ	10.1a	4.5b	2.8b	2.9b
4–5 mm	EZ	5.6a	2.8ab	2.4b	1.6b
>5 mm		3.5a	2.0ab	1.6b	1.2b

2.2 Aluminium accumulation and radial transport in roots

Aluminium is accumulated by roots with a rapid initial phase and a lower rate thereafter (Zhang and Taylor, 1989, 1990). The primary binding site of Al^{3+} is likely the pectic matrix of the cell walls with its negatively charged carboxylic groups having a particularly high affinity for Al^{3+} (Blamey et al., 1990). Short-term Al accumulation by roots is closely related to the pectin content and may explain the differences in Al contents between apical root sections of *Zea mays* and *Vicia faba* (Fig. 2). It appears that the Al content and thus the binding of Al to the pectic matrix are closely positively correlated to Al-induced callose formation and thus Al sensitivity (see below).

The role of the pectin content for Al accumulation and Al sensitivity has been further substantiated using different approaches (long-term treatment with a cellulose synthesis inhibitor, medium-term NaCl pre-treatment, short-term treatment with pectin-degrading enzymes) modifying the pectin content of intact maize plants (Horst et al., 1999) and maize cell-suspension cultures (Schmohl and Horst, 2000).

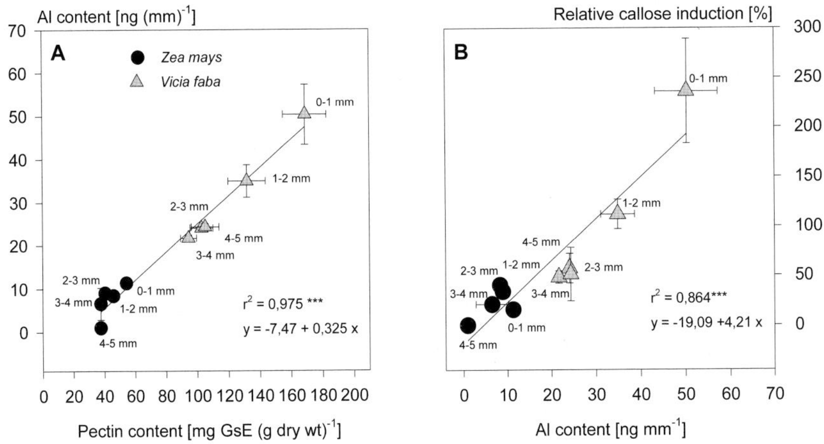

*Fig. 2. Relationships between pectin and Al content (A) and Al content and relative callose induction (digitonin = 100) (B) of root sections of maize and faba bean. Roots were incubated for 3 h in nutrient solution ± 50 μM Al or 10 μM digitonin at pH 4.3. ***significant at P<0.001.*

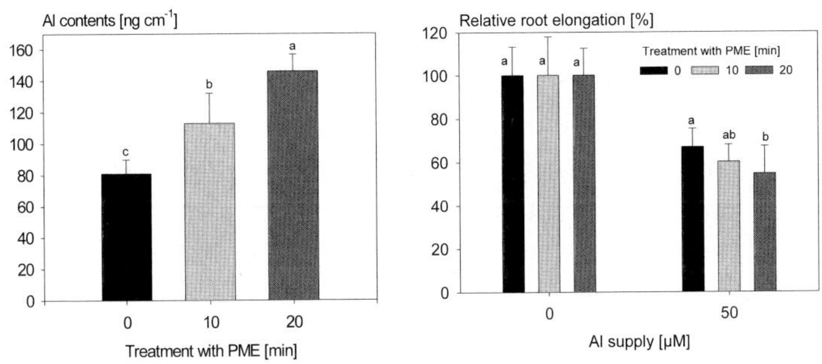

Fig. 3. Effect of treatment of maize root tips (1 cm) with PME on Al content (A) and relative root elongation (B). After PME treatment, roots were incubated in nutrient solution ± 50 μM Al, pH 4.3 for 3 h. Different letters indicate significant differences at P<0.05 (Tukey test).

In fact, the factor responsible for Al binding to pectin is not the pectin content alone but its negative charge determined by its degree of methylation (DM) which is controlled by pectin methylesterase (PME) (Gerendás (this issue)). Schmohl et al. (2000) provided evidence that Al accumulation and Al sensitivity of maize cell-suspension cells is modulated by the DM of their cell walls. Also short-term treatment of intact maize roots with PME enhanced Al accumulation and Al-induced inhibition of root elongation (Fig. 3). A modulating role of the DM of root cell-walls in Al resistance is supported by the comparison of potato transformants differing in the expression of PME from *Petunia inflata* (Schmohl and Horst 2000; Fig. 4): Transformants with higher PME expression accumulated more Al, produced more callose, and were more inhibited in root growth when exposed to Al than the wild-type.

Due to the strong Al-binding capacity of the cell wall it is not surprising that particularly after shorter periods of Al treatment which induce Al toxicity, most of the Al in the root is located in the cell walls of the outer cortical cells (Delhaize et al., 1993; Marienfeld and Stelzer, 1993, Godbold et al., 1995; Godbold and Jentschke, 1998). In this respect radial and cellular distribution of Al in the DTZ of maize was markedly different from the distribution of the stable isotopes of Ca, Mg, K after 1 h of application as revealed by SIMS (Fig. 5). The transport from the rhizodermis to the endodermis was shown to be time-dependent (10 s to 3 h) and no transport of Al into the central cylinder through the fully differentiated endodermis could be observed within 3 h of Al treatment in *Zea mays* and *Vicia faba* (Marienfeld et al. 2000). An accumulation of Al in the cell wall can also be expected on the basis of the measured low rates of transport of Al through the plasma membrane into the symplast of the model plant *Chara corallina* (Rengel and Reid, 1997, Taylor et al., 2000). However, both studies as well as those by Marienfeld et al. (2000) in maize also show, that a rapid transfer of Al from the apoplast to the symplast does occur. Using different techniques Tice et al., (1992), Lazof et al. (1994), and Vazquez et al., (1999) demonstrated the accumulation of Al in the symplast in wheat, soybean, and maize, respectively, leading to a rather uniform cellular distribution of Al or even accumulation of Al in the symplast at the expense

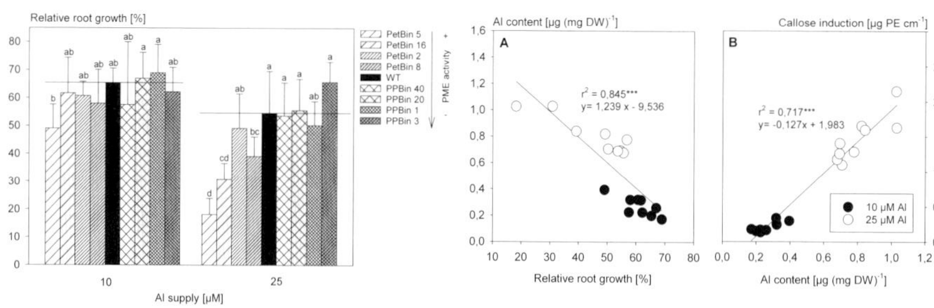

Fig. 4. Effect of Al supply on relative root growth (-Al = 100) of transgenic potato lines, with different expression of PME (left). Different letters indicate significant differences between lines at P<0.05 (Tukey test)., Right: relationships between relative root growth (- Al = 100) of potato genotypes (wild-type and PME transformants) and Al contents (A), and Al and Al-induced callose contents (B) in root apices. Treatment of intact plants in nutrient solution for 24 h at pH 4.3.

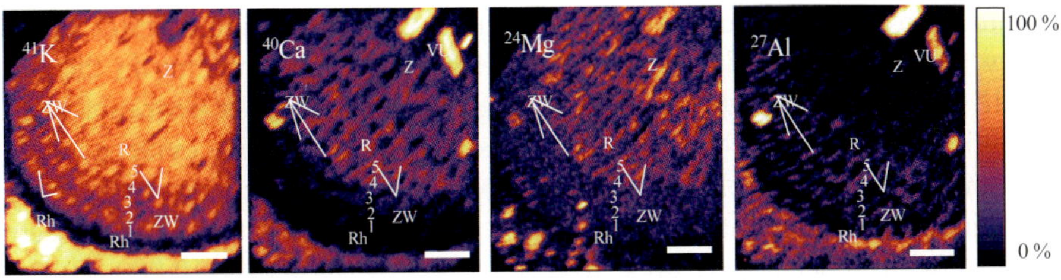

Fig. 5. Distribution of the isotopes ^{41}K, ^{40}Ca, ^{24}Mg and ^{27}Al in cryosections of the 1–2 mm root zone of the Al sensitive maize genotype Lixis after Al treatment (90 μM Al, pH 4.3 in nutrient solution in 0.6% agarose, 1 h). Bar = 50 μm. R = cortex, Rh = rhizodermis, Vu = contamination, Z = stele, ZW = cell wall, numbers 1–5 = cortex layers.

of the cell wall. However, we have challenged the results based on cellular Al localization using morin, because morin is unable to bind to cell wall-bound Al (Eticha et al., 2005b). The rapid uptake of Al, transfer to the central cylinder, transport to the shoots and accumulation of Al in the vacuoles of the leaves is a typical feature of Al accumulator plant species such as *Hydrangea* (Ma et al. 1997, Naumann and Horst, 2003) and buckwheat (Ma et al., 1998). The reasons for the difference in mobility of Al between Al excluders (most plant species) and includers are not yet understood.

2.3 Apoplastic solute flow in the roots

Aluminium modifies cell-wall composition and properties. Cellulose synthesis was inhibited in favour of callose synthesis in barley (Teraoka et al., 2002) and the contents of pectin hemicellulose increased in squash root-apices (Van et al., 1994), thus enhancing the binding of Al in the cell wall which was interpreted by the authors as an Al tolerance mechanism. However, as shown above, binding of Al to pectins appears to be more closely related to Al sensitivity in maize. Cross linking of pectins by Al^{3+}, thus reducing cell-wall extensibility, has been claimed responsible for the inhibition of root elongation by Al (Schildknecht and Vidal, 2002). But as shown above, direct application of Al to the elongation zone was not inhibitory to root elongation in maize. Therefore, in maize a direct interference of Al with the process of cell elongation as primary target of Al^{3+} appears less probable.

The Al-induced modifications of structural properties of the cell wall (Schildknecht and Vidal, 2002) affect the mobility of solutes in the apoplast of the root cortex. Schmohl and Horst (2002) demonstrated that the release of proteins in general and acid phosphatase specifically, and pectins was inhibited by Al in maize root apices and maize cell-suspensions cultures which they attributed to a reduction of cell-wall porosity by Al-binding in the cell wall. However, these results can also be explained by a lower permeability of the plasma membrane for macromolecules. A more convincing evidence of an inhibition of the apoplastic solute bypass-flow in maize root apices was provided by Sivaguru et al., (2006). They showed accumulation of fluorescent probes with molecular weights from 524 (HPTS) and 3,000, 10,000, 40,000 (neutral dextran Texas Red conjugates) in the outer cortical cells, especially in the DTZ and inhibition of transfer of these solutes to the xylem and finally the shoot (Fig. 6). Water flow was not affected in contrast to the expectations expressed by Blamey et al., (1993) showing a strong reduction by Al of water flow through an artificial pectate membrane.

Since the inhibition of callose synthesis by pre-treatment of the roots with 2-deoxy-D-glucose (DDG) prior to Al treatment partially alleviated the Al-induced inhibition of solute bypass-flow, it is assumed that callose deposition does not only contribute to the inhibition of cell to cell trafficking of solutes through plasmodesmata (Sivaguru et al., 2000b) but also of apoplastic bypass flow in root cortical cell walls.

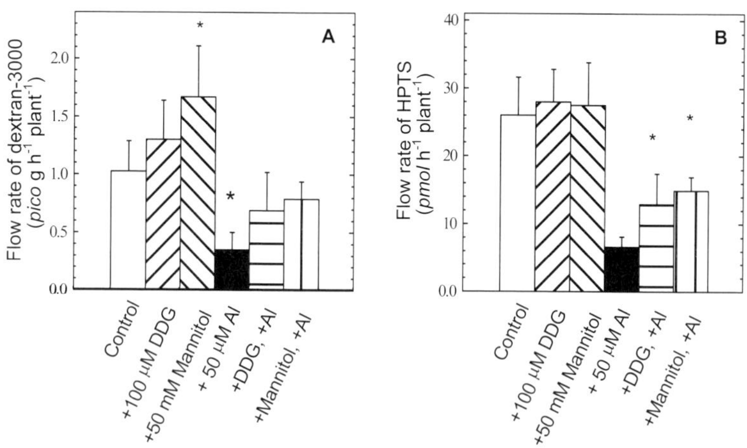

*Fig. 6. Effect of Al (50 µM), mannitol (50 mM) and DDG (100 µM) pre-treatment on the flow rates of dextran-3,000-Texas Red (A) or HPTS (B) in root xylem-exudates of maize cv. Lixis. DDG was applied 1 h prior to Al treatment. Al, mannitol, and fluorescent tracers were applied to the 1 cm root apex for two hours including the 1 h exudate collection-period. Values are means of eight to ten independent replicates SE. Analysis of covariance (A and B) coupled with Tukey honest significant difference test shows *if significant compared to the control at $P<0.05$. The influence of either mannitol or DDG over Al treatments was significant at the $P<0.05$ level.*

2.4 Impairment of membrane functions and related physiological disorders

Aluminium not only rapidly affects cell-wall but also plasma-membrane properties (Ishikawa and Wagatsuma, 1998). Interaction of Al with membrane lipids and proteins (Akeson et al., 1989; Caldwell, 1989; Jones and Kochian, 1997) induces modifications of its structural properties such as fluidity (Vierstra and Haug, 1978). Such structural change in membrane properties is one of the prerequisites in addition to an increase in the cytosolic Ca^{2+} activity for the induction of callose synthesis (Kauss et al., 1990) a most sensitive response of root apices to Al (see above). Binding of Al to the plasma membrane alters its electrical properties. In *Curcubita pepo* Ahn et al. (2001) found a reduction of the plasma-membrane surface negativity. Also, in most studies Al supply rapidly induced membrane depolarization (Lindberg et al., 1991, Olivetti et al., 1995) specifically in the most Al-sensitive root zone (DTZ) (Sivaguru et al., 1999). This may be related to an inhibition of the H^+-APTase activity (Ahn et al., 2001) which may lead to a disturbance of the H^+ homeostasis in the cytosol (Lindberg and Strid, 1997; Plieth et al., 1999). As a result of the changes in plasma-membrane properties by Al, its ion-transport properties are affected. In *Glycine max* Al treatment led to a rapid decrease of K^+ efflux without changing K^+ influx (Horst et al., 1992, Staß and Horst, 1995). In *Triticum aestivum* the Al-enhanced release of malate was charge-balanced by a release of K^+ (Ryan et al., 1995) which is in agreement with our results in *Zea mays* where Al-induced citrate release through an anion channel was observed without affecting the K^+ outward rectifier (Kollmeier et al., 2001). Al-induced impairment of membrane functions may be related to Al-enhanced oxidative stress through the formation of reactive radicals leading to lipid peroxidation (Yamamoto et al., 1997, 2002) and protein oxidation (Boscolo et al., 2003). Among the identified genes that are expressed after Al treatment, are particularly oxidative stress genes (Richards et al., 1998). Transformation of *Arabidopsis thaliana* with such genes conferred Al resistance (Ezaki et al., 2001). However, oxidative stress in roots appears not to be the primary cause for Al-induced inhibition of root elongation, because in most cases it could be observed only after prolonged Al treatment (Cakmak and Horst, 1991; Maltais and Houde, 2002; Boscolo et al., 2003). But sustained Al resistance may require protection mechanisms against oxidative stress.

In spite of these changes in plasma-membrane structure und functions it needs to be stressed that there is no indication that at physiological Al concentrations a severe disruption of plasma-membrane functions are a prerequisite for inhibition of root elongation and callose formation (Horst et al., 1992). It appears that Al triggers signal transduction pathways leading to the observed symplastic physiological disorders. In this regard the effect of Al on cytosolic Ca^{2+} seems to play a crucial role (Rengel and Zhang, 2003). An increase in cytosolic Ca^{2+} as immediate response to Al treatment has been demonstrated in different plants (Jones, et al., 1998; Zhang and Rengel, 1999). The source of Ca^{2+} is likely the apoplastic Ca^{2+} pool, because Ca^{2+} bound in the apoplast is liberated by Al^{3+} and the change of the plasma-membrane potential results in an activation of Ca^{2+} channels. However, the triggering by Al of a release of Ca^{2+} from symplastic Ca^{2+} pools cannot be excluded (Rengel and Zhang, 2003). Increasing cytosolic Ca^{2+} can explain two cellular distortions: callose formation and disorganisation of the cytoskeleton (Rengel and Zhang, 2003). An increase in cytosolic Ca^{2+} is one of the prerequisites for the induction of callose synthesis by different elicitors (Kauss et al., 1990). Aluminium-induced alterations of the cytoskeleton have been reported by Blancaflor et al. (1998), Sivaguru et al. (1999), Sivaguru et al. (2000a), and Schwarzerova et al. 2002). Although a direct effect of cytosolic Al on the cytoskeleton cannot be ruled out, an interaction of apoplastic Al with the cell-wall – plasma membrane – cytoskeleton continuum appears more likely (Horst et al., 1999; Sivaguru et al., 2000c).

3. ALUMINIUM RESISTANCE

As shown above, Al binds readily to binding sites in the apoplast and the plasma membrane in the most Al-sensitive sites of the root apex. Since this may lead to enhanced transport of Al to the symplast and/or to impairment of root growth and functions (see above), it cannot only be expected, but has to be postulated that reduced binding of Al in the apoplast is a prerequisite for Al resistance. Kinraide et al. (1992) were able to explain inhibition of root elongation by Al^{3+} in the presence of competing cations, including protons, on the basis of the computed cation distribution on a negatively charged root membrane surface. Blamey et al. (1993) and Grauer and Horst (1992) came to comparable conclusions based on similar but conceptually different approaches. A lower root cation-exchange capacity as a measure of cell-wall negativity has been reported in plant species adapted to acid soils with high Al supply (Blamey et al., 1990; Büscher et al., 1990). However, across a large range of plant species a clear relationship between root CEC and Al resistance does not exist (Grauer, 1992). There are several reports showing that Si nutrition enhances Al resistance of plants (Hodson and Evans, 1995). Both, *ex planta* and *in planta* effects are involved, but the latter effects are only poorly understood (Cocker et al., 1998). In maize, Kidd et al. (2001)

provided evidence for an Al-induced enhanced release of phenolics by Si-preteated plants thus detoxifying Al in the apoplast. We, too, attributed Si-enhanced Al resistance of maize clearly to an *in planta* effect (Wang et al., 2004). However, we could not relate Si-enhanced Al resistance to an enhanced release of phenolics or organic acid anions, but rather to a modification of the Al binding capacity of the cell wall. Al supply greatly changed the Si accumulation in the cell walls of the root apex (Fig. 7). Cell wall Si reduced the most labile exchangeable Al fraction in the cell wall (Wang et al., 2004). In addition to the cell wall, also the plasma membrane contributes to the negativity of the apoplast. Wagatsuma and Akiba (1989) and Wagatsuma et al. (1991) related differences in Al resistance between plant species to the plasma membrane negativity of protoplasts, and Yermiyahu et al. (1997) ascribed the higher Al sensitivity of the *Triticum aestivum* cultivar Scout to its higher plasma-membrane negativity compared to the Al-resistant cultivar Atlas.

An even more effective way to reduce the impact of Al on apoplastic functions is the release of Al complexing solutes particularly organic acid anions from the Al-sensitive apical root zone (Ma et al., 2001, Ryan et al., 1997). Using the patch-clamp technique, it is now well established that the Al-induced release of malate in *T. aestivum* (Ryan et al., 1997; Zhang et al., 2001) and citrate in *Z. mays* (Kollmeier et al., 2001; Pineros and Kochian, 2001) is mediated by plasma-membrane anion-channels. In Al-resistant cultivars the frequency and magnitude of the Al-induced anion currents were greater (Zhang et al., 2001; Fig. 8). In wheat the expression level of a gene encoding an Al-activated malate transporter (ALMT1) has been identified as decisive for Al resistance (Sasaki et al., 2004). Transgenic Al-sensitive barley expressing ALMT1 acquired Al resistance (Delhaize et al., 2004).

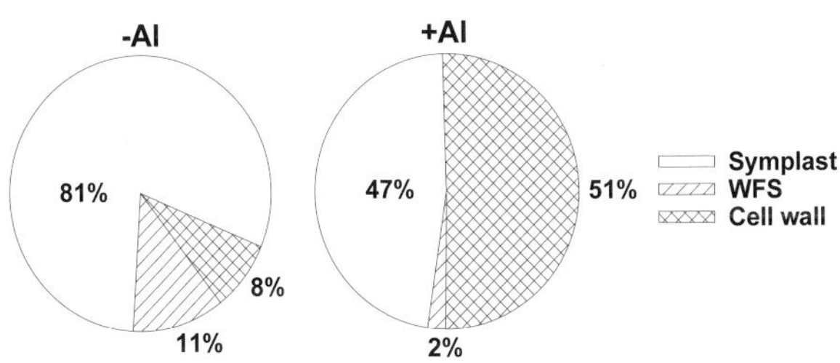

Fig. 7. Distribution of Si between symplast, water free space (WFS) and cell walls of root tips of maize as affected by Al. Plants were pre-cultured for 36 h in 1.4 mM Si, and then treated with 0 or 25 μM Al for 12 h in the presence of Si. From Wang et al., 2004.

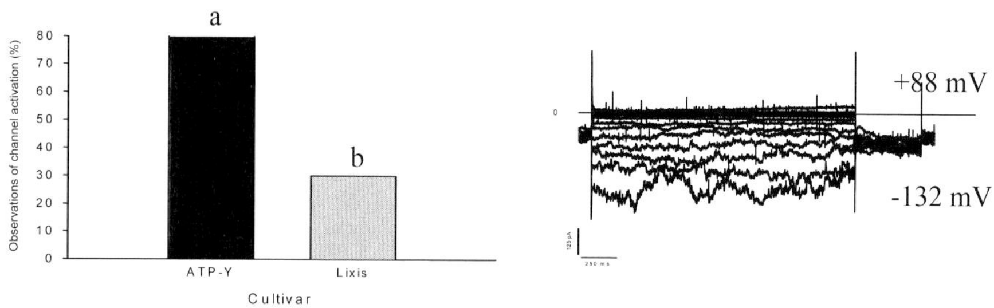

Fig. 8. Application of the patch-clamp technique to protoplasts isolated from the apical root cortex. Pre-incubation of intact roots with 90 μM Al for 1 h induced a citrate and malate-permeable, large conductance anion channel in protoplast isolated from the distal elongation zone (DTZ) but not of the elongation zone (EZ). Anion-channel activity was induced in 80% of the DTZ protoplasts from the Al-resistant maize cultivar ATP-Y, but only in 30% from the Al-sensitive cultivar Lixis. After Kollmeier et al., 2001.

The role of the metabolism of organic acids in Al resistance is still a matter of discussion. In most studies a clear relationship between the root content and release of organic acids did not exist (Ryan et al., 2001). Also, the activities of enzymes involved in the synthesis of organic acids did not differ significantly between Al resistant and Al-sensitive genotypes. However, based on a detailed study on release from (Fig. 9A) and content of (Fig. 9B) specific 1 mm apical root sections, we showed that the Al-sensitive cultivar is not capable of maintaining the level of citrate in the apical root sections in spite of a lower citrate release-rate (Kollmeier and Horst, 2001). This was in agreement with a general trend of Al-enhanced activities of enzymes involved in citrate synthesis such as NAD malate-dehydrogenase (MDH) and phosphoenolpyruvate decarboxylase (PEPC) (but not citrate synthase (CS)) in the Al-resistant cultivar and of citrate degrading aconitase in the Al-sensitive cultivar (Kollmeier and Horst, 2001). The strongest evidence for an involvement of organic acid synthesis in Al resistance comes from studies using transgenic plants with modified organic acid metabolism. Fuente et al. (1997) reported on transgenic tobacco with greatly enhanced citrate accumulation and exudation though expression of a bacterial CS in the cytoplasm which conferred Al resistance to these plants. However, attempts by other authors to enhance citrate synthesis and to improve Al resistance of tobacco using the same gene were not successful (Delhaize et al., 2001). Although not only the overexpression of citrate synthase but also of MDH (Tesfaye et al., 2001) and of PEPC (Osaki et al.,

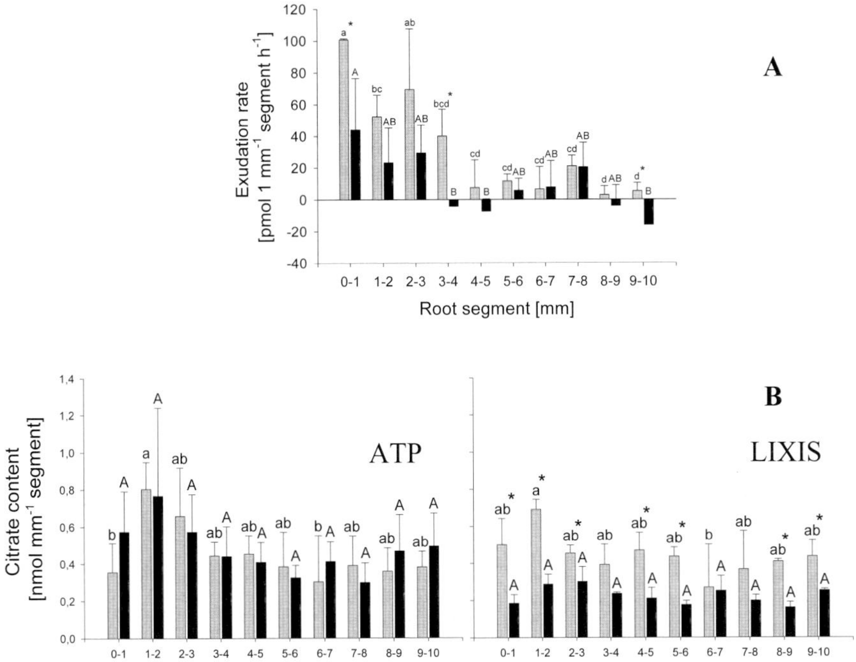

Fig. 9. A Aluminium-induced citrate release from excised 1-mm root segments of maize cultivars (Black colums cv Lixis, grey colums cv ATP). Al treatment 50 μM Al for 3 h. B Citrate tissue contents in apical 1-mm segments of 2 maize cultivars. Intact roots were treated in agarose gels for 2h ± 90 μM A (+Al black colums). Means with standard deviation, n = 4, means with the same letter are not significantly different between segments and stars denote significant differences between cultivars (A) or Al treatments(B) at P<0.05 (Tukey test).

2001) were reported to enhance Al resistance of plants it appears that the key to Al-enhanced release of organic acid anions and thus Al resistance is the expression, activity, and control of plasma-membrane anion channels rather than the synthesis of organic acids.

In addition to organic acid anions there are indications that the release of polypeptides (Taylor et al. 1997) and phenols (Heim et al., 1999; Kidd et al., 2001) may also be involved in genotypic Al resistance in wheat and maize, respectively. Among the most Al-resistant plant species are Al includers, such as *Camellia sinensis* and *Hydrangea macrophylla*. In these plants Al resistance is attributed to the detoxification of Al by organic acids (Ma et al., 1997, 1998) particularly in the vacuoles, where Al may precipitate in association with silicate and or phosphate (Naumann et al., 2001). There is also increasing evidence that phenols may be involved in the internal sequestration of Al (Stoutjesdijk et al. 2001).

4. CONCLUSIONS

The apoplast of the most Al-sensitive root zone plays an important role in Al toxicity and resistance in maize. Although symplastic lesions of Al toxicity cannot be excluded, the protection of the apoplast appears to be a prerequisite for Al resistance. An in-depth molecular characterisation of Al-induced apoplastic reaction in the most sensitive root zone is urgently required.

REFERENCES

Ahn, S.J., Sivaguru, M., Osawa, H., Ch Chung, G. and Matsumoto, H. (2001). Aluminum inhibits the H^+-ATPase activity by permanently altering the plasma membrane surface potentials in squash roots. *Plant Physiol., 126,* 1381–1390.

Akeson, M.A., Munns, D.N. and Buran, R.G. (1989). Adsorption of AL3+ to phosphathidyl choline vesicles. *Biochem. Biophys. Acts, 986,* 33–40.

Bennet, RJ., Breen, C.M. and Fey, M.V. (1985). Aluminium uptake sites in the primary root of *Zea mays* L. *S. Afr. J. Plant Soil, 2,* 1–7.

Blamey, F.P.C., Asher, C.J., Kerven, G.L. and Edwards, D.G. (1993). Factors affecting aluminum sorption by calcium pectate. *Plant Soil, 149,* 87–94.

Blamey, F.P.C., Edmeades, D.C. and Wheeler, D.M. (1990). Role of root cation-exchange capacity in differential aluminum tolerance of Lotus species. *J. Plant Nutr., 13[6],* 729–744.

Blancaflor, E.B., Jones, D.L., and Gilroy, S. (1998). Alterations in the cytoskeleton accompany aluminum-induced growth inhibiton and morphological changes in primary roots of maize. *Plant Physiol., 118,* 159–172.

Boscolo, P.R.S., Menossi, M. and Jorge, R.A. (2003). Aluminum-induced oxidative stress in maize. *Phytochemistry, 62,* 181–189.

Büscher, P., Koedam, N. and Speybroeck, Van D. (1990). Cation exchange properties and adaption to soil acidity in bryophytes. *New Phytol., 115,* 177–186.

Cakmak, I. and Horst, W.J. (1991). Effect of aluminium on lipid peroxidation, superoxide dismutase, catalase and peroxidase activities in root tips of soybean (*Glycine max* L.). *Plant Physiol., 83,* 463–468.

Caldwell, C.R. (1989). Analysis of aluminum and divalent cation binding to wheat root plasma membrane proteins using terbium phosphorescence. *Plant Physiol., 91,* 233–241.

Cocker, K.M., Evans, D.E. and Hodson, M.J. (1998). The amelioration of aluminium toxicity by silicon in higher plants: solution chemistry or an in planta mechanism? *Physiol. Plant., 104,* 608–614.

Collet, L. and Horst, W.J. (2001). Characterisation of maize cultivars in their adaptation to acid soils on the single plant level. In: W.J., Horst, et al. (eds) *Plant Nutrition – Food Security and Sustainability of Agro-ecosystems* (pp. 86–87). The Netherlands. Kluwer Academic Publishers, Dordrecht, ISBN 0-7923-7105-4.

Delhaize, E. and Ryan, P.R. (1995). Aluminum toxicity and tolerance in plants. *Plant Physiol., 107,* 315–321.

Delhaize, E., Craig, S., Beaton, C.D., Bennet, R.J., Jagadish, V.C. and Randall, P.J. (1993). Aluminum tolerance in wheat (*Triticum aestivum* L.). I. Uptake and distribution of aluminum in root apices. *Plant Physiol., 103,* 685–693.

Delhaize, E., Hebb, D.M. and Ryan, P.R. (2001). Expression of as Pseudomonas aeruginosa citrate synthase gene in tobacco is not associated witth either enhanced citrate accumulation or efflux. *Plant Physiol., 125,* 2059–2067.

Delhaize E., Ryan P.R., Hebb D.M., Yamamoto Y., Sasaki T. and Matsumoto H. (2004). Engineering high-level aluminum tolerance in barley with the ALMT1 gene. *Proc. Natl. Acad. Sci. USA, 101,* 15249–15254.

Eticha, D., Thé, C., Welcker, C., Narro, L., Staß, A. and, Horst, W.J. (2005a). Aluminium-induced callose formation in root apices: inheritance and selection trait for adaptation of tropical maize to acid soils. *Field Crops Res., 93,* 252–263.

Eticha, D., Staß, A. and Horst, W.J. (2005b). Localization of aluminium in the maize root apex: Can morin detect cell wall-bound aluminium? *J. Exp. Botany, 56,* 1351–1357.

Ezaki, B., Katsuhara, M., Kawamura, M. and Matsumoto, H. (2001). Different Mechanisms of four aluminum (Al)-resistant transgenes for Al toxicity in Arabidopsis. *Plant Physiol., 127,* 918–927.

Fuente, J.M.d.l., Ramirez-Rodriguez, V., Cabrera-Ponce, J.L. and Herrera-Estrella, L. (1997). Aluminum tolerance in transgenic plants by alteration of citrate synthesis. *Science, 276,* 1566–1568.

Godbold, D.L., Jentschke, G. and Marschner, P. (1995). Solution pH modifies the response of Norway spruce seedlings to aluminium. *Plant Soil., 171,* 175–178.

Godbold, D.L. and Jentschke, G. (1998). Aluminium accumulation in root cell walls coincides with inhibition of root growth but not with inhibition of magnesium uptake in norway spruce. *Physiol. Plant., 102,* 553–560.

Grauer, U.E. and Horst, W.J. (1992). Modelling cation amelioration of aluminium phytotoxicity. *Soil Sci. Soc. Am. J., 56,* 166–172.

Grauer, U.E. and Horst, W.J. (1992). Modeling cation amelioration of aluminum phytotoxicity. *Soil Sci. Soc. Am. J., 56[1],* 166–172.

Hasenstein, K.H. and Evans, M.L. (1988). Effects of cations on hormone transport in primary roots of *Zea mays*. *Plant Physiol., 86,* 890–894.

Heim, A., Luster, J., Brunner, I., Frey, B. and Frossard, E. (1999). Effects of aluminium treatment on Norway spruce roots: Aluminium binding forms, element distribution, and release of organic substances. *Plant Soil, 216,* 103–116.

Hodson, M.J. and Evans, D.E. (1995). Aluminium/silicon interactions in higher plants. *J. Exp. Bot., 46[283],* 161–171.

Horst, W.J., Asher, C.J., Cakmak, I., Szulkiewicz, P. and Wissemeier, A.H. (1992). Short-term responses of soybean roots to aluminium. *J. Plant Physiol., 140,* 174–178.

Horst, W.J. (1995). The role of the apoplast in aluminium toxicity and resistance of higher plants: a review. *Z.Pflanzenernähr.Bodenk., 158,* 419–428.

Horst, W.J., Püschel, A-K. and Schmohl, N. (1997). Induction of callose formation is a sensitive marker for genotypic aluminium sensitivity in maize. *Plant Soil, 192,* 23–30.

Horst, W.J., Schmohl, N., Kollmeier, M., Baluska, F. and Sivaguru, M. (1999). Does aluminium affect root growth of maize through interaction with the cell wall – plasma membrane – cytoskeleton continuum? *Plant Soil, 215,* 163–174.

Ishikawa, S. and Wagatsuma, T. (1998). Plasma membrane permeability of root-tip cells following temporary exposure to Al ions is a rapid measure of Al tolerance among plant species. *Plant Cell Physiol., 39,* 516–525.

Jones, D.L. and Kochian, L.V. (1997). Alumium interaction with plasma membrane lipids and enzyme metal binding sites and its potential role in Al cytotoxicity. *FEBS, 400,* 51–57.

Jones, D.L., Gilroy, S., Larsen, P.B., Howell, S.H. and Kochian, L.V. (1998). Effect of aluminum on cytoplasmic Ca^{2+} homeostasis in root hairs of *Arabidopsis thaliana* (L.). *Planta, 206,* 378–387.

Kauss, H., Waldmann, T. and Quader, H. (1990). Ca^{2+} as a signal in the induction of callose synthesis: In Ranjeva, R. and Baudet, A. (eds), *Signal Perception and Transduction in Higher Plants*. Springer Verlag Heidelberg.

Kerven, G.L., Edwards, D.G., Asher, C.J., Hallman, P.S. and Kokot, S. (1989). Aluminium determination in soil solution. I. Evaluation of existing colorimetric and separation methods for the determination of inorganic monomeric aluminium in the presences of organic ligands. *Aust. J. Soil Res., 27,* 79–90.

Kidd, P.S., Llugany, M., Poschenrieder, C., Gunsé, B. and Barcelo, J. (2001). The role of root exudates in aluminum resistance and silicon-induced amelioration of aluminum toxicity in three varieties of maize (*Zea mays* L.). *J. Exp. Bot., 52[359],* 1339–1352.

Kinraide, T.B. (1991). Identity of the rhizotoxic aluminium species. *Plant Soil, 134,*167–178.

Kinraide, T.B., Ryan, P.R. and Kochian, L.V. (1992). Interactive effects of Al^{3+}, H^{+}, and other cations on root elongation considered in terms of cell-surface electrical potential. *Plant Physiol., 99,* 1461–1468.

Kochian, L.V. (1995). Cellular mechanisms of aluminum toxicitiy and resistance in plants. *Annu. Rev. Plant Physiol. Plant Mol. Biol., 46,* 237–260.

Kochian L.V., Hoekenga O.A. and Piñeros M.A. (2004) How do crop plants tolerate acid soils? Mechanisms of aluminum tolerance and phosphorus efficiency. *Ann. Rev. Plant Biol.,* 55, 459–493.

Kollmeier, M. and Horst, W.J. (2001). Aluminium-induced exudation of citrate from the root tip of *Zea mays* (L.): Are differential impacts of Al on citrate metabolism involved in genotypic differences? In:

W.J., Horst et al. (eds), *Plant Nutrition – Food Security and Sustainability of Agro-ecosystems* (pp 492–493). The Netherlands. Kluwer Academic Publishers, Dordrecht, ISBN 0-7923-7105-4.
Kollmeier, M., Felle, H.H. and Horst, W.J. (2000). Genotypical differences in Al resistance of Zea mays (L.) are expressed in the distal part of the transition zone. Is reduced basipetal auxin flow involved in inhibition of root elongation by Al? *Plant Physiol., 122,* 945–956.
Kollmeier, M., Dietrich, P., Bauer, C.S., Horst, W.J. and Hedrich, R. (2001). Aluminum activates a citrate-permeable anion channel in the Al-sensitive zone of the maize root apex: a comparison between an Al-sensitive and an Al-resistant cultivar. *Plant Physiol., 126,* 397.
Lazof, D.B., Goldsmith, J.G., Rufty, T.W. and Linton, R.W. (1994). Rapid uptake of aluminum into cells of intact soybean root tips. *Plant Physiol., 106,* 1107–1114.
Lindberg, S. and Strid, H. (1997). Aluminium induces rapid changes in cytosolic pH and free calcium and potassium concentrations in root protoplasts of wheat *(Triticum aestivum). Physiol. Plant., 99,* 405–414.
Lindberg, S., Szynkier, K. and Greger, M. (1991). Aluminium effects on transmembrane potential in cells of fibrous roots of sugar beet. *Physiol. Plant., 83,* 54–62.
Llugany, M., Poschenrieder, C. and Barcelo, J. (1995). Monitoring of aluminium-induced inhibition of root elongation in four maze cultivars differing in tolerance to aluminium and proton toxicity. *Physiol. Plant., 93,* 265–271.
Ma, J.F., Hiradate, S., Nomoto, K., Iwashita, T. and Matsumoto, H. (1997). Internal detoxification mechanism of Al in hydrangea. Identification of Al form in leaves. *Plant Physiol., 113,* 1033–1039.
Ma, J.F., Hiradate, S. and Matsumoto, H. (1998). High aluminum resistance in buckwheat. II. Oxalic acid detoxifies aluminum internally. *Plant Physiol., 117,* 753–759.
Ma, J.F., Ryan, P.R. and Delhaize, E. (2001). Aluminium tolerance in plants and the complexing role of organic acids. *Trends Plant Sci., 6[6],* 273–278.
Maltais, K. and Houde, M. (2002). A new biochemical marker for aluminium tolerance in plants. *Physiol. Plant., 115,* 81–86.
Marienfeld, S. and Stelzer, R. (1993). X-ray microanalyses in roots of Al-treated *Avena sativa* plants. *J. Plant Physiol., 141,* 569–573.
Marienfeld, S., Schmol, N., Klein, M., Schröder, W.H., Kuhn, A.J. and Horst, W.J. (2000). Localization of aluminium in root tips of Zea mays and Vicia faba. *J. Plant Physiol., 156*, 666–671.
Matsumoto, H. (2000). Cell biology of aluminum toxicity and tolerance in higher plants. *Intern. Rev. Cytol., 200,* 1–46.
Naumann, A. and Horst, W.J. (2003). Effect of aluminium supply on aluminium uptake, translocation, and blueing of *Hydrangea macrophylla* (Thunb.) Ser. cultivars in a peat-clay substrate. *J. Hort. Sci. Biotechnol., 78,* 463–469.
Naumann, A., Kunz, U., Lehmann, H., Stelzer, R. and Horst, W.J. (2001). Effect of aluminium on root morphology of *Hydrangea macrophylla*. In: W.J., Horst et al. (eds), *Plant Nutrition – Food Security and Sustainability of Agro-ecosystems.* (pp 516–517). The Netherlands. Kluwer Academic Publishers, Dordrecht, ISBN 0-7923-7105-4.
Olivetti, G.P., Cumming, J.R. and Etherton, B. (1995). Membrane potential depolarization of root cap cells precedes alumium tolerance in snapbean. *Plant Physiol., 109,* 123–129.
Osaki, M., Nursyams, D., Begum, H.H. and Watanabe, T. (2001). Study on aluminium resistance in relation to organic-acid anion exudation from roots of PEPC transgenic rice plants. In: W.J., Horst et al. (eds). *Plant Nutrition – Food Security and Sustainability of Agro – ecosystems*. (pp 516–517). The Netherlands. Kluwer Academic Publishers, Dodrecht, ISBN 0-7923-7105-4.
Pineros, M.A. and Kochian, L.V. (2001). A patch-clamp study on the physiology of aluminum toxicity and aluminum tolerance in maize. Identification and characterization of Al^{3+}-induced anion channels. *Plant Physiol., 125,* 292–305.
Plieth, C., Sattelmacher, B., Hansen, U.P. and Knight, M.R. (1999). Low-pH-mediated elevations in cytosolic calcium are inhibited by aluminium: apotential mechanism for aluminium toxicity. *Plant J., 18[6],* 643–650.
Rengel, Z. (1996). Uptake of aluminium by plant cells. *New Phytol, 134.,* 389–406.
Rengel, Z. and Reid, R.J. (1997). Uptake of Al across the plasma membrane of plant cells. *Plant Soil, 192,* 31–35.
Rengel, Z. and Zhang, W.H. (2003). Role of dynamics of intracellular calcium in aluminium-toxicity syndrome. *New Phytol., 159,* 295–314.
Richards, K.D., Schott, E.J., Sharma, Y.K., Davis, K.R. and Gardner, R.C. (1998). Aluminum induces oxidative stress genes in *Arabidopsis thaliana*. *Plant Physiol*., 116, 409–418.

Ryan, P.R., DiTomaso, J.M. and Kochian, L.V. (1993). Aluminum toxicity in roots: An investigation of spatial sensitivity and the role of the root cap. *J. Exp. Bot., 44[259],* 437–446.

Ryan, P.R., Delhaize, E. and Randall, P.J. (1995). Characterisation of Al-stimulated efflux of malate from the apices of Al-tolerant wheat roots. *Planta, 196,* 103–110.

Ryan, P.R., Skerrett, M., Findlay, G.P., Delhaize, E. and Tyerman, S.D. (1997). Aluminum activates an anion channel in the apical cells of wheat roots. *Proc. Natl. Acad. Sci., 94,* 6547–6552.

Sasaki, T., Yamamoto, Y., Ezaki, B., Katsuhara, M., Ahn, S.J., Ryan, P.R., Delhaize, E. and Matsumoto, H. (2004). A wheat gene encoding an aluminum-activated malate transporter. *Plant J., 37,* 645–653

Schildknecht, P.H.P.A. and Campos Vidal, B.A (2002). Role for the cell wall in Al^{3+} resistance and toxicity: crystallinity and availability of negative charges. LifeXY, 1, 1087–1095.

Schmohl, N. and Horst, W.J. (2000). Cell wall pectin content modulates aluminium sensitivity of *Zea mays* (L.) cells grown in suspension culture. *Plant, Cell Environ., 23,* 735–742.

Schmohl, N., Pilling, J., Fisahn, J. and Horst, W.J. (2000). Pectin methylesterase modulates aluminium sensitivity in *Zea mays* and *Solanum tuberosum*. *Physiol. Plant., 109,* 419–427.

Schmohl, N. and Horst, W.J. (2002). Effect of aluminium on the activity of acid phosphatase and the exudation of macromolecules by roots and suspension-culture cells of *Zea mays* L. *J. Plant Physiol., 159,* 1213–1218.

Schreiner, K.A., Hoddinott, J. and Taylor, G.J. (1994). Aluminum-induced deposition of (1,3)-β-glucans (callose) in *Triticum aestivum* L. *Plant Soil ,162,* 273–280.

Schwazerova, K., Zelenkova, S., Nick, P. and Opatrny, Z. (2002). Aluminum- Induced rapid changes in the microtubular cytoskeleton of tobacco cell lines. *Plant Cell Physiol., 43,* 207–216.

Sivaguru, M. and Horst, W.J. (1998). The distal part of the transition zone is the most aluminum-sensitive apical root zone of maize. *Plant Physiol., 116,* 155–163.

Sivaguru, M., Baluska, F., Volkmann, D., Felle, H. and Horst, W.J. (1999). Impacts of aluminum on the cytoskelelton of the maize root apex. Short-term effects on the distal part of the transition zone. *Plant Physiol., 119,* 1073–1082.

Sivaguru, M., Yamamoto, Y. and Matsumoto, H. (2000a). Differential impacts of aluminium on microtubule organisation depends on growth phase in suspension-cultured tobacco cells. *Physiol. Plant., 107,* 110–119.

Sivaguru, M., Fujiwara, T., Samaj, J., Baluska, F., Yang, Z., Osawa, H., Maeda, T., Mori, T., Volkmann, D. and Matsumoto, H. (2000b). Aluminum-induced (1,3)-β-glucan inhibits cell-to-cell trafficking of molecules through plasmodesmata. A new meachnism of aluminum toxicity in plants. *Plant Physiol., 124,* 991–1005.

Sivaguru, M., Matsumoto, H. and Horst, W.J. (2000c). Control of the response to aluminium stress. In: P., Nick (ed.), *Plant Microtubules: Potential for Biotechnology*. Springer Verlag, Berlin, ISBN 3-540-67105-6, 103–120.

Sivaguru, M, Horst, W.J. Eticha, D. and Matsumoto, H. (2006). Aluminium inhibits apoplastic flow of high molecular weight solutes in root apices of *Zea mays* L. *J. Plant Nutr. Soil Sci., 169,* 679–690.

Staβ, A. and Horst, W.J. (1995). Effect of aluminium on membrane properties of soybean (*Glycine max*) cells in suspension culture. *Plant Soil, 171,* 113–118.

Stoutjesdijk, P.A., Sale, P.W.G. and Larkin, P.J. (2001). Posible involvement of condensed tannins in aluminium tolerance of *Lotus pedunculatus*. *Aust. J. Plant Physiol., 28,* 106–107.

Taylor, G.J., Basu, A., Basu, U., Slaski, J.J., Zhang, G. and Good, A.G. (1997). Al-induced, 51-kilodalton, membrane-bound proteins are associated with resistance to al in a segregating population of wheat. *Plant Physiol., 114,* 363–372.

Taylor, G.J. (1991). Current views of the aluminum stress response; the pyhsiological basis of tolerance. *Current topics in Plant Biochem. Physiol.: 10,* 57–93.

Taylor, G.J., McDonald-Stephens, J.L., Hunter, D.B., Bertsch, P.M., Elmore, D., Rengel, Z. and Reid, R.J. (2000). Direct measurement of Aluminum uptake and distribution in single cells of *Chara corallina*. *Plant Physiol., 123,* 987–996.

Teraoka, T., Kaneko, M., Mori, S. and Yoshimura, E. (2002). Aluminum rapidly inhibits cellulose synthesis in roots of barley and wheat seedlings. *J. Plant Physiol., 159,* 17–23.

Tesfaye, M., Temple, S.J., Allan, D.L., Vance, C.P. and Samac, D.A. (2001). Overexpression of malate dehydrogenase in transgenic alfalfa enhances organic acid synthesis and confers tolerance to aluminum. *Plant Physiol., 127,* 1836–1844.

Tice, K.R., Parker, D.R. and DeMason, D.A. (1992). Operationallly defined apoplastic and symplastic aluminum fractions in root tips of aluminum-intoxicated wheat. *Plant Physiol., 100,* 309–318.

Uexküll, H.R.v. and Mutert, E. (1995). Global extent, development and economic impact of acid soils. *Plant Soil, 171,* 1–15.

Van, H.L., Kuraishi, S. and Sakurai, N. (1994). Aluminum-induced rapid root inhibition and changes in cell-wall components of squash seedlings. *Plant Physiol., 106,* 971–976.

Vázquez, M.D., Poschenrieder, C., Corrales, I. and Barcelo, J. (1999). Change in apoplastic aluminum during the initial growth response to aluminum by roots of a tolerant maize variety. *Plant Physiol., 119,* 435–444.

Vierstra, R. and Haug, A. (1978). The effect of Al^{3+} on the physical properties of membrane lipids in *Thermoplasma acidophilum. Biochem. Biophys. Res. Comm. 84(1),* 138–143.

Vierstra, R. and Haug, A. (1989). The effect of Al^{3+} on the physical properties of membrane lipids in *Thermoplasma acidophilum. Biochem. Biophys. Res. Comm., 84[1],* 138–143. 1978.

Wagatsuma, T. and R. Akiba. (1989). Low surface negativity of root protoplasts from aluminum-tolerant plant species. *Soil Sci. Plant Nutr., 35,* 443–452.

Wagatsuma, T., Nakashima, T. and Twaraya, K. (1991). Identification of aluminium-tolerant protoplasts in the original root protoplast population from several plant species differing in aluminium tolerance. In: Wright, R.J., et al. (eds) *Plant–Soil Interactions at Low pH*, pp. 789–793. Kluwer Academic Publishers, Dordrecht.

Wang, Y., Staß, A. and Horst, W.J. (2004). Apoplastic binding of aluminum is involved in silicon-induced amelioration of aluminum toxicity in *Zea mays. Plant Physiol., 136,* 3762–3770.

Wissemeier, A.H., Klotz, F. and Horst, W.J. (1987). Aluminium induced callose synthesis in roots of soybean (*Glycine max* L.). *J. Plant Physiol. 129,* 487–492.

Wissemeier, A.H., Diening, A., Hergenröder, A., Horst, W.J. and Mix-Wagner, G. (1992). Callose formation as parameter for assessing genotypical plant tolerance of aluminium and manganese. *Plant Soil,146,* 67–75.

Wissemeier, AH. and Horst, W.J. (1995). Effect of Calcium supply on aluminium-induced callose formation, its distribution and persistence in roots of soybean (*Glycine max* (L.) Merr.). *J. Plant Physiol., 145,* 470–476.

Yamamoto, Y., Hachiya, A. and Matsumoto, H. (1997). Oxidative damage to membranes by a combination of aluminum and iron in suspension-cultured tobacco cells. *Plant, Cell Physiol., 38,* 1333–1339.

Yamamoto, Y., Kobayashi, Y., Devi, S.R., Rikiishi, S. and Matsumoto, H. (2002). Aluminum toxicity is associated with mitochondrial dysfunction and the production of reactive oxygen species in plant cells. *Plant Physiol., 128*, 63–72.

Yermiyahu, U., Brauer, D.K. and Kinraide, T.B. (1997). Sorption of aluminum to plasma membrane vesicles isolated from roots of scout 66 and atlas 66 cultivars in wheat. *Plant Physiol., 115,* 1119–1125.

Zhang, G. and Taylor, G.J. (1989). Kinetics of aluminum uptake by excised roots of aluminum-tolerant and aluminum-sensitive cultivars of *Triticum aestivum* L. *Plant Physiol., 91,* 1094–1099.

Zhang, G. and Taylor, G.J. (1990). Kinetics of aluminum uptake in *Triticum aestivum* L. *Plant Physiol., 94,* 577–584.

Zhang, W.H. and Rengel, Z. (1999). Aluminium induces an increase in cytoplasmic calcium in intact wheat root apical cells. *Aust. J. Plant Physiol., 26,* 401–409.

Zhang, W.H., Ryan, P.R. and Tyerman, S.D. (2001). Malate-permeable channels and cation channels activated by aluminum in the apical cells of wheat roots. *Plant Physiol., 125,* 1459–1472.

SIGNIFICANCE OF POLYAMINES FOR PECTIN-METHYLESTERASE ACTIVITY AND THE ION DYNAMICS IN THE APOPLAST

J. GERENDAS
Institute for Plant Nutrition and Soil Science, University Kiel, Germany, jgerendas@plantnutrition.uni-kiel.de

Abstract. Activity of pectin methyl esterase (PME), controlling the degree of methylation of the carboxyl groups of pectins, strongly affects the cation exchange capacity of the cell wall and thus the ion dynamics in this compartment. PMEs have been identified in various isoforms, which differ with respect to optimal pH and other properties. Activity of PME is depended on the electrostatic potential of the cell-wall, as well as on the concentration and composition of salts in the apoplastic fluid. Di- and polyamines (merely putrescine, spermidine, spermine) are protonated at physiological pH and are often increased in response to ammonium supply, K deficiency, salt, water and other stresses. They have also been found in the plant apoplast and were shown to influence PME activity. However, in the present investigation N form and K supply had only marginal effects on PME activity of Vicia faba, which could not be related to apoplastic polyamine concentrations. Further investigations using potato tuber PME revealed no coherent scheme of the influence of polyamines on PME activity. However, the pH strongly affected PME activity, which was more than 8-fold higher at pH 7 as compared to pH 5. In view of its strong pH dependence and substantial alterations of apoplastic pH it is suggested that this feature may be more relevant under *in vivo* conditions than the regulation via polyamines. However, more detailed studies on the interaction of polyamines and apoplastic pH with respect to PME activity are required before final conclusions can be drawn.

Key words: cell wall pH, nitrogen form, pectin methyl esterase, polyamines, putrescine

B. Sattelmacher† and W.J. Horst (eds.), The Apoplast of Higher Plants: Compartment of Storage, Transport and Reactions, 67–83.

1. INTRODUCTION

The activity of pectin methyl esterase (PME), an enzyme responsible for the de-esterification of methylated polygalacturonic acid (pectin), has a significant impact on the cation exchange-capacity in the leaf apoplast and thus for the ion dynamic in cell walls. It has been shown that the PME activity (EC 3.1.1.11) is modulated, at least in vitro, by polyamines (Charney et al., 1992). Furthermore, the influence of N form and K supply on the polyamine contents and turnover is well documented (Smith et al., 1982; Gerendás, 1992). In the following the significance of the polyamine status and the interrelationship with the PME activity and thus the ion dynamics in the apoplast will be discussed.

2. SIGNIFICANCE OF PME FOR CELL-WALL PROPERTIES

2.1 Composition of the plant cell-wall and significance of PME

The plant cell-wall is heterogeneous both in time and space and consists of polysaccharides (mostly cellulose, crosslinking glycans and pectins) associated with structural proteins (Carpita and Gibeaut, 1993). In addition numerous enzymes, mostly hydrolases and some oxidases (Noat et al., 1991) are present in the cell wall and catalyse the many reactions responsible for the function of the cell wall for plant growth and metabolism, like the rate and direction of cell expansion, the response to cell wall-degrading pathogenic enzymes, the water and ion binding properties of the cell wall, and the production of oligosaccharides that represent wall fragments with signalling roles (Fry, 1995; Messiaen and van Cutsem, 1999; Pilling et al., 2000). A coherent picture of the role played by metal ions, pH and the PME in cell-wall extension was developed based on the observation that cell-wall growth leads to a decrease of local proton concentrations because the electrostatic potential difference ($\Delta\psi$) of the wall decreases. This in turn activates pectin methylesterase, which restores the initial $\Delta\psi$ value (Moustacas et al., 1991). The increase of metal-ion concentration resulting from the increased attraction of metal ions in the polyanionic cell-wall matrix, also results in the activation of wall-loosening enzymes (see below [2.2] for further details on ion interactions).

Two more recent examples highlight the significance of PME. The expression of an inducible gene encoding PME was found to be tightly correlated, both spatially and temporally, with border cell separation in pea root caps (Wen et al., 1999). The results are in agreement with the hypothesis that the demethylation of pectin by PME plays a key role in cell-wall metabolism. In ripening grape berries mRNAs coding for PME accumulate from one week before the onset of ripening until complete maturity, indicating that this transcript represents an early marker of veraison (i.e. inception of ripening) and could be involved in berry softening (Barnavon et al., 2001). The degree of methyl esterification of insoluble pectins decreased throughout the development from 68% in green stages to less than 20% for the ripe berries, and this observation is consistent with the induction of PME mRNAs during ripening. For a detailed account of cell wall composition see Carpita (this volume, pp 17–22, and references therein).

Pectins consist of rhamnogalacturonanes and polygalacturonic acid, whose uronic acid groups are either methylated or otherwise esterated, or left as free carboxyl groups representing cation exchange sites. These are mostly saturated with Ca^{2+} (Carpita and Gibeaut, 1993), but also with Mg^{2+} and other cations to some extent (Grignon and Sentenac, 1991). The cation saturation is also of significance for the stability of the gels formed. The rhamnogalacturonanes may be subdivided into rhamnogalacturonan I and II. Rhamnogalacturonan II is very heterogeneous in structure and accounts for a minor proportion of the total pectin (Carpita and Gibeaut, 1993).

The degree of pectin methylation und thus the cation-exchange properties of plant cell-walls is adjusted by a group of extra cellular enzymes, the pectin methylesterases (PMEs; EC 3.1.1.11), which catalyse the de-esterification of pectins and thus are of particular significance for cell-wall metabolism (Rexova and Markovic, 1976; Tieman et al., 1992).

PME has been identified as neutral, alkaline and acidic isoforms (Bordenave and Goldberg, 1994; Richard et al., 1994). In sweet cherries four different pectin esterase (PE) isoforms, with different isoelectric points, have been detected, and the PE isoforms differ in optimum pH and thermal stability (Alonso et al., 1996). In soybean seeds seven isoforms of PME exhibiting isoelectric points (pI values) between 6.0 and 9.5 in 1 M NaCl-extracts of total homogenates were identified (Markovic and Obendorf, 1998). Whereas the bound enzyme exhibited more basic PME isoforms (pl 8-9.5), the soluble enzyme had more active bands at pI 6.5, 7.0 and 7.5. Despite different pI values the pH optimum was 7.8 for both soluble and bound PE (Markovic and Obendorf, 1998). In the pericarp of the tomato fruit three main forms (A-C) of PME were identified that exhibit different kinetic properties and isoelectric points (Warrilow and Jones, 1995). This was attributed to their different physiological functions. The isoforms showing only weak activation by Ca^{2+}, a small inhibition by low pH and polygalacturonate are particular suited to continue working during the processes of disruption and degradation that typify fruit ripening, while the other isoforms, whose activity is under continuous fine control by local ionic and pH conditions in the wall, and are more likely candidates for a role in the modulation of cell-wall growth (Warrilow and Jones, 1995). All these results suggest that the different kinetic properties of the three forms of PE reflect different physiological functions.

PME has been detected in all higher-plant species as well as in a number of plant pathogenic fungi and bacteria (Rexova-Benkova and Markovic, 1976; Bordenave, 1996). Many cDNAs and some genomic sequences coding for PMEs have been published for several higher-plant species such as *Arabidopsis thaliana*, *Petunia inflata* and *Lycopersicon esculentum* (Hall et al., 1994; Mu et al., 1994; Richard et al., 1994; Gaffe et al., 1997; Micheli et al., 1998).

2.2 Significance of PME activity for ion relationships in the apoplast

The contents of pectins and their degree of methylation determines the cation exchange properties of plant cell-walls to large extent (Fry, 1995), and thus the storage capacity of the apoplast for ions as well as their dynamics in this plant compartment. The PMEs that catalyse de-esterification of pectins and thus control the degree of methylation modify the ion-exchange characteristics of the plant cell-wall to suit their particular functions (Tieman et al., 1992). The significance of the CEC of the wall has been already stressed by Noat et al. (1991), who pointed out that the CEC represents binding sites for positively charged (enzyme) proteins and metabolites, which has far-reaching consequences for the activity and efficiency of wall enzymes in general. This may be mediated either (1) by reducing the mobility of enzymes and/or substrates within the wall, or (2) by affecting the conformation of the enzyme and thus its catalytic properties. They showed that the parietal electrostatic potential plays a central role in the regulation of the activity of the wall enzymes. As this potential is based upon PMEs an autoregulation mechanism was suggested: the PME increases the wall electrostatic potential and this leads to its own inhibition favouring at the same time the activity of the glycosidases, which take part in growth. Thus, the wall potential slows down again and the PME is reactivated. In the normal conditions a steady state is attained, which is strictly dependent on the ion concentrations surrounding the enzyme. It was concluded that particularly the electrostatic interactions in the wall induce important regulatory properties in enzymic systems, which are not considered when studying these isolated enzymes outside their natural environment. PME activity showed a strong interaction with type ($NaCl$, $MgCl_2$, $CaCl_2$) and concentration of salts. Maximum activity was achieved at different salts concentrations ($CaCl_2$ < $MgCl_2$ < NaCl) and each curve showed a strong dose response (Noat et al., 1991). Maximum PME activity may be obtained in relatively alkaline medium (pH 8) and the optimum salt concentration is increased as the pH falls. The inhibition of PME activity by reduced salt concentrations below the optimum was attributed to the strong binding of PME to its substrate. Above the optimum salt concentration binding of the enzyme to the hydrolysable sites is prevented and hydrolysis is inhibited (Noat et al., 1991). In agreement with the previous statements Goldberg et al. (1992) stressed that PME activity shows a strong dependence on the ion milieu, which has been related to modifications of its substrate pectin. In sweet cherries all four isoforms identified were activated by calcium and sodium cations (Alonso et al., 1996). They were inactivated by D-galacturonic acid, which was discussed in relation to the sharp increase of pectin degradation by polygalacturonases during ripening (Barrett and Gonzalez, 1994). PME reduced the degree of esterification of pectins, rendering them susceptible to degradation by polygalacturonases.

A coherent picture of the role played by metal ions and pH in cell-wall extension of soybean cells was depicted by Moustacas et al. (1991). Metal ions, by attraction in the polyanionic cell-wall matrix, amplify the mechanism based on the activation of PME due to a decrease of local proton concentrations following a decrease of the electrostatic potential difference ($\Delta\psi$) of the wall. The increased activation of PME restores the initial $\Delta\psi$ value, and amplification by the attraction of metal ions in the polyanionic cell-wall matrix is basically due to the release of enzyme molecules that were initially bound to "blocks" of carboxy groups. Moreover, the apparent "inhibition" of pectin methylesterase by high salt concentrations may be considered as a device which prevents the electrostatic potential from becoming too high.

PME was also shown to be responsible for the metal-cation distribution in ripening tomato (Tieman and Handa, 1994). In transgenic tomato fruits that show more than 10-fold reduction in PME activity due to antisense technology the low PME activity in the transgenic fruit pericarp modified both accumulation and partitioning of cations between soluble and bound forms and selectively impaired accumulation of Mg^{2+} over other major cations. Decreased PME activity was associated with a 30–70% decrease in bound Ca^{2+} and Mg^{2+} in transgenic pericarp. Levels of soluble Ca^{2+} increased 10–60%, whereas levels of soluble Mg^{2+} and Na^+ were reduced by 20–60% in transgenic pericarp, indicating that PME plays a key role in regulating cation binding to the cell wall with possible determining consequences for tissue integrity during fruit senescence (Tieman and Handa, 1994).

3. INTERACTION OF POLYAMINES WITH THE ACTIVITY OF PME

3.1 Factors influencing the polyamine concentration in the apoplast

Polyamines represent a group of small molecules that are protonated at physiological pH and thus are (poly-) cationic in nature (Bachrach, 1970). The most important ones are the diamines diaminopropane (DAP), putrescine (1,4-diaminobutane) and cadaverine (1,5-diaminopentane) as well as the higher polyamines spermidine (N-aminopropyl putrescine) and spermine (N,N'-diaminopropyl putrescine). Putrescine is either synthesised from arginine via agmatine (by arginine decarboxylase, EC 4.1.1.19) or ornithine (by ornithine decarboxylase, EC 4.1.1.17) (Fig. 1, Slocum et al., 1984). The higher polyamines are synthesised by successive coupling with aminopropyle residues derived from decarboxylated S-adenosyl methionine (SAM). Since this metabolite is also the precursor of ethylene, physiological interactions between polyamines on the one hand and phytohormones and senescence processes on the other hand have been observed (Apelbaum et al., 1985; Grossmann et al., 1993; Nunn et al., 2002). Polyamines not only occur as free metabolites, but are also conjugated with cinnamic acid derivates and other compounds (Slocum et al., 1984; Smith, 1985; Neves et al., 2002; Puga et al., 2003).

Breakdown of polyamines is accomplished by oxidation by diamine (Federico and Angelini, 1987) and polyamine oxidases (Kaur-Sawhney et al., 1981) localised in the cell wall, which generate pyrroline derivates and H_2O_2 (Smith, 1985). The di- and polyamine oxidases show a close relationship to the activity of peroxidases in the cell wall (Angelini et al., 1993) and were found to be mostly associated with vascular tissue (Kaur-Sawhney et al., 1981; Smith, 1985). The H_2O_2 generated during the course of polyamine oxidation may serve as substrate for apoplastic peroxidases and thus support lignification (Federico and Angelini, 1987).

An accumulation of polyamines has been observed under various stress conditions, namely salt (Friedman et al., 1989), water (Turner and Stewart, 1986), acid (Young and Galston, 1983) and oxidative stress (Langebartels et al., 1991). Ammonium nutrition (Flores et al., 1985; Gerendás and Sattelmacher, 1995) and mineral cation deficiency (Smith, 1984), particularly K deficiency (Smith et al., 1982; Gerendás et al., 1997) increases the polyamine contents, especially the putrescine concentration. This has been discussed in view of the cationic nature of the polyamines, serving as organic cations to maintain the ion balance within the tissue (Murty et al.,

1971). However, the contribution of polyamines to the maintenance of the ion balance, as discussed in several reviews (e.g. Slocum et al., 1984) has been questioned due to limited information on their subcellular distribution (Gerendás et al., 1997). The increase of polyamine contents observed under stress conditions is often mediated by an increased activity of arginine decarboxylase, while ornithine decarboxylation is often associated with metabolic regulation during cell growth and development.

The mobility of polyamines within the plant has not been investigated to large extent, although polyamine oxidases are located in the apoplast (see above) and several polyamines have been identified in phloem and xylem sap of sunflower, mung bean and other plant species (Friedman et al., 1989). In maize plants grown with nitrate or ammonium at different N and K supply a strong increase of the total free putrescine concentration was observed when grown with high ammonium supply (Table 1). This agrees well with the frequent observation that nutrient stress results in a substantial accumulation of putrescine, while higher polyamines are affected to a much smaller degree.

Putrescine could also be detected in the xylem sap, and concentrations correlated well with those found in the roots (Fig. 2). Diamino propane,

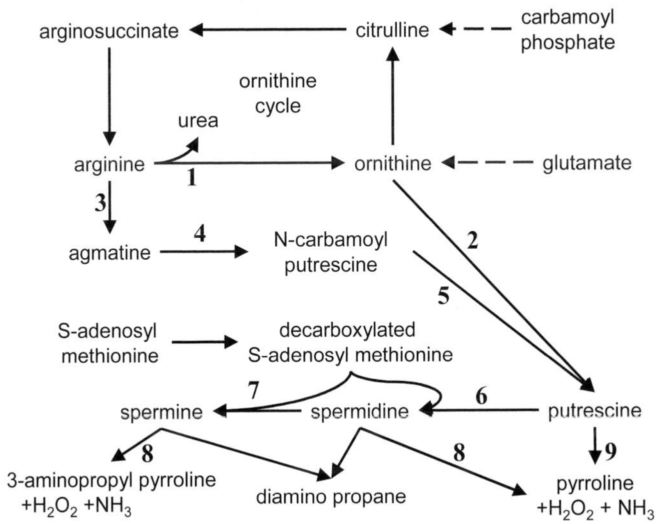

Fig. 1. Pathways of polyamine metabolism. 1 = arginase, 2 = ornithine decarboxylase, 3 = arginine decarboxylase, 4 = agmatine iminohydrolase, 5 = N-carbamoylputrescine amidohydrolase, 6 = putrescine aminopropyl transferase, 7 = spermidine aminopropyl transferase, 8 = polyamine oxidase, 9 = diamine oxidase (after Slocum et al., 1984).

Table 1. Influence of N form, N and K supply on polyamine concentrations in roots, shoots and the xylem exudate of maize plants. (Plants were grown for 6 days in complete nutrient solution using the N and K supplies as indicated. NH_4^+-containg nutrient solutions were buffered with $CaCO_3$. Plants were grown in a growth chamber (25°C, 70% relative humidity, 250 µmol m^{-2} s^{-1} PPFD for 14 h per day). Polyamines were extracted using 0.2 M perchloric acid and analysed according to Smith and Davies (1987). Means followed by the same letter are not significantly different (Tukey test, = 0.05, n = 3).

N supply	NO_3		NH_4	
	1.6 mM K	0.1 mM K	1.6 mM K	0.1 mM K
	Putrescine concentration of shoots (nmol g^{-1} DW)			
5 mM	457 f	2165 c	745 de	3525 b
10 mM	409 f	2252 c	909 d	9604 a
20 mM	487 ef	2134 c	2006 c	14790 a
	Putrescine concentration of roots (nmol g^{-1} DW)			
5 mM	559 e	2611 ab	1144 d	3383 ab
10 mM	525 e	2436 bc	1376 cd	2538 b
20 mM	604 e	2667 ab	3904 ab	4555 a
	Putrescine concentration of xylem sap (nmol ml^{-1})			
5 mM	2.12 d	4.21 c	3.17 d	8.19 b
10 mM	2.26 d	7.79 cd	5. 59 cd	21.53 bc
20 mM	2.31 d	8.20 cd	47.03 ab	144.77 a

produced by polyamine oxidases, was found in traces, and spermine, although present in low concentrations only, increased in plants grown with high ammonium or low potassium supply. This may suggest an influence of the xylem transport on polyamine concentrations in the leaf apoplast, which was indicated also by other studies investigating the significance of N supply on apoplastic polyamine concentrations (C. Langebartels, personal communication). Apoplastic washing fluids of *Vicia faba* leaves were obtained after infiltration with 50 mM MES, pH 6 (Mühling and Sattelmacher, 1995). Results showed a strong accumulation of free putrescine in the apoplast of ammonium-grown plants, which was not observed in plants grown with nitrate, and showed a similar trend as the total putrescine contents. A slight accumulation of apoplastic spermidine levels during ammonium nutrition was also observed, while total contents were largely unaffected. Apoplastic spermine was not detected and N supply had no effect on total spermine contents.

The impact of N form and K supply on polyamine levels was also investigated using *Ricinus communis* and *Vicia faba*. Plant were grown in complete nutrient solutions containing either 5 mM nitrate or ammonium at different K supplies (0.1 vs. 1.6 mM). Nutrient solution pH was maintained using a pH-stat equipment. However, with *Ricinus communis* the collection of apoplastic washing fluid (Mühling and Sattelmacher, 1995) proved difficult and cytosolic marker enzymes indicated substantial contamination.

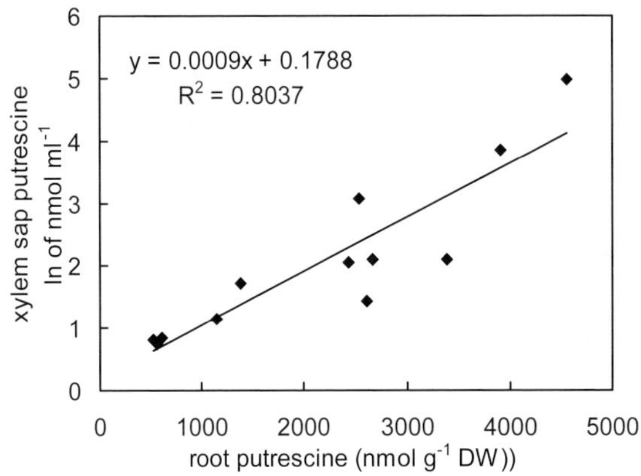

Fig. 2. Relationship between the putrescine contents of roots and the concentration in the xylem sap. Plants were grown as in Table 1 (n = 3).

From *Vicia faba* leaves polyamines were extracted using 0.2 M perchloric acid and analysed as their FMOC derivatives (Huhn et al., 1995). Briefly, the crude extract was spiked with the internal standard diaminohexan, mixed with a borate buffer and derivatised using 9-fluorenylmethyl chloroformate. Polyamines were separated on an ODS II column (Hypersil, 3 µm) by binary gradient elution using an acetate-acetonitrile gradient and recorded by fluorescence detection (excitation 265 nm, emission 316 nm). In agreement with the statement made above, the total content of polyamines in the leaves showed a differential response to the treatments imposed (Fig. 3). The

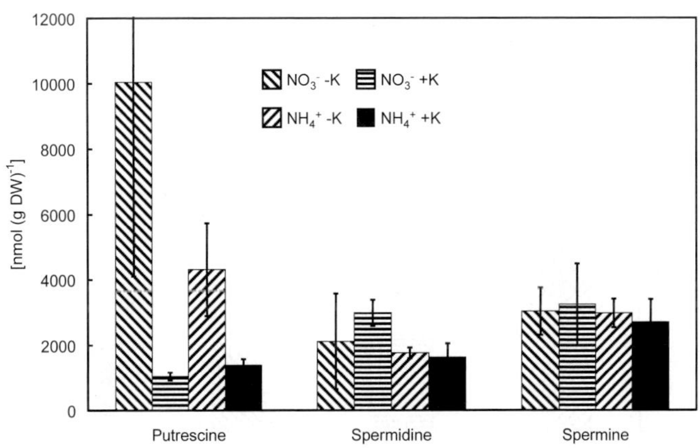

Fig. 3. Influence of N form and K supply on the polyamine contents of leaves of Vicia faba. Plants were grown for four weeks in complete nutrient solution using 5 mM N and 0.1 (-K) vs. 1.6 mM K (+K). Plants were grown using a pH stat equipment in a growth chamber (25°C, 70% relative humidity, 250 µmol m^{-2} s^{-1} PPFD for 14 h per day, n=3).

influence was strongest on putrescine, while spermine contents were not affected. When grown with high (normal) K supply the putrescine values for ammonium-grown plants were only slightly higher than for nitrate-grown controls, while K deficiency substantially increased the putrescine level, particularly in nitrate-grown plants.

The influence of N and K supply on apoplastic polyamine concentration was also investigated (Fig. 4). Apoplastic washing fluid (AWF) was obtained using 50 mM MES, pH 6 (Mühling and Sattelmacher, 1995) and analysed for polyamines as before. Only putrescine was present in detectable amounts (not corrected for dilution). The concentration was highest in the AWF obtained from NO_3^--grown K-deficient leaves, while the other treatments reached a lower level. The putrescine concentration in the AWF obtained from NH_4^+ -K plants was surprisingly low. Thus the polyamine concentration in the AWF did not follow the total content.

The results presented indicate that not only the total, but also the apoplastic polyamine contents of plants are influenced by the form und concentration of N and K supplied.

3.2 Behaviour of polyamines in the plant cell-wall

Due to the cationic nature of the polyamines they exhibit a strong affinity to polyanionic structures like pectins, which was first characterised by D'Orazi and Bagni (1987). Formation constants with pectin were in the order of 10^5 for putrescine and spermidine, and 10^6 for spermine. Calcium

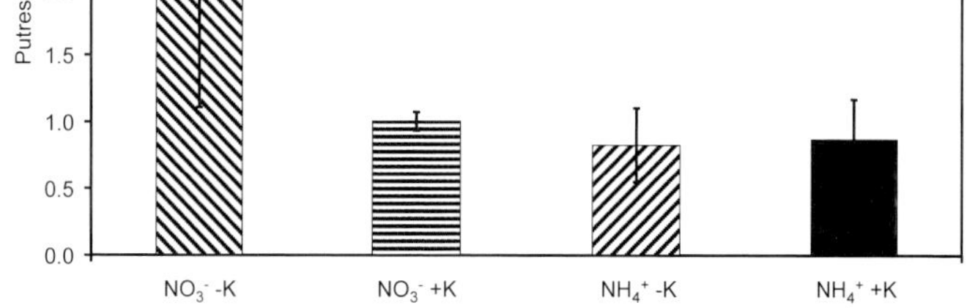

Fig. 4. Influence of N form and K supply on the putrescine concentration in the AWF of leaves of Vicia faba. Plants were grown as in Fig. 3 (n = 3).

ions were shown to compete weakly with spermine by lowering the formation constant by a factor of 4-5. The interaction of polyamines and Ca^{2+} was later on investigated in detail by Messiaen et al. (1997) using purified carrot cell-walls. The ion selectivity sequence of the walls for polyamines followed the sequence spermine^{4+} > spermidine^{3+} $\approx$ Ca^{2+} > putrescine^{2+}. The polyamines were subjected only to electro-selectivity and probably did not induce any favorable supramolecular conformation of pectin like the one induced by Ca^{2+}. The lower selectivity of the cell wall for putrescine was partly attributed to its inability to access and displace Ca^{2+} from higher affinity sites within dimerized pectin substances, while spermidine and spermine were able to remove Ca^{2+} completely from both the walls and the pure polygalacturonates (Messiaen and van Cutsem, 1999). In other words the polyamines behave more ideally as cations of different charge and thus seem not to induce favourable conformation of pectin fragments such as egg boxes in the presence of Ca^{2+} (Morris et al., 1982; Powell et al., 1982). This structure is essential for the pectic signalling cascade which requires recognition of the pectic fragments by the plant cell, initiating morphological and defence responses (Messiaen and van Cutsem, 1994). It was shown that due to the binding of higher polyamines to pectin fragments these structures were not recognized by the 2F4 anti-pectin monoclonal antibody (Messiaen and van Cutsem, 1999). The authors concluded that polyamines can act on plant-cell physiology by modulating the transduction of the pectic signal.

3.3 Influence of polyamines on PME dynamics

The interaction between polyamines and pectin methylesterase activity was studied in detail by Charnay et al. (1992) showing that polyamines are of crucial importance for the regulation of apoplastic pectin metabolism. Both activity and transcription of pectin methylesterase were regulated by polyamines to some extent. The higher polyamines spermidine and spermine were more efficient than the inorganic salts tested and the reaction rate versus the polyamine concentration exhibited a bell-shaped curve. As mentioned already polyamines show a high affinity to polyglacturonic acids, which affects the formation of enzyme substrate complexes (Moustacas et al., 1991). Apparently, the regulatory effect of polyamines is effective in concentration ranges typically fond in the apoplast of non-stressed plant tissues (Charnay et al., 1992). Thus, polyamines may act as efficient regulators of the cell-wall pH via the control of the electrostatic cell-wall potential.

The PME activity is typically measured by the release of protons, which can either be followed as a change of pH (Hagermann and Austin, 1986) or as the amount of alkali required to maintain the pH using a pH stat titrator (Charnay et al., 1992). Initially we determined the PME activity in leaves of *Vicia faba* after homogenisation in 1 M NaCl. The assay was based on the acidification brought about by the hydrolysis of methylated pectin, which is monitored by the change of absorption at 620 nm of the pH indicator bromothymol blue. But extracts showed a strong darkening, which was most likely due to the presence of polyphenole oxidases. Inclusion of insoluble poly(vinylpolypyrrolidone) (McMillan and Péromelon, 1995) circumvented the problem, but the influence of the nutrient treatments (5 mM nitrate vs. ammonium at 0.1 mM or 1.6 mM K) did not induce significant differences in PME activity (Fig. 5).

In order to characterise the influence of polyamines on the PME activity in vitro, the model plant potato was used as a more suitable source of PME. Its activity in tuber extracts was measured in the presence of different polyamine concentrations after gel-filtering of extracts to remove endogenous polyamines. To overcome sensitivity restrictions encountered with the previous titration method in terms of pH changes per unit time, the PME activity was determined by pH monitoring, which allowed more sensitive detection of pH changes (Fig. 6). As this would not consider the buffering capacity of the assay medium a well-defined hydroxide addition induced a pH shift, which can be derived from the parallel shift of the two regression lines, providing an estimate of the buffering capacity of the medium. Using this approach the rate of pectin demethylation can be calculated accurately.

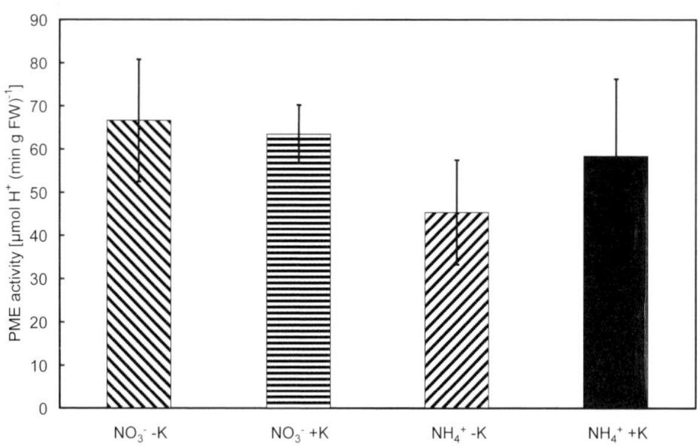

Fig. 5. Influence of N form and K supply on the total PME activity in leaves of Vicia faba. Plants were grown as in Fig. 3 (n = 3).

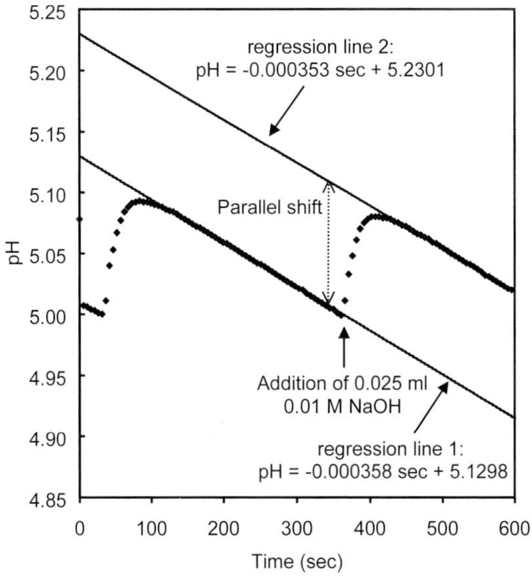

Fig. 6. Determining pectin methylesterase activity by pH monitoring. The regression lines allow to determine the trend of pH changes accurately, and the parallel shift induced by the hydroxide addition is a measure of the buffering capacity. Taking both into account allows calculating the H^+ release accurately.

The influence of increasing polyamine concentrations (0, 0.01, 0.1, 1 and 10 mM) was tested under in vitro conditions (Fig. 7). At pH 7 spermidine and spermine had no influence on in vitro PME activity extracted from potato tubers, and cadaverine, which is not usually observed in potatoes (Fig. 3), induced a moderate increase in PME activity only when supplied at 0.1 mM at pH 7. Putrescine stimulated PME activity at 1 mM, but such levels were not found in the AWF (even when correcting for dilution). At pH 5 PME activity was not substantially altered by spermidine or cadaverine, while spermine seems to induce a differential response. However, further measurements indicated that this response is not consistent. The results presented here are in agreement with those of Leiting and Wicker (1997) who also found that the polyamines had a negligible stimulating effect, which does not support the view expressed by Charnay et al. (1992) that the PME activity is highly controlled by the presence of polyamines. In agreement with previous reports the pH strongly influenced the PME activity (McMillan and Péromelon, 1995; Saénz et al., 2000), which in the absence of polyamines was 0.68±0.08 at pH 7 and 0.08±0.01 at pH 5.

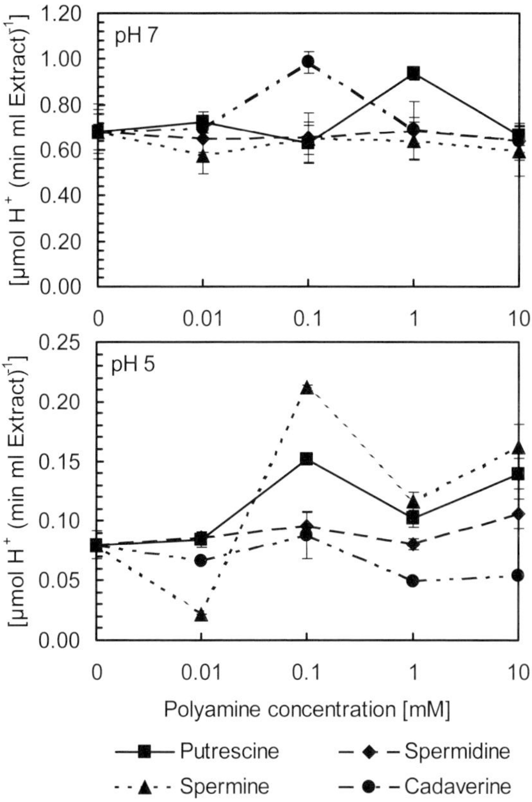

Fig. 7. Influence of the polyamine concentration on the PME activity of potato tuber extracts at pH 7 (top) and pH 5 (bottom) determined in vitro.

4. CONCLUSIONS

It is well accepted that the PME activity is of fundamental importance for the ion dynamics in the apoplast. Previous in-vitro studies have shown that PME activity may be modulated by polyamines, but results presented here suggest that the interaction is of limited significance in relation to the polyamine concentrations expected in apoplastic fluids. N form and potassium supply are known for their influence on total polyamine contents (Slocum et al., 1984; Smith, 1985), but the moderate influence of N form and K supply on apoplastic polyamine concentrations observed here reduces the significance of the interaction between PME and polyamines even further. Overall the influence of polyamines on the PME activity determined in vitro is rather small.

The strong dependence of activity of PME on the surrounding pH is striking and in view of numerous reports on the effect of nutrient supply, particularly N form (Mühling and Läuchli, 2001; Kosegarten et al., 2001; Mattsson and Schjoerring, 2002), on apoplastic pH it is concluded that this mechanism may have more substantial influence on the PME activity. In the present study the apoplastic pH was not considered, since suitable tools to study its dynamics (e.g. fluorescent ratio imaging or micro electrodes) were not available. More detailed studies on the interaction between PME activity, polyamines, and apoplastic pH are required before final conclusions can be drawn.

REFERENCES

Alonso, J., Rodriguez, M.T. and Canet, W. (1996). Purification and characterization of four pectinesterases from sweet cherry (*Prunus avium* L.). *J. Agric. Food Chem.*, 44, 3416–3422.

Angelini, R., Bracaloni, M., Federico, R., Infantino, A. and Porta-Puglia A. (1993). Involvement of polyamines, diamine oxidase and peroxidase in resistence of chickpea to Ascochyta rabiei. *J. Plant Physiol.*, 142, 704–709.

Apelbaum, A., Goldlust, A. and Icekson, J. (1985). Control by ethylene of arginine decarboxylase activity in pea seedlings and its implication for hormonal regulation of plant growth. *Plant Physiol.*, 79, 635–640.

Bachrach, U. (1970). Metabolism and function of spermine and related polyamines. *Ann. Rev. Microbiol.*, 24, 109–134.

Barnavon, L., Doco, T., Terrier, N., Ageorges, A., Romieu, C. and Pellerin, P. (2001). Involvement of pectin methyl-esterase during the ripening of grape berries: Partial cDNA isolation, transcript expression and changes in the degree of methyl-esterification of cell wall pectins. *Phytochem.*, 58, 693–701.

Barrett, D.M. and Gonzalez, C. (1994). Activity of softening enzymes during cherry maturation. *J. Food Sci.*, 59, 574–577.

Bordenave, M. (1996). Analysis of pectin methyl esterases. In: Linskens, H.F., Jackson, J.F. (eds) *Plant Cell Wall Analysis*. Springer, Berlin, pp 165–180.

Bordenave, M. and Goldberg, R. (1994). Immobilized and free apoplastic pectinmethylesterases in mung bean hypocotyl. *Plant Physiol.*, 106, 1151–1156.

Carpita, N.C. and Gibeaut, D.M. (1993). Structural models of primary cell walls in flowering plants: consistency of molecular structure with the physical properties of the walls during growth. *Plant J.*, 3, 1–30.

Charnay, D., Nari, J. and Noat, G. (1992). Regulation of plant cell wall pectin methyl esterase by polyamines. Interactions with the effects of metal ions. *Europ. J. Biochem.*, 205, 711–714.

D'Orazi, D. and Bagni, N. (1987). In vitro interactions between polyamines and pectic substances. *Biochem. Biophys. Res. Commun.*, 148, 1259–1263.

Federico, R. and Angeline, R. (1987). On the physiological role of diamines and polyamine oxidases in the cell wall. The Phytochemical Society of Europe, Symposium on Amines in Plants. Oxford University Press.

Friedman, R.A., Altman, A. and Levin, N. (1989). The effect of salt stress on polyamine biosynthesis and content in mung bean plants and in halophytes. *Physiol. Plant.*, 76, 295–302.

Fry, S. (1995). Polysaccharide-modifying enzymes in the plant cell wall. *Ann. Rev. Plant Physiol. Plant Mol. Biol.*, 46, 497.

Gaffe, J., Tiznado, M.E. and Handa, A.K. (1997). Characterization and functional expression of a ubiquitously expressed tomato pectin methylesterase. *Plant Physiol.*, 114, 1547–1556.

Gerendás, J., (1992). Einfluß von Form und Konzentration des Stickstoffangebotes auf Wachstum und Physiologie junger Maispflanzen (*Zea mays* L.). *Schriftenreihe des Instituts für Pflanzenernährung und Bodenkunde, Universität Kiel.* Band 20.

Gerendás, J. and Sattelmacher, B. (1995). Einfluß des Ammoniumangebotes auf Wachstum, Mineralstoff- und Polyamingehalt junger Maispflanzen. Zeitschrift für Pflanzenernährung und Bodenkunde, 158, 299–305.

Gerendás, J. Zhu, Z., Bendixen, R., Ratcliffe, R.G. and Sattelmacher, B. (1997). Physiological and biochemical processes related to ammonium toxicity in higher plants. Zeitschrift für Pflanzenernährung und Bodenkunde, 160, 239–251.

Goldberg, R., Pierron, M., Durand, L. and Mutaftschiev, S. (1992). In vitro and in situ properties of cell wall pectin methylesterases from Mung bean hypocotyls. *J. Exp. Bot.*, 43, 41–46.

Grignon, C. and Sentenac, H. (1991). pH and ionic conditions in the apoplast. *Ann. Rev. Plant Physiol. Plant Mol. Biol.*, 42, 102–125.

Grossmann, K., Siefert, F., Kwiatkowski, J., Schraudner, M., Langebartels, C. and Sandermann, H. Jr. (1993). Inhibition of ethylene production in sunflower cell suspensions by the plant growth retardant BAS 111..W: Possible relations to changes in polyamine and cytokinin contents. *J. Plant Growth Regul.*, 12, 5–11.

Hagermann, A.E. and Austin, P.J. (1986). Continous spectrophotometric assay for plant pectin methylesterase. *J. Agric. Food Chem.*, 34, 440–444.

Hall, L.N., Bird, C.R., Picton, S., Tucker, G. A., Seymour, G.B. and Grierson, D. (1994). Molecular characterisation of cDNA clones representing pectin esterase isozymes from tomato. *Plant Mol. Biol.*, 25, 313–318.

Huhn, G., Mattusch, J. and Schulz, H. (1995). Determination of polyamines in biological materials by HPLC with 9-fluorenylmethyl chloroformate. *Fresenius J. Anal. Chem.*, 351, 563–566.

Kaur-Sawhney, R., Flores, H.E. and Galston, A.W. (1981). Polyamine oxidase in oat leaves: a cell wall-localized enzyme. *Plant Physiol*, 68, 494–498.

Kosegarten, H., Hoffmann, B. and Mengel, K. (2001) The paramount influence of nitrate in increasing apoplastic pH of young sunflower leaves to induce Fe deficiency chlorosis, and the re-greening effect brought about by acidic foliar sprays. *J. Plant Nutr. Soil Sci.*, 164, 155–163.

Langebartels, C., Kerner, K., Leonardi, S., Schraudner. M., Trost, M., Heller, W. and Sandermann, H. Jr (1991). Biochemical plant responses to ozone: I. Differential induction of polyamine and ethylene biosynthesis in tobacco. *Plant Physiol.*, 95, 882–889.

Leiting, V.A. and Wicker, L. (1997). Inorganic cations and polyamines moderate pectinesterase activity. *J. Food Sci.*, 62, 253–255.

Markovic, O. and Obendorf, R.L. (1998). Soybean seed pectinesterase. *Seed Sci. Res.*, 8, 455–461.

Mattsson, M. and Schjoerring, J.K. (2002). Dynamic and steady-state responses of inorganic nitrogen pools and NH_3 exchange in leaves of *Lolium perenne* and *Bromus erectus* to changes in root nitrogen supply. *Plant Physiol.*, 128, 742–750.

McMillan, G.P. and Péromelon, M.C.M. (1995). Purification and characterization of a high pI pectin methyl esterase isoenzyme and its inhibitor from tubers of *Solanum tuberosum* subsp. *tuberosum* cv. Katahdin. *Physiol. and Mol. Plant Pathol.*, 46, 413–427.

Messiaen, J. and van Cutsem, P. (1994). Pectic signal transduction in carrot cells: Membrane, cytosolic and nuclear responses induced by oligogalacturonides. *Plant Cell Physiol.*, *35*, 677–689.

Messiaen, J. and van Cutsem, P. (1999). Polyamines and pectins. II. Modulation of pectic signal transduction. *Planta,* 208, 247–256.

Messiaen, J., Cambier, P. and van Cutsem, P. (1997). Polyamines and pectins. I. ion exhange and selectivity. *Plant Physiol.*, 113, 387–395.

Micheli, F., Holliger, C., Goldberg, R. and Richard, L. (1998). Characterization of pectin methylesterase-like gene AtPME3: a new member of a gene family comprising at least 12 genes in *Arabidopsis thaliana*. *Gene,* 220, 13–20.

Morris, E.R., Powell, D.A., Gidley, M.J. and Rees, D.A. (1982). Conformations and interactions of pectins. I. Polymorphism between gel and solid states of calcium polygalacturonate. *J. Mol. Biol.*, 155, 507–516.

Moustacas, A. M., Nari, J., Borel, M., Noat, G. and Ricard, J. (1991). Pectin methylesterase, metal ions, and plant cell-wall extension. *Biochem. J.*, 279, 351–354.

Mu, J.H., Stains, J.P. and Kao, T.H. (1994). Characterization of a pollen-expressed gene encoding a putative pectin esterase of *Petunia inflata*. *Plant Mol. Biol.*, *25*, 539–544.

Mühling, K.H. and Läuchli A. (2001). Influence of chemical form and concentration of nitrogen on apoplastic pH of leaves. *J. Plant Nutr.*, 24, 399–411.

Mühling, K.H. and Sattelmacher, B. (1995). Apoplastic ion concentration of intact leaves of fieldbean (*Vicia faba* L.) as influenced by ammonium and nitrate nutrition. *J. Plant Physiol.*, 147, 81–86.
Murty, K.S., Smith, T.A. and Bould, C. (1971). The relation between the putrescine content and potassium status of black currant leaves. *Ann. Bot.*, 35, 687–695.
Neves, C., Santos, H., Vilas, B.L., Amancio, S. (2002). Involvement of free and conjugated polyamines and free amino acids in the adventitious rooting of micropropagated cork oak and grapevine shoots. *Plant Physiol. Biochem.*, 40, 1071–1080.
Noat, G., Nari, I., Moustacas, A.M. and Borel M. (1991). Properties of the plant cell wall and interactions between the matrix and the bound enzymes. *Bull. Soc. Bot. France Act. Bot.*, 138, 263–277.
Nunn, A.J., Reiter, I.M., Haeberle, K.H., Werner, H., Langebartels, C., Sandermann, H., Heerdt, C., Fabian, P. and Matyssek, R. (2002). "Free-air" ozone canopy fumigation in an old-growth mixed forest: Concept and observations in beech. *Phyton,* 4, 105–119.
Pilling, J., Willmitzer, L. and Fisahn, J. (2000). Expression of a Petunia inflata pectin methyl esterase in Solanum tuberosum L. enhances stem elongation and modifies cation distribution. *Planta*, 210, 391–399.
Powell, D.A., Morris, E.R., Gidley, M.J. and Rees, D.A. (1982). Conformations and interactions of pectins. II. Influence of residue sequence on chain association in pectate gels. *J. Mol. Biol.*, 155, 517–531.
Puga, H.M.I., Gallardo, M. and Matilla, A.J., (2003). The zygotic embryogenesis and ripening of *Brassica rapa* seeds provokes important alterations in the levels of free and conjugated abscisic acid and polyamines. *Physiol. Plant.*, 117, 279–288.
Rexova, B.L and Markovic, O. (1976). Pectic enzymes. *Adv. Carbohydrate Chem. Biochem.*, 33, 323–385.
Richard, L., Qin, L.X., Gadal, P. and Goldberg, R. (1994). Molecular cloning and characterisation of a putative pectin methylesterase cDNA in *Arabidopsis thaliana* (L). *FEBS Letters,* 355, 135–139.
Saénz, J.M., Tellez A., de la Garza, H., de la Luz Reyes, M., Contreras-Esquivel, J.C. and Aguilar, C.N. (2000). Purification and some properties of pectinesterase from potato (*Solanum tuberosum* L.) alpha cultivar. *Braz. Arch. Biol. Technol.,* 43, 393–398.
Slocum, R.D., Kaur-Sawhney, R. and Galston, A.W. (1984). The physiology and biochemistry of polyamines in plants. *Arch. Biochem. Biophys*., 235, 283–303.
Smith, T.A. (1985) Polyamines. *Ann. Rev. Plant Physiol.*, 36, 117–143.
Smith, G.S., Lauren D.R., Cornforth, I.S. and Agnew, M.P. (1982). Evaluation of putrescine as a biochemical indicator of the potassium requirements of lucerne. *New Phytol.,* 91, 419–428.
Tieman, D.M and Handa, A.K. (1994). Reduction in pectin methylesterase activity modifies tissue integrity and action levels in ripening tomato (*Lycopersicon esculentum* Mill.) fruits. *Plant Physiol.*, 106, 429–436.
Tieman, D.M., Harriman, R.W., Ramamohan, G. and Handa, A.K. (1992). An antisense pectin methyl esterase gene alters pectin chemistry and soluble solids in tomato fruit. *Plant Cell*, 4, 667–679.
Turner, L.B. and Stewart, G.R. (1986). The effect of water stress upon polyamine levels in barley (*Hordeum vulgare* L.) leaves. *J. Exp. Bot.*, 37, 170–177.
Warrilow, A.G.S. and Jones, M.G. (1995). Different forms of tomato pectinesterase have different kinetic properties. *Phytochem*., 39, 277–282.
Wen, F., Zhu, Y. and Hawes, M.C. (1999). Effect of pectin methylesterase gene expression on pea root development. *The Plant Cell,* 11, 1129–1140.
Young, N.D. and Galston A.W. (1983). Putrescine and acid stress. *Plant Physiol.*, 71, 767–771.

Section 2

The Root Apoplast – Implication for Ion Acquisition and Short-Distance Transport

THE APOPLAST: A KINETIC PERSPECTIVE

An ESR study of manganese binding in plant tissue

A.D.M. GLASS
Department of Botany, University of British Columbia, Vamcouver, Canada,
aglass@interchange.ubc.ca

1. INTRODUCTION

The apoplast was defined by Münch (1930) as the non-living (extracellular) component of the plant, in contrast to the living (cytoplasmic) symplast. As such the apoplast was considered to make up a physical continuum through which water and solutes might freely move by diffusive or bulk flow throughout the plant. The cell walls that constitute the apoplast may be made up of a number of chemically and physically distinct layers. The middle lamella is the first-formed layer, deposited following cell division, and is rich in galacturonic acid residues. The dissociation of the uronic acid carboxyl groups generates the negatively charged carboxylates that are responsible for the cation-binding properties of cell walls. Upon the middle lamellae are layered the primary and secondary cell walls that are composed of cellulose microfibrils embedded in a matrix of pectins, a diverse group of xyloglucans (the hemicelluloses) together with structural proteins (Cosgrove, 1997). The cell-wall matrix, in contrast to the cellulose microfibrils, provides numerous porous regions where the hydraulic conductivity is high compared to more ordered semi-crystalline regions within the cellulose microfibrils. Traditional wisdom has suggested that the bulk flow or diffusion of water is reduced to various extents by regions of the apoplast where cell walls are impregnated with non-permeable substances such as suberin and lignin, as for example in the endodermis. Nevertheless recent measurements documented in the paper by Steudle and

B. Sattelmacher† and W.J. Horst (eds.), The Apoplast of Higher Plants: Compartment of Storage, Transport and Reactions, 87–96.

Ranathunge (this volume, pp. 119–130) suggests that even in the presence of putative apoplastic barriers (e.g. the Casparian band and suberin lamellae) the permeability of the apoplast is higher than formerly anticipated. Moreover, switching between the apoplastic and symplastic pathways of water flow may provide a mechanism for regulating the root's hydraulic conductivity. This viewpoint is in part substantiated in the paper by Schreiber et al. (this volume, pp. 109–118) who demonstrate that the extent of suberization of the endodermis or hypodermis is highly variable according to both species and environmental conditions. However, they also conclude that suberization per se does not necessarily lead to complete impermeability of the apoplast to water or solute movement. The biochemical nature of the cell wall, particularly of the middle lamella with its abundant carboxyl groups, represents an electrostatic (Donnan) barrier to free diffusion throughout the cell wall. Thus, as a water and solute permeable pathway from soil to the shoot (the major biosynthetic region of the plant), the apoplast typically consists of the extracellular regions of the root only as far as the endodermis. Because of the electrostatic properties of the cell wall with regard to charged species the apoplast is generally considered to include a Water Free Space (WFS) and a Donnan Free Space (DFS). These are not considered to be spatially distinct (see Sattelmacher (2000) for discussion). Water and solute flux from this cortical region of the apoplast might theoretically lead vertically to the shoot by a route that is external to the stele, either through the apoplast or the symplast, but clearly the low resistance provided by tracheary elements within the stele makes entry into the stele the preferred pathway. To enter the stele water and solute must bypass the impermeable endodermal barrier, briefly entering the endodermal symplasm before rejoining the apoplast within the stele. The apoplast within the endodermis includes the stelar regions of the root and stem, including xylem vessels and tracheids and cell walls of parenchyma cells. Within leaves the apoplast continues beyond the stele to the cell walls of the mesophyll and epidermis. Clearly, close to the root tip there may be no endodermal development and thus the apoplast is continuous with the differentiating vascular elements. Nevertheless, as a proportion of the whole root, except in very young germinating seeds, this pathway represents only a very small proportion of the total cross-sectional area available for water and solute entry.

Throughout a long period of botanical study, much attention and considerable discussion has focused on defining the details of radial water and solute movement from soil to stele, via the apoplast as far as the endodermis and hence via the symplast to the tracheary elements, or via entry to the symplasm at the epidermis and ongoing movement via the cortical symplasm or some combinations of these two pathways. In this Section I will address only solute fluxes and generally consider these as they may be understood using kinetic methods of analysis. More comprehensive reviews of apoplastic transport and particularly the relevance of the apoplast for mineral nutrition of plants may be found in Sattelmacher (2000).

2. RADIAL PATHWAY OF SOLUTE FLUX IN ROOTS

Although many introductory textvolumes state that solute movement across the cortex typically occurs within the apoplast as far as the endodermis, where solutes must enter the symplasm, bypassing the more impermeable Casparian band of the endodermis, studies have demonstrated that under natural conditions, particularly when plant demand is high and available soil solution ion concentration is low, substantial depletion zones develop around the root where ion concentrations are considerably diminished. For example Barber (1968) grew corn plants in soil that was initially uniformly labeled with ^{86}Rb, to show by means of autoradiography that extensive regions depleted in this tracer developed over time. Likewise Bagshaw et al. (1972) reported on the gradual depletion of inorganic phosphate around onion roots as the growing season progressed. Ion fluxes up to the root surface across these depletion zones depend upon diffusion for monovalent cations such as potassium or ammonium and even divalent anions such as orthophosphate that are readily bound to charged soil particles, and by bulk flow of soil solution in the case of monovalent anions such as nitrate or chloride (Barber, 1962). Indeed, these regions of nutrient depletion provide a convincing rationale for extending the absorptive regions of the root beyond these depletion zones by means of root hairs and mycorrhizae. Interestingly, where plant demand is low and ion concentration is high the opposite situation might develop and ion concentration may build up at the root surface generating a gradient for diffusion away from the root.

Nevertheless, especially under natural conditions, by virtue of low ion concentrations at the root surface, it has been suggested that only the outermost regions of the root epidermis and outer cortex could possible participate in nutrient uptake into the symplasm because the rapid uptake of ions by these outermost cells would so deplete apoplastic ion concentration that further transport within the apoplast to the endodermis would be virtually insignificant to the nutrition of the plant. Thus a more realistic picture of nutrient uptake might involve absorption from the apoplast in outermost cells (including root hairs and mycorrhizae) and onward travel within the symplasm across the endodermis into the apoplast of the stele. Clearly the realization of this prediction must depend upon soil solution strength and plant demand for the particular nutrient. Using available estimates for rates of K^+ uptake by barley roots, Bange (1973) calculated that if bulk solution K^+ concentration were 1 mM or above, then K^+ might permeate within the apoplast up to the endodermis and in this case the whole of the root cortex, rather than only the outermost layers, could serve for ion uptake. Similar conclusions might be derived for other nutrients. Available literature estimates of soil-solution concentrations show enormous geographic, regional and seasonal variations (Reisenauer, 1966; Wolt, 1994). Even in agricultural soils, Wolt (1994) recorded soil-solution concentrations of nitrate, ammonium, potassium and other ions that varied over orders of magnitude, from ~0.01 mM to ~5 mM and above. Thus conclusions regarding the two pathways must depend upon prevailing conditions and perhaps both pathways operate according to these conditions. It is reasonable to imagine that soil-solution concentrations of ions such as potassium, ammonium, nitrate and phosphate, which are typically supplied to soils in large amounts as inorganic fertilizers under agricultural conditions, might be quite high early in the season. In this case the whole of the cortical apoplast might be viewed as a source for ion uptake. The situation probably undergoes gradual change during the season, increasingly favouring absorption from outermost layers and root hairs and mycorrhizae, and travel within the symplasm, later on in the season. Most plant physiologists have advanced understanding of root ion absorption by use of hydroponically grown roots in which diffusion barriers and depletion zones are diminished by aeration and thorough mixing of nutrients. The paper by Bücking et al. (this volume, pp 99-110) examines the role of the apoplast in facilitating both the uptake and exchange of nutrients between plant and fungal partners. By use of fluorescent dyes they reveal that the fungal sheath is not a barrier to the entry of nutrients into the mycorrhizal root complex.

3. THE APOPLAST AS A NUISANCE TO PLANT PHYSIOLOGISTS

The general consensus suggests that the apoplast is a cation binding matrix, by virtue of the presence of large numbers of dissociated carboxyl residues that are common in the pectic fraction of the cell wall. Clearly the extent of these ion binding sites will be reduced by methylation and increased by demethylation (Sattelmacher, 2000). Nevertheless, as described below, kinetic evidence suggests that anions, too, may be bound to the root apoplast. For the majority of plant physiologists who study ion uptake by plant roots, the primary interest is in the absorption of ions across the plasma membrane. The apoplast is then simply a nuisance that complicates the measurement of ion uptake because of its capacity to bind ions during the period of ion-uptake measurement. Where experimenters employ radioactive or stable isotopes to measure ion uptake, these must be washed out of the apoplast, or else the measure of uptake will be an overestimate. This has been dealt with in the past by a period of desorption in non-labelled, but otherwise chemically identical solution after the uptake period. Many experimenters have paid little regard to the dynamics of ion binding in the apoplast in designing experimental protocols, particularly the duration of the desorption period and the temperature of the desorption solution. Cram (1968) made the situation abundantly clear when he argued that without a prior knowledge of the half-lives of apoplastic and cytosolic isotopic exchange the ion-influx might be underestimated by loss from the cytosol during desorption if the duration of desorption was too long and overestimated if too short by virtue of not effectively removing tracer from the apoplast. Although ecologists often find it counterintuitive that ions that are commonly in short supply and that typically limit plant growth should efflux from roots, the evidence is very clear. Even plants maintained on low external concentrations of ions such as ammonium or nitrate may efflux these ions at rates that increase from ~10% of influx at low μM concentrations to ~30% or higher as external ion concentrations approach 1 mM. For example, rice plants grown in 2 μM ammonium solutions were found to efflux ammonium at a rate that was 10% of influx (Wang et al., 1993). As a simple rule of thumb, the influx period should be chosen to be a small proportion of the half-life of cytoplasmic exchange so that there is insufficient time for the tracer concentration to rise within the cytoplasm to a level that can bring about a significant efflux. As the duration of the influx period is extended, the cytoplasm will approach a quasi-steady state, while efflux may rise. Under such conditions the flux that is measured will increasingly approximate net flux (influx minus efflux), rather than influx. Likewise, the duration of the period selected to desorb ions from the cell wall should be of a sufficient length that all tracer is washed from the apoplasm. However, if too long then significant quantities of tracer will be removed from the cytoplasm and this efflux will reduce the estimate of influx. Clearly it is necessary to compromise here because the two separate goals take us in opposite directions. In order to properly arrive at the appropriate influx and desorption times it is necessary to have a clear definition of the half-lives of exchange of the cytoplasm and the apoplast. Typically this is achieved through the use of compartmental analysis. No influx measurements should be made in ignorance of these parameters. The methodology has been described in detail in numerous sources (Pitman, 1963; Cram, 1968; MacRobbie, 1971; Clarkson, 1974) and these sources should be consulted. In brief, the method involves growing the plant system under steady state conditions (i.e. external solutions are not allowed to deplete significantly by use of peristaltic pumps that continuously replace ions being depleted), then exposing the plant system (e.g. intact plant roots) to tracer for a time period that is guesswork in the first assay, but ideally will bring the cytoplasmic compartment to the same specific activity as that of the external solution. When the duration of this exposure is five times the half-life for cytoplasmic exchange then this goal is achieved. Next, the tracer solution is replaced by chemically identical but non-labelled solution and at intervals of time (often very short initially and longer as time elapses) the solution is completely replaced and analyzed to determine tracer content. In studies of efflux from roots of white spruce labelled with $^{13}NH_4^+$ we used intervals of 5 s (2X) followed by 10 s (2), 15 s (6X), 30 s (4X), 1 min (4X) and 2 min (7X) (Kronzucker et al., 1995a). When the log of counts for each interval are plotted against time, one obtains a plot such as is shown in Fig 1. By regression of the lines that are assumed to correspond to different structural compartments of the root one can obtain half-life values for exchange (k/0.693) where k is the slope of regression. Verification that the lines generated by regression correspond to real entities of the apoplast or symplast (see next section) has been obtained independently (Siddiqi et al., 1991; Kronzucker et al., 1995b).

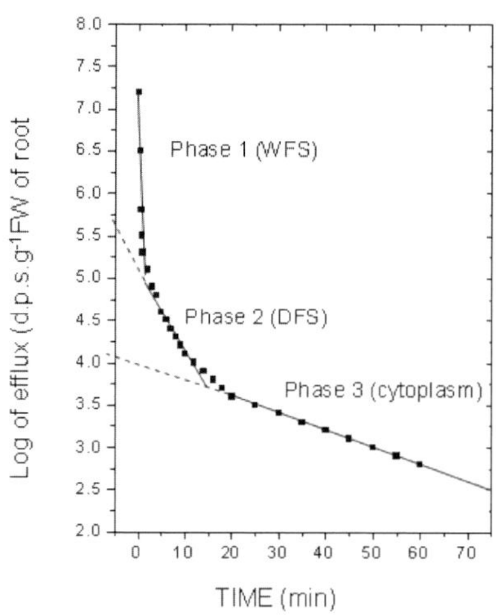

Fig. 1. Typical form of plot for efflux of ions from roots pre-labelled with radioactive ions after transfer to non-labelled but otherwise chemically identical solution. Regression of the linear phases gives phases that correspond to efflux from the Water Free Space (Phase 1), the Donnan Free Space (Phase 2) and the cytoplasm (Phase 3).

4. WHAT IS THE ROLE OF THE APOPLAST IN ION UPTAKE?

It is generally considered that there are two components within the apoplast, namely the Water Free Space (WFS) and the Donnan Free Space (DFS). Respectively, these components are equated with the regions within which non-charged molecules e.g. water and sucrose would pass freely, and those that represent charged entities (principally pectic carboxyl groups) that interact with charged species. Detailed analysis of the dynamics of tracer efflux from roots labelled with $^{13}NH_4^+$ has revealed the presence of two kinetically distinct components with half-lives that are in the range of 3 s and 0.5–1 min, respectively in spruce and rice (Wang et al., 1993; Kronzucker et al., 1995a). Intuitively these would be expected to correspond to the WFS and DFS. Treatments designed to destroy the integrity of the root plasma membranes such as heat treatment or detergent substantially reduced tracer accumulation in the putative cytoplasmic compartment but had little effect on putative WFS and DFS properties. Likewise, treatments designed to impact upon the DFS such as elevated concentrations of Ca^{2+} resulted in a reduced concentration of ammonium in the DFS but failed to impact upon the putative WFS or the cytoplasmic compartments (Kronzucker et al., 1995b). Thus we conclude that the two components, WFS and DFS have half-lives of exchange of ~3 and ~30 to 60 s, respectively (Wang et al, 1993; Kronzucker et al., 1995a). An important question regarding ion transport across the plasma membranes of root cells is the role if any of the DFS. Since ion binding is generally considered to be a preliminary to traversing the lipid bilayer how do we distinguish electrostatic binding to uronic acid residues from binding to transporter proteins? In short-term (<1 min) measurements of ion uptake using tracers such as ^{13}N that are detected with great sensitivity we have found very rapid accumulation of ions (Unkles et al., 2004) that shows no indication of a gradual rise in tracer accumulation that might be anticipated if ion binding to transporter proteins occurred with a half-life ≥30 s. With regard to binding to the DFS and the direct role of the latter in transport across the lipid bilayer it is possible by means of a simple computer model to compute the expected rise in tracer specific activity in the WFS, the DFS, and the cytoplasm, and the anticipated fluxes of NH_4^+ across the plasma membrane using the half-lives for tracer exchange obtained experimentally as described above. The results show (unpublished) that if ions were to be first bound to the apoplast DFS ($t_{0.5} \geq 30$ s) and then transferred across the plasma membrane, then in ion uptake studies there should be a gradual rise in the measured transmembrane influx over time to a steady value after about 2–3 min as the apoplast DFS came to equilibrium with bulk solution. However, real measurements show no such gradual rise. In other words, the apoplast compartment from which ions are transferred across the plasma membrane must very rapidly achieve equilibrium with the external solution. This leads to the only viable conclusion, namely that ion uptake across the plasma membrane proceeds from the WFS and that binding in the apoplast is not directly related to ion transfer across the plasma membrane. Ions in solution that enter the root apoplast pass rapidly to the plasma membrane surface while simultaneously and with a much longer half-life bind to charged sites of the apoplast DFS.

5. ION BINDING IN THE APOPLAST DFS

The rationale that Donnan binding of ions in the apoplast was associated with negatively charged pectic carboxyl groups has long been recognized. Indeed, the cation exchange capacity (CEC) of plant tissues has been widely used to emphasize the capacity of plant tissues to retain cations in the apoplast. In our analyses of the ion concentrations of the various kinetically defined components derived from efflux analysis using roots of plants maintained under steady state conditions we were able to confirm that the NH_4^+ concentration of the apoplast was significantly higher than that of the bulk solution, signifying additional binding over and above that associated with the WFS (Wang et al., 1993; Kronzucker et al., 1995a). Therefore it came as a surprise to find that nitrate concentrations were also computed to be higher than those of the bulk solution (Siddiqi et al., 1991; Kronzucker et al., 1995c), when the dogma leads us to believe that DFS binding pertains exclusively to cations. This was found to be true for roots of barley and spruce. It is possible that there are other explanations of this finding, but clearly many studies with microbial systems have identified periplasmic proteins that participate in responses to nutrient deprivation. For example, Davies et al. (1994) have documented the synthesis of a number of periplasmic proteins in *Chlamydomonas reinhardtii* in response to sulfate deprivation. One of these is an arylsulfatase whose activity scavenges sulfate by hydrolyzing aromatic sulfates. The roles of the other periplasmic proteins remain to be resolved. Likewise, periplasmic phophatases induced by phosphate stress were reported in *Spirodela oligorrhiza* (Bieleski and Johnson, 1974. The existence in higher plant cell walls of ion binding proteins in addition to uronic acid residues seems likely although such proteins appear not to have been documented. The review by Sattelmacher (2000) examines the limited evidence for nutrient ion storage in the DFS of plant roots. This is suggestive but not nearly convincing support for this concept. By contrast, the high concentration of potentially toxic ions such as Mn or Cd in root and leaf cell-walls suggest that the DFS may serve an important protective role in reducing potential toxicity of such ions (Hart et al., 1998; Wang, 2003). The opposite hypothesis, namely that a high concentration of apoplast-bound Al may result in increased toxicity was advanced by Schmohl and Horst (2000) and Horst et al., (this volume, pp 51-68). These authors observed increased Al binding and sensitivity to Al in maize suspension cells after increasing the pectin content by various treatments and speculated that the pectic fraction may play a role in determining the sensitivity to Al. However, as the authors acknowledged, the increased sensitivity to Al might have been due to increased entry of Al into the cytosol that was associated with the increased Al-binding. Likewise, in a study of Mn permeation into maize roots, Bacic et al. (1993) concluded that Mn-binding in the extracellular space of the tissue seemed to be the rate limiting step in entry across the root cell membranes. The resolution of this question as with others related to the function of the apoplast awaits further study.

REFERENCES

Bacic, G., Schara, M., Ratkovic, S. (1993). An ESR study of manganese binding in plant tissue. *Gen. Physiol. and Biophys.*, 12, 49–54.

Bagshaw, R., Vaidyanathan, L.V. and Nye, P.H. (1972). The supply of nutrient ions to plant roots in soil. *V. Plant Soil.*, 37, 617–626.

Bange, G.G.J. (1973). Diffusion and absorption of ions in plant tissue Part 3. The role of the root cortex cells in ion absorption. *Acta Bot. Neerl.*, 22, 529– 542.

Barber, S.A. (1962). A diffusion and mass flow concept of nutrient availability. *Soil Sci.*, 93, 39–49.

Barber, S.A. (1968). Mechanism of potassium absorption by plants, pp. 293–310. In: Kilmer (ed.), *The Role of Potassium in Agriculture*, Madison, USA.

Bieleski, R.L. and Johnson, P.N. (1974). The external location of phosphatase in phosphorus deficient *Spirodela oligorrhiza. Aust. J. Biol. Sci.*, 25, 707–720.

Clarkson, D.T. (1974). *Ion Transport and Cell Structure in Plants*. McGraw Hill, London.

Cosgrove, D.J. (1997). Assembly and enlargement of the primary cell wall in plants. *Annu. Rev. Cell Develop. Biol.*, 13, 171–201.

Cram, W.J. (1968). Compartmentation and exchange of chloride in carrot root tissue. *Biochim. Biophys. Acta.*, 163, 339–353.

Davies, J.P., Yildiz, F. and Grossman, A.R. (1994). Mutants of Chlamydomonas with aberrant responses to Sulfur deprivation. *Plant Cell.*, 6, 53–63.

Hart, J.J., Welch, R.M., Norvell, W.A., Sullivan, L.A. and Kochian, L.V. (1998). Characterization of cadmium binding, uptake, and translocation in intact seedlings of bread and durum wheat cultivars. *Plant Physiol.*, 116, 1413–1420.

Kronzucker, H.J. Siddiqi, M.Y. and Glass, A.D.M. (1995a). Compartmentation and flux characteristics of ammonium in spruce. *Planta*, 196, 691–698.

Kronzucker, H.J., Siddiqi, M.Y. and Glass, A.D.M. (1995b). Analysis of $^{13}NH_4^+$ Efflux in Spruce: A test case for phase identification in compartmental analysis. *Plant Physiol.*, 109, 481–490.

Kronzucker, H.J., Siddiqi, M.Y., and Glass, A.D.M. (1995c). Compartmentation and flux characteristics of nitrate in spruce. *Planta*, 196, 674–682.

MacRobbie, E.A.C. (1971) Fluxes and compartmentation in plant cells. *Annu. Rev. Plant Physiol.*, 22, 75–96.

Münch, E. (1930). Die Stoffbewegung in der Pflanze. Jena. Germany. Fischer

Pitman, M.G. (1963) The determination of the salt relations of the cytoplasmicphase in cells of beet root tissue. *Aust. J. Biol. Sci.*, 16, 647–668.

Reisenauer, H.M. (1966). Mineral nutrients in soil solution. In: *Environmental Biology*. P.L. Altman and D.S. Dittmer (eds). *Fed. Amer. Soc. Exp. Biol. Bethesda*. pp. 507–508.
Sattelmacher, B. (2000). The apoplast and its significance for plant mineral nutrition. *New Phytol.*, 149, 167–192.
Schmohl, N. and Horst, W.J. (2000). Cell wall pectin content modulates aluminium sensitivity of *Zea mays* (L.) cells grown in suspension culture. *Plant, Cell and Environ.*, 23, 735–742.
Siddiqi, M.Y., Glass, A.D.M. and Ruth, T.J. (1991). Studies of the uptake of nitrate in barley. III. Compartmentation of NO_3^-. *J. Exp. Bot.*, 42, 1455–1463.
Unkles, S.E., Wang, R., Wang, Y., Glass, A.D.M., Crawford, N. and Kinghorn, J.R. (2004). Nitrate Reductase is Required for Nitrate Uptake into Fungal but not Plant Cells. *J. Biol. Chem.*, 279, 28182–28186.
Wang, J.J. (2003). Kinetics of manganese uptake by excised roots of sensitive and tolerant tobacco genotypes. *J. Plant Nutr.*, 26, 1439–1450.
Wang, M., Siddiqi, M.Y., Ruth, T.J. and Glass, A.D.M. (1993). Ammonium uptake by rice roots. I. Fluxes and subcellular distribution of $^{13}NH_4^+$. *Plant Physiol.*, 103, 1249–1258.
Wolt, J.D. (1994). *Soil Solution Chemistry: Applications to Environmental Science and Agriculture*. New York, N.Y. Wiley.

THE APOPLAST OF ECTOMYCORRHIZAL ROOTS – SITE OF NUTRIENT UPTAKE AND NUTRIENT EXCHANGE BETWEEN THE SYMBIOTIC PARTNERS

H. BÜCKING[1], R. HANS[2] and W. HEYSER[2]
[1]*State University of New Jersey, Biology Department, USA, bucking@camden.rutgers.edu*
[2]*University Bremen, Center for Environmental Research and Technology, Germany.*

Abstract. Between 80 and 90% of all known plant species live in close interaction with mycorrhizal fungi in a mutalistic interaction, the mycorrhizal symbiosis. Mycorrhizal root tips with their extramatrical mycelium increase the absorbing surface area of mycorrhizal roots and contribute significantly to the nutrient uptake of plants. The following paper deals with the role of the apoplast for nutrient uptake and nutrient exchange between both partners. Investigations by use of fluorescent dyes as apoplastic tracers showed that the fungal sheath of the ectomycorrhizal roots does not act as an effective apoplastic barrier for the entry of nutrients into the mycorrhizal root cortex. However, nutrients such as P can be absorbed by hyphae of the extramatrical mycelium or the fungal sheath and the transfer to the host plant is controlled by the fungal symplast. The results indicate that the uptake of P by the extramatrical mycelium and the transfer across the interfacial apoplast to the mycorrhizal host plant is not primarily regulated by the host plant demand for P, but by the flux of carbohydrates from the mycorrhizal host plant to the fungal symbiont. A model system shows how the carbohydrate and P exchange between both symbiotic partners is possibly linked.

Key words: apoplast, ectomycorrhiza, interface, nutrient exchange, P transport

1. INTRODUCTION

In general, a mycorrhizal infection can enhance nutrient uptake and thereby plant growth by (1) an increase of the absorbing surface area, (2) the mobilization of sparingly available nutrients and (3) highly efficient uptake

B. Sattelmacher† and W.J. Horst (eds.), The Apoplast of Higher Plants: Compartment of Storage, Transport and Reactions, 97–108.

systems. This positive effect of an ectomycorrhizal infection is well established for the macronutrients P and N and some trace elements such as Cu and Zn (Bücking and Heyser, 1994; Smith and Read, 1997). By contrast, knowledge of the effect of the mycorrhizal infection on the uptake of nutrients such as K, Mg and Ca is limited and the results of different studies are not consistent (Marschner and Dell, 1994).

The efficiency of a mycorrhizal infection for the P nutrition of a host plant is not due to a higher P absorption under all supply conditions but to the capability to accumulate P under high external supply and to remobilize this storage pool under stress and to maintain a continuous flux of P to the mycorrhizal host also under P-limiting conditions (Harley and Smith, 1983; Bücking and Heyser, 2000).

At sufficient and supraoptimal P supply a mycorrhizal infection has no positive effect on the P absorption (Bücking and Heyser, 2000) and the growth of arbuscular mycorrhizal plants is reduced (Peng et al., 1993). This can be explained by the carbon costs of a mycorrhizal infection for the host plant. In return for the beneficial effect on nutrient supply, around 20 % of the assimilated carbon from the plant is translocated to the fungal symbiont for the formation, maintenance, and function of mycorrhizal structures (Finlay and Söderström, 1992).

Presently, only little is known about the mechanisms involved in the regulation and the polarization of transfer processes occuring between both symbiotic partners. The aim of the current investigations was to examine the uptake of nutrients by the ectomycorrhizal symbiosis and to learn more about the exchange of nutrients between both symbiotic partners and the possible regulation of these transfer processes.

2. THE ROLE OF THE APOPLAST FOR NUTRIENT UPTAKE OF MYCORRHIZAL ROOTS

An ectomycorrhizal root is generally characterized by two prominent structures, a fungal sheath completely surrounding the mycorrhizal root and the Hartig net, an intercellular network of hyphae within the root cortex, where the main exchange processes between the symbiotic partners are localized (Fig. 1). If the fungal sheath is impermeable to nutrient ions, the underlying root tissue would be isolated from the soil solution and an enclosed shared apoplastic nutrient exchange compartment will be created in which the conditions (e.g. pH, ion composition) can be controlled by both symbiotic partners efficiently. However, the presence of a barrier to apoplastic transport in the mantle would necessitate the uptake of nutrients from the soil solution

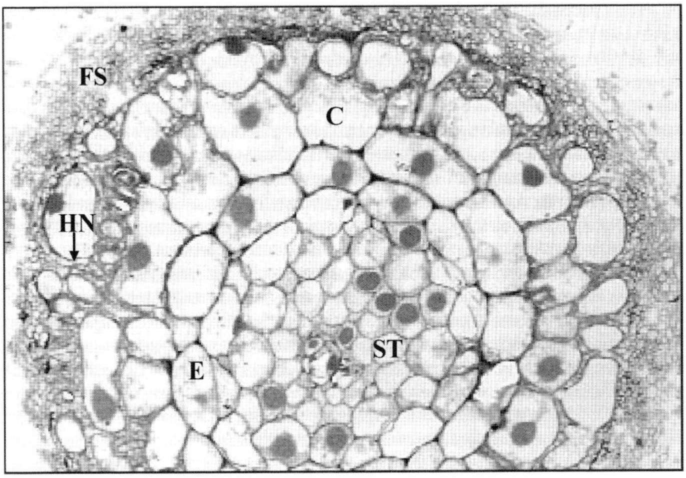

Fig. 1. Cross-section of an ectomycorrhizal root (Pinus sylvestris/Suillus bovinus). Abbreviations: C – cortex, E – endodermis, FS – fungal sheath, HN – Hartig net, ST – stele.

into the fungal symplast and therefore, the mycorrhizal root would be completely dependent on the nutrient translocation by the fungal symbiont.

The permeability of an ectomycorrhizal fungal sheath has been examined by several authors using fluorochromes as apoplastic tracers, but the results were controversial. Ashford and coworkers (1989) for example showed that the sheath of *Pisolithus tinctorius* ectomycorrhizas is impermeable to the fluorescent dye cellufluor. However, it can be assumed, that the effect of an ectomycorrhizal sheath on apoplastic movement may differ with the fungal species, structure and properties of the mantle and possibly also with the growth conditions of the mycorrhizal root tips. Vesk et al. (2000) reported that the permeability of the fungal sheath of *P. tinctorius* ectomycorrhizas grown on a wet agar surface differed from those grown in air.

Microscopic investigations after the use of sulphorhodamine G as an apoplastic tracer showed that the fungal sheaths of various ectomycorrhizal fungal species grown in a Petri dish system (Bücking and Heyser, 1994) did not prevent penetration of this fluorescent dye into the mycorrhizal root cortex of pines (Fig. 2). In all mycorrhizal associations the fluorescent dye could be detected within the cortical apoplast. In contrast to the fungal sheath, the endodermis with its Casparian band of these mycorrhizal short roots acts an effective apoplastic barrier. A labeling was only found in the radial cell walls of the endodermis up to the Casparian band and not within the stele of these short roots.

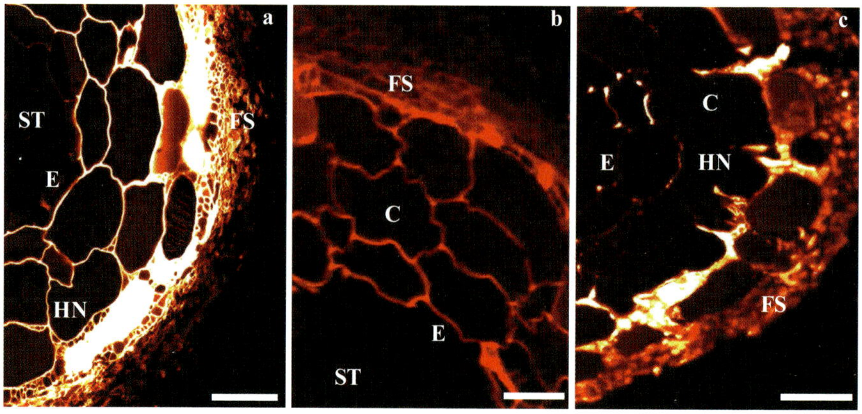

Fig. 2 Movement of sulphorhodamine G across the fungal sheath of ectomycorrhizal roots inoculated with Suillus bovinus (a), Pisolithus tinctorius (b) and Rhizopogon roseolus (c) C – cortex, E – endodermis, FS – fungal sheath, HN – Hartig net, ST – stele (scale bar: 15 μM).

However, movement and diffusivity of large uncharged or negatively charged dyes might be quite different from that of hydrated inorganic ions and especially of ions, that are potential nutrients for the fungal symbiotic partner. Investigations after labeling with the stable isotopes ^{25}Mg, ^{41}K and ^{44}Ca using the laser microprobe mass analysis (in collaboration with A. J. Kuhn and W. H. Schröder, F. Z. Jülich, Germany) showed that the movement of these ions across the fungal sheath of *Suillus bovinus* is retarded compared to non-mycorrhizal roots. Confirming the results after the use of fluorescent dyes, the endodermis in non-mycorrhizal and mycorrhizal pine roots acted as an efficient apoplastic diffusion barrier, limiting the apoplastic movement of the divalent cations Ca and Mg into the stele (Bücking et al., 2002). By contrast, Kuhn et al. (2000) reported a significant labeling of the apoplast within the stele of mycorrhizal spruce roots after exposure to ^{25}Mg and ^{44}Ca. The radial movement of the divalent cations, Ca and Mg, is closely correlated to the endodermal differentiation and the highest uptake of these elements can be found in regions of the root, where the endodermis is not completely differentiated (Häussling et al., 1988). The development of the Casparian band and the differentiation of the primary stage of the endodermis, however, start in fine roots of pine close to the root tip (Behrmann, 1995).

The investigations with laser microprobe mass analysis also showed that an ectomycorrhizal infection can affect the exchangeability of cations within the cortical apoplast of pines (Bücking et al., 2002). The exchangeability of cations in roots inoculated with *Suillus bovinus* was higher than in non-mycorrhizal pine roots. Modifications of the host cell-wall as a consequence

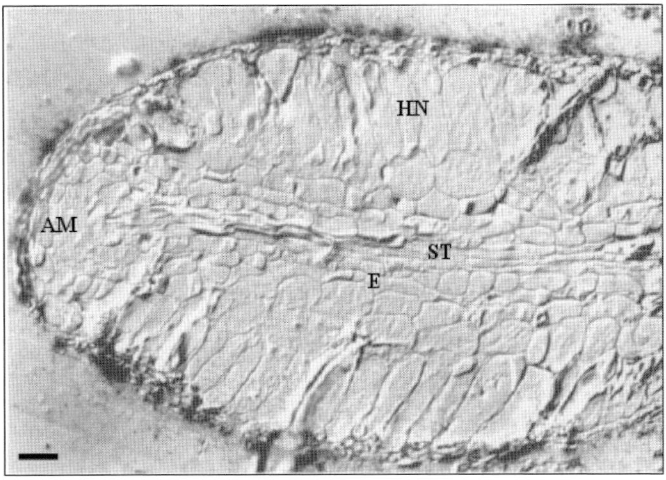

Fig. 3. Microautoradiograph of an ectomycorrhizal root of Populus alba x P. tremula after application of ^{33}P-orthophosphate for 2 d (scale bar – 100 μm, Bücking and Heyser 2001). Abbreviations: AM – apical meristeme, C – cortex, E – endodermis, FS – fungal sheath, HN – Hartig net, ST – stele.

of an infection of pine roots by *Suillus bovinus* were also reported by Duddridge and Read (1984).

However, the possibility of an apoplastic nutrient transport across the fungal sheath in ectomycorrhizal roots does not mean that under normal soil conditions, where depletion zones around ectomycorrhizal roots occur and the main nutrient acquisition of mycorrhizal systems takes place at the extraradical mycelium, the fungal symplast does not compete with its host plant for limited nutrient resources. Microautoradiographic studies after $^{33}P_i$ application gave evidence that P was rapidly accumulated in the ectomycorrhizal sheath and was only slowly translocated via the Hartig net to the cortical cells (Fig. 3). An apoplastic movement of P across the fungal sheath does not occur and it can be inferred, that P has to pass the fungal symplast, before it can be absorbed by the mycorrhizal host plant.

3. THE LOCALIZATION OF CARBON AND PHOSPHATE TRANSFER IN THE SYMBIOSIS

The compatible interaction between mycorrhizal fungi and plants is based on a bi-directional transfer of nutrients and carbohydrates across an interface, whose structure and development vary between different types of mycorrhizal associations. Since there is no direct symplastic continuity between partners, nutrients must pass an interfacial apoplast before they can be absorbed (for review see: Peterson and Bonfante, 1994). Models of transfer processes across this mycorrhizal interface generally involve (1) the passive efflux of P and carbohydrates through the fungal and plant plasma membrane into the interfacial apoplast and (2) the active absorption of nutrients by both partners driven by an H^+-ATPase (Smith et al., 1994b). This protein operates as a primary transporter by pumping protons into the interfacial apoplast, thereby generating a proton-motive force, which can be used for an active transport through the plasma membrane.

Based on the H^+-ATPase activity a bidirectional transfer across the same interface structure was proposed for ectomycorrhizal associations by Lei and Dexheimer (1988). The protocol used for detection of H^+-ATPase activity, however, is now considered to be unreliable and there are indications that in fungal hyphae the proton pumps are spatially segregated from the proton-coupled transport systems (Harold, 1994) and might not always represent the capability of membranes for an active absorption of nutrients. Jones et al. (1991) reported that the P acquisition efficiency was higher in younger than in older mycorrhizal associations and they concluded, that in mycorrhizal roots the maximum fluxes of P and carbohydrates are temporally separated.

To get further information on the localization of exchange processes in ectomycorrhizal associations, microautoradiographic investigations after $^{14}CO_2$ assimilation and $^{33}P_i$ supply were carried out (Bücking and Heyser, 2001; Bücking et al., 2001). The investigations revealed, (1) that the fungal partner is a strong sink for photosynthates, (2) that similar to the situation for P most of the carbohydrate transfer takes place in median parts of mycorrhizal roots, (3) that carbohydrates absorbed by the mycorrhizal fungus are translocated to the fungal sheath and are homogeneously distributed and (4) that in the main exchange zone the nuclei of the cortical cells showed also a high sink strength indicating their increased metabolic activity (Fig. 4, Bücking and Heyser, 2001).

Based on the fact that for the P transfer as well as for the carbohydrate transfer a main exchange zone in median parts of the ectomycorrhizal root can be detected, it can be assumed that in ectomycorrhizal roots the exchange processes are localized at the same interface structure (Bücking and Heyser, 2001). It can be inferred that the varying efficiency for the carbohydrate and P translocation in different zones along the mycorrhizal axis is due to different developmental stages of the Hartig net. In apical zones the differentiation of the Hartig net is not completed, in older, basal parts of the mycorrhiza the number of mitochondria is reduced indicating a lower metabolic activity of these cells (Smith and Read, 1997).

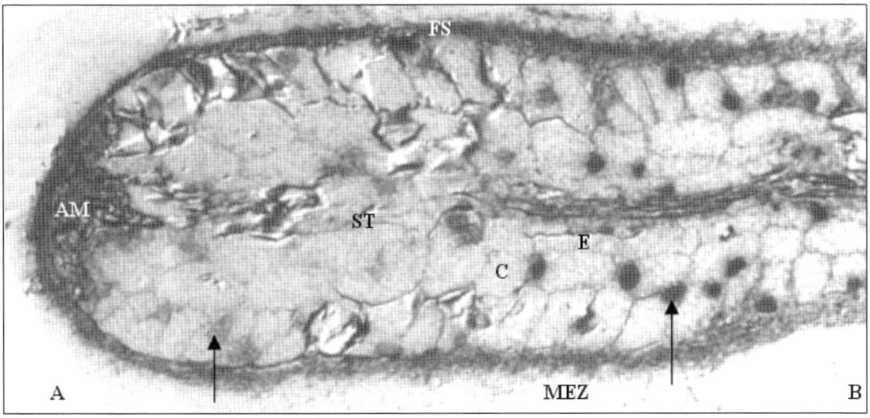

Fig. 4. Microautoradiograph of an ectomycorrhizal poplar root after assimilation of $^{14}CO_2$ (Bücking and Heyser, 2001). Abbreviations: A – apical, AM – apical meristeme, B – basal, C – cortex, E – endodermis, FS – fungal sheath, HN – Hartig net, MEZ – main exchange zone, ST – stele. The arrows mark nuclei in different zones of the mycorrhiza.

The high metabolic activity of the cortical cells in the main exchange zone for P and carbohydrates in ectomycorrhizal associations indicates a possible increase in protein biosynthesis in this region of the root. The development of a mycorrhizal infection is accompanied by molecular changes in both partners leading to alterations in protein biosynthesis (up- and down-regulation) and the expression of mycorrhiza-specific genes. Some of these processes are associated with numerous morphological and anatomical changes during the development of an ectomycorrhizal association (Tarkka et al., 1998) and some are obviously involved in the nutrient transfer across the interface (Smith et al., 2001; Nehls et al., 2001; Chalot et al., 2002). Rosewarne et al. (1999) reported for example that LePT1, a P transporter of *Lycopersicon esculentum*, which is assumed to be mainly involved in the P uptake from the interface in arbuscular mycorrhizal roots, is highly expressed especially in arbuscule-containing cortical cells.

The labeling of the hyphae in the fungal sheath was higher than that of the hyphae of the Hartig net indicating that photosynthates are accumulated in this part of the mycorrhiza (Fig. 4). Similar to P, which is translocated by cortical cells to sinks in the plant to maintain the concentration gradient across the interface to allow a further transport, absorbed photosynthates are converted by the fungal symbiont to fungal sugars such as trehalose, mannitol and glycogen (e.g. Lewis and Harley, 1965).

4. THE REGULATION OF THE EXCHANGE PROCESSES BETWEEN SYMBIOTIC PARTNERS

Although carbon allocation to the roots increases during mycorrhizal associations, the amounts of carbon estimated to leak out of intact root cells into the interfacial apoplast are thought to be insufficient for the extensive hyphal growth, which can be observed. In addition, the flow of P across the interface and the estimated P influx rate into plant cells has been observed to be considerably larger than the efflux from the fungal hyphae in axenic cultures (Smith et al., 1994a). Therefore, it has been suggested by several authors (e.g. Smith et al., 1994b) that in the mycorrhizal interface conditions exist which lead to an enhanced efflux or a decrease in the level of competing uptake systems.

It has been suggested, that the P transfer to the host plant is regulated by the intracellular P content in the hyphae of the Hartig net (Bücking and Heyser, 2000). Transfer of P across the interface would (1) reduce the intracellular P content in hyphae of the Hartig net, (2) stimulate the flux of P from the extramatrical hyphae to the Hartig net, and (3) therefore increase the P absorption by the extramatrical mycelium. It has been concluded that the P demand of the host plant regulates the P uptake by a mycorrhizal fungus (Cairney and Smith, 1992). The regulation by the host would ensure that the mycorrhizal root absorbs P with the greatest efficiency when the plant is under P limitation (Thomson et al., 1990).

However, the P flow across the interfacial apoplast to the mycorrhizal host plant is not only be regulated by the P deficiency of the plant. Investigations with poplar and pine seedlings, placed in the dark in advance of a P supply, showed even if the plant was P-deficient, the P absorption and transfer by the mycorrhizal fungus to its host plant was reduced when the photosynthetic activity and the transfer of carbohydrates across the mycorrhizal interface to the fungal symbiotic partner was low (Bücking and Heyser, 2001; Bücking and Heyser, 2003). Under these conditions the cytoplasmic P content of the Hartig net was reduced and, instead, a high number of polyphosphate (polyP) granules could be detected within the hyphae. The function of polyP in mycorrhizal fungi and other microorganism are (1) P storage in vacuoles, (2) regulation of the levels of ATP and other nucleoside triphosphates, (3) homeostasis of cations, and (4) participation in membrane transport-processes, cell-wall formation and gene expression (Kulaev et al., 1999). The presence of polyP granules in the fungal sheath and the Hartig net indicated that the decrease of the P absorption and transfer to the host plant, which was observed, was not due to a P and ATP deficiency of the mycorrhizal fungus (Bücking and Heyser, 2003). A reduction in P uptake and transfer by a lower carbon transfer by the host was recently also shown in arbuscular mycorrhizal associations (Bücking and Shachar-Hill, 2005).

Investigations with axenic cultures of different ectomycorrhizal fungi showed that the P efflux across the fungal plasma membrane is affected by the carbohydrate supply to the mycorrhizal fungus. The disaccharid sucrose which was assumed to be the carbohydrate translocated through the plant plasma membrane into the interfacial apoplast (Smith and Read, 1997), induces a higher P efflux through the plasma membrane of ectomycorrhizal fungi.

The decrease of P absorption and P transfer from a mycorrhizal fungus to its host plant when the carbohydrate flow from the host plant to the mycobiont is reduced indicates that both exchange processes are possibly linked. A model, in which a possible interaction between the carbohydrate and P flow in a mycorrhiza is illustrated, is shown in Fig. 5.

According to this model P_i, absorbed by the extramatrical mycelium by an active process can (1) replenish the cytoplasmic metabolically active fungal P_i pool, (2) be transferred to the host plant in form of short chain polyP or (3) be accumulated in fungal vacuoles in form of polyP. The P uptake by the extramatrical mycelium and the P efflux into the interfacial apoplast is regulated by the intracellular P_i concentration within the hyphae. P_i, transferred through the fungal plasma membrane into the interfacial

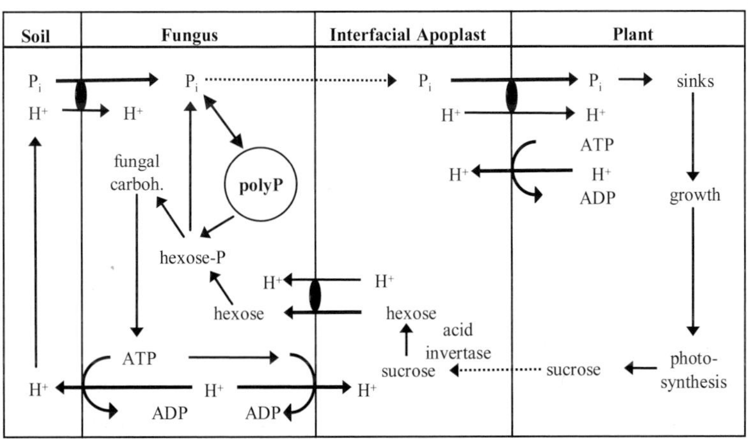

Fig. 5. Model showing a possible interaction between carbohydrate and P transfer in a mycorrhizal symbiosis. Further development of existing models by Smith and Smith (1990) and Smith et al. (1994b). Bold arrows – active processes, thin, dotted arrows – passive efflux processes. Abbreviations: P_i – inorganic P, PolyP – polyphosphates.

apoplast can be absorbed by the host plant by an active process driven by an H^+-ATPase located in the plant plasma membrane. A continous flow of P_i across the interfacial apoplast to the mycorrhizal host plant and the transfer to the sinks will (1) improve plant growth and photosynthetic activity of the plant, (2) enhance the transfer of carbohydrates to the mycorrhizal roots and (3) promote an efflux of sucrose through the plant plasma-membrane into the interfacial apoplast. A plant acid invertase in the interfacial apoplast hydrolyzes the sucrose into the hexoses glucose and fructose, which can be absorbed by the mycorrhizal fungus by an active process. The activity of this invertase is regulated by the pH (Salzer and Hager, 1993) and thereby stimulated by the activity of the H^+-ATPases located in both plasma membranes. The uptake of hexoses by the mycorrhizal fungus and the consequent conversion of these hexoses to fungal carbohydrates, such as mannitol and trehalose via hexose-phosphates driven by a polyP-hexokinase, will (1) enhance the remobilization of polyphosphates, (2) increase the intracellular P_i concentration within the hyphae and, thereby, promote the efflux of P_i through the fungal plasma membrane into the interfacial apoplast. The transfer of carbohydrates by the host plant to the mycorrhizal fungus will promote (1) the supply of the mycobiont with ATP needed for active uptake processes such as P uptake and (2) the extension of the extramatrical mycelium and, therefore, the access of the mycorrhizal fungus to further P resources within the soil compartment.

5. CONCLUSIONS

The apoplast of ectomycorrhizal roots is involved in both nutrient uptake and nutrient exchange between both symbiotic partners. Even if the apoplast of the fungal sheath is not completely impermeable for the movement of fluorescence dyes and/or ions into the mycorrhizal root cortex, nutrients such as P are taken up from the apoplast by the fungal symplast, and the transfer to the host plant is controlled by the mycorrhizal fungus. P is only transferred across the interfacial apoplast to the host plant, when the fungus is supplied with carbohydrates by its host plant in return. Our present knowledge about the conditions in the interfacial apoplast that are involved in the regulation of nutrient exchange between the symbiotic partners is limited. We still need to get more information about the molecular and biochemical processes that are involved in polarizing and controlling the flux of nutrients across the mycorrhizal interface.

ACKNOWLEDGEMENTS

The authors wish to thank Simone Sikora for technical assistance and Yair Shachar-Hill (Michigan State University, USA) for critical reading the manuscript. The investigations with the Laser microprobe mass analysis were carried out in collaboration with A. J. Kuhn and W. H. Schröder from the FZ Jülich, Germany. Plants of Populus alba x P. tremula were kindly provided by R. Nitschke and H. Rennenberg (University Freiburg, Germany).

REFERENCES

Ashford, A. E., Allaway, W. G., Peterson, C. A. and Cairney, J. W. G. (1989). Nutrient transfer and the fungus-root interface. *Aust. J. Plant Physiol.*, 16, 85–97.

Behrmann, P. (1995). Entwicklung, Struktur und Funktion der Endodermis in mykorrhizierten und nicht-mykorrhizierten Baumwurzeln unter besonderer Berücksichtigung der Kiefer (*Pinus sylvestris* L.). Ph.D. thesis, University of Bremen, Germany.

Bücking, H. and Heyser, W. (1994). The effect of ectomycorrhizal fungi on Zn uptake and distribution in seedlings of *Pinus sylvestris* L. *Plant Soil*, 167, 203–212.

Bücking, H. and Heyser, W. (2000). Subcellular compartmentation of elements in non-mycorrhizal and mycorrhizal roots of *Pinus sylvestris*: an X-ray microanalytical study. I. The distribution of phosphate. *New Phytol.*, 145, 311–320.

Bücking, H. and Heyser, W. (2001). Microautoradiographic localization of phosphate and carbohydrates in mycorrhizal roots of *Populus tremula* x *Populus alba* and the implications for transfer processes in ectomycorrhizal associations. *Tree Physiol.*, 21, 101–107.

Bücking, H. and Heyser, W. (2003). Uptake and transfer of nutrients in ectomycorhizal associations: interactions between photosynthesis and phosphate nutrition. *Mycorrhiza*, 13, 59–68.

Bücking, H., Warner, J., Hespe, C. and Heyser, W. (2001). Autoradiographische und cytochemische Untersuchungen zum Assimilattransfer in der ektotrophen Mykorrhiza. In R. Langenfeld-Heyser, A. Polle and E. Fritz (eds), Schriften aus der Forstlichen Fakultät der Universität Göttingen und der Niedersächsischen Forstlichen Versuchsanstalt. Band 131: Neues zum Stofftransport in Bäumen. *J. D. Sauerländer's Verlag*. Frankfurt am Main, pp. 108–120.

Bücking, H. and Shachar-Hill, Y. (2005). Phosphate uptake, transport and transfer by the arbuscular mycorrhizal fungus *Glomus intraradices* is stimulated by increased carbohydrate availability. *New Phytol.*, 165, 899–912.

Bücking, H., Kuhn, A. J., Schröder, W. H. and Heyser, W. (2002). The fungal sheath of ectomycorrhizal pine roots: an apoplastic barrier for the entry of calcium, magnesium and potassium into the root cortex? *J. Exp. Bot.*, 53, 1659–1669.

Cairney, J. W. G. and Smith, S. E. (1992). Influence of intracellular phosphorus concentration on phosphate absorption by the ectomycorrhizal basidiomycete *Pisolithus tinctorius*. *Mycol. Res.*, 96, 673–676.

Chalot, M., Javelle, A., Blaudez, D., Lambilliote, R., Cooke, R., Sentenac, H., Wipf, D. and Botton, B. (2002). An update on nutrient transport processes in ectomycorrhizas. *Plant Soil,* 244, 165–175.

Duddridge, J. A. and Read, D. J. (1984). The development and ultrastructure of ectomycorrhizas. II. Ectomycorrhizal development on pine in vitro. *New Phytol.*, 96, 575–582.

Finlay, R. and Söderström, B. (1992). Mycorrhiza and carbon flow to the soil. In M. J. Allen (ed.), *Mycorrhizal Functioning*. Chapman and Hall, New York, pp. 134–160.

Harley, J. L. and Smith, S. E. (1983). Mycorrhizal symbiosis. *Academic Press*, London.

Harold, F. M. (1994). Ionic and electrical dimensions of hyphal growth. In J. G. H. Wessels and F. Meinhardt (eds), *The Mycota I. Growth, Differentiation and Sexuality*. Springer-Verlag, Berlin. pp. 89–109.

Häussling, M., Jorns, C. A., Lehmbecker, G., Hecht-Buchholz, C. and Marschner, H. (1988). Ion and water uptake in relation to root development in norway spruce (*Picea abies* (L.) Karst.). *J. Plant Physiol.*, 133, 486–491.

Jones, M. D., Durall, D. M. and Tinker, P. B. (1991). Fluxes of carbon and phosphorus between symbionts in willow ectomycorrhizas and their changes with time. *New Phytol.*, 119, 99–106.

Kuhn, A. J., Schröder, W. H. and Bauch, J. (2000). The kinetics of calcium and magnesium entry into mycorrhizal spruce roots. *Planta*, 210, 488–496.

Kulaev, I., Vagabov, V. and Kulakovskaya, T. (1999). New aspects of inorganic polyphosphate metabolism and function. *J. Biosci. Bioeng.*, 88, 111–129.

Lei, J. and Dexheimer, J. (1988). Ultrastructural localization of ATPase activity in the *Pinus sylvestris*/*Laccaria laccata* ectomycorrhizal association. *New Phytol.*, 108, 329–334.

Lewis, D. H. and Harley, J. L. (1965). Carbohydrate physiology of mycorrhizal roots of beech. I. Identity of endogenous sugars and utilization of exogenous sugars. *New Phytol.*, 64, 224–237.

Marschner, H. and Dell, B. (1994). Nutrient uptake in mycorrhizal symbiosis. *Plant Soil*, 159, 89–102.

Nehls, U., Mikolajewski, S., Magel, E. and Hampp, R. (2001). Carbohydrate metabolism in ectomycorrhizas: gene expression, monosaccharide transport and metabolic control. *New Phytol.*, 150, 533–541.

Peng, S., Eissenstat, D. M., Graham, J. H., Williams, K., Hodge, N. C. (1993) Growth depression in mycorrhizal citrus at high-phosphorus supply. *Plant Physiol.*, 101, 1063–1071.

Peterson R. L. and Bonfante, P. (1994). Comparative structure of vesicular-arbuscular mycorrhizas and ectomycorrhizas. *Plant Soil*, 159, 79–88.

Rosewarne, G. M., Barker, S. J., Smith, S. E., Smith, F. A. and Schachtman, D. P. (1999). A *Lycopersicon esculentum* phosphate transporter (LePT1) involved in phosphorus uptake from a vesicular-arbuscular mycorrhizal fungus. *New Phytol.*, 144, 507–516.

Salzer, P. and Hager, A. (1993). Characterization of wall bound invertase isoforms of *Picea abies* cells and regulation by ectomycorrhizal fungi. *Physiol. Plant.*, 88, 52–59.

Smith, S. E. and Read, D. J. (1997). *Mycorrhizal Symbiosis* (2nd ed). Academic Press, London.

Smith, S. E. and Smith, F. A. (1990). Tansley review No. 20. structure and function of the interfaces in biotrophic symbioses as they relate to nutrient transport. *New Phytol.*, 114, 1–38.

Smith, S. E., Dickson, S., Morris, C. and Smith, F. A. (1994a). Transfer of phosphate from fungus to plant in VA mycorrhizas: calculations of the area of symbiotic interface and of fluxes of P from two different fungi to *Allium porrum* L. *New Phytol.* 127, 93–99.

Smith S. E., Gianinazzi-Pearson, V., Koide, R. and Cairney, J. W. G. (1994b). Nutrient transport in mycorrhizas: structure, physiology and consequences for efficiency of the symbiosis. *Plant Soil*, 159, 103–113.

Smith, S. E., Dickson, S. and Smith, F. A. (2001). Nutrient transfer in arbuscular mycorrhizas: how are fungal and plant processes integrated? *Aust. J. Plant Physiol.*, 28, 683–694.

Tarkka, M., Niini, S. S. and Raudaskoski, M. (1998). Developmentally regulated proteins during differentiation of root system and ectomycorrhiza in Scots pine (*Pinus sylvestris*) with Suillus bovinus. *Physiol. Plant.*, 104, 449–455.

Thomson, B. D., Clarkson, D. T. and Brain, P. (1990). Kinetics of phosphorus uptake by the germ-tubes of the vesicular-arbuscular mycorrhizal fungus, Gigaspora margarita. *New Phytol.*, 116, 647–653.

Vesk, P. A., Ashford, A. E., Markovina, A. -L. and Allaway, W. G. (2000). Apoplasmic barriers and their significance in the exodermis and sheath of *Eucalyptus pilularis – Pisolithus tinctorius* ectomycorrhizas. *New Phytol.*, 145, 333–346.

CHEMICAL COMPOSITON OF APOPLASTIC TRANSPORT BARRIERS IN ROOTS

Quantification of suberin depositions in endodermal and hypodermal root cell walls

L. SCHREIBER, R. FRANKE and K. HARTMANN

IZMB – Institute for Cellular and Molecular Botany, Department of Ecophysiology, University of Bonn, Germany, lukas.schreiber@uni-bonn.de

Abstract. The lipophilic biopolymer suberin is deposited to endodermal and hypodermal root cell walls forming apoplastic transport barriers. Comparing 10 different species, it becomes evident that suberization of apoplastic barriers of roots is strongly species-dependent and can vary by more than 2 orders of magnitude. In response to environmental stress factors, suberization of apoplastic barriers can significantly increase (salt stress) or decrease (nutrient deficiency). Radial hydraulic conductivity in the apoplast of corn roots decreased as a result of an increased suberization of the apoplastic barriers. Based on the suberin determination in apoplastic barriers in roots of different species, it must be concluded that (i) there is a large variability in the degree of suberization of apoplastic barriers in roots due to internal and external factors and (ii) suberization per se does not necessarily lead to complete impermeability of the apoplast for water and dissolved solutes.

Key words: endodermis, hydraulic conductivity, hypodermis, salt stress, suberin

1. INTRODUCTION

Plant roots are designed to take up water and dissolved nutrients from the soil solution, but at the same time they have to be protected from desiccation under drought conditions and from infection by soil-borne pathogens (Marschner, 1995). In order to cope with these demands, the extracellular biopolymer suberin is deposited to the apoplast of specific root tissues. Both, the endodermis and the hypodermis of roots are characterized by the

B. Sattelmacher† and W.J. Horst (eds.), The Apoplast of Higher Plants: Compartment of Storage, Transport and Reactions, 109–117.

occurrence of the lipophilic biopolymer suberin (Schreiber et al., 1999). Using histochemical techniques to stain cell-wall depositions with lipophilic dyes and analyzing them using light microscopy, suberin depositions have already been visualized in the past (Wilson and Peterson, 1983). Only recently, the direct analysis of cell-wall depositions employing modern analytical techniques such as gas chromatography and mass spectrometry has enabled the detailed characterization of suberized plant-root cell-walls (Zeier and Schreiber, 1997).

Using these techniques offers the opportunity to get qualitative as well as quantitative data on suberin composition in root cell-walls. These data on suberin amounts and on its composition has been correlated with functional properties of apoplastic barriers in plant roots such as the hydraulic conductivity and the reaction towards environmental stress factors like drought, nutrient deficiency, or salt stress. In the following, the techniques isolating suberized plant-root cell-walls and analyzing them by gas chromatography and mass spectrometry will be described. Results on suberin composition of apoplastic barriers in roots of 10 different species are summarized as well as effects of different environmental factors on suberization. The correlation between the degree of suberization and radial hydraulic flow of water through the root apoplast of corn roots is discussed.

2. CELL WALL ISOLATION AND SUBERIN ANALYSIS

2.1 Enzymatic cell wall isolation

Plant roots cut in segments of several cm length were immersed in enzyme solutions containing cellulase and pectinase. By this approach "normal", chemically unmodified carbohydrates of cell walls of most tissues were digested, but chemically modified cell walls like xylem vessels, endodermal cell walls and hypodermal cell walls resisted the enzymatic attack (Schreiber et al., 1994). Following the enzymatic treatment of root samples, different root tissues had to be separated by manual manipulation using forceps and a binocular.

The approach provided purified endodermal and hypodermal cell wall samples. Preparing the hypodermis from different species, it was observed that in most cases the root hypodermis could be only isolated together with the epidermis (rhizodermis). For this reason, preparations of apoplastic barriers isolated from the outer root surface are called RHCW (rhizodermal and hypodermal cell walls), whereas the endodermal cell wall samples are called ECW (endodermal cell walls). Prior to chemical analysis, isolated cell-wall preparations were washed several times in deionized water and stored at room temperature over silica gel.

2.2 Suberin analysis using gas chromatography

Before analysing the chemical composition, enzymatically isolated cell wall samples were extensively extracted using a mixture of chloroform and methanol. The extracts, containing monomeric lipophilic material were stored for the subsequent analysis. Extracted cell walls were degraded applying different techniques specific for the analysis of suberin (Kolattukudy and Agrawal, 1974), lignin (Lapierre et al., 1995), sugars (Blakeney et al., 1983) and amino acid composition after acidic hydrolysis. In the following, suberin analysis will be described in more detail, since this lipophilic biopolymer is the most important constituent of apoplastic barriers in the root.

Suberin is a polyester of hydroxy fatty acids of different chain lengths. It can be degraded by trans-esterification (Kolattukudy and Agrawal, 1974; Zeier and Schreiber, 1997). One milligram of the isolated and extracted cell-wall material was immersed in 1 ml of borontrifluoride/methanol at 70°C for 16 h. By this treatment, carboxyl groups were methylated and polymers split into monomers. Released monomers were extracted from the reaction mixture using chloroform, whereby dotriacontane was added as an internal standard for the chromatographic analysis. Before analyzing samples by gas chromatography, hydroxyl groups were derivatized for 40 min at 70°C using BSTFA (N,N-bis-trimethylsilyltrifluoroacetamide) giving the trimethylsilyethers.

One microliter of suberin samples was injected onto a fused silica capillary column (inner diameter 0.32 mm; film thickness 0.1 mm) using a HP 5890 gas chromatograph with an on-column injector. Peaks were quantified by FID (flame ionization detector), and compounds identified by mass spectrometry. (HP 5971A mass selective detector).

3. SUBERIN COMPOSITION OF HYPO- AND ENDODERMAL CELL WALLS

Extracts obtained from isolated cell-wall samples did not contain significant amounts of aliphatic compounds. In some cases, trace amounts of membrane steroids and primary fatty acids were detected (Zeier et al., 1999). However, it must be assumed that these compounds were sorbed to the cell-wall samples during enzymatic isolation. During enzymatic isolation of cell walls, different kinds of compounds from the cells were released to the enzyme solution, and lipophilic compounds may have been adsorbed by the lipophilic suberin polymer, at least to some extent.

In isolated and completely extracted ECW and RHCW, a series of different structural biopolymers were detected. Besides carbohydrates still forming the largest fraction of these cell-wall samples, amino acids derived from structural cell-wall proteins, lignin and suberin were present in significant amounts (Schreiber et al., 1999). The most abundant suberin monomers released from all ECW and RHCW samples were ω-hydroxyacids, 1, ω-diacids, primary carboxylic acids, primary alcohols, and 2-hydroxyacids (Schreiber et al., 1999). Depending on the species, the chain lengths of these suberin monomers varied from C_{16} up to C_{32}. The most prominent monomers in ECW and RHCW samples, which have been described as being diagnostic markers for the occurrence of suberin, were the C_{18}-unsaturated ω-hydroxyacid (18-hydroxy-octadec-9-enoic acid) and 1,ω-diacid (octadec-9-ene-1,18-dioic acid).

Absolute amounts of suberin monomers detected in RHCW and ECW have been referred to the surface area of the isolated cell wall samples (Table 1). They varied over a large range. They were obviously dependent on the respective species analyzed. The highest amounts of suberin were detected in RHCW of *I. germanica*, and lowest in ECW of *M. deliciosa*. Suberin amounts in RHCW of the 10 investigated species varied by a factor of 232 between the species with the lowest and the highest amount (Table 1). In

Table 1. Suberin amounts in apoplastic barriers isolated from the roots of 10 different plant species. a) Zeier and Schreiber (1998); b) Schreiber et al. (1999); c) Effinger (2002); d) Zeier et al. (1999); e) Hartmann (2002); f) multi = multilayered hypodermis; g) RHCW = rhizodermal and hypodermal cell walls; h) III = tertiary endodermis with U-shaped cell wall thickening; i) I = primary endodermis with Casparian bands; k) II = secondary endodermis with suberin lamella; l) I-III = three developmental states over the length of the corn root.

Species	Outer root surface suberin amount [μg cm^{-2}]	Anatomy	Endodermis suberin amount [μg cm^{-2}]	Anatomy
Iris germanica [a)]	95.0	multi [f)]	20.8	III [h)]
Monstera deliciosa [a)]	53.7	multi [f)]	0.23	I [i)]
Aspidistra eliator [a)]	12.6	multi [f)]	2.10	III [h)]
Agapanthus africanus [a)]	11.0	multi [f)]	2.89	III [h)]
Clivia miniata [b)]	5.42	multi [f)]	0.69	I [i)]
Oryza sativa [c)]	4.53–9.58	RHZW [g)]	9.61–17.2	III [h)]
Pisum savtivum [d)]	3.31	RHZW [g)]	7.54	I [i)]
Ricinus communis [e)]	2.81–10.6	RHCW [g)]	22.5	II [k)]
Zea mays (Helix) [e)]	0.51–2.43	RHCW [g)]	0.61–1.24	I–III [l)]
Cicer arietinum [d)]	0.41	RHCW [g)]	22.4	II [k)]

ECW, variability between the species having the highest and the lowest suberin amount was also very pronounced (factor of 90; Table 1).

Due to the large variability in cell-wall suberization of RHCW and ECW, we postulate that there are also large differences in the efficiency of apoplastic transport barriers in roots depending on species. Just from the occurrence of suberin in RHCW and ECW, which can be detected in traces due to the high sensitivity of the analytical techniques applied, it cannot be simply concluded that these cell walls are always perfectly sealed with suberin in the apoplast.

4. EFFECTS OF ENVIRONMENTAL FACTORS ON SUBERIN COMPOSITION

Histochemically, it has been readily shown that plants react on environmental stress factors by strengthening their apoplastic transport barriers (North and Nobel, 1995; Reinhardt and Rost, 1995). Applying our quantitative analytical techniques, we were able to correlate an increase or decrease of suberization in roots directly with environmental stress factors. Salt stress significantly increased suberization in ECW and RHCW of *Ricinus communis* roots. However, nitrate deficiency resulted in a significant decrease (Figs. 1 and 2).

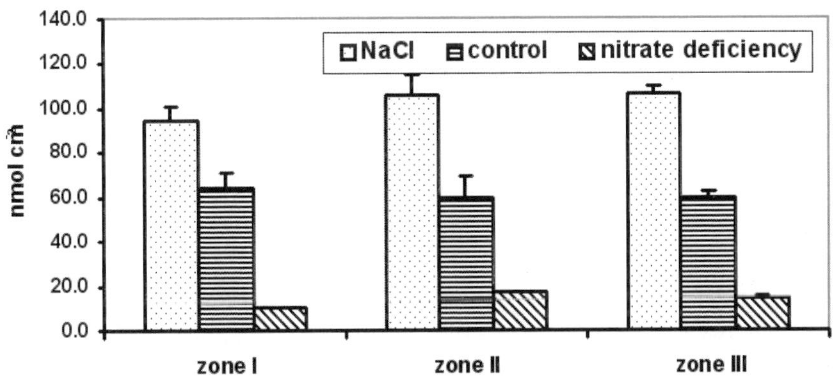

Fig. 1. Effect of environmental stress factors on aliphatic suberin amounts [nmol cm^{-2}] in ECW (endodermal cell walls) of hydroponically grown, 30 days-old Ricinus communis plants; zone I to III = different root zones increasing in age; NaCl = 100 mmol l^{-1} NaCl; nitrite deficiency = 0 mmol l^{-1} NO_3.

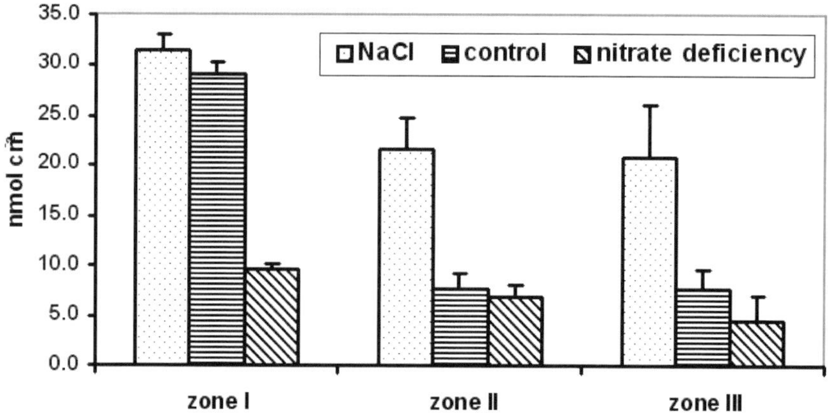

Fig. 2. Effect of environmental stress factors on aliphatic suberin amounts [nmol cm^{-2}] in RHCW (rhizodermal and hypodermal cell walls) of hydroponically grown, 30 days-old Ricinus communis plants; zone I to III = different root zones increasing in age; NaCl = 100 mmol l^{-1} NaCl; nitrate deficiency = 0 mmol l^{-1} NO_3.

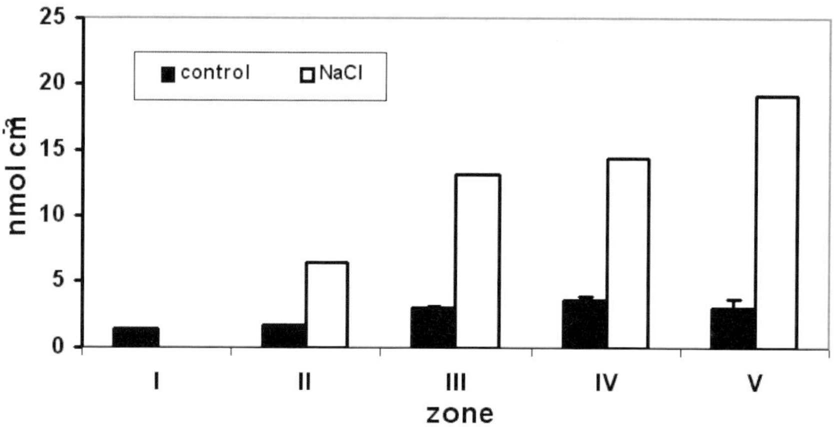

Fig. 3. Effect of NaCl stress on aliphatic suberin amounts [nmol cm^{-2}] in ECW (endodermal cell walls) of hydroponically grown 14 days-old Zea mays seedlings; zone I to V = different root zones increasing in age; NaCl = 100 mmol l^{-1} NaCl.

In case of salt stress, it is argued that plants increased their apoplastic barriers in roots in order to either reduce apoplastic uptake of sodium chloride or water losses. The opposite may be true during nutrient deficiency. Here, one could argue that external nutrient limitation led to a decrease in suberization of apoplastic barriers in order to facilitate an apoplastic passage of nutrients increasing the surface area for subsequent active uptake.

In part, effects of salt stress on suberization of ECW of corn (*Zea mays*) were comparable to those of *R. communis*. Upon salt stress, corn roots reacted by a strong increase in suberization of apoplastic transport barriers in ECW (Fig. 3). However, different from *R. communis*, an increase in suberization could not be measured in RHCW of corn (Fig. 4). Both species seem to have different strategies to react to salt stress in terms of a suberization of their apoplastic transport barriers. *R. communis* increased suberization of both ECW and RHCW. In corn, suberization was increased only in ECW.

5. CORRELATION BETWEEN CELL-WALL SUBERIZATION IN ROOTS AND WATER TRANSPORT ACROSS THE APOPALST

Experimental determination of the hydraulic conductivity in corn roots revealed that the radial hydraulic conductivity in the root apoplast correlated with the degree of suberization of apoplastic transport barriers in the root (Zimmermann et al., 2000). Compared to hydroponically grown corn roots,

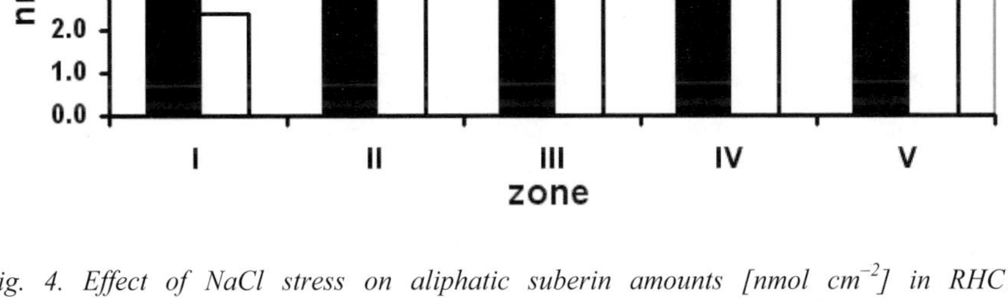

Fig. 4. Effect of NaCl stress on aliphatic suberin amounts [nmol cm^{-2}] in RHCW (rhizodermal and hypodermal cell walls) of hydroponically grown 14 days-old Zea mays seedlings; zone I to V = different root zones increasing in age; NaCl = 100 mmol l^{-1} NaCl.

suberization in RHCW of aeroponically grown corn roots was increased by a factor 2.4. In response to this increase of suberization, radial hydraulic conductivity in the apoplast significantly decreased by a factor between 1.5 to 3.6 (Zimmermann and Steudle, 1998; Zimmermann et al., 2000). From these data two things can be concluded: (i) Plants obviously adapt the degree of suberization in the root apoplast in response to environmental stimuli and (ii) there is always a certain amount of water transported across the suberized apoplast .

6. CONCLUSIONS AND FUTURE APPROACHES

Using modern analytical techniques, suberin composition and amount in specific root tissues (endodermis and hypodermis) forming apoplastic transport barriers can be analyzed with a high resolution. Thus it became possible to show the effect of different environmental stress factors (salt stress, nutrient deficiency) on the differentiation of the root apoplast and to correlate apoplastic water flow with the degree of suberization.

To date, there is virtually nothing known about biosynthesis of aliphatic suberin in roots. Hence, it will be a main future task to detect genes and enzymes involved in the synthesis of suberin in roots using molecular biological techniques. More insights into the molecular aspects of suberin formation in roots may result in new important biotechnological approaches such as the design of crops with increased suberization of roots rendering them more resistant towards drought or pathogen attack.

REFERENCES

Blakeney, A.B., Harris, P.J., Henry, R.J. and Stone, B.A. (1983) A simple and rapid preparation of alditol acetates for monosaccharide analysis. *Carbohyd. Res.*, 113, 291–299.

Effinger, N. (2002) Apoplastische Barrieren in den Wurzeln von Reis und Mais: chemische Zusammensetzung und Einfluss auf die radiale hydraulische Leitfähigkeit. Diploma Thesis, University of Würzburg.

Hartmann, K. (2002) Struktur, Funktion und chemische Zusammensetzung suberinisierter Transportbarrieren im Apoplasten Höherer Pflanzen. Doctoral Thesis, University of Würzburg.

Kolattukudy, P.E. and Agrawal, V.P. (1974) Structure and composition of aliphatic constituents of potato tuber skin (suberin). *Lipids*, 9, 682–691.

Lapierre, C., Pollet, B. and Rolando, C. (1995) New insights into the molecular architecture of hardwood lignins by chemical degradative methods. *Res. Chem. Intermed.*, 21, 397–412.

Marschner, H. (1995) *Mineral Nutrition of Higher Plants*. London: Academic Press.

North, G.B. and Nobel, P.S. (1995) Hydraulic conductivity of concentric root tissues of Agave deserti Engelm. under wet and drying conditions. *New Phyt.*, 130, 47–57.

Reinhardt, D.H. and Rost, T.L. (1995) Salinity accelerates endodermal development and induces an exodermis in cotton seedling roots. *Environ. Exp. Bot.*, 35, 563–574.

Schreiber, L., Breiner, H.W., Riederer, M., Düggelin, M. and Guggenheim R. (1994) The Casparian strip of *Clivia miniata* Reg. roots: isolation, fine structure and chemical nature. *Bot. Acta*, 107, 353–361.

Schreiber, L., Hartmann, K., Skrabs, M. and Zeier, J. (1999) Apoplastic barriers in roots: chemical composition of endodermal and hypodermal cell walls. *J. Exp. Bot.*, 50, 1267–1280.

Wilson, C.A. and Peterson, C.A. (1983) Chemical composition of the epidermal, hypodermal, endodermal and intervening cortical cell walls of various plant roots. *Ann. Bot.*, 51, 759–769.

Zeier, J. and Schreiber, L. (1997) Chemical composition of hypodermal and endodermal cell walls and xylem vessels isolated from *Clivia miniata*: identification of the biopolymers lignin and suberin. *Plant Phys.*, 113, 1223–1231.

Zeier, J. and Schreiber, L. (1998) Comparative investigation of primary and tertiary endodermal cell walls isolated from the roots of five monocotyledoneous species: chemical composition in relation to fine structure. *Planta*, 206, 349–361.

Zeier, J., Goll, A., Yokoyama, M., Karahara, I. and Schreiber L. (1999) Structure and chemical composition of endodermal and rhizodermal/hypodermal walls of several species. *Plant Cell Environ.*, 22, 271–279.

Zimmermann, M.H. and Steudle, E. (1998) Apoplastic transport across young maize roots: effect of the exodermis. *Planta*, 206, 7–19.

Zimmermann, M.H., Hartmann, K., Schreiber, L. and Steudle, E. (2000) Chemical composition of apoplastic transport barriers in relation to radial hydraulic conductivity of corn roots (*Zea mays* L.). *Planta*, 210, 302–311.

APOPLASTIC WATER TRANSPORT IN ROOTS

E. STEUDLE and K. RANATHUNGE
Lehrstuhl Pflanzenökologie, Universität Bayreuth, Germany, ernst.steudle@uni-bayreuth.de

Abstract. Results obtained from combined measurements at the cell and root levels (cell and root pressure probes) indicate an important role of apoplastic water transport in roots, even in the presence of apoplastic barriers (Casparian bands and suberin lamellae in the endo- and exodermis). The composite transport model (CTM) of the root explains the variable root hydraulic conductivity (Lp_r) and its physiological benefits, as well as low root reflection coefficients, and the switching between pathways depending on the water demand from the shoot (apoplastic *vs.* cell-to-cell transport). Switching between pathways provides a coarse, and changes in aquaporin activity a fine regulation of root Lp_r. In experiments with excised corn roots (cell and root levels), the extent of suberisation of roots was varied but successfully tested the CTM. Recent measurements of the hydraulics of rice roots also supported the CTM, and the view that apoplastic barriers exhibit water permeabilities of greater than usually assumed. In the rice experiments, the hydraulic conductivity through the apoplastic passage could be modified by blocking apoplastic pores with ink particles or insoluble precipitates of copper ferrocyanide, analogous to a Pfeffer cell.

Key words: apoplastic barriers, composite-transport model, hydraulic conductivity, rice, root, water transport

1. INTRODUCTION

Different from ions, water flow across roots involves no active pumping. Both, across the root cylinder and in xylem vessels along the root, water flow is down-hill following gradients in free energy (water potential) or pressure. Water uptake by plant roots can be described by simple force/flow relations analogous to Ohm's law, and is characterised by hydraulic conductances or resistances. The latter parameters are known to be highly

B. Sattelmacher† and W.J. Horst (eds.), The Apoplast of Higher Plants: Compartment of Storage, Transport and Reactions, 119–130.

variable. This affects the water status of plants. At a given rate of transpiration, the water supply by roots determines the water status of the shoot and its ability to assimilate carbon dioxide. The regulation of water input by roots is as important as that of the output (stomata). For technical reasons, much is known about the regulation of the latter, but little about the regulation of water uptake. Evidence collected over the past decade shows that the phenomenon of variable root hydraulics is not only related to the permeability of root cell-membranes for water (as it largely is for nutrient ions), but also depends on its apoplastic passage. The presence of root apoplastic barriers is important (Casparian bands and suberin lamellae in the endo- and exodermis). The anatomical complexity of the root dictates that the flow of water through it will also be complex. Flow is nicely described by a composite transport model (CTM), which allows for differences in movement through membranes of individual cells and along the apoplast, as well as through various tissues (see reviews: Hose et al., 2001; Steudle, 2000a,b, 2001, 2002a,b; Steudle and Frensch, 1996; Steudle and Heydt, 1997; Steudle and Peterson, 1998). In the following, recent findings are summarised, which relate to apoplastic water flow in roots. Results have been obtained using cell and root-pressure probes (Steudle, 1993) and different types of pressure chambers and pressure perfusion-techniques for herbaceous (maize, sunflower, bean, onion, barley) and woody (oak, spruce, beech) plants. Some recent results of rice have been added (Ranathunge et al., 2003; Ranathunge et al., 2004; Ranathunge et al., 2005a, b). Because of differences in the structure of roots of wetland plants, they allow a more detailed view on root hydraulics and tests of current models as do others. Hence, data from rice rather than from corn are used in this review to substantiate the role of apoplastic transport of water and of composite transport (CT) even in the presence of apoplastic barriers in roots.

2. WATER FLOW IN ROOTS IS VARIABLE: FLOWS AND FORCES

The numerous factors, which affect the capability of roots to take up water can be classified in two groups, (i) the hydraulic resistance of roots, and (ii) the different forces that move water in roots. The hydraulic resistance depends on the size of root systems, i.e. on the area available for water uptake. However, it also highly depends on the specific hydraulic conductivity, which is a measure of conductance per unit surface area (sometimes also given per unit root length) and per unit driving force (Pa or MPa). Hydraulic conductivities (Lp_r in m s^{-1} MPa^{-1}) rather than conductances are used for a quantitative comparison of different root species or roots at different developmental stages of a given species.

Usually, the radial transport of water across the root cylinder rather than the axial transport within the root xylem controls the overall water uptake by plant roots (Steudle and Peterson, 1998). When referred to unit area of outer root surface, the radial flow of water across a root (J_{Vr} in m^3 m^{-2} s^{-1}) is given by:

$$J_{Vr} = Lp_r\,[P_r - \sigma_{sr}(\pi_x - \pi_m)] \qquad (1)$$

The expression in brackets represents the driving force, which comprises the pressure difference between root xylem and soil solution (root-xylem pressure, P_r referred to reference atmospheric pressure), and the difference in osmotic pressure between the two compartments (xylem, π_x, and root medium, π_m). In the case of a transpiring plant, P_r would be negative, but may be also positive in the absence of transpiration (root pressure). According to Eq. (1), the osmotic component is modified by the root reflection coefficient, σ_{sr}. This indicates that roots may not behave like an ideal semipermeable osmometer and $\sigma_{sr} < 1$. The reflection coefficient is a measure of the passive selectivity of roots to solutes such as nutrient ions or others present in xylem sap and soil solution.

3. PATHWAYS FOR WATER IN ROOTS

In the root cylinder, water uses the pathway(s) of lowest resistance. Usually, this is the apoplast, i.e. the passage around protoplasts. During root development this may change by the formation of Casparian bands and suberin lamellae in the endo- and exodermis. Casparian bands are formed in one or more layers of the hypodermis (exodermis) or in the endodermis by the deposition of suberin and lignin in anticlinal cell walls. They may largely interrupt water and ion flow, and enforce a transmembrane movement of these compounds. Depending on the species, but also on the root's developmental and physiological state, root Lp_r comprises different components. It clearly incorporates the hydraulic conductivity of root cell-membranes (Lp), which, in turn, has a symplastic (mediated by plasmodesmata) and a transmembrane component. The latter is dominated by aquaporins or water channels, which are subjected to some "gating" by different internal or external parameters such as drought, high salinity, temperature, nutrient status, or diurnal rhythm. Aquaporins are thought to be the molecular basis of water relations (Clarkson et al., 2000; Henzler et al., 1999; Javot and Maurel, 2002; Maurel, 1997; Maurel and Chrispeels, 2001; Schäffner, 1998; Steudle, 2000a,b, 2001; Steudle and Henzler, 1995;

Tyerman et al., 1999; Tyerman et al., 2002). To date, the symplastic and transmembrane components cannot be separated experimentally, and are, hence, summarised as a cell-to-cell component which is measurable using the cell pressure probe (Steudle, 1993). Pressure probes also permit measurement at the root level (individual roots and root systems; Hose et al., 2000). Comparison of data allowed conclusions about pathways and transport models (see reviews).

4. DRIVING FORCES DURING WATER UPTAKE AND LOSSES: COMPOSITE TRANSPORT

According to Eq. (1), there are two distinct driving forces for water. Assuming that the reflection coefficient is unity (semipermeable osmotic barrier), they can be summarised by an overall difference in water potential. The hydrostatic driving force (P_r) should dominate in transpiring plants where transpiration causes tensions in the root xylem and the xylem osmotic pressure is rather small. On the other hand, there should be positive root pressures in its absence, when osmotic forces dominate. From Eq. (1), one would expect that root Lp_r is the same regardless of a pressure (hydrostatic) or osmotic diriving force (hydrostatic or osmotic water flow). However, this is not so. Differences can be as large as a few orders of magnitude in woody, and up to a factor of ten in herbaceous species (see reviews). The reason for the deviations is that, in the presence of osmotic forces across the root cylinder, the driving force along the apoplast is small due to the low selectivity (reflection coefficient) of the cell-wall compartment ($\sigma_{cw} \approx 0$). Along the cell-to-cell passage, this is different ($\sigma_{cc} \approx 1$). Overall, the contribution of the apoplast to osmotic water uptake should be small. By contrast, presence of hydrostatic pressure gradients across a root effectively drags water along both the apoplastic and cell-to-cell pathways.

Differences are well understood in terms of the CTM. The CTM explains the contribution of different pathways to overall root Lp_r and the switching between pathways depending on conditions. The term "composite transport" results from irreversible thermodynamics where it has been used to explain the overall permeability of patchy membranes in terms of permeabilities of individual arrays (see reviews). The CT of roots explains the findings of (i) variable root Lp_r, (ii) low reflection coefficients, (iii) differences between osmotic and hydrostatic root Lp_r (i.e. the switching in the transport model), and differences between species (e.g herbaceous *vs.* woody; Steudle and Frensch, 1996; Steudle and Peterson, 1998). The parallel arrangement of apoplastic and cell-to-cell pathways is the most relevant feature of the CTM. Composite transport is more efficient in explaining variable root Lp_r, and the other findings as do older models, which assume a variation in driving force or of root membrane permeability (Fiscus, 1975; Weatherley, 1982).

5. COMPOSITE TRANSPORT IN ROOTS OF HERBACEOUS PLANTS

Most of the evidence in favour of the CTM has been derived from pressure probe work (cell and root level) with excised roots of herbaceous plants such as corn, bean, onion, or sunflower. Some of the evidence is summarised in the following.

Early work with the root pressure probe indicated that roots behave like osmometers, though not like ideal ones (low root σ_{sr}; see reviews). Results indicated an apoplastic bypass for both water and small solutes, which also referred to the endo- and exodermal Casparian bands. However, the

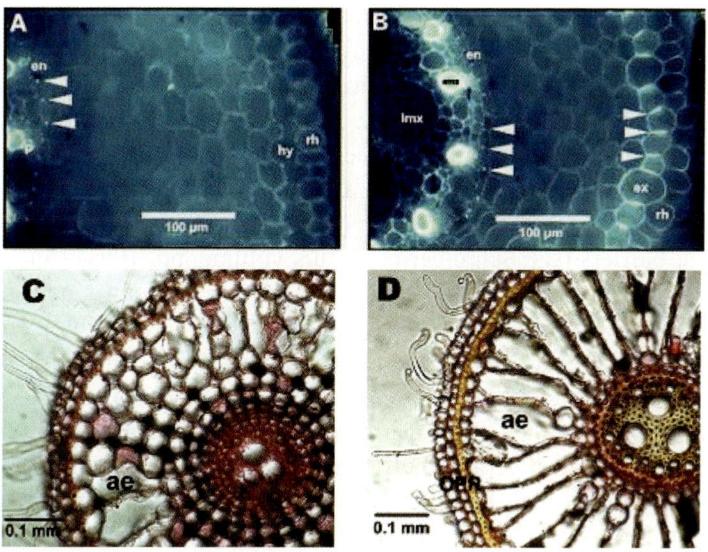

Fig. 1. A, B Freehand cross-sections of roots of 8 day-old maize plants stained with berberine-aniline blue and viewed under an epifluorescence microscope. Lignified vessels and Casparian bands appear bright. In A (hydroponic culture), the distance from the tip was 80 mm. rh=rhizodermis, hy=hypodermis without Casparian bands; en=endodermis (primary state) with Casparian bands (arrowheads); p=mature protoxylem. In B (aeroponic culture), the distance from the root tip was 50 mm. ex=mature exodermis, (secondary state) with Casparian bands (arrowheads). C, D Root cross-sections of 30 day-old rice plants stained with Sudan Red 7B. Development of aerenchyma at 20 (C) and 100 mm (D) from the root tip. ae=aerenchyma, OPR=outer part of the root.

permeability of apoplastic barriers depends on the species, growth conditions, and developmental state of these structures (Frensch et al., 1996; Barrowclough et al., 2000; Hose et al. 2001).

For example, in roots of corn and sunflower, even rather big ABA molecules could be swept from the root medium into the xylem by solvent drag (Freundl et al., 1998, 2000). The fact that the solvent drag was bigger in roots of corn than in those of sunflower may indicate that the "porosity" of Casparian bands is higher in the former. It turned out that reflection coefficients for ABA (as estimated from solvent drag) depended on pH. At slightly alkaline pH = 8, when ABA was present in anionic form, the reflection coefficient was bigger than at slightly acid (pH = 4.8) when part of the ABA was present in undissociated form. As a general rule, uncharged solutes are more permeable through the apoplast than are charged nutrient ions.

In corn roots, the effect of Casparian bands in the exodermis was tested for by growing roots with and without an exodermis. The existence of an exodermis caused a decrease of root hydrostatic Lp_r by a factor of 3.6 at constant root membrane Lp (Zimmermann and Steudle, 1998). There was no change in the presence of low osmotic gradients as expected from the CTM. Results were correlated with chemical analyses in different root zones, which indicated that aliphatic suberin in Casparian bands of the exodermis caused the decrease in root Lp_r (Zimmermann et al., 2000)

Despite these results, there is still a lack of information of how different growth conditions, stresses and other treatments affect the permeability of apoplastic barriers (see Schreiber et al, this volume, pp. 109–118; Schreiber et al., 2005). From the work of Schreiber's group in Bonn much is known about the chemistry of apoplastic barriers (Schreiber et al., 1999). However, there is still considerable uncertainty about the actual porosity and submicroscopic structure of these barriers (Hose et al., 2001; Ma and Peterson, 2003). As they form, permeability is reduced, but rarely to zero. Passive selectivity (reflection coefficients) increases as the cell-to-cell path becomes more important for water and solutes. Unlike the Casparian bands, there is much less quantitative evidence about the role of suberin lamellae, which coat the inner cell-wall surface of endo- and exodermis.

6. PHYSIOLOGICAL BENEFITS OF COMPOSITE TRANSPORT

Composite transport of water in roots is beneficial for the plant. It provides a mechanism to regulate water uptake according to the needs of the shoot. In the presence of high rates of transpiration, demands for water are high. Under these conditions, water flow across the root is hydraulic in nature and root Lp_r high. Both, the apoplastic and the cell-to-cell pathways are used. When transpiration is switched off, the cell-to-cell passage is left which has a relatively high resistance. This prevents water losses to a dry soil during the night. This type of a physical adjustment due to a switching on or off of the apoplastic passage has been termed a "coarse regulation". It allows rapid changes of root Lp_r according to the needs of the plant. It would explain most of the variability in root Lp_r. Under conditions of water shortage, plants develop roots, which are heavily suberized and have a low root Lp_r because of substantial apoplastic barriers. These barriers prevent water losses to the dry soil (back flow), but should have a negative effect on water uptake, when conditions become more favorable. The disadvantage can be compensated for by the existence of water channels which are under metabolic control and are inhibited/gated by factors such as heavy metals, hypoxia, nutrient deficiency, low temperature, drought and high salinity (see reviews). Water-channel activity may provide a "fine regulation" of water uptake and the only way to take up water under harsh conditions.

Apoplastic barriers are much discussed in relation to water uptake. However, they are probably important as barriers preventing excessive water losses to the dry soil under conditions of water shortage in the absence of a sufficient transpirational force for water uptake (Stavosky and Peterson, 1993; Steudle, 2000b; Taleisnik et al., 1999). Although it is well documented that roots suberise in response to drought and other stresses, there are only few measurements of changes in root Lp_r and in root cell Lp. These measurements are badly needed for the comparison between cell and root level to work out contributions of pathways (Azaizeh et al., 1992). Besides the interplay between the physical switching between pathways and water-channel activity, there are of course other means for plants to regulate water uptake such as by root growth under favorable and root death under adverse conditions. However, these responses are on a much longer time scale compared to those mentioned above.

7. COMPOSITE TRANSPORT IN RICE ROOTS

Wetland plants such as paddy rice develop root systems which grow into hypoxic substrates at the risk that oxygen delivered from the shoot down to root tips would be largely lost by radial diffusion across the outer part of the roots (OPR; Ranathunge et al., 2003, 2004). Hence, there should be barriers such as Casparian bands and suberin lamellae to prevent radial oxygen loss. In Fig. 1D, it is shown that the OPR of rice roots has a well-developed exodermis with Casparian bands and suberin lamellae. This may cause problems for the water uptake, when the apoplastic passage is blocked by Casparian bands and the cell-to-cell passage affected by suberin lamellae. Accordingly, it has been shown that rice plants, even though rooting in a wet substrate, may suffer from water shortage (Hirasawa et al., 1992).

For the first time, we studied the hydraulics of rice roots in some detail using pressure chambers and probes as well as a new perfusion technique (Fig. 2). The results indicated the importance of apoplastic water transport, at least in the OPR. Miyamoto et al., (2001) showed that root Lp_r and σ_{sr} were rather low, which was interpreted as a major resistance at the endodermis and some apoplastic bypass of water and solutes despite suberisation. When the small size of rice root systems is taken into account, the water supply to the shoot must indeed be quite limited.

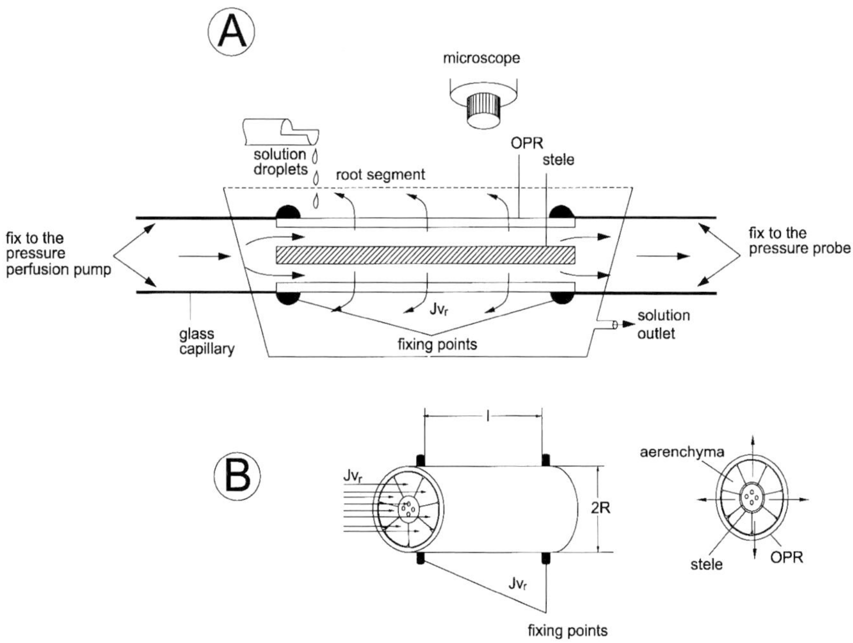

Fig. 2. (A) Pump perfusion setup: A syringe was mounted on a Braun-Melsungen pump (not shown) that created pump rates between 1.7 X 10^{-9} and 1.1 X 10^{-7} mm^3 s^{-1}. One end of the root segment was used as an inlet. This was fixed to the syringe by a narrow and rigid Teflon tube. The other end was connected to a pressure probe to measure resulting steady state pressures while perfusing aerenchyma with the nutrient solution. (B) Schematic diagram to show radial water flow across the outer part of the root segment during pressure perfusion. At a given pump rate, stationary pressure was established where the volume flow provided by the pump equalled the radial volume flow across the outer part of the root (OPR).

*Table 1. (A) Hydraulic conductivity (Lp_r) of individual rice (*Oryza sativa*, L.) roots, whole root systems and the outer part of roots (Lp_{OPR}), which were grown in hydroponics with bubbled air for 31–40 days, as measured by the root pressure probe (single roots), pressure chamber techniques (root systems) and pressure perfusion technique (outer part of roots), respectively. (B) Calculated treatment / control ratios of Lp_{OPR} after blocking the apoplast of the OPR either with China ink particles or $Cu_2[Fe(CN)_6]$ precipitates (Ppt.), or closing water channels treating with 50 μM $HgCl_2$ for 20 min. Ratio of bulk (P_{fOPR}) and diffusional (P_{dOPR}) water permeability of the OPR are also given for two different distances from the root tip. (C) Reflection coefficients of single adventitious roots ($_{sr}$) and outer part of roots ($_{sOPR}$) as measured with root pressure probe or pump perfusion technique. Osmotic water flow was induced by increasing the osmotic pressure of the medium by adding different osmotica.*

A)

	Hydraulic conductivity (Lp) × 10^{-8} m s^{-1} MPa^{-1}				References
	Whole root systems	Individual roots	Outer part of root (OPR)		
			20–50 mm	50–100 mm	
Hydrostatic	4.0 ± 1.7	3.8 ± 0.6	150 ± 60	110 ± 30	Ranathunge et al., 2003
	5.6 ± 2.7	5.0 ± 2.5	–	–	Miyamoto et al., 2001
Osmotic	3.1 ± 0.9	1.1 ± 0.5	–	–	Ranathunge et al., 2003
(NaCl)	4.2 ± 2.5	9.2 ± 3.0	–	–	Miyamoto et al., 2001

B)

Distance from root tip (mm)	Treatment /control ratio of Lp_{OPR}			Ratio of bulk (P_{fOPR}) and diffusional (P_{dOPR}) water permeability of the OPR	References
	$HgCl_2$/control	Ink/control	Ppt./control		
20–50	0.90 ± 0.10	0.75 ± 0.09	0.28 ± 0.06	620	Ranathunge et al., 2004, 2005a
50–100	0.92 ± 0.03	0.67 ± 0.13	0.31 ± 0.05	1200	

C)

	Reflection coefficient (σ_{sr}) of excised roots		References
	Ethanol	NaCl	
Whole root	0.04 ± 0.02	0.18 ± 0.06	Ranathunge et al., 2003
	0.09 ± 0.01	0.28 ± 0.11	Miyamoto et al., 2001
Distance from the root tip (mm)	Reflection coefficient (σ_{sOPR}) of outer part of the root		References
	Mannitol	NaCl	
20–50	0.13 ± 0.04	0.09 ± 0.02	Ranathunge et al., 2003
50–100	0.13 ± 0.04	0.11 ± 0.03	

In subsequent measurements, the perfusion technique was used to measure the water permeability of OPR separately, and to get more detailed information as to whether the exo- or the endodermis was limiting water uptake. Despite suberisation of the exodermis, the Lp_{OPR} was larger by a factor of 30 than the overall root Lp_r (Table 1A), and σ_{OPR} siginifcantly smaller than the overall σ_{sr} (Table 1C). So, despite suberisation, there was a considerable bypass of water and small solutes in the exodermis as proposed for other species (see above). A dominating apoplastic transport component was also indicated by the fact that the diffusional water permeability (measured with heavy water) was smaller than the bulk-water permeability by a factor of as large as 600–1200 (Table 1B). The partial blockage of the apoplastic pathway by China ink particles or by the precipitation of the insoluble salt, copper ferrocyanide analogous to Wilhelm Pfeffer's famous experiments on osmosis (for details, see Ranathunge et al., 2005a, b) resulted in a substantial decrease of the root's Lp_r. Effects of precipitation were more pronounced than those of the blockage of apoplastic pores by China ink particles (diam. $\approx$50 nm) and even more pronounced as compared with the inhibition of water channels along the transmembrane path using the water channel blocker, $HgCl_2$. The results strongly support the idea of a dominating apoplastic water flow. Similar results were obtained during the blockage of the apoplast of corn roots using the same precipitation technique. The experiments with rice roots showed that there was an apoplastic transport of water even in the presence of a well-developed exodermis with apoplastic barriers. Anatomical studies did not reveal any kind of patchiness or of other irregularities of the exodermis.

8. CONCLUSIONS

The results reviewed here indicate that, depending on conditions, the apoplastic water flow may substantially contribute to the overall rate of uptake of water by plant roots. A composite-transport model applies which explains the variability of the hydraulic conductivity of roots. The model provides a useful adaptation of plants to the water needs of the shoot and the availability of water in the soil. Both, apoplastic barriers in the root as well as aquaporins in root cell-membranes contribute to the overall regulation of the hydraulic conductivity of roots.

ACKNOWLEDGEMENTS

This work was supported by the Deutsche Forschungsgemeinschaft, Schwerpunktprogramm "Apoplast". Publications obtained within this progam are indicated by an asteriks (*). The work on the paper indicated by two asteriks (**) was published during a subsequent DFG project on apoplastic water flow in rice roots (Ste 319/3-4).

REFERENCES

Azaizeh, H., Gunse, B. and Steudle, E. (1992). Effects of NaCl and $CaCl_2$ on water tranport across root cells of maize (*Zea mays* L.) seedlings. *Plant Physiol.*, 99, 886–894.

*Barrowclough, D.E., Peterson, C.A. and Steudle, E. (2000). Radial hydraulic conductivity along developing onion roots. *J. Exp. Bot.*, 51, 547–557.

*Clarkson, D.T. Carvajal, M., Henzler, T., Waterhouse, R.N., Smyth, A., Cooke D.T. and Steudle, E. (2000). Root hydraulic conductance: diurnal aquaporin expression and the effects of nutrient stress. *J. Exp. Bot.*, 51, 61–70.

Fiscus, E. (1975). The interaction between osmotic- and pressure-induced water flow in plant roots. *Plant Physiol.*, 55, 752–759.

Frensch, J., Hsiao, T.C. and Steudle, E. (1996). Water and solute transport along developing maize roots. *Planta*, 198, 348–355.

*Freundl, E., Steudle, E. and Hartung, W. (1998). Water uptake by roots of maize and sunflower affects the radial transport of absciic acid and its concentration in the xylem. *Planta*, 207, 8–19.

*Freundl, E., Steudle, E. and Hartung, W. (2000). Apoplastic transport of abscisic acid through roots of maize: effect of the exodermis. *Planta*, 210, 222–231.

*Henzler, T., Waterhouse, R.N., Smyth, A.J., Carvajal, M., Cooke, D.T., Schäffner, A.R., Steudle, E. and Clarkson, D.T., (1999). Diurnal variations in hydraulic conductivity and root pressure can be correlated with the expression of putative aquaporins in the roots of *Lotus japonicus*. *Planta*, 210, 50–60.

Hirasawa, T., Tsuchida, M. and Ishibara, K. (1992). Relationship between resistance to water transport and exudation rate and the effect of the resistance on the midday depression of stomatal aperture in rice plants. *Jap. J. Crop Sci.*, 61, 145–152.

*Hose, E., Steudle, E. and Hartung, W. (2000). Abscisic acid and the hydraulic conductivity of roots: a study using cell- and root pressure probes. *Planta*, 211, 874–882.

*Hose, E., Clarkson, D.T., Steudle, E., Schreiber, L. and Hartung, W. (2001) The exodermis: a variable apoplastic barrier. *J. Exp. Bot.,* 52, 2245–2264.

Javot, H. and Maurel, C. (2002). The role of aquaporins in root water uptake. *Annals of Bot.*, 90, 301–313.

Ma, F. and Peterson, C.A. (2003). Current insights into the development, structure, and chemistry of the endodermis and exodermis of roots. *Can. J. Bot.*, 81, 405–421.

Maurel, C. (1997). Aquaporins and water permeability of plant membranes. *Ann. Rev. Plant Physiol. Plant Mol. Biol.*, 48, 399–429.

Maurel, C. and Chrispeels, M.J. (2001). Aquaporins. A molecular entry into plant water relations. *Plant Physiol.*, 125, 135–138.

*Miyamoto, N., Steudle, E., Hirasawa, T. and Lafitte, R. (2001). Hydraulic conductivity of rice roots. *J. Exp. Bot.*, 52, 1–12.

*Ranathunge, K., Steudle, E. and Lafitte, R. (2003). Control of water uptake by rice (*Oryza Sativa* L.): role of the outer part of the root. *Planta*, 217, 193–205.

*Ranathunge, K., Kotula, L., Steudle, E. and Lafitte, R. (2004) Water permeability and reflection coefficient of the outer part of young rice roots are differently affected by closure of water channels (aquaporins) or blockage of apoplastic pores. *J. Exp. Bot.*, 55, 433–447.

*Ranathunge, K., Steudle, E. and Lafitte, R. (2005a). Blockage of apoplastic bypass-flow of water in rice roots by insoluble salt precipitates analogous to a Pfeffer cell. *Plant Cell and Environ.,* 28, 121–133.

**Ranathunge, K., Steudle, E. and Lafitte, R. (2005b). A new precipitation technique provides evidence for the permeability of Casparian bands to ions in young roots of corn (*Zea mays* L.) and rice (*Oryza sativa* L.). *Plant Cell Environ.,* 28, 1450–1462.

Schäffner, A.R. (1998). Aquaporin function, structure, and expression: there are still surprises to come up in water channels. *Planta*, 204, 131–139.

Schreiber, L., Hartmann. K.D., Skrabs, M. and Zeier, J. (1999). Apoplastic barriers in roots: chemical composition ofendodermal and hypodermal cell walls. *J. Exp. Bot.*, 50, 1267–1280.

*Schreiber L., Franke, R., Hartmann, K.D., Ranathunge, K. and Steudle, E. (2005). The chemical composition of suberin in apoplastic barriers affects radial hydraulic conductivity differently in the roots of rice (*Oryza sativa* L. cv. IR64) and corn (*Zea mays* L. cv. *Helix*). *J. Exp. Bot.,* 56, 1427–1436.

Stavosky, E. and Peterson, C.A. (1993). Effects of drought and subsequent rehydration on the structure, vitality and permeability of *Allium cepa adventitious* roots. *Can. J. Bot.*, 71, 700–707.

Steudle, E. (1993). Pressure probe techniques: basic principles and application to studies of water and solute relations at the cell, tissue, and organ level. In: Smith, J.A.C. and Griffith, H. (eds), *Water Deficits: Plant Responses from Cell to Community* (pp. 5–36). Oxford: Bios Scientific Publishers.

* Steudle, E. (2000a). Water uptake by roots: an integration of views. *Plant Soil*, 226, 45–56.

*Steudle, E. (2000b) Water uptake by roots: effects of water deficit. *J. Exp. Bot.*, 51, 1531–1542.

*Steudle, E. (2001). The cohesion-tension mechanism and the acquisition of water by plant roots. *Ann. Rev. Plant Physiol. Plant Mol. Biol.*, 52, 847–875.

*Steudle, E. (2002a). Transport of water in plants. *Environ. Contr. Biol.*, 40(1), 29–37.

*Steudle, E. (2002b). Aufnahme und Transport des Wassers in Pflanzen. *Nova Acta Leopoldina* NF, 85, 251–278.

Steudle, E. and Frensch, J. (1996). Water transport in plants: role of the apoplast. *Plant Soil*, 87, 67–79.

Steudle, E. and Henzler, T. (1995). Water channels in plants: do basic concepts of water transport change? *J. Exp. Bot.*, 46, 1067–1076.

Steudle, E. and Heydt, H. (1997). Water transport across tree roots. In: Rennenberg, H., Eschrich, W. and Ziegler, H. (eds). *Trees –Contributions to Modern Tree Physiology* (pp. 239–255). Backhuys Publishers, Leiden, The Netherlands.

*Steudle, E. and Peterson, C.A. (1998). How does water get through roots? *J. Exp. Bot.*, 49, 775–788.

Taleisnik, E. and Peyrano, G., Cordoba, A. and Arias, C. (1999). Water retention capacity in root segments differing in the degree of exodermis development. *Ann. Bot.*, 83, 19–27.

*Tyerman, S.D., Bohnert, H.J., Maurel, C., Steudle, E., Smith, J.A.C. (1999). Plant aquaporins: their molecular biology, biophysics and significance for plant water relations. *J. Exp. Bot.*, 50, 1055–1071.

Tyerman, S.D., Niemietz, C.M., Bramley, H. (2002). Plant aquaporins: multifunctional water and solute channels with expanding roles. *Plant Cell Environ.*, 25, 173–194.

Weatherley, P.E. (1982). Water uptake and flow into roots. In: Lange, O.L., Nobel, P.S., Osmond, C.B., Ziegler, H. (eds), *Encyclopedia of Plant Physiology*, Vol. 12B (pp. 79–109). Springer-Verlag, Berlin.

*Zimmermann, H.M., Hartmann, K.D., Schreiber, L. and Steudle, E. (2000). Chemical composition of apoplastic transport barriers in relation to radial hydraulic conductivity of roots of maize (*Zea mays* L.). *Planta*, 210, 302–311.

*Zimmermann, H.M. and Steudle, E. (1998), Apoplastic transport across young maize roots: effect of the exodermis. *Planta*, 206, 7–19.

Section 3

Ion Uptake from and Loading into the Apoplast: Characterization of Channel Properties and Relevance for the Nutrition of Plants

LONG DISTANCE TRANSPORT IN PLANTS: TOWARDS ANALYSES OF REGULATORY INTERACTIONS BETWEEN MEMBRANE TRANSPORT SYSTEMS AND CELL WALL IONIC ATMOSPHERE IN VASCULAR TISSUES

H. SENTENAC
Biochimie et Physiologie Moléculaire des Plantes, France, sentenac@ensam.inra.fr

In terrestrial plants, the autotrophic status requires long distance transport of water, inorganic ions and organic solutes, in the xylem and phloem vasculatures, between the roots which take up water and mineral nutrients, and the aerial parts which harvest CO_2 and light. Long distance transport ensures translocation to the various organs, sources and sinks, of water and nutrients required for growth. It also plays a role in root–shoot signaling since, for instance, evidence is available that recycling of inorganic nutrient ions from the shoots to the roots via the phloem sap plays a role in the regulation of nutrient acquisition from the soil (Kochian and Lucas, 1988; White, 1997; Tillard et al., 1998).

Preservation of the hydromineral homeostasis of growing organs under fluctuating environmental conditions, e.g., changes in soil water and nutrient ion availability, air humidity or light intensity, requires tight control of long-distance transport. The regulation mechanisms are very complex, integrated at the whole plant level, under hormonal control, and still poorly understood. On the other hand, the membrane transport systems responsible for uptake/secretion of solutes from/into the apoplast of the vascular tissues can be looked at as terminal effectors of the signaling and regulation pathways. The present section is dedicated to the molecular identification and functional characterization of such systems, and their interaction/regulation with/by the apoplast ionic atmosphere. The papers in the section mainly

B. Sattelmacher† and W.J. Horst (eds.), The Apoplast of Higher Plants: Compartment of Storage, Transport and Reactions, 133–136.

concern K^+ channels and transporters (by Abshagen-Keunecke and Hansen, 177–192, Ache and Deeken, 163–176, and Fromm and Hedrich, 149–162), anion channels (paper by Köhler and Raschke, 193–214) and H^+-ATPases (Fromm and Hedrich, 149–162, Köhler and Raschke, 192–214). Aspects regarding regulation by external concentrations and interactions with the apoplastic ionic atmosphere concern H^+ and Ca^{2+} ions.

The electrical properties of the apoplast affect the ionic composition of the medium that bathes the cell membrane (Grignon and Sentenac, 1991; Sattelmacher, 2001). Together with sugars, uronic acids are major constituents of the cell-wall matrix. Dissociation of their caroboxylic groups results in high concentrations of fixed anionic charges, up to several hundred of mmol l^{-1} (Sentenac and Grignon 1981). The resulting "Donnan" potential accumulates cations (and excludes anions). Cations are accumulated as free hydrated ions but can also be directly associated to the anionic groups. The latter phenomenon mainly concerns H^+ (i.e., dissociation/association of the uronic carboxylic groups: intrinsic pK close to 3.2) and Ca^{2+} (dissociation constant in the 10^{-2}–10^{-1} mol l^{-1} range) (Sentenac and Grignon, 1981). Compared to H^+ and Ca^{2+}, cations like K^+, Na^+ or Mg^{2+} do not seem to significantly associate with cell wall anionic groups (Haynes, 1980; Sentenac and Grignon, 1981). Thus H^+ and Ca^{2+} are endowed with a particular status in the apoplast. The consequences of this special status on the ionic atmosphere of a closed apoplast (e.g., the root stele) in planta are not clear. However, it can be speculated that the cell-wall pH and free Ca^{2+} concentration are well buffered, and probably poorly dependent on the concentrations of other ions. Interestingly, the external concentrations of H^+ and Ca^{2+} are used by the cell to control the activity of transport systems at work on its membrane, as discussed in the following sections. Indeed, it has been shown that *Shaker* K^+ channels expressed in phloem tissues (members of the *Shaker* subgroup 2) in many plant species are sensitive to both Ca^{2+} and H^+ (voltage-dependent block; papers in this volume by Ache and Deeken, pp. 151–164, and Fromm and Hedrich, 137–150). Also, anion channels characterized in maize leaf in bundle-sheath cells (Abshagen-Keunecke and Hansen, this volume, pp. 165–180) and H^+-ATPase activity observed in barley root stelar tissues (Köhler and Raschke, this volume, pp. 181–200) are sensitive to pH. Interestingly, correlated with seasonal cambial activity and expression of K^+ channels in cambial region, high K^+ and Ca^{2+} contents in cambial region (in the apoplast?) stimulate wood formation in poplar (Fromm and Hedrich, this volume, pp. 137–150).

The following four papers do concern pioneering research in plant biology. Indeed, we are only at the beginning of the analysis of the molecular mechanisms and regulation of long distance transport in plants. So far, the regulatory interactions between membrane transport activity and ionic conditions in the apoplast have been poorly investigated. Furthermore, the transport activities of the cell membrane themselves are still poorly characterized at the molecular and electrophysiological levels in vascular tissues (when compared with the guard-cell membrane; see the discussion/conclusion section of the paper by Abshagen-Keunecke and Hansen, this volume, pp. 165–180). The reason of this situation is that, until recently, the different cell types present in vascular tissues appeared as hardly accessible to electrophysiological techniques since too numerous and buried in the heart of the organs, as in the root, the stem and even the leaf. However, isolation procedures have been adapted (Wegner and Raschke, 1994; Roberts and Tester, 1995; Keunecke et al., 1997; see the papers from Abshagen-Keunecke and Hansen, this volume, pp. 165–180, Köhler and Raschke, this volume, pp. 181–200) and transgenic constructs expressing GFP under control of specific promoters can be used to sort vascular cells (Ivashikina et al., 2003). It is eagerly wished that such tools, along with knock-out mutant lines (Gaymard et al. 1998; Deeken et al., 2002), will allow rapid progress in our molecular understanding of long distance transport in plants.

REFERENCES

Deeken, R., Geiger, D., Fromm, J., Koroleva, O., Ache, P., Langenfeld-Heyser, R., Sauer, N., May, ST. and Hedrich, R. (2002) Loss of the *AKT2/3* potassium channel affects sugar loading into the phloem of *Arabidopsis*. *Planta*, 216, 334–344.

Gaymard, F., Pilot, G., Lacombe, B., Bouchez, D., Bruneau, D., Boucherez, J., Michaux-Ferriere, N., Thibaud, J.-B. and Sentenac, H. (1998) Identification and disruption of a plant shaker-like outward channel involved in K^+ release into the xylem sap. *Cell.*, 94, 647–655.

Grignon, C. and Sentenac, H. (1991) pH and ions in the apoplast. *Annu. Rev. Plant Physiol. Plant Mol. Biol.*, 42, 103–128.

Haynes, R.J. (1980) Ion exchange properties of roots and ionic interactions within the root apoplasm: their role in ion accumulation by plants. *Bot. Rev.*, 46, 75–99.

Ivashikina, N., Deeken, R., Ache, P., Kranz, E., Pommerring, B., Sauer, N. and Hedrich, R, (2003) Generation of AtSUC2 promoter-GFP-marked companion cells for patc-clamp studies and expression profiling. *Plant J.*, 36, 931–45.

Keunecke, M., Sutter, J.U., Sattelmacher, B. and Hansen, U.P. (1997) Isolation and patch-clamp measurements of xylem contact cells for the study of their roles in the exchange between apoplast and symplast of leaves. *Plant Soil*, 196, 239–244.

Kochian, L.V. and Lucas, W.J. (1988) Potassium transport in roots. *Adv. Bot. Res.*, 15, 136–151.

Roberts SK and Tester M (1995) Inward and outward K^+-selective currents in the plasma membrane of protoplasts from maize root cortex and stele. *Plant J.*, 8, 811–825.

Sattelmacher, B. (2001) The apoplast and its significance for plant mineral nutrition. *New Phytol.*, 149, 167–192.

Sentenac, H. and Grignon, C. (1981) A model for predicting ionic equilibrium concentrations in cell walls. *Plant Physiol.*, 68, 415–419.

Tillard, P., Passama, L. and Gojon, A. (1998) Are phloem amino acids involved in the shoot to root control of NO_3^- uptake in *Ricinus communis* plants? *J. Exp. Bot.*, 49, 1371–1379.

Wegner, L.H. and Raschke, K. (1994). Ion channels in the xylem parenchyma of barley roots: a procedure to isolate protoplasts from this tissue and a patch-clamp exploration of salt passageways into xylem vessels. *Plant Physiol.*, 105, 799–813.

White, P.J. (1997) The regulation of K^+ influx into roots of rye (*Secale cereale* L.) seedlings by negative feedback via the K^+ flux from shoot to root in the phloem. *J. Exp. Bot.*, 48, 2063–2073.

THE ROLE OF POTASSIUM IN WOOD FORMATION OF POPLAR

J. FROMM[1] and R. HEDRICH[2]
[1]*Fachgebiet Angewandte Holzbiologie der TU München, Germany, fromm@wzw.tum.de;*
[2]*Julius-von-Sachs-Institut für Biowissenschaften, Germany*

Abstract. The influence of potassium (K) supply on wood formation and the molecular mechanisms of K^+-dependent xylogenesis were studied in poplar, one of the most important tree species in the field of wood biotechnology. Structural analyses revealed that the wood elongation zone as well as the vessel lumina were significantly reduced upon K^+ starvation. In contrast to plants grown under optimal K^+ supply, plants grown under limiting K^+ concentrations also showed low and equally distributed K^+ contents in vessels, fibres and cambial cells. To study the molecular basis of K^+ transport during wood formation, the EST database from the cambial region of poplar was searched for sequence homologies to known K^+ transporters from *Arabidopsis*. By quantitative RT-PCR we found that especially the *P. tremula* outward rectifying K^+ channel (PTORK) and *P. tremula* K^+ channel 2 (PTK2) correlated with seasonal cambial activity. The transcripts of both channels coincided with the seasonal K^+ variations in the wood formation zone. The investigation of the biophysical properties showed that PTORK mediates K release upon membrane depolarization, while PTK2 is almost voltage-independent and might play a role in K^+ uptake. By using immunofluorescence microscopy PTORK was localised in the plasma membrane of sieve elements and xylem ray-cells indicating a function in K^+ release from these cells. In addition, a PM H^+-ATPase which generates the necessary H^+-gradient for the uptake of K^+ and other nutrients into cambial cells could be localised in active cambial and differentiating xylem cells as well as in ray cells surrounding vessels. Since the ion channels are involved in the regulation of wood formation an important biotechnological challenge would be the generation of transgenic trees with modified K^+ channels in order to change wood structure and thereby optimise wood properties.

Key words: electrophysiolgy, immunolocalisation, K^+ channels, poplar, wood production

B. Sattelmacher† and W.J. Horst (eds.), The Apoplast of Higher Plants: Compartment of Storage, Transport and Reactions, 137–149.

1. INTRODUCTION

Up to these days wood has never stopped being the most important natural and endlessly renewable source of raw material. Its major role is not only pointed on the provision of paper, constructing material and wood-based products, but worldwide mainly on the provision of energy. Woody plants are gradually gaining more scientific and economic importance and poplar is, apart from Eucalyptus and Pine spp., one of the main model tree species for basic and applied wood research. Because of its ease of vegetative propagation, its suitability for genetic transformation, its rather small genome size and the availability of genetic maps, poplar has become the common used model tree species in Europe and the United States.

Since nutrient availability is an important extrinsic factor in wood growth (Fig. 1), the effect of various minerals on xylogenesis is worthy to investigate.

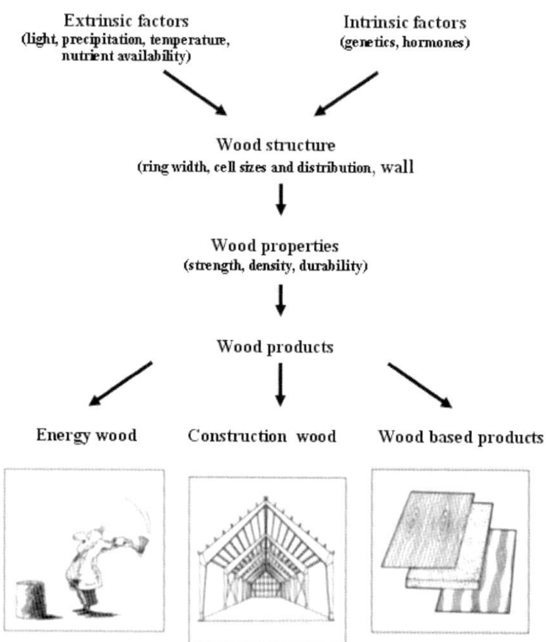

Fig. 1. Modification of wood structure, properties and products by extrinsic and intrinsic growth factors.

K homeostasis in higher plants in general and in trees in particular, is dependent on the availability of nutrients, the state of mycorrhizal association, as well as the physiological condition of the plant. During wood formation K^+ ions are involved in various aspects of xylogenesis (Dünisch and Bauch, 1994a, b; Kuhn et al., 1997), especially during cell expansion when the symplastic K^+ content increases (Dünisch et al., 1998). Differentiating vessels represent a strong sink for K because this osmolyte seems to be the driving force for cell expansion during primary-wall formation in wood. Confirming this suggestion, fertilized spruce stands showed wider tree rings and prolonged cambial activity compared to non-fertilized stands (Dünisch and Bauch, 1994a, b). Seasonal changes in the cambial K and calcium content of the balsam poplar (*P. trichocarpa*) could be detected via EDXA and revealed that the reactivation of the cambium in springtime is accompanied by high concentrations of K^+ and Ca^{2+} (Arend and Fromm, 2000). When the cambium starts developing latewood during July the K concentration still remains at a high level, whereas calcium decreases shortly after cambial reactivation. So far, there is some information on seasonal variation in K content within trees available.

Knowing that K is transported within plant organs, tissues and even cells, its uptake and transport mechanisms have been studied extensively within different cell types of the root, shoot and leaf (Hedrich and Roelfsema, 1999). Once K is taken up by the roots, it is transported via the xylem towards the shoots and the leaves. After leaving the xylem a high percentage of the K content can be transported via the phloem to various sinks like differentiating wood, young leaves, seeds, fruits or growing roots (Ache et al., 2001, Deeken et al., 2000, Eschrich et al., 1988, Fromm and Bauer, 1994). K uptake channels of the AKT1- and AtKC1-type (*Arabidopsis thaliana* K^+ channel) as well as carriers of the high-affinity K^+ transporter (HKT1) and K^+ uptake (KUP) family are responsible for the K uptake from the soil (Brüggemann et al., 1999, Gassman et al., 1996, Hirsch et al., 1998, Ivashikina et al., 2001, Kim et al., 1998, Reintanz et al., 2002, Rodriguez-Navarro, 2000, Schroeder and Fang, 1991, Spalding et al., 1999). Loss-of-channel function mutants could prove the necessity of three root *Shaker*-like K^+channels for K uptake into and its transport within roots (Gaymard et al., 1998, Hirsch et al., 1998, Reintanz et al., 2002). Within the *Arabidopsis* genome there are two K efflux channels of the *Shaker*-type, stelar K^+ outward rectifier (SKOR, Gaymard et al., 1998) and guard cell outward rectifying K^+ channel (GORK, Ache et al., 2000), of which SKOR plays an important role in loading K into the xylem. The expression of this channel is induced by K^+ and repressed by abscisic acid. In contrast to SKOR, AtKUP1 as the high-affinity K carrier of *A. thaliana*, is enhanced by K^+ depletion (Fu and Luan, 1998, Kim et al., 1998). Measurements on the phloem using the aphid stylet technique have further shown that the membrane potential of the phloem is dominated by K^+ conductance (Ache et al., 2001). Based on these results, specifically light-regulated corresponding *AKT 2/3*-like K^+ channels were identified in several species such as *Arabidopsis*, maize and broad bean (Ache et al., 2001, Bauer et al., 2000, Deeken et al., 2000 and 2002, Marten et al., 1999).

To give a description of the fundamental processes involved in K-dependent xylogenesis molecular and biophysical techniques have been used analysing K^+ transporters of poplar. One K^+ uptake channel (KPT1), one K^+ release channel (PTORK) and one weak voltage-independent channel (PTK2) as well as a broadly expressed KUP-type of K^+ transporter from a poplar wood EST library (Sterky et al., 1998) were isolated and transporter functions were verified by heterologous expression in *Xenopus oocytes* or *Escherichia coli.* Their properties as well as expression patterns will be discussed in the context of wood formation in Section 4.

2. THE IMPORTANCE OF POTASSIUM NUTRITION; STORAGE AND REMOBILIZATION FOR WOOD GROWTH

Within the various mineral nutrients essential for tree growth K but also magnesium as well as calcium are important for basic developmental processes like cell division and differentiation and therefore wood formation (Larson, 1967; Wardrop, 1981; Eklund and Eliasson, 1990). Old growth spruce stands fertilized with K, Ca and Mg show 30% more biomass, a significant increase in periclinal cell divisions in the cambium and higher K, Ca and Mg contents in different tree fractions than those of the controls (Dünisch and Bauch, 1994a). Uptake and transport of mineral nutrients depend on their availability in the soil as well as on the physiological state of the trees (Van de Geijn and Petit, 1979). The path of K, Ca and Mg during uptake into the root and long-range transport into the shoot was followed by multiple stable isotope labelling in intact spruce trees (Kuhn et al., 1995), showing about 60–70% of the Mg and Ca and 30% of the K content in the xylem cell walls originating from labelling solutions. By using optical emission spectrometry Kuhn et al. (1997) further studied the distribution of K, Ca and Mg in spruce stems and found highest contents of K and Mg in the cambium, whereas Ca was dominant in the developing phloem. In more detail the same authors investigated the distribution and incorporation of the three elements in intact de-rooted plants exposed to labelling solutions before analysed by EDXA, LAMMA and SIMS. Results point to a radial element exchange between xylem, cambium and phloem as well as to a relevant bidirectional longitudinal transport. High contents of K indicate rays as being the major reloading point in trees.

During the period of leaf senescence in fall, evidence for a rapid K export out of the leaf lamina into the sieve tubes of the subtending stem could be demonstrated in beech (Eschrich et al., 1988). K, Mg and P are retrieved from the leaf blade prior to shedding, and deposited mainly in cortex and pith tissues of the stem. Until the vegetative dormancy K^+ ions accumulate within rays, from where they can be re-mobilized in spring to be used for wood formation. In spring, the increase in metabolic activity especially in the buds is accompanied by an increase of K^+ and a decrease in inorganic P in both, diffuse-porous and ring-porous trees (Fromm and Eschrich, 1986). During this time action potentials based on K, Ca and Cl-fluxes can be evoked in the phloem (Fromm and Spanswick, 1993). In addition, we found a seasonal variation in K and Ca levels in the poplar cambium with high K levels in summer and a strong reduction of K in winter. Seasonal changes in cambial K content, osmotic potential, and cambial activity correlated strongly throughout the season, increasing from spring to summer and decreasing from summer to autumn (Wind et al., 2004). When we analysed the distribution of K in actively growing twigs, we found the highest K concentrations in the cambium/xylem differentiation zone and lower levels in the mature xylem as well as the phloem (Langer et al., 2002). This pattern was most pronounced in plants grown in nutrient solution with 10 mM K, rather than with those supplied with 0.05 mM K. Thus, changes in K content between different tissues appear to depend on the K supply. A similar result was obtained in differentiating xylem cells; plants grown at low K concentrations showed no significant differences in K content between differentiating fibres, vessels and cambial cells, while those with optimal K supply had the highest K levels in the vessels (Langer et al., 2002). Furthermore, the zone of expanding xylem cells was threefold larger when trees were grown in 10 mM K versus 0.05 mM K. Coinciding with the narrow xylem elongation zone in the K^+ starved trees was an earlier initiation of secondary cell walls.

To prove a possible effect of K supply on cell enlargement rooted cuttings of *P. trichocarpa* were cultivated either on 1 mM or 11 mM K during the time of active cambial growth. The different levels of K supplied in the nutrient solution were reflected by different K levels in the cambial region as measured by EDXA (Langer et al., 2002). The effect of K on the expansion of xylem cells derived from the cambium was investigated by measurements of the lumen area of newly-formed vessels and fibres. Vessels showed a distinct tendency to have an increased lumen area with increasing K levels in the nutrient solution. After treatment of the stem with TEA, a K^+ channel blocker, the vessel-lumen area did not expand as much as untreated stems. In contrast to vessels, the lumen area of newly-formed fibres was not affected by K nutrition or TEA treatments (Langer et al., 2002). Due to the key role that K plays in the action of the osmo-hydraulic system, the strong vacuolisation of active cambial cells may be a structural prerequisite for effective expansion of differentiating vessels (Arend and Fromm, 2003). A possible effect of abscisic acid (ABA) on K dependent cambial growth was investigated and revealed that ABA strongly decreased the K content within the cambial zone and reduced cambial activity, as well as the number of expanding cambial cell derivatives (Wind et al., 2004).

Concerning the role of K in phloem transport we found that the electrical properties of the SE/CC complex are dominated by a K^+ conductance. In search for a respective K^+ channel involved in phloem transport we cloned VfK1 (*Vicia faba* K^+ channel, Ache et al., 2001). Since fructose rather than sucrose or glucose feeding via the petiole induced VfK1 gene activity, we measured the fructose sensitivity of the sieve tube potential through cut aphid stylets and found that the potential was dominated by a K^+ conductance. To investigate the role of K in phloem transport of poplar the aphid stylet technique will also be applied to this tree species in future.

3. MOLECULAR ANALYSIS OF POPLAR K^+ TRANSPORTERS

After searching the EST database from the cambium of *P. tremula* x tremuloides (Sterky et al., 1998) for sequence homologies to known K^+ transporters we identified the following DNA fragments with homologies to the channels and carriers of Arabidopsis: SKOR, the gene for the outward rectifier expressed in endodermis and xylem parenchyma cells (Gaymard et al., 1998), the phloem channel AKT2/3 (Ache et al., 2001, Deeken et al., 2000, Lacombe et al., 2000, Marten et al., 1999), the guard-cell channel of the KAT1-type (Anderson et al., 1992), and AtTrh which mediates K^+ transport (Rigas et al., 2001). The corresponding full-length cDNAs were cloned and the homologues were named PTORK (*P. tremula* outward rectifying K^+ channel), PTK2 (*P. tremula* K^+ channel 2), KPT1 and PtKUP1 (*P. tremula* K^+ uptake transporter), respectively. The deduced proteins showed all structural features of members of the "green" *Shaker* channel family (Hedrich and Becker, 1994). While KPT1 is associated with K^+ uptake during stomatal opening and bud development (Langer et al., 2004), PTORK and PTK2 are involved in xylem and phloem K^+ transport of young poplar twigs. PTK2 was predominantly found in the phloem, PTORK was detected in both phloem and xylem fractions. Therefore, we focused on their functional properties during wood formation (Langer et al., 2002).

Using the double-electrode voltage-clamp technique gene-products of PTORK and PTK2 cRNA's were analysed 3–5 days after injection into *Xenopus oocytes* (cf. Ache et al., 2000, Geiger et al., 2002). In PTORK-expressing *oocytes* we found that membrane depolarization elicited an outward rectifying current with a slow sigmoidal activation kinetic (Langer et al., 2002). Measurements further showed that PTORK is under control of the membrane potential and external K^+ concentration. It shared closest functional similarities with SKOR and enables K^+ release in a voltage- and K-dependent manner. In contrast to PTORK PTK2 shared the basic features of its related *Arabidopsis Shaker* family channel, the weakly voltage-dependent AKT2/3. Like its counterpart in *Arabidopsis* PTK2 is able to mediate both uptake and release of K^+ in response to changes in membrane potential, calcium and pH (Langer et al., 2002). PtKUP1 was expressed in a K uptake-deficient *Escherichia coli* strain, where it rescued K-dependent growth.

In order to localize the site of PTORK, PTK2 and PtKUP1 expression, mRNA was isolated from various tissues for quantitative RT-PCR analyses. Highest amounts of PTORK and PTK2 transcripts were detected in the petioles and the phloem while PtKUP1 seems to be expressed ubiquitous at low levels (Langer et al., 2002). Furthermore, to link the different K^+ transporters to seasonal changes of wood formation, we compared the expression profiles throughout the year. All transcript levels were low during cambial dormancy in winter. In contrast to PtKUP1 which remained at low levels throughout the seasons suggesting a housekeeping function, PTORK and PTK2 expression was induced at temperatures above 10–15°C which reactivate the cambium and initiate wood formation. Since PTORK and PTK2 transcripts accompany the annual K^+ changes in poplar branches, they should be involved in the regulation of K^+-dependent wood formation.

To study the biophysics of poplar ion-channels *in vivo* patch-clamp studies were performed on isolated protoplasts from PTORK and PTK2 expressing suspension cultures (Langer et al., 2002). Therefore, poplar branches were induced to build callus and the resulting meristematic tissues were used to generate suspension cultures. Protoplasts were isolated and the plasma membrane K^+ conductances were compared to the electrical properties of *Xenopus oocytes* expressing PTORK and PTK2 individually. Concerning PTORK it was shown that the properties of this channel are similar in both experimental systems and also to other plant depolarization-activated K^+ release channels (Ache et al., 2000, Gaymard et al., 1998). With regard to PTK2, its voltage dependency in protoplasts differed from that measured in PTK2 expressing *oocytes* where inward rectification was weak in contrast to PTK2-expressing poplar cells. The fact that functional Shaker K^+ channels are formed by four alpha subunits (MacKinnon, 1991), and that members of different subfamilies are able to form hetero-tetramers (Daram et al., 1997, Dreyer et al., 1997, Ehrhardt et al., 1997) might explain the opposite features of PTK2. We, therefore, suggest that poplar suspension cells might express an additional K^+ channel alpha subunit which transforms PTK2 into an inward rectifier.

4. LOCALISATION OF POPLAR ION TRANSPORTERS

Using a specific antibody against PTORK on tissue sections of poplar branches we were able to localise this channel by fluorescence microscopy (Arend et al., 2005). The specifity of the antibody against PTORK was checked by western-blot analysis of *Xenopus oocytes* expressing the channels. PTORK appears in the plasma membrane of sieve elements of the phloem as well as in differentiating fibres and xylem rays where it is mainly active when a contact cell is adjacent to a vessel. The activity is restricted to the period of cambial activity and is not detectable in winter, due to reduced K^+ transport during dormancy. Since the appearance of PTORK is correlated to the period of wood formation, the function of PTORK could be a release of K^+ from xylem ray parenchyma and sieve elements.

In addition, the plasma membrane H^+-ATPase which also might play a fundamental role in the physiology of cambial growth was localised in the poplar stem by using monoclonal as well as polyclonal antibodies (Arend et al., 2002, Arend et al., 2004). In coincidence with the activity of the K^+ channels strong immunoreactivity was visualised by the monoclonal antibody and was restricted to the active growth period in spring and summer. Only a slight activity was found in cross sections and tissue homogenates in autumn and winter. The PM H^+-ATPase was localised in the cambial zone, in differentiating xylem cells and in ray cells surrounding vessels in the mature xylem. Immunogold labeling showed that the enzyme accumulated in the plasma membrane of the labeled cells. Auxin applied to dormant plants induced the formation of this PM H^+-ATPase in the wood formation zone which correlates to elder findings showing that auxin is known to stimulate H^+ excretion (Talbott et al., 1988, Lüthen et al., 1990, Hager et al., 1991). In addition, since "auxin genes" have been found during cell elongation (Abel et al., 1994, Abel and Theologis, 1996) and high concentrations of endogenous auxin are present in young shoots and twigs (Little and Savidge, 1987) as well as in the cambium (Sundberg et al., 2000), the activation of dormant plants by auxin points to a genetic upregulation of PM H^+-ATPase.

Using a polyclonal antibody, further PM H^+-ATPases were localised in the inner phloem and in cells of the cortex parenchyma (Arend et al., 2002). The different labeling pattern given by both antibodies indicated that different isoforms of the enzyme are present in poplar. Molecular studies also show that the PM H^+-ATPase is encoded by a multigene family in plants, indicating that isoforms exhibit a distinct pattern of tissue-specific expression (Palmgren, 2001). Under conditions of low K supply an increased abundance of PM H^+-ATPases occurred, and immunolabelling experiments showed that this increase was restricted to vessel-associated cells (VACs) of the wood ray parenchyma (Arend et al., 2004). Measurements of extracellular H^+ concentrations using non-invasive, H^+-selective microelectrodes revealed a lowering of the pH at the surface of VACs and an enhancement of net efflux of H^+ in plants grown with low K^+ supply. These results indicate an upregulation of specific isoforms of the PM H^+-ATPase in VACs under K-deficient conditions and suggest a key role for these PM H^+-ATPases in unloading K^+ from the xylem stream.

Taken together, the occurrence of PM H^+-ATPase in active twigs indicates that the enzyme plays a fundamental role in growth and differentiation of wood. It generates the necessary proton-motive force for the uptake of K and nutrients into expanding cells, a process that seems to be regulated by auxin.

5. CONCLUSIONS AND PERSPECTIVES

Being a renewable ressource of highly economic significance, wood and its utilisation is in these days of ecological awareness gaining more and more importance. According to an analysis done by the Food and Agriculture Organisation (FAO) there will be a worldwide 25% increase of the need for wood between 1996 and 2010 (Whiteman and Brown, 1999). Apart from the traditional fields of use, like wood as construction material, source of energy, and for the pulp and paper industry, new fields of application are developing, e.g. wood-based ceramics (Hofenauer et al., 2003) or biogenic adhesives (Roffael et al., 2000).

Because of its highly complex chemical composition (cellulose, hemicellulose and lignin) wood can also be used for the production of plastic and other sorts of chemical materials, as well as for the production of food and also textiles. Therefore, wood is an ideal raw material for a developing ligno-chemical industry able to replace products of the petro-chemical industry like plastics in future (Plomion et al., 2001).

One of the most intensively studied wood components is lignin which provides rigidity to the cell walls, causes water impermeability to vessels and forms a chemical barrier against microbial attack (Monties, 1989, Northcote, 1989, Moerschbacher et al., 1990). Its distribution in wood cell walls is best to be determined by high-resolution electron micoscropic techniques (Fromm et al., 2003). However, for the pulp and paper industry it is considered as a negative factor, because residual lignin in wood fibres causes discolouration of the pulp and reduces pulp quality (Chiang et al., 1988). Since the chemical pulping methods are known to be environmently unfriendly and need a high energy input, it was a big task to improve wood towards the paper making process by generating transgenic trees, which either have a lower lignin content or a modified lignin composition that is easier to be seperated from cellulose during Kraft-pulping (Chen et al., 2001). Apart from these developments another biotechnological challenge is to genetically transform trees in order to get a specific wood structure and so be able to modify distinct wood properties (strength, density) as required for various products (Fig. 1). We were able to show that nutrient availability in general, and K nutrition especially, does affect the structure of wood (Langer et al., 2002, Wind et al., 2004). Therefore the generation of transgenic trees with modified K^+ channels could be a promising strategy to alter wood structure in order to improve the individually demanded properties of wood. For example, in order to optimise construction wood trees with overexpressed PTORK channels presumably have more cell wall material per area, leading to a higher wood density and strength.

REFERENCES

Abel, S., Oeller, P.W. and Theologis, A. (1994). Early auxin-induced genes encode short-lived nuclear proteins. *Proc. Natl. Acad. Sci. USA*, 91, 326–330.

Abel, S. and Theologis, A. (1996). Early genes and auxin action. *Plant Physiol.*, 111, 9–17.

Ache, P., Becker, D., Ivashikina, N., Dietrich, P., Roelfsema, M.R.G. and Hedrich, R. (2000). GORK, adelayed outward rectifier expressed in guard cells of *Arabidopsis thaliana*, is a K^+-selective, K^+-sensing ion channel. *FEBS Lett.*, 486, 93–98.

Ache, P., Becker, D., Deeken, R., Dreyer, I., Weber, H., Fromm, J. and Hedrich, R. (2001). VFK1, *a Vicia faba* K channel involved in phloem unloading. *Plant J.*, 27, 571–580.

Anderson, J.A., Huprikar, S.S., Kochian, L.V., Lucas, W.J. and Gaber, R.F. (1992). Functional expression of a probable *Arabidopsis thaliana* potassium channel in *Saccharomyces cerevisae*. *Proc. Natl. Acad. Sci. USA*, 89, 3736–3740.

Arend, M. and Fromm, J. (2000). Seasonal variation in the K, Ca and P content and distribution of plasma membrane H^+-ATPase in the cambium of *Populus trichocarpa*. In: Savidge, R.A., Barnett, J.R. and Napier, R. (eds). *Cell and Molecular Biology of Wood Formation*, pp.67–70. BIOS Scientific Publishers Ltd., Oxford.

Arend, M. and Fromm, J. (2003). Ultrastructural changes in cambial cell derivatives during xylem differentiation in poplar. *Plant Biol.*, 5, 255–264.

Arend, M., Weisenseel, M.H., Brummer, M., Osswald, W. and Fromm, J. (2002). Seasonal changes of plasma membrane H^+-ATPase and endogenous ion current during growth in poplar plants. *Plant Physiol.*, 129, 1651–1663.

Arend, M., Monshausen, G., Wind, C., Weisenseel, M.H. and Fromm, J. (2004). Effect of potassium deficiency on the plasma membrane H^+-ATPase of the wood ray parenchyma in poplar. *Plant, Cell Environ.*, 27, 1288–1296.

Arend, M., Stinzing, A., Wind, C., Langer, K., Latz, A., Ache, P., Fromm, J. and Hedrich, R. (2005). Polar-localised poplar K^+ channel capable of controlling electrical properties of wood-forming cells. *Planta*, 223, 140–148.

Bauer, C.S., Hoth, S., Haga, K., Phillipar, K., Aoki, N. and Hedrich, R. (2000). Differential expression and regulation of K^+ channels in the maize coleoptile: molecular and biophysical analysis of cells isolated from the cortex and vasculature. *Plant J.*, 24, 139–145.

Brüggemann, L., Dietrich, P., Becker, D., Dreyer, I., Palme, K. and Hedrich, R. (1999). Channel-mediated high-affinity K^+ uptake into guard cells from *Arabidopsis*. *Proc. Natl. Acad. Sci. USA*, 96, 3298–3302.

Chen, C., Baucher, M., Christensen, J.H. and Boerjan, W. (2001). Biotechnology in trees: towards improved paper pulping by lignin engineering. *Euphytica*, 118, 185–195.

Chiang, V.L., Puumala, H., Takeuchi, H. and Eckert, R.E. (1988). Comparison of softwood and hardwood kraft pulping. *Tappi J.*, 71, 173–176.

Daram, P., Urbach, S., Gaymard, F., Sentenac, H. and Cherel, I. (1997). Tetramerization of the AKT1 plant potassium channel involves its C-terminal cytoplasmic domain. *EMBO J.*, 16, 3455–3463.

Deeken, R., Sanders, C., Ache, P. and Hedrich, R. (2000). Developmental and light-dependent regulation of a phloem-localised K^+ channel of *Arabidopsis thaliana*. *Plant J.*, 23, 285–290.

Deeken, R., Fromm, J., Koroleva, O., Ache, P., Langenfeld-Heyser, R., Geiger, D., Sauer, N., May, S.T. and Hedrich, R. (2002). Loss of the AKT2/3 potassium channel affects sugar loading into the phloem of *Arabidopsis*. *Planta*, 216, 334–344.

Dreyer, I., Antunes, S., Hoshi, T., Müller-Röber, B., Palme, K., Pongs, O., Reintanz, B. and Hedrich, R. (1997). Plant K^+ channel a-subunits assemble indiscriminately. *Biophys. J.*, 72, 2143–2150.

Dünisch, O. and Bauch, J. (1994a). Influence of mineral elements on wood formation of old growth spruce (*Picea abies* L. *Karst.*). *Holzforschung*, 48, 5–14.

Dünisch, O. and Bauch, J. (1994b). Influence of soil substrate and drought on wood formation of spruce (*Picea abies* L. *Karst.*) under controlled conditions. *Holzforschung*, 48, 447–457.

Dünisch, O., Bauch, J., Müller, M. and Greis, O. (1998). Subcellular quantitative determination of K and Ca in phloem, cambium and xylem cells of spruce (*Picea abies* L. *Karst.*) during earlywood and latewood formation. *Holzforschung*, 52, 582–588.

Ehrhardt, T., Zimmermann, S. and Müller-Röber, B. (1997). Association of plant K^+ in channels is mediated by conserved C-termini and does not affect subunit assembly. *FEBS Lett.*, 409, 166–170.

Eklund, L. and Eliasson, L. (1990). Effects of calcium ion concentration on cell wall synthesis. *J. Exp. Bot.*, 41, 863–867.

Eschrich, W., Fromm, J. and Essiamah, S. (1988). Mineral partitioning in the phloem during autumn senescence of beech leaves. *Trees*, 2, 73–83.

Fromm, J. and Bauer, T. (1994). Action potentials in maize sieve tubes change phloem translocation. *J. Exp. Bot.*, 45, 463–469.

Fromm, J. and Eschrich, W. (1986). Changes of adenine nucleotide and orthophosphate concentrations in buds of deciduous trees during spring activation. *Trees*, 1, 42–46.

Fromm, J. and Spanswick, R. (1993). Characteristics of action potentials in willow (*Salix viminalis*). *J. Exp. Bot.*, 44, 1119–1125.

Fromm, J., Rockel, B., Lautner, S., Windeisen, E. and Wanner, G. (2003). Lignin distribution in cell walls determined by TEM and backscattered SEM techniques. *J. Structural Biol.*, 143, 77–84.

Fu, H.H. and Luan, S. (1998). AtKUP1: a dual-affinity K^+ transporter from *Arabidopsis*. *Plant Cell*, 10, 63–73.

Gassman, W., Rubio, F. and Schroeder, J.I. (1996). Alkali cation selectivity of the root high-affinity potassium transporter HKT1. *Plant J.*, 10, 852–869.

Gaymard, F., Pilot, G., Lacombe, B., Bouchez, D., Bruneau, D., Boucherez, J., Michaux-Ferriere, N., Thibaud, J.-B. and Sentenac, H. (1998). Identification and disruption of a plant *shaker*-like outward channel involved in K^+ release into the xylem sap. *Cell*, 94, 647–655.

Geiger, D., Becker, D., Lacombe, B. and Hedrich, R. (2002). Outer pore residues control H^+- and K^+-sensitivity of the *Arabidopsis* potassium channel AKT3. *Plant Cell*, 14, 1859–1968.

Hager, A., Debus, G., Edel, H.G., Stransky, H. and Serrano, R. (1991). Auxin induces exocytosis and the rapid synthesis of a high-turnover pool of plasma-membrane proton ATPase. *Planta*, 185, 527–537.

Hedrich, R. and Becker, D. (1994). Green circuits - The potential of plant specific ion channels. *Plant Mol. Biol.*, 26, 1637–1650.

Hedrich, R. and Roelfsema, M.-R.G. (1999). Plant ion transport. In: *Encyclopedia of Life Sciences*. Available: http://www.els.net; Macmillan Reference Ltd.

Hirsch, R., Lewis, B.D., Spalding, E.P. and Sussmann, M.R. (1998). A role for the AKT1 potassium channel in plant nutrition. *Science*, 280, 4806–4810.

Hofenauer, A., Treusch, O., Gahr, M., Schmidt, J., Tröger, F., Krenkel, W., Wegener, G. and Fromm, J. (2003). Dense reaction infiltrated SiSiC-ceramics derived from wood-based materials. *Advanced engineering materials*, 5(11), 794–799.

Ivashikina, N., Becker, D., Ache, P., Meyerhoff, O., Felle, H.H. and Hedrich, R. (2001). K^+ channel profile and electrical properties of Arabidopsis root hairs. *FEBS Lett.*, 25503, 1–7.

Kim, E.J., Kwak, J.M., Uozumi, N. and Schroeder, J.I. (1998). AtKUP1: an *Arabidopsis* gene encoding high-affinity potassium transport activity. *Plant Cell*, 10, 51–62.

Kuhn, A.J., Bauch, J. and Schröder, W.H. (1995). Monitoring uptake and contents of Mg, Ca and K in Norway spruce as influenced by pH and Al, using microprobe analysis and stable isotope labelling. *Plant Soil*, 168–169, 135–150.

Kuhn, A.J., Schröder, W.H. and Bauch, J. (1997). On the distribution and transport of mineral elements in xylem, cambium and phloem of spruce (*Picea abies* L. *Karst.*). *Holzforschung*, 51, 487–496.

Lacombe, B., Pilote, G., Michard, E., Gaymard, F., Sentenac, H. and Thibaud, J.B. (2000). A *shaker*-like K^+ channel with weak rectification is expressed in both source and sink phloem tissues of *Arabidopsis*. *Plant Cell*, 12, 837–851.

Langer, K., Ache, P., Geiger, D., Stinzing, A., Arend, M., Wind, C., Regan, S., Fromm, J. and Hedrich, R. (2002). Poplar potassium transporters capable of controlling K^+ homeostasis and K^+-dependent xylogenesis. *The Plant J.*, 32, 997–1009.

Langer, K., Levchenko, V., Fromm, J., Geiger, D., Steinmeyer, R., Lautner, S., Ache, P. and Hedrich, R. (2004). The poplar K^+ channel KPT1 is associated with K^+ uptake during stomatal opening and bud development. *The Plant J.*, 37, 828–838.

Larson, P.R. (1967). Assessing wood quality of fertilized coniferous trees. Forest Fertilization Symposium, Gainsville, 275–280.

Little, C.H.A. and Savidge, R.A. (1987). The role of plant growth regulators in forest tree cambial growth. *Plant Growth Regul.*, 6, 137–169.

Lüthen, H., Bidgon, M. and Böttger, M. (1990). Reexamination of the acid growth theory of auxin action. *Plant Physiol.*, 93, 931–939.

MacKinnon, R. (1991). Determination of the subunit stoichiometry of a voltage-activated potassium channel. *Nature*, 350, 232–235.

Marten, I., Hoth, S., Deeken, R., Ache, P., Ketchum, K.A., Hoshi, T. and Hedrich, R. (1999). AKT3, a phloem-localized K^+ channel, is blocked by protons. *Proc. Natl. Acad. Sci. USA*, 96, 7581–7586.

Moerschbacher, B.M., Noll, L., Gorrichon, L. and Reisener, H.-J. (1990). Specific inhibition of lignification breaks hypersensitive resistance of wheat to stem rust. *Plant Physiol.*, 93, 465–470.

Monties, B. (1989). Lignins. In: P.M. Dey and J.B. Harborne (eds), *Methods in Plant Biochemistry: Plant Phenolics*, Vol. 1, pp. 113–157. Academic Press, New York.

Northcote, D.H. (1989). Control of plant cell wall biosynthesis: an overview. In: N.G. Lewisand M.G. Paice (eds), *Plant Cell Wall Polymers* (ACS Symposium Series), Vol. 399, pp. 1–15. American Chemical Society, Washington, D.C.

Palmgren, M.G. (2001). Plant plasma membrane H^+-ATPases: powerhouses for nutrient uptake. *Annu. Rev. Plant Physiol. Plant Mol. Biol.*, 52, 817–845.

Plomion, C., Leprovost, G. and Stokes A. (2001). Wood formation in trees. *Plant Physiol.*, 127, 1513–1523.

Reintanz, B., Szyroki, A., Ivashikina, N., Ache, P., Godde, M., Becker, D., Palme, K. and Hedrich, R. (2002). AtKC1, a silent *Arabidopsis* potassium channel a-subunit modulates root hair K^+ influx. *Proc. Natl. Acad. Sci. USA*, 99, 4079–4084.

Rigas, S., Debrosses, G., Haralampidis, K., Vicente-Agullo, F., Feldmann, K., Grabov, A., Dolan, L. and Hatzopoulos, P. (2001). Trh1 encodes a potassium transporter required for tip growth in *Arabidopsis* root hairs. *Plant Cell*, 13 (1), 139–151.

Rodriguez-Navarro, A. (2000). Potassium transport in fungi and plants. *Biochim. Biophys. Acta*, 1469, 1–30.

Roffael, E., Dix, B. and Okum, J. (2000). Use of spruce tannin as a binder in particleboards and medium density fiberboards (MDF). *Holz als Roh- und Werkstoff*, 58, 301–305.

Schroeder, J.I. and Fang, H.H. (1991). Inward rectifying K^+ channels in guard cells provide a mechanism for low-affinity K^+ uptake. *Proc. Natl. Acad. Sci. USA*, 88, 11583–11587.

Spalding, E.P., Hirsch, R.E., Lewis, D.R., Qi, Z., Sussmann, M.R. and Lewis, B.D. (1999). Potassium uptake supporting plant growth in the absence of AKT1 channel activity. *J. Gen. Physiol.*, 113, 909–918.

Sterky, F., Regan, S., Karlsson, J. et al. (1998). Gene discovery in the wood-forming tissues of poplar: analysis of 5692 expressed sequence tags. *Proc. Natl. Acad. Sci. USA*, 95,13330–13335.

Sundberg, B., Uggla, C. and Tuominen, H. (2000). Cambial growth and auxin gradients. In: Savidge, R.A., Barnett, J.R. and Napier, R. (eds), *Cell and Molecular Biology of Wood Formation*, pp. 169–188. Oxford: BIOS Scientific Publishers Ltd.

Talbott, L.D., Ray, P.M. and Roberts, J.K.M. (1988). Effect of indole acetic acid- and fusicoccin-stimulated proton extrusion on internal pH of pea internode cells. *Plant Physiol.*, 87, 211–216.

Van de Geijn, S.C. and Petit, C.M. (1979). Transport of divalent cations. Cation exchange capacity of intact xylem vessels. *Plant Physiol.,* 64, 954–958.

Wardrop, A.B. (1981). Lignification and xylogenesis. In: J.R. Barnett, (ed.) *Xylem Cell Development.* Castle House Publ. Ltd., London. 115–155.

Whiteman, A. and Brown, C. (1999). The potential role of forest plantations in meeting future demands for industrial wood products. *International Forestry Review,* 1(3), 143–152.

Wind, C., Arend, M. and Fromm, J. (2004). Potassium-dependent cambial growth in poplar. *Plant Biol.*, 6, 30–37.

TRANSPORT CHARACTERISTICS OF ION CHANNELS AS INFLUENCED BY APOPLASTIC PROPERTIES

K^+ channels of the phloem

P. ACHE and R. DEEKEN

Molekulare Pflanzenphysiologie und Biophysik, Germany, ache@botanik.uni-wuerzburg.de

Abstract. Our studies demonstrate that members of the *AKT2/3* subfamily exhibit unique features among the plant *Shaker*-like K^+ channels. They show weak voltage dependence and rectification, are blocked by calcium or protons and all members investigated so far are phloem located. *AKT2/3*-type channels are involved in the control of sugar and potassium (K) translocation in the phloem from source to sink. The membrane potential (E_m) is controlled by the channels via a voltage clamp near the equilibrium potential for K^+ (E_K), to maintain sugar/H^+ co-transport. This function could be demonstrated by co-expression of *AKT2/3*-type channels with sucrose transporters in *Xenopus* oocytes and confirmed the findings in *planta* performed by the aphid stylectomy technique and analysis of an *AKT2/3*-loss-of-function mutant. Furthermore, channel regulation *in vivo* is accomplished by changes in the number and activity of K^+ channels in response to changes in the sink/source relation, extarcellular sugar concentration and pH. Some members of this channel subfamily are highly expressed in the source phloem, while others are restricted to the sink, indicating specialised functions in different plants, e.g. rosette plants (*Arabidopsis*), leguminoses (*Vicia faba*) or perennial plants like poplar.

Key words: AKT 2/3, electrophysiology, K^+ channels, phloem loading, sucrose transport

1. INTRODUCTION

1.1 Plant potassium channels

The family of *Shaker*-like K^+ channels from *Arabidopsis thaliana* consists of six transmembrane domains (S1 to S6) and a pore region (P) located between S5 and S6 (Hedrich and Roelfsema, 1999). Sequence

B. Sattelmacher† and W.J. Horst (eds.), The Apoplast of Higher Plants: Compartment of Storage, Transport and Reactions, 151–163.

alignments of the nine K^+ channel genes, present in *Arabidopsis*, resulted in a phylogenetic tree that branches into five subfamilies with different electrical and physiological properties (Mäser et al., 2001). Subfamilies were named after the first *Arabidopsis* member identified (*KAT1*, *AKT1*, *AKT2/3*, *AtKC1*, and *SKOR*). There are three major functional groups within the five subfamilies of Shaker-like K^+ channels:

a) The K^+-uptake channels (6 members), represented by *KAT1*, *KAT2*, *AKT1*, *AKT5*, *SPIK* and *AtKC1*, which are voltage-dependent and, with the exception of *AtKC1*, acid-activated, provide a molecular pathway for K uptake (Schachtman et al., 1992, Hedrich et al., 1995, Müller-Röber et al., 1995, Mäser et al., 2002, Pilot et al., 2001, Szyroki et al., 2001, Moulin et al., 2002, Reintanz et al., 2002). They open in response to membrane hyperpolarisation.
b) The K^+-release channels (2 members), represented by *GORK* and *SKOR*, are activated upon depolarisation in a K-dependent manner (Gaymard et al., 1998, Ache et al., 2000, Ivashikina et al., 2001)
c) The *AKT2/3* family (1 member) of largely voltage-independent channels can act as a K^+-uptake or K^+-release systems. In contrast to the inward rectifiers, this type of channel is blocked by extracellular protons and Ca^{2+}. All of the channels so far investigated in different plant species are phloem-located (Marten et al., 1999, Philippar et al., 1999, Bauer et al., 2000, Deeken et al., 2000, Ache et al., 2001, Langer et al., 2002)

In the following we focus on the biophysical and physiological properties of the *AKT2/3* channel type.

1.2 Phloem features and K^+ transport

The phloem presents a network for assimilate allocation and retrieval of minerals (Pate and Jeschke, 1995, Marschner et al., 1996), as well as a pathway for chemical and electrical communication within the plant (Fromm and Bauer, 1994). K^+ is the major cation in the phloem and stimulates sugar loading into the phloem sap (Peel and Rogers, 1982, Fromm and Eschrich, 1989). Phloem loading coincides with an increase in symplastic K^+ concentration, likely to maintain electrical neutrality which is required for creating a pH gradient (Marschner et al., 1997). In addition, the K^+ concentration in the sieve tube may affect the volume flow rate in the phloem (Mengel, 1980). The membrane potential that transiently changes during phloem-propagating action potentials is possibly repolarised on K^+ release from the sieve tube (Fromm and Bauer, 1994). In search for K^+

channels involved in the control of sugar loading into, or unloading from the sieve element/companion-cell complex (SE/CC), we identified members of the *AKT2/3* family expressed in the vascular system of *Arabidopsis thaliana*, *Zea mays*, *Vicia faba* and *Populus tremula* x tremuloides (Marten et al., 1999, Philippar et al., 1999, Ache et al., 2001, Langer et al., 2002). Out of the 11 *Arabidopsis thaliana Shaker*-like K^+ channels only KAT1 (in cv. Wassilewskija, Ivashikina et al. 2003) or KAT2 (in cv. Columbia, Pilot et al., 2001) were together with *AKT2/3* remarkably expressed in *Arabidopsis* phloem protoplasts.

2. LOCALISATION OF *AKT2/3* TYPE CHANNELS

We have localized the plant K^+ channel *AKT2/3* in the phloem of *Arabidopsis thaliana*, using whole mount in situ hybridisation (Marten et al., 1999). Promoter–GUS studies of *AKT2/3* have confirmed these findings (Deeken et al., 2000, Lacombe et al., 2000). *ZMK2* transcripts and currents were identified from a phloem protoplast fraction of *Zea mays* seedlings, derived from vascular strands of mesocotyl and coleoptile (Bauer et al., 2000). VFK1, a *Vicia faba* K^+ channel, was detected in the phloem by in situ hybridisation and in situ RT-PCR (Ache et al., 2001). Finally, quantitative RT-PCR of phloem-enriched fractions of poplar stems revealed *PTK2* gene expression predominately in the phloem (Langer et al., 2002). The role of this phloem K^+ channel family in phloem physiology was further studied with respect to transcriptional regulation under different physiological conditions.

3. GENE REGULATION

Since the *Arabidopsis thaliana AKT2/3* transcripts were predominantly found in the phloem of the green parts of the shoot, including sepals of the flower, a source-specific function was proposed for this K^+ channel, implying a role in phloem loading (Deeken et al., 2000). During the light period, transcript levels gradually increased, peaked around noon, and dropped again in the afternoon and night. This light induction was CO_2-dependent, indicating that photosynthate induces *AKT2/3* transcription. Further evidence that the expression of *AKT2/3* is activated in light-treated tissues only came from studies where one half of the rosette was illuminated and the other half shaded, or alternatively rosettes were completely illuminated while the inflorescence stalk was shaded. The fact that *AKT2/3* transcription is exclusively induced in the phloem of illuminated parts of the plant provides evidence for local gene activation rather than a signal travelling to shaded parts via the sieve elements. In contrast to the source-specific regulation of the *AKT2/3* gene, transcripts of the orthologs *VFK1* from *Vicia faba* and *ZMK2* from *Zea mays* accumulate under sink conditions, indicating a role in phloem unloading (Ache et al., 2001, Philippar et al., 2003). Upon experimental source-sink transitions and during the filling process of cotyledons with nutrients, the *Vicia faba* phloem K^+ channel was induced. *ZMK2* expression was restricted to growing young leaves and the base of mature leaves. Transcripts accumulated during the dark period, pointing to a function of this phloem channel in sugar unloading. In order to reveal the characteristics of the *AKT2/3*-type K^+ channels in phloem physiology we have analysed the electrical properties of the different proteins.

4. ELECTRICAL PROPERTIES

Biophysical properties of *AKT2/3*-like K^+ channels were analysed using heterologous expression and mutagenesis analysis, while the role of K^+ channel transport was investigated in wild-type and mutant *Arabidopsis* plants. The voltage-dependent gating of *AKT2/3* differs markedly from strong inward rectifiers which activate negative of −80 mV only and thus do not conduct K^+ in the depolarised state (Marten et al., 1999). In contrast, the voltage-dependence of *AKT2/3* in the range of +50 to −170 mV is much less pronounced, leaving a significant fraction of *AKT2/3* channels open at depolarised voltages. As a result, this channel is able to conduct both inward and outward currents. The instantaneous and time-dependent K^+ current components of the *AKT2/3* channel protein expressed in *Xenopus* oocytes were sensitive to K^+ channel blockers and correspond to two different gating modes (Dreyer et al., 2001). Instead of the membrane potential, modifiers of *AKT2/3* K^+ transport capacity are extracellular calcium and protons (Marten et al., 1999). Cell elongation and phloem transport are accompanied by changes of H^+ and Ca^{2+} concentrations in the apoplast. In exudation saps of various mono- and dicotyledons, the apoplastic Ca^{2+} concentration can reach millimolar ranges (up to 5 mM). A rise of external Ca^{2+} up to this physiological range inhibited AKT2/3-conducted K^+ currents at voltages negative to −150 mV (Marten et al., 1999). The voltage dependence of the Ca^{2+} block indicates that the Ca^{2+} binding site may be located within the pore of AKT2/3. In addition to Ca^{2+}, protons modulate the AKT2/3 channel. Extracellular acidification reduced the macroscopic K^+ conductance level, whereas the K^+ inward rectifiers of guard cells from potato (*KST1*), *Arabidopsis* (*KAT1*), and the channel from broad bean are activated by extracellular acidification. In every case, the pH regulation was observed in the physiological range (Hedrich et al., 1995, Hoth et al., 1997, Brüggemann et al., 1999, Marten et al., 1999, Ache et al., 2001). In guard cells which express several different *Shaker* channel types, a calcium-sensitive and highly caesium-sensitive inward rectifier represents the dominant inward K^+ conductance (Szyroki et al., 2001). K^+ currents of guard cells, however, lacking the *AKT2/3* subunit were no longer blocked by external calcium ions, indicating that *AKT2/3* subunits provide for sensitivity towards this channel inhibitor (Ivashikina et al., 2005).

In addition to *AKT2/3* from *Arabidopsis*, all other phloem-localised K^+ channels characterised by heterologous expression in *Xenopus* oocytes and in *planta*, have shared most of the electrophysiological characteristics. *VFK1* from *Vicia faba*, *ZMK2* from *Zea mays*, and *PTK2* from *Populus tremula* x tremuloides exhibit the same basic properties (Philippar et al., 1999, Bauer et al., 2000, Ache et al., 2001, Langer et al., 2002). The localisation in the phloem and the electrical properties of the *AKT2/3* family (Table 1) point to their role in phloem physiology.

Table 1. Features of AKT2/3 type channels compared to KAT1.

	AKT2/3	*VFK1*	*ZMK2*	*PTK2*	*KAT1*
phloem localisation	yes	yes	yes	yes	no
preferred expression	source	sink	sink	source/sink	guard cell
voltage dependency	weak	weak	weak	weak	strong
apoplastic Cs^+ block	yes	n.d.	yes	yes	yes
apoplastic Ba^{2+} block	yes	yes	yes	yes	yes
apoplastic H^+ block	yes	yes	yes	yes	no
apoplastic Ca^{2+} block	yes	yes	yes	yes	no

5. POST TRANSLATIONAL REGULATION

Compared to the weak voltage-dependence of *PTK2* injected oocytes, K^+ channels of *PTK2*-expressing poplar cells exhibited strong inward rectification (Langer et al., 2002). The voltage-dependent calcium block and high sensitivity to caesium were observed also with *AKT2/3*. This feature suggests that functional *Shaker* K^+ channels are formed by four alpha subunits, and members of different subfamilies are able to form hetero-tetramers (Dreyer et al., 1997). Baizabal-Aguirre et al. (1999) confirmed the observations that the voltage-dependence of heteromers was dominated by the strong inward rectifier *KAT1* when it was co-expressed with *AKT2/3* in *Xenopus* oocytes. In addition to the membrane potential, external protons, and Ca^{2+} ions, the *AKT2/3* channel activity further depends on cytoplasmic factors as indicated by the rundown behaviour following patch excision (Marten et al., 1999). Two-hybrid screens have shown, that *AKT1* is able to interact with *AKT2/3* or *AtKC1* (Pilot et al., 2003), and the protein phosphatase, *At*PP2CA with *AKT2/3* (Vranová et al., 2001). When co-expressed with *AKT2/3*, this phosphatase promotes inward rectification of *AKT2/3* homomers (Cherel et al., 2002). In order to understand the molecular basis for the different electrical properties of *AKT2/3*, the protein structure in relation to its function was studied in more detail.

6. STRUCTURE AND FUNCTION RELATIONS

The phloem-localised *AKT2/3* is a structural homolog of the K^+ uptake channels *AKT1* and *KAT1*, showing 60% amino acid identity in the S1–S6 region (Ketchum and Slayman, 1996). A histidine residue conserved in the outer pore region of all cloned inward-rectifying K^+ channels was identified to be the major component of the acid activation mechanism of the potato guard-cell K^+ channel *KST1*, but not of *KAT1* from *Arabidopsis* (Hoth et al., 1997, Hoth and Hedrich, 1999). In addition to the pore histidine, another extracellular histidine in the S3–S4 linker and an aspartate in the pore region contribute to pH sensing in *KST1*. In contrast, pH sensitivity in *KAT1* depends on residues other than histidines (Hoth and Hedrich, 1999). The vascular associated K^+ inward rectifier *KZM1* from *Zea mays* represents the first plant K^+ channel insensitive to external pH changes (Philippar et al., 2003). When compared to *KST1*, *KZM1* possesses an arginine residue instead of histidine in the S3-S4 linker. If *KST1* is mutated accordingly, this channel also becomes pH-insensitive (Hoth et al., 1997), indicating the importance of this extracellular histidine in pH sensing among different plant species. The inverted pH sensitivity of the phloem expressed *AKT2/3*-type channel suggests a different molecular mechanism which accounts for proton sensitivity. Using site-directed mutagenesis followed by heterologous characterization in *Xenopus* oocytes, Geiger et al. (2002) showed that the pH-mediated decrease in single-channel conductance observed in *AKT2/3*-WT is lost in *AKT2/3* mutants. Thus, a histidine residue at position 228 in the linker domain between S5 and the pore and a serine at position 271 in the ascending loop of the pore could be identified to be involved in proton sensing of *AKT2/3*. In addition to pH sensitivity, the serine at position 271 confers K sensitivity to the phloem expressed K^+ channel, indicating that protonation of these peripheral amino acids in the outer pore controls K^+-dependent K^+ currents. In addition to the regulation of phloem channel protein activity by extra-cellular factors we studied the effect of a loss-of-channel mutation on phloem sugar-transport.

7. REGULATION OF SUGAR TRANSPORT

Sucrose/H^+ symporters mediate electrogenic, proton-coupled transport of sucrose into the SE/CC complex (Stadler et al., 1996) and are voltage-dependent as shown by Zhou et al. (1997) and Boorer et al. (1996). The proton gradient and voltage drop across the membrane is generated by H^+ pumps (Langhans et al., 2001), which coexist with the sucrose/H^+ symporter (Stadler and Sauer, 1996) and K^+ channels (Bauer et al., 2000, Deeken et al., 2000, Pilot et al., 2001) in the plasma membrane of phloem cells. Sugar uptake into the phloem of barley leafs led to a strong depolarisation of about 30 mV (Fromm and Eschrich, 1989). In the phloem, *AKT2/3* channels seem to counteract sucrose-induced depolarisation. At potentials positive to the activation threshold of strong inward rectifiers (> –80 mV, *KAT1* or *KAT2*), *AKT2/3* maintains the K^+-dependent membrane potential when challenged with sucrose. In an in vitro experiment we have demonstrated that only *AKT2/3* – but neither of the strong rectifiers *KAT1* or *GORK*, was able to stabilise the membrane potential during sucrose uptake (Deeken et al., 2002). When the phloem-specific sucrose/H^+ symporter, *AtSUC2* from *Arabidopsis* was expressed in *Xenopus leavis* oocytes alone, a sucrose-induced depolarisation of the membrane was detected. But in oocytes, which expressed both, the phloem K^+ channel *AKT2/3*, and the sucrose/proton symporter, *AtSUC2*, the drop in membrane potential during uptake of sucrose was prevented. In contrast to *AKT2/3*, *KAT2* which is also expressed in the phloem behaved almost like a K^+ electrode and did not prevent membrane depolarisation upon sucrose uptake. Complete blocking of *AKT2/3* with Ba^{2+} or TEA^+, however, recovered the sucrose-dependent depolarisation (Deeken et al., 2002). This indicates that *in vivo* the phloem K^+ channel *AKT2/3* re-polarises the membrane rather than catalysing the bulk flow of K^+. The latter function could be provided by the strong inward rectifier, *KAT2*. Similar results were obtained by characterisation of the sugar-import machinery of maize. Philippar et al. (2003) showed that ZmSUT1 represents a sucrose/H^+ symporter that acts under the voltage control of the *AKT2/3* ortholog *ZMK2*. As was found with *KAT2,* the inward rectifier *KZM1* of *Zea mays* was not able to prevent depolarisation of the membrane potential during sugar uptake of *Xenopus* oocytes.

In contrast to *AKT2/3,* but in agreement with *ZMK2*, the ortholog *VFK1* from *Vicia faba* is found in sink tissues as well as during the transition of tissues from source to sink suggesting a role in phloem unloading (Ache et al., 2001, Philippar et al., 2003). In sink tissues at the sites of phloem unloading, sucrose is released. In most plants, an apoplastic invertase converts the disaccharide into glucose and fructose, which are the substrates of monosaccharide transporters for uptake into the sink cells. Using the aphid stylectomy technique (Wright and Fischer, 1981, Pritchard, 1996) we were able to correlate the phloem K^+ conductance with the rate of sugar loading and unloading in *Arabidopsis thaliana* and *Vicia faba* (Ache et al., 2001, Deeken et al., 2002). In *Vicia faba*, depolarisation in response to a 10-fold change in the apoplastic K^+ concentration indicated that K^+ transporters represent a major ionic conductance in the SE/CC complex. Under *VFK1*-inducing conditions, mimicking unloading by the presence of fructose, however, K^+ channels predominantly controlled the phloem membrane-conductance.

8. LOSS OF PHLOEM K^+ CHANNEL MUTANTS

In order to unravel the regulatory role of the *AKT2/3*-type of phloem-expressed K^+ channels in loading and long-distance transport of photoassimilates, we characterised the *akt2/3-1* loss-of-function mutant of *Arabidopsis thaliana* (Deeken et al., 2002). Application of the aphid stylectomy technique revealed differences in membrane potential of the SE/CC complex between *akt2/3-1* and wild-type plants. The resting potential of the mutant phloem membrane-complex was more positive than that of wild-type plants and the change in K^+ conductance was reduced upon a 10-fold increase of K^+ applied through the apoplast. As a consequence, the sugar content of the *akt2/3-1* phloem sap was only half of that measured in wild-type plants, indicating a defect in phloem sucrose-loading. The impaired carbon allocation of the *akt2/3-1* mutant was visualised by the distribution of ^{14}C-labelled fixation products, which were detected by microautoradiography. In cross sections of inflorescence stalks lacking the *AKT2/3* phloem channel, ^{14}C-labelled photoassimilates were predominately localised in chlorenchyma cells, whereas in wild-type plants they were restricted to the phloem. It has been demonstrated that during long-distance transport of sucrose from source to sink, photosynthates tend to leak away from the sieve tubes (loss = 6% per cm; Patrick and Turvey, 1981, Minchin and Thorpe, 1987). Steady retrieval is therefore required to maintain the photosynthate concentrations at levels sufficient to drive the pressure flow and to nourish sink tissues. In the phloem K^+ channel mutant, allocation of sugars is less efficient because of impaired retrieval of sugars along the long distance transport pathway. As a consequence, the high assimilate demand of strong sink tissues such as fast growing inflorescence stalks, flowers and seeds cannot be satisfied. In line with our interpretation, *akt2/3-1* plants generated a reduced number of inflorescence stalks in *akt2/3-1* mutant plants as compared to the wild-type. Furthermore the mutant stalks were shorter but had an increased diameter. In addition, the number of rosette leaves after three weeks in continuous light or ten weeks under short-day conditions was decreased. These findings suggest that the phloem K^+ channel *AKT2/3* could be important for charge balance and osmotic adaptation during assimilate loading and might stimulate sugar loading due to membrane repolarisation.

9. CONCLUSION

Comparing the findings on phloem-expressed K^+channels with data from the literature, we finally proposed a model for the contribution of K^+ channels to phloem transport processes (Fig. 1, Ache et al., 2001). In *Vicia faba* different apoplastic and intracellular K^+ concentrations or SE/CC membrane potentials (E_m) are found at sites of phloem sugar loading and unloading (Mühling and Sattelmacher, 1997, Walker et al., 2000, van Bel, 1993). The equilibrium potential for K^+ (E_K) is around –60 mV at the source and –120 mV at the sink site and the membrane potentials are around –150 mV (source) and –80 mV (sink). Under these conditions, K is taken up by *VFK1* into the phloem at the source (E_m negative to E_K) and released at the sink site (E_m positive to E_K). Due to the proton symport mechanism, massive sugar allocation, both at the sink or the source lead to strong depolarisation of the phloem E_m (up to 30 mV, Fromm and Eschrich, 1989 and own unpublished results) and a related pH increase in the apoplast. At the loading site, sucrose is taken up by proton symport, whereas at the sink site, it is unloaded by a so far unknown mechanism and in most plants cleaved to fructose and glucose by an acid invertase. Subsequently, uptake of the resulting hexoses into the adjacent sink cells is also accomplished by H^+ symporters, which leads also to a H^+ loss in the apoplast. The elevated pH causes activation of *AKT2/3* type channels, which now will clamp the membrane potential close to E_K and thus guarantee the membrane potential-dependent sugar allocation. In *Vicia faba*, the continuous presence of fructose induces *VFK1* expression to support higher K^+ channel density in the sink compared to the source.

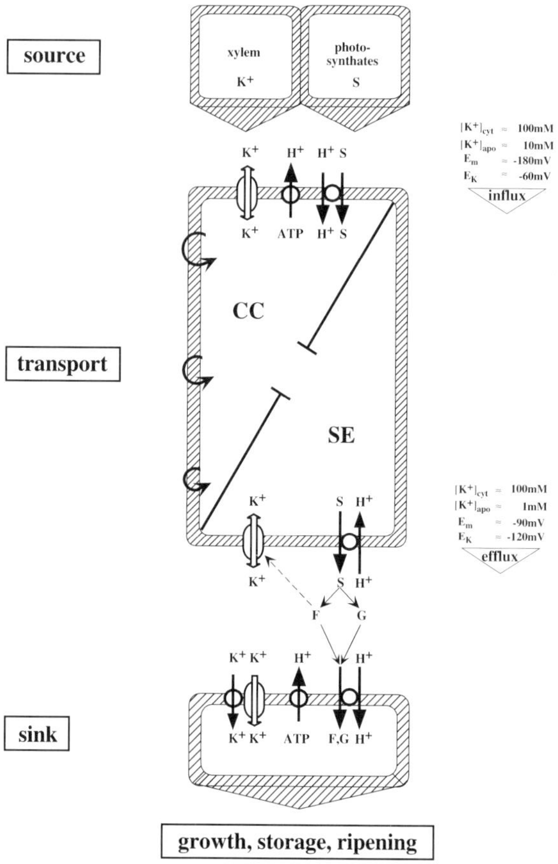

Fig. 1 (Ache et al. *2001). Model: Contribution of K^+ channels to phloem transport processes. The source site is characterized by K^+ concentrations of about 100 mM within the cytoplasm of phloem cells ($[K^+]_{cyt}$) and 10 mM in the apoplast ($[K^+]_{apo}$), resulting in an E_K of around 60 mV. The resting potential (E_m) of phloem cells at the source site is negative to E_K. At the sink site, the K^+ gradient is two orders of magnitude and E_K consequently at around 120 mV. E_m at the sink site is around 90 mV. The dotted line indicates that the number of K^+ channels is increased via fructose-dependent VFK1 expression. Abbreviations: CC=Companion Cell, SE=Sieve Element, F=Fructose, G=Glucose, S=Sucrose.*

Symbols: VFK1-channel transporter, pump, antiporter symporter

REFERENCES

Ache, P., Becker, D., Ivashikina, N., Dietrich, P., Roelfsema, M.R.G. and Hedrich, R. (2000). GORK, a delayed outward rectifier expressed in guard cells of *Arabidopsis thaliana*, is a K^+-selective, K^+-sensing ion channel. *FEBS Letters*, 486, 93–98.

Ache, P., Becker, D., Deeken, R., Dreyer, I., Weber, H., Fromm, J. and Hedrich, R. (2001). VFK1, a *Vicia faba* K^+ channel involved in phloem unloading. *Plant J.*, 27, 571–580.

Baizabal-Aguirre, V.M., Clemens, S., Uozumi, N. and Schroeder, J.I. (1999). Suppression of inward-rectifying K^+ channels KAT1 and AKT2 by dominant negative point mutations in the KAT1 α-subunit. *J. Membr. Biol.*, 167, 119–125.

Bauer, C.S., Hoth, S., Haga, K., Philippar, K., Aoki, N. and Hedrich, R. (2000). Differential expression and regulation of K^+ channels in the maize coleoptile: molecular and biophysical analysis of cells isolated from cortex and vasculature. *Plant J.*, 24, 139–145.

Boorer, K.J., Loo, D.D.F., Frommer, W.B. and Wright, E.M. (1996). Transport mechanism of the cloned potato H^+/Sucrose cotransporter StSUT1. *J. Biol. Chem.*, 41, 25139–25144.

Brüggemann, L., Dietrich, P., Becker, D., Dreyer, I., Palme, K. and Hedrich, R. (1999). Channel-mediated high-affinity K^+ uptake into guard cells from *Arabidopsis*. *Proc. Natl. Acad. Sci. USA*, 96, 3298–3302.

Cherel, I., Michard, E., Platet, N., Mouline, K., Alcon, C., Sentenac, H. and Thibaud, J.-B. (2002). Physical and functional interaction of the *Arabidopsis* K^+ channel AKT2 and Phosphatase AtPP2CA. *Plant Cell*, 14, 1–14.

Deeken, R., Sanders, C., Ache, P. and Hedrich, R. (2000). Developmental and light-dependent regulation of a phloem-localised K^+ channel of *Arabidopsis thaliana*. *Plant J.*, 23, 285–290

Deeken, R., Fromm, J., Koroleva, O., Ache, P., Langenfeld-Heyser, R., Geiger, D., Sauer, N., Bennett, M. J. and Hedrich, R. (2002). Loss of the AKT2/3 Potassium channel affects sugar loading into the phloem of *Arabidopsis*. *Planta*, 216, 334–344.

Dreyer, I., Michard, E., Lacombe, B. and Thibaud, J.-B. (2001). A plant *Shaker*-like K^+ channel switches between two distinct gating modes resulting in either inward-rectifying or "leak" current. *FEBS lett.*, 505, 233–239.

Dreyer, I., Antunes, S., Hoshi, T., Müller-Röber, B., Palme, K., Pongs, O., Reintanz, B. and Hedrich, R. (1997). Plant K^+ channel α-subunits assemble indiscriminately. *Biophys. J.*, 72, 2143–2150.

Fromm, J. and Bauer, T. (1994). Action potentials in maize sieve tubes change phloem translocation. *J. Exp. Bot.*, 45, 463–469.

Fromm, J. and Eschrich, W. (1989). Correlation of ionic movements with phloem unloading and loading in barley leaves. *Plant Physiol. Biochem.*, 27, 577–585.

Gaymard, F., Pilot, G., Lacombe, B., Bouchez, D., Bruneau, D., Boucherez, J., Michaux-Ferriere, N., Thibaud, J.-B. and Sentenac, H. (1998) Identification and disruption of a plant *Shaker*-like outward channel involved in K^+ release into the xylem sap. *Cell.*, 94, 647–655.

Geiger, D., Becker, D., Lacombe, B. and Hedrich, R. (2002). Outer pore residues control the H^+ and K^+ sensitivity of the *Arabidopsis* potassium channel AKT3. *Plant Cell*, 14(8), 1859–1868.

Hedrich, R. and Roelfsema, M.-R. G. (1999). Plant ion transport. In: *Encyclopedia of Life Sciences*, Macmillan Reference Ltd.

Hedrich, R., Moran, O., Conti, F., Busch, H., Becker, D., Gambale, F., Dreyer, I., Küch, A., Neuwinger, K. and Palme, K. (1995). Inward rectifier potassium channels in plants differ from their animal counterparts in response to voltage and channel modulators. *Eur. Biophys. J.*, 24, 107–115.

Hoth, S. and Hedrich, R. (1999). Distinct molecular bases for pH sensitivity of the guard cell K^+ channels KST1 and KAT1. *J. Biol. Chem.*, 274(17), 11599–11603.

Hoth, S., Dreyer, I. and Hedrich, R. (1997). Mutational analysis of functional domains within plant K^+ uptake channels. *J. Exp. Bot.*, 48, 415–420.

Ivashikina, N., Becker, D., Ache, P., Meyerhoff, O., Felle, H. and Hedrich, R. (2001). K^+ channel profile and electrical properties of *Arabidopsis* root hairs. *FEBS Lett.*, 508, 463–469.

Ivashikina, N., Deeken, R., Ache, P., Kranz, E., Pommerrenig, B., Sauer, N. and Hedrich, R. (2003). Isolation of AtSUC2:GFP-marked companion cells for patch-clamp studies and expression profiling. *Plant J.*, 36, 931–945.

Ivashikina, N., Deeken, R., Fischer, S., Ache, P. and Hedrich, R. (2005). AKT2/3 subunits render guard cell K^+ channels Ca^{2+} sensitive. *J. Gen. Physiol.*, 125, 483–492.

Ketchum, K.A. and Slayman, C.W. (1996). Isolation of an ion channel gene from *Arabidopsis thaliana* using the H5 signature sequence from voltage-dependent K^+-channels. *FEBS Lett.*, 378, 19–26.

Lacombe, B., Pilot, G., Michard, E., Gaymard, F., Sentenac, H. and Thibaud, J.B. (2000). A *Shaker*-like K^+ channel with weak rectification is expressed in both source and sink phloem tissues of *Arabidopsis*. *Plant Cell*, 12, 837–851.

Langer, K., Ache, P., Geiger, D., Stinzing, A., Arend, M., Wind, C., Regan, S., Fromm, J. and Hedrich, R. (2002). Poplar potassium transporters capable of controlling K^+ homeostasis and K^+-dependent xylogenesis. *Plant J.*, 32, 997–1009.

Langhans, M., Ratajczak, R., Lützelschwab, M., Michalke, W., Wächter, R., Fischer-Schliebs, E. and Ullrich, C.I. (2001). Immunolocalization of plasma-membrane H^+-ATPase and tonoplast-type pyrophosphatase in the plasma membrane of the sieve tube element-companion cell complex in the stem of *Rhizinus communis* L. *Planta*, 213, 11–19.

Marschner, H., Kirkby, E.A. and Cakmak, I. (1996). Effect of mineral nutritional status on shoot-root partitioning of photoassimilates and cycling of mineral nutrients. *J. Exp. Bot.*, 47, 1255–1263.

Marschner, H., Kirkby, E.A. and Engels, C. (1997). Importance of cycling and recycling of mineral nutrients within plants for growth and development. *Bot. Acta*, 110, 265–273.

Marten I., Hoth, S., Deeken, R., Ache, P., Ketchum, K.A., Hoshi, T. and Hedrich, R. (1999). AKT3, a phloem-localised K^+ channel, is blocked by protons. *Proc. Natl. Acad. Sci. USA*, 96, 7581–7586.

Mäser, P., Thomine, S., Schroeder, J.I., Ward, J.M., Hirschi, K., Sze, H., Talke, I.N., Amtmann, A., Maathuis, F.J., Sanders, D., Harper, J.F., Tchieu, J., Gribskov, M., Persans, M.W., Salt, D.E., Kim, S.A. and Guerinot, M.L. (2001). Phylogenetic relationships within cation transporter families of *Arabidopsis*. *Plant Physiol.*, 126, 1646–1667.

Mengel, K. (1980). Effect of potassium on the assimilate conduction to storage tissue. *Ber. Deutsch. Bot. Ges.*, 93, 353–362.

Minchin, P.E.H. and Thorpe, M.R. (1987). Measurement of unloading and reloading of photoassimilate within the stem of bean. *J. Exp. Bot.*, 38, 211–220.

Mouline, K., Very, A.A., Gaymard, F., Boucherez, J., Pilot, G., Devic, M., Bouchez, D., Thibaud, J.B. and Sentenac, H. (2002). Pollen tube development and competitive ability are impaired by disruption of a *Shaker* K^+ channel in Arabidopsis. *Genes Dev.*, 16, 339–350.

Mühling, K.H. and Sattelmacher, B. (1997). Determination of apoplastic K^+ in intact leaves by ratio imaging of PBFI fluorescence. *J. Exp. Bot.*, 48, 1609–1614.

Müller-Röber, B., Ellenberg, J., Provart, N., Willmitzer, L., Busch, H., Becker, D., Dietrich, P., Hoth, S. and Hedrich, R. (1995). Cloning and electrophysiological analysis of KST1, an inward-rectifying K^+ channel expressed in potato guard cells. *EMBO J.*, 14, 2409–2416.

Pate, J.S. and Jeschke, W.D. (1995). Role of stems in transport, storage and circulation of ions and metabolites by the whole plant. In: Gartner, B., (ed.), *Plant Stems Physiology and Functional Morphology* New York: Academic Press, pp. 177–204.

Patrick, J.W. and Turvey, P.M. (1981). The pathway of radial transfer of photosynthate in decapitated stems of *Phaseolus vulgaris* L. *Ann. Bot.*, 47, 611–621.

Peel, A.J. and Rogers, S. (1982). Stimulation of sugar loading into sieve elements of willow by potassium and sodium salts. *Planta*, 154, 94–96.

Philippar, K., Fuchs, I., Lüthen, H., Hoth, S., Bauer, C.S., Haga, K., Thiel, G., Ljung, K., Sandberg, G., Böttger, M., Becker, D. and Hedrich, R. (1999). Auxin-induced K^+ channel expression represents an essential step in coleoptile growth and gravitropism. *Proc. Natl. Acad. Sci. USA*, 96, 12186–12191.

Philippar, K., Büchsenschütz, K., Abshagen, M., Fuchs, I., Geiger, D., Lacombe, B. and Hedrich, R. (2003). The K^+ channel KZM1 is capable to mediate potassium uptake into the phloem and guard cells of the C4 grass *Zea mays*. *J. Biol. Chem.*, 278, 16973–16981.

Pilot, G., Lacombe, B., Gaymard, F., Cherel, I., Boucherez J., Thibaud, J.-B. and Sentenac, H. (2001). Guard cell inward K^+ channel activity in *Arabidopsis* involves expression of the twin channel subunits KAT1 and KAT2. *J. Biol. Chem.*, 276, 3215–3221.

Pilot, G., Gaymard, F., Mouline, I., Cherel, I. and Sentenac, H. (2003). Regulated expression of *Arabidopsis Shaker* K^+ channel genes involved in K^+ uptake and distribution in the plant. *Plant Mol. Biol.*, 51, 773–787.

Pritchard, J. (1996). Aphid stylectomy reveals an osmotic step between sieve tube and cortical cells in barley roots. *J. Exp. Bot.*, 47, 1519–1524.

Reintanz, B., Ivashikina, N., Szyroki, A., Ache, P., Godde, M., Becker, D., Palme, K. and Hedrich, R. (2002). AtKC1, a silent *Arabidopsis* potassium channel α-subunit modulates root hair K^+ influx. *Proc. Natl. Acad. Sci. USA*, 99, 4079–4048.

Schachtman, D.P., Schroeder, J.I., Lucas, W.J. anderson, J.A. and Gaber, R.F. (1992). Expression of an inward-rectifying potassium channel by the *Arabidopsis* KAT1 cDNA. *Science*, 258, 1654–1658.

Stadler, R. and Sauer, N. (1996). The *Arabidopsis thaliana* AtSUC2 gene is specifically expressed in companion cells. *Bot. Acta*, 109, 299–306.

Szyroki, A., Ivashikina, N., Dietrich, P., Roelfsema, M.R., Ache, P., Reintanz, B., Deeken, R., Godde, M., Felle, H., Steinmeyer, R., Palme, K. and Hedrich, R. (2001). KAT1 is not essential for stomatal opening. *Proc. Natl. Acad. Sci. USA*, 98, 2917–2921.

van Bel, A.J.E. (1993). II. The transport phloem. Specifics of its functioning. *Progr. Bot.*, 54, 134–150.

Vranová, E., Tähtiharju, S., Sriprang, S., Willekens, H., Heino, P., Palva, E.T., Inzé, D. and Van Camp, W. (2001). The AKT3 potassium channel protein interacts with the AtPP2CA protein phosphatase 2C. *J. Exp. Bot.*, 52, 181–182.

Walker, N.A., Zhang, W.-H., Harrington, G., Holdaway, N. and Patrik, J.W. (2000). Effluxes of solutes from developing seed coats of *Phaseolus vulgaris* L. and *Vicia faba* L.: locating the effect of turgor in a coupled chemiosmotic system. *J. Exp. Bot.*, 48, 1047–1055.

Wright, J.P. and Fisher D.B. (1981). Measurement of the sieve tube membrane potential. *Plant Phys.*, 67, 845–848.

Zhou, J.-J., Theodoulou, Sauer, N., Sanders, D. and Miller, A.J. (1997). A kinetic model with ordered cytoplasmic dissociation for SUC1, an Arabidopsis H^+/Sucrose cotransporter expressed in *Xenopus oocytes*. *J. Membrane Biol.*, 159, 113–125.

ION UPTAKE FROM THE XYLEM INTO THE SYMPLASM OF THE MAIZE LEAF

M. ABSHAGEN-KEUNECKE and U.-P. HANSEN
Center of Biochemistry and Molecular Biology, University of Kiel, Germany, uphansen@zbm.uni-kiel.de

Abstract. Whereas the uptake of ions into the xylem roots is well understood, less is known about the uptake from the xylem to the symplast of the leaves. LAMMA experiments using $^{85}Rb^+$ as a tracer showed that ions move over the cell membrane between xylem and bundle-sheath cells of small veins of *Zea maize*. A new isolation technique was developed that enabled the separation of mesophyll cells and bundle-sheath cells of maize leaves. Whole-cell patch-clamp experiments revealed three types of channel, MB-1 and MB-2 in bundle-sheath cells and MM-2 in mesophyll cells. Stimulation by acidic lumenal medium suggested that MB-1 and MB-2 take care of charge balancing during uptake of anions, whereas stimulation at high pH would provide MM-2 with a fail-safe function by clamping membrane potential to E_K, if the H^+ pump is inactive. The distinction between MB-1 and MB-2 was based on current density, temporal behaviour and different conductivity with Rb^+ at 100 mM KCl at the lumenal side. MB-2 requires ATP for activation. Freshly prepared bundle-sheath cells are either of MB-1- or of MB-2-type. The isolation procedure developed for the patch clamp experiments was employed to separate c-DNA from vascular strands from that of mesophyll cells and of epidermis cells. Physiological behaviour, especially pH-dependence, indicated that KZM1 found in the vascular strands originated from the phloem. Thus the genetic identification of MB-1 and MB-2 is still an open question.

Key words: bundle sheath cells, ion uptake, mesophyll cells, pH, rubidium, tracer

1. INTRODUCTION

Nutrient transport from the soil to the symplasm of leaves occurs through a long apoplastic compartment: the xylem; these vascular strands

B. Sattelmacher† and W.J. Horst (eds.), The Apoplast of Higher Plants: Compartment of Storage, Transport and Reactions, 165–180.

extend from the roots to the leaves. The transport through this apoplastic compartment is still a matter of debate. Solvent flow through the xylem powered by evaporation from the stomata is assumed in the cohesion theory (Tyree, 1997; Steudle, 2002, Section 2), whereas Zimmermann et al. (2002, section 4) postulated an important role of active transport processes.

At either end, membrane transport has to be involved in the movement of nutrients out of and into the symplast of the surrounding cells. Expression of the involved transporters is subject to tissue-specific and time-dependent regulation (Butt et al., 1997; Czempinski et al., 1999; Yamada and Bohnert, 2000; Furbank et al., 2001; Shi et al., 2002). Increasing awareness of the importance of the apoplast (Sattelmacher, 2001) stimulated tissue-specific analysis in order to understand how transporters fulfil the functions imposed by the contact with a special apoplastic compartment. In roots, uptake into the xylem has been studied by several workers. They assigned a crucial role to xylem parenchyma cells for the translocation of cations and anions from the symplast to the xylem (Amtmann et al., 1999; Köhler and Raschke, 2000; Wegner et al., 1994 and Section 4, this volume) and considered the participation of multiple gene families in the response to cation stress (Maathuis et al., 2003). The ion-motive forces for all of these transport processes are established by a H^+-ATPase (Jahn et al., 1998; Kinoshita and Shimazaki, 2002, and Sections 3 and 5, this volume).

At the other end of the vascular strands, in the leaves, research has been focussed on phloem transport. Especially, cation channels like AKT2/3 in *Arabidopsis* (Deeken et al., 2002) or KZM1 in *Zea mays* (Philippar et al., 2003), and sucrose/H^+ symports like ZmSUT1 in *Zea mays* (Aoki et al., 1999; Bauer et al., 2000; Philippar et al., 2003) involved in loading of photosynthates have been investigated in great physiological and genetic detail. (Further details on phloem transport are given in Section 3 and 5, this volume).

A less clear picture is available for the transfer from the xylem to the symplast of the leaf. First results dealing with cation channels were obtained in *Zea mays*. A C4 plant with Krantz antomy provides special benefits for these studies as the apoplast of the vascular strands and that of the bulk leaf (embedding the mesophyll cells and guard cells) are strictly separated by a suberin layer (Evert et al., 1985, 1996; Canny, 1995) surrounding the bundle-sheath cells (see Fig. 1). However, this suberin layer is also a severe obstacle that has prevented the study of the transport properties of bundle-sheath cells. This problem has been overcome by a new isolation procedure which enables the separation of mesophyll cells and bundle-sheath cells (Keunecke et al., 1997).

2. VASCULATURE OF THE MAIZE LEAF

2.1 Organisation of the vascular apoplast in maize leaves

Servicing cells engaged in various leaf processes is achieved by the vascular system of the leaf (Russell and Evert, 1985; Nelson and Dengler, 1997). The highly ordered, striate array of veins consists of different types of veins with different functions. The large veins transport photosynthates out of the leaf blade, small and intermediate bundles collect these photosynthase and are engaged in phloem loading. Transverse veins transport photosynthates from the smaller to the larger veins (Fritz et al., 1983; Altus and Canny, 1985).

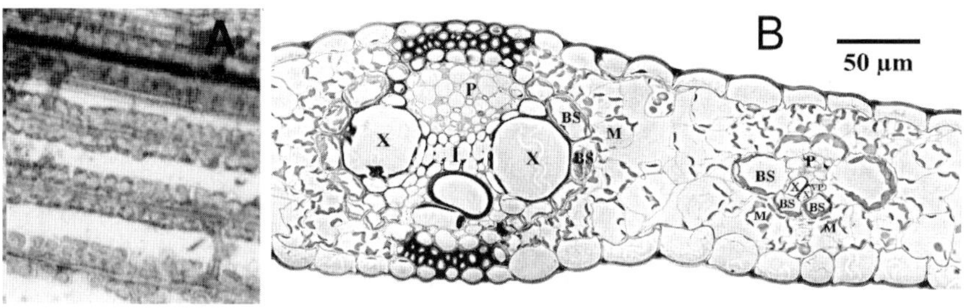

Fig. 1. (A). Vascular system of a maize leaf after the mesophyll cells have been removed by the method of Keunecke et al. (1997) showing a large vein at the top, three small veins and transverse veins between them. (B) Section of a maize leaf showing a large vein and a small vein. (micrograph made by Inge Dörr, Botany, Kiel). X-xylem vessel, P = phloem, I = intermediate cells, BS = bundle-sheath cells, M = mesophyll cells, VP = vascular parenchyma cells. (B) shows the typical arrangement of cells in and outside a minor (right-hand side) and a large (left-hand side) vein. The apoplasts of the veins are surrounded by a ring of bundle-sheath cells (BS). Inside the ring are the xylem (X), the phloem (P) and intermediate cells (I). The arrangement of mesophyll cells (M) around a ring of bundle-sheath cells (BS) is typical of C4 grasses. The bundle-sheath cells are surrounded by an impermeable suberin layer resulting in a complete separation of the apoplasts in the veins and of the apoplast of the mesophyll cells (Evert et al., 1977, 1985, 1996, Hattersley and Browning, 1981). Differentiation of photosynthetic functions in bundle-sheath cells and mesophyll cells provides C4 plants, living in hot and arid regions, with the benefit of "turbo loading" of CO_2. In the mesophyll cells, CO_2 is bound to phosphoenolpyruvate. The bound CO_2 is delivered via different biochemical carriers (Heldt, 1999) to the Calvin cycle in the bundle-sheath cells. The resulting requirement of lower CO_2 concentrations allows partial closing of the stomata and reduces water loss. The CO_2 carriers and the photosynthates can flow through plasmodesmata (Weiner et al., 1988) between mesophyll and bundle-sheath cells. Thus, also nutrients entering the bundle-sheath cells can proceed to the symplast of the leaf via these plasmodesmata.

Fig. 1A shows the veins of a maize leaf which was subject to the isolation procedure of Keunecke et al. (1997). The mesophyll cells are washed out and the frame work of veins become visible.

3. LOCATING THE UPTAKE FROM THE VASCULAR APOPLAST

3.1 First attempts using dyes

Fig. 1B shows that the xylem (X) has direct contact to bundle-sheath cells (BS). However, also cells inside the ring of BS are in contact with the xylem. They, too, may be involved in the uptake of nutrients from the xylem. The first estimate of where nutrients enter the symplast was obtained from the distribution of dyes. Altus and Canny (1985) and Canny (1990a, b, c; 1993) concluded from the accumulation of the dye sulphorhodamine G "on the inner tangential cell walls of the parenchyma-sheath cells of small and intermediate veins" that, there, water had entered the symplast of maize leaves and had left the dye behind (Canny, 1990a, b, c). In dicotyledon leaves, bundle-sheath cells of the smallest veins are named explicitely as the locus of entry to the symplast (Canny, 1993). The same was found by Evert et al. (1985) for Prussian blue crystals and lanthanum. Especially, the heavy deposit of lanthanum in the cell walls of bundle-sheath cells in small and intermediate veins indicated uptake at this location. Interestingly, GUS-staining of PIP (plasma membrane intrinsic protein) aquorin promoter MipA labeled water channels in minor, but not in major veins of tobacco (Yamada and Bohnert, 2000).

Accumulation of ions (K^+, Cl^-, Ca^{2+}) in small veins of dicotyledon leaves measured by cryo-analytical scanning electron microscopy was less than that of the impermeant dye. This led Canny (1993) to the suggestion that ions were taken up with the water. Sugar, in contrast, traced by ^{14}C sucrose entered the vascular parenchyma and then the thick-walled sieve tubes, but did not accumulate in the bundle-sheath cells (Fritz et al., 1983). The results of Canny (1990a, b, c) and Fritz et al. (1983) are not a contradiction. Using the sed1 mutant of maize, Russin et al. (1996) showed that uptake of nutrients in mutants can occur without functional plasmodesmata between bundle-sheath cells and vascular parenchyma cells, but export of carbohydrates from the bundle-sheath cells required functional plasmodesmata. This indicates different pathways for import and export.

3.2 Tracing ion uptake in maize by $^{85}Rb^+$

K^+ is the most abundant ion in the xylem sap (Canny 1995a; Mühling and Sattelmacher, 1995), and thus K^+ is the first choice when locus and mechanism of nutrient uptake from the xylem to the symplast is to be investigated.

Rb^+ has a long tradition as a tracer for K^+ fluxes in plants (Hirsch et al., 1998; MacRobbie, 1995). In animal cells, it was used to monitor K^+ fluxes not only through channels (Salva and Matteson, 1991), but also through cotransporters (KCC, Shen et al., 2000) and pumps (Na^+/K^+, Gaur et al., 2000). Nevertheless, it had to be tested whether Rb^+ enters the symplast of bundle-sheath cells via K^+ channels, especially as Hedrich et al. (1995) have found a K^+ uptake channel in corn coleoptiles that is not permeant to Rb^+. Also blocking of K^+ channels by Rb^+ is known from many inward rectifier K^+ channels (Thompson and Begenisich, 2001). Nevertheless, the experimental results shown below justify the use of Rb^+ as a tracer of K^+ (Fig. 5).

Fig. 2 presents the results of laser microprobe mass analysis (LAMMA) for measuring local ion concentrations with a resolution of 1 μm. Measurements were done when the front of the $^{85}Rb^+$ flow was reaching the investigated small veins of maize 22.5 min after immersion of the leaf into a 20 mM $^{85}Rb^+$ solution. In small veins (Fig. 2A), the high accumulation in the cell walls between xylem and bundle-sheath cells (X-BS) and between adjacent bundle-sheath cells (BS-BS) indicates that in the case of small veins

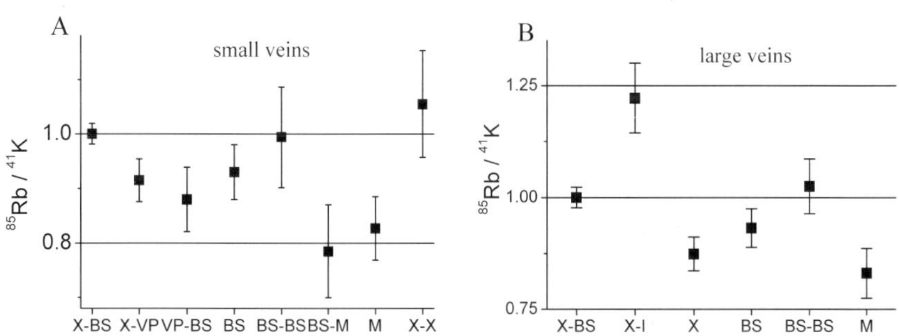

Fig. 2. Relative concentration of $^{85}Rb^+$ at different loci in (A) small and (B) large veins of maize, 22.5 min. after the leaf was dipped into a solution of 20 mM RbCl + 1 mM KCl, pH 7.2. Single letters denote the center of the related compartment (Fig. 1) and letters connected with a hyphen denote the cell wall between two compartments. The ratio in X-BS is normalized to 1.0 (Keunecke et al., 2001). The value of X is low because the aqueous interior is lost.

of maize there is a direct transport via the xylem/bundle-sheath cell interface, thus confirming the suggestions of Canny (1990a). In large veins (Fig. 2B), $^{85}Rb^{+}$ is also enriched in BS cells before it reaches mesophyll cells. However, a very high value is found in the wall of intermediate cells (X-I). It may be speculated that this accumulation is related to phloem loading for recycling of cations to the root (Marschner et al., 1997; Engels and Kirkby, 2001) or to the cambrium (Canny, 1995b).

4. CHANNELS AT THE INTERFACE XYLEM – BUNDLE-SHEATH CELL

4.1 Physiological identification of K^{+}-channels for uptake from the xylem

Bundle-sheath cells and also mesophyll cells (for comparison) were isolated by the method described by Keunecke et al. (1997) and Keunecke and Hansen (2000). Briefly, a dissected leaf was put into a syringe filled with 10 ml tap water and exposed to positive pressure (which worked better than negative pressure). After infiltration, the leaf was cut longitudinally and exposed to 1.5% cellulase, 0.5% macerozyme (Yakult Honsha), 2.0% pectinase (Sigma), 0.5% BSA (Sigma), 5 mM $CaCl_2$, 0.5 Mol sorbitol, puffered to pH 6.5 (Tris/Mes). After 6 h at room temperature or 2 h at 37°C, the leaf was stretched using needles; like this they stayed for another 30 min in the enzyme solution. Bundle-sheath cells remaining at the vascular bundles and free mesophyll cells were separated by a washing procedure.

Whole-cell patch clamp experiments on bundle-sheath cells and mesophyll cells revealed three types of channels. Identification was based on their temporal behaviour and on pH dependence as shown in Fig. 3.

Whole-cell recordings like those in Fig. 3 represent the sum of currents from several transporters. However, the dependence on K^{+} concentration, as shown in Fig. 4A for MB-2, indicates that the major component results from a K^{+} channel. At 10 mM KCl, the Nernst potentials for K^{+} and Cl^{-} are zero, and the reversal potential is also zero as expected. Increasing the K^{+} concentration on the cytosolic side to 100 mM makes the reserval potential shift by about 40–50 mV. This is less than the expected 60 mV (Nernst potential). Thus, it has to be assumed that in whole-cell experiments the conductivity of unidentified other channels, probably chloride channels, contributes to the IV curves besides the obvious K^{+} channel. However, also in single-channel experiments, a slope of less than 60 mV/decade was

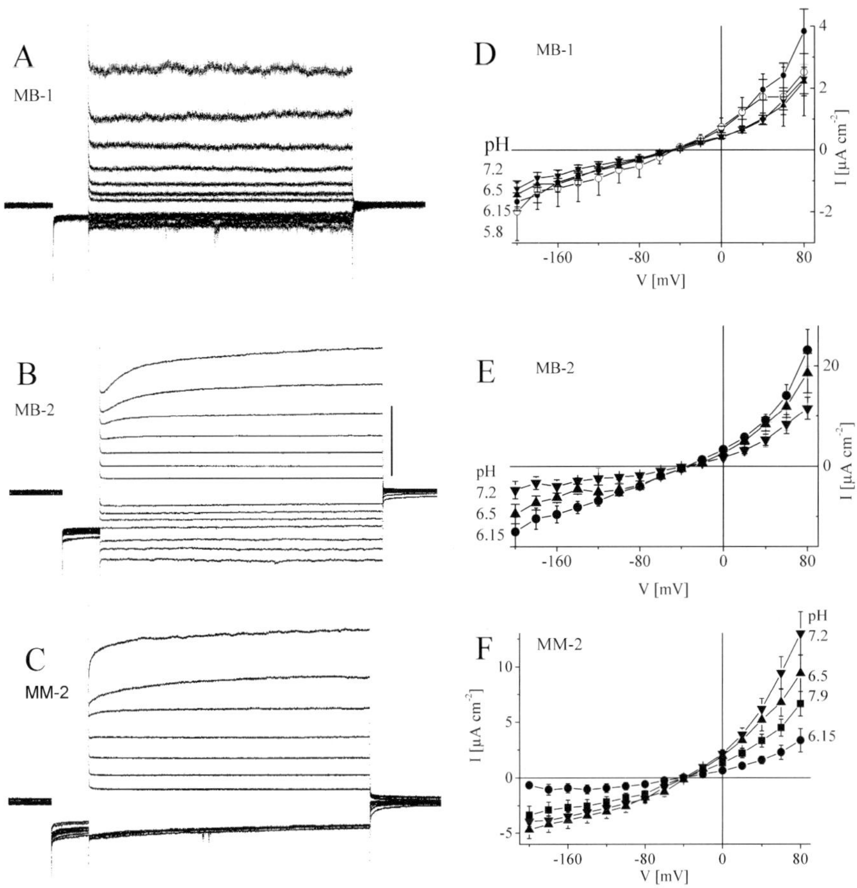

Fig. 3. Time series (left column) and pH-dependencies (right column) of MB-1, MB-2 and MM-2. The bathing medium was: 10 mM KCl, 2 mM $MgCl_2$, 5 mM $CaCl_2$, 450 mOsm (sorbitol), pH 5.8 to 7.9 (Tris/Mes), and the pipette medium was: 100 mM KCl, 2 mM $MgCl_2$, 5.6 mM $CaCl_2$, 10 mM EGTA, 2 mM MgATP, 450 mOsm (Sorbitol), pH 7.2 (Tris/MES).

found (Keunecke et al., 1997) indicating that the channels may also conduct other ions.

Experiments on MB-2 often showed a run-down phenomenon, i.e. current decreases several minutes after perfusing the cytoplasm with pipette medium. Fig. 4B shows that this resulted from ATP depletion. The requirement of ATP raises the question of whether MB-2 is connected to a SUR transporter (Leonhardt et al., 1997). This ABC transporter often confers ATP sensitivity to K^+ channels.

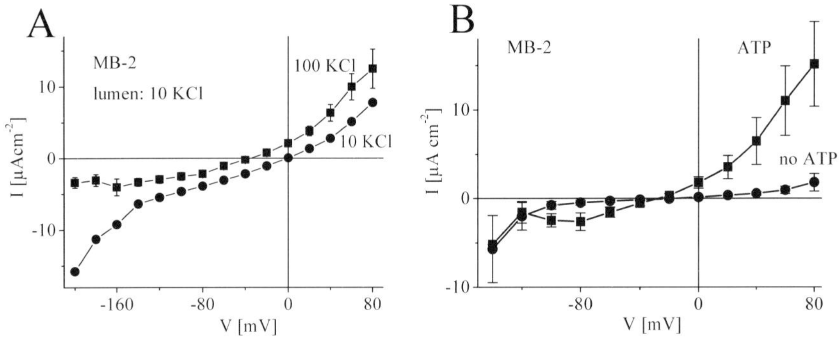

Fig. 4. (A) Dependence of IV-curves of MB-2 on cytosolic K^+ concentration. IV-curves obtained at two different cytosolic K^+ concentrations at pH 7.2. (B) Dependence of MB-2 on cytosolic ATP.

4.2 Manifesting the distinction between MB-1 and MB-2

Keunecke and Hansen (2000) based the distinction between MB-1 and MB-2 mainly on different temporal behaviour and on current density. More convincing support for this distinction came from experiments with Rb^+, which were done in order to check whether it was legitimate to use Rb^+ as a tracer of K^+ in the LAMMA-studies in Fig. 2.

The effect of replacing 10 mM K^+ by 10 mM Rb^+ was tested with MB-1, MB-2 and MM-2 at cytosolic concentrations of 10 and 100 mM KCl. In MB-2 and MM-2 there was no effect on the reversal potential. Fig. 5C and D show that in MB-2 the current was increased, and in MM-2 it was decreased. Similar results were obtained at cytosolic 10 mM KCl (not shown). The interesting effects occurred in MB-1. At a cyctosolic concentration of 10 mM KCl, Rb^+ reduced the current, but also here there was no effect on the reversal potential (Fig. 5A). With 100 mM KCl on the cytosolic side, a dramatic shift in reversal potential occurred (Fig. 5B). The origin of the shift is not quite clear. A shift towards the reversal potential of Rb^+ (plus infinity) may indicate a change in selectivity from K^+ to Rb^+. However, the reversal potential $E_r = E_{K,Rb} = E_{Cl} = 0$ mV in Fig. 5A, and the move towards E_{Cl} (+60 mV) in Fig. 5B seem to indicate a Rb^+-induced closure of the K^+-channel (Hedrich et al., 1995) and an opening of the chloride channel. Whatever the mechanism is, the results in Fig. 5 clearly demonstrate that MB-1 and MB-2 are really different.

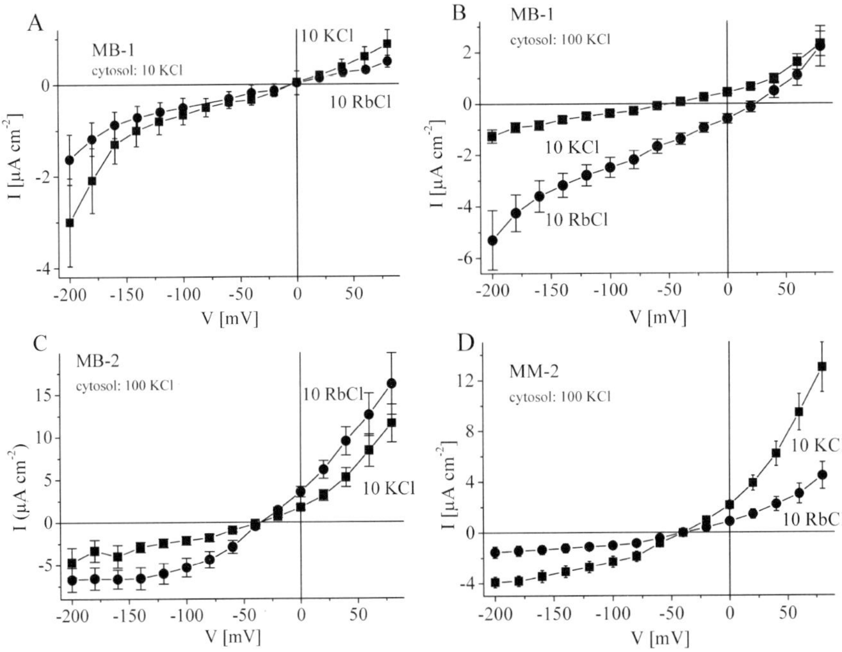

Fig. 5. Influence of 10 mM RbCl on reversal potential of MB-1 at cytosolic concentration of (A) 10 mM KCl and (B) 100 M KCl. Other ingredients as given in the legend of Fig. 1.

4.3 Suggested physiological function of MB-1, MB-2 and MM-2

The most intriguing finding in Fig. 3 is the inverse pH-dependence of the channels in bundle-sheath cells (MB-1 and MB-2) and in mesophyll cells (MM-2). We supposed that this is strongly related to their physiological function. The border bundle-sheath cells/xylem is the locus of the uptake of salts comprising an anion (preferably NO_3^-) and a cation (preferably K^+). The anion can reach the negative interior of the bundle-sheath cell only by means of a $2H^+$/anion-cotransporter (Sanders and Hansen, 1981). Acidification in the veins as a prerequisite of proton-driven uptake has been shown by Canny (1987) and Mühling and Läuchli (2000). As illustrated in the scheme of Fig. 6, a P-ATPase has to transfer $2H^+$ to supply the cotransporter with H^+. For the sake of charge balance, 1 K^+ has to enter via a K^+ channel, driven by membrane potential. Thus, 1 KNO_3 is taken up. The establishment of a driving force for the cotransporter leads to an acidification of the xylem. It is an efficient strategy to open the channel for charge balance only during transport activity of the cotransporter. Constant activity would otherwise be an unneccessary load on the proton pump and would waste energy. Fig. 3 shows that MB-1 and MB-2 fulfill these requirement as they open when acidification signals cotransport activity.

The involved cotransporter is still unknown. The only cotransporter that has been genetically identified in maize cells so far, is ZmSUT1 (Aoki et al., 1999), a H^+/sugar symporter in the phloem. However, the actual type working in bundle-sheath cells remains to be identified. Maybe, it is similar to the H^+/anion cotransporter involved in xylem loading in maize roots (McClure et al., 1990).

In mesophyll cells, the situation seems to be different. Similar to phloem-localized AKT3 (Marten et al., 1999), protons block the MM-2. Obviously, there is little need in mesophyll cells for anion uptake by cotransport from the apoplast. Thus, MM-2 has another function. It is long known that the P-ATPase has periods of low and high activities. In the case of low activity, absolute membrane potential has to be kept at a minimum value, usually E_K. Thus, opening of MM-2 could fulfill a fail-safe function when alkalinisation of the apoplast signals low pump activity. A similar function is suggested for the ZMK2 (Philippar et al., 1999, 2003) during phloem loading.

4.4 Single-channel experiments

Nernst potential E_K and reversal potential of the IV curves in Figs. 3 and 4 did not coincide, indicating that the characteristics of the K^+ channels are partially hidden by the presence of other transporters in whole-cell experiments. Thus, inside-out excided patches were used to study single-channel behavior. In the first set of experiments (Keunecke et al., 1997), two types of channel were found in bundle-sheath cells, one channel with an

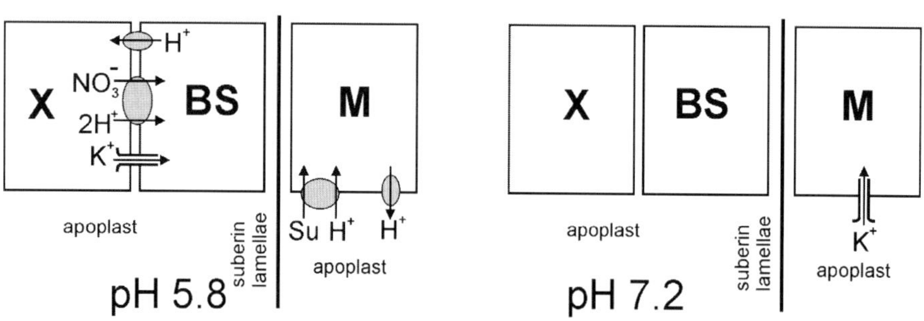

Fig. 6. Suggested physiological functions of bundle sheath (BS) and mesophyll (M) transporters depending on pH (Keunecke and Hansen, 2001).

Ohmic IV curve and one inward rectifier (IR). Single-channel conductivity determined by fit-per eye or an automatic level detector (Riessner et al., 2002) resulted in single-channel conductivities of about 20 pS.

In order to find a correspondence between whole-cell experiments and single-channel experiments, both kinds of experiments were done sequentially at the same protoplast. The single-channel currents in MB-1 and MB-2-type patches were identical within the statistical limits (MB-1: at −100 mV: -1.1 ± 0.1 pA and at +60 mV: 1.1 ± 0.2 pA; MB-2: at −100 mV –1.2 ± 0.1 pA and at + 60 mV 1.1 ± 0.1 pA, n = 3). Also, no significant difference in open propability was found. The only difference that may explain the higher currents of MB-2 vs MB-1 in Figs. 3 and 5 is a higher number of channels in MB-2 patches. However, a final statement needs further investigations.

4.5 Molecular biology

Table 1 shows the K^+ channels of maize that have been cloned up to now. Since KAT1-like channels are stimulated by external acidification (Hoth et al., 1997, 2001; Pilot et al., 2001; Pratelli et al., 2002), and because KZM1 is always present at all locations and ages of the maize leaf (as expected for MB-1 and MB-2), it was investigated whether KZM1 might correspond to MB-1 or MB-2 described above. The method of Keunecke (Keunecke et al., 1997; Keunecke and Hansen, 2000) was modified to separate leaves of 5-week old maize plants into epidermis, mesophyll cells and vascular strands, and total

Table 1. Potassium channels of maize cloned so far. IR = Inward rectifier. References (ref): 1= Philippar et al., 1999, 2 = Philippar et al., 2003; 3 = Büchsenschütz et al., 2005, 4 = Su et al., 2005.

	Tissue	Arabi-dopsis	Type	Effect of external acidification		ref
ZMK1	non-vascular	AKT1	IR	Activated	Growth	1
ZMK2	vascular tissue	AKT2/3	IR	Deactivated	Sink	1
KZM1	vascular/bundle sheath strands and guard-, subsidiary cells	KAT2	IR	Insensitive	Perma nent	2
KZM2	guard cells	KAT 1				3
ZMK2,1	vascular/bundle sheath strands	KAT1	IR	Activated		4

RNA was isolated using the Plant RNeasy Extraction kit (Quiagen, Hilden, Germany). First-strand cDNA synthesis and RT-PCR (Szyroki et al., 2001) using a light cycler (Roche, Mannheim, Germany) revealed that KZM1 was predominantly expressed in the vascular strand fraction and to a much lower extend in epidermis cells. There was no detectable expression in mesophyll cells (Philippar et al., 2003). The other channel also found in vascular tissue is ZMK2 (Philippar et al., 1999; 2003).

Expression of the cDNA in *Xenopus oocytes* revealed the physiological characteristics of these channels as shown in Table 1. It is obvious that ZMK2 and KZM1 do not show the most characteristic feature of MB-1 and MB-2 (and of KAT1-channels), namely, the stimulation by acidic external (xylem) medium (Fig. 3D-F). Further, KZM1 was found to have a lower conductivity for Rb^+ than for K^+, whereas MB-2 showed higher conductivity for Rb^+. Thus it has to be concluded that both ZMK2 and KZM1 found in vascular strands originate from phloem cells and not from bundle-sheath cells. Hence the genetic characterization of MB-1 and MB-2 is still an open question.

Revealing the genetic identity of MB-1 and MB-2 may enable the use of dyes targeted to MB-1 or MB-2 to solve the question which could not be answered above, namely, are MB-1 and MB-2 expressed in different cells as suggested by Keunecke and Hansen (2000), and are they expressed only in the membrane adjacent to the xylem or uniformly in the bundle-sheath cells?

5. DISCUSSION

The isolation technique of Keunecke et al. (1997) has opened the access to the study of the impact of the vascular apoplast on transport to the symplast of the leaves. However, the results reported above are just an initial step. The first major goal should be the integration of the transporters in a thermodynamically coupled network as outlined by Gradmann (2001) and Gradmann and Hoffstadt (1998) in order to understand the mutual interaction with variable compositions of the xylem sap (Sattelmacher, 2001). However, even very elementary information is still missing, e.g., the value of the free running membrane potential as a crucial component of the driving force, the evidence for an anion-induced depolarisation as an indication for cotransport. The gap becomes obvious when the state of the research on guard cells is considered. Cation (Dietrich et al., 2001) and anion-channels (Schroeder, 1995; Hugouvieux et al., 2002), H^+-ATPases (Fricke and Willmer, 1990; Becker et al., 1993) and aquaporins (Sarda et al., 1997) have been succesfully characterized. The next step would be the inclusion of control functions. Also here, guards cells show the pie in the sky dealing with the influence of Ca^{2+} (McAinsh et al., 1997; Blatt, 2000), G-proteins (Kelly et al., 1995; Wang et al., 2001) or ABC-transporters (Leonhardt et al., 1999; 2001), phytohormones (Assmann and Shimazaki, 1999; Wilkinson and Davies, 2002; Hugouvieux et al., 2002) or CO_2 (Assmann, 1999; Roelfsema et al., 2002; Dietrich et al., 2001).

The increasing interest in the leaf apoplast (Sattelmacher, 2001) may help to draw the attention to the neglected transport properties of bundle-sheath cells. Understanding control by different apoplastic compositions (Schneider et al., 1994; Prima and Botton, 1998; Lohaus et al., 2000) and phytohormones (Blatt and Thiel, 1993; Wilkinson and Davies, 2002) would have a great impact on the understanding of plant nutrition.

REFERENCES

Altus, D.P. and Canny, M.J. (1985) Water pathways in wheat leaves I. The division of fluxes between different vein types. *Aust. J. Plant Physiol.*, 12, 173–181.

Amtmann, A., Jelitto, T.C. and Sanders, D. (1999) K^+-selective inward-rectifying channels and apoplastic pH in barley roots. *Plant Physiol.*, 120, 331–338.

Aoki, N., Hirose, T., Takahashi, S., Ono, K., Ishimaru, K. and Ohsugi, R. (1999) Molecular cloning and expression analysis of a gene for a sucrose transporter in maize (*Zea mays* L.). *Plant Cell Physiol.*, 40, 1072–1078.

Assmann, S.M. (1999) The cellular basis of guard cells sensing of rising CO_2. *Plant Cell Environ.*, 22, 629–637.

Assmann, S.M. and Shimazaki, K. (1999) The multisensory guard cell. Stomatal responses to blue light and abscisic acid. *Plant Physiol.*, 119, 809–815.

Bauer, C., Hoth, S., Haga, K., Philippar, K., Aoki, N. and Hedrich, R. (2000) Differential expression and regulation of K^+ channels in the maize coleoptile: Molecular and biophysical analysis of cells isolated from cortex and vasculature. *Plant J.*, 24, 139–145.

Becker, D., Zeilinger, C., Lohse, G., Depta, H. and Hedrich, R. (1993) Identification and biochemical characterization of the plasma membrane H^+ -ATPase in guard cells of *Vicia faba* L. *Planta*, 190, 44–50.

Blatt, M.R. (2000) Ca^{2+} signalling and control of guard-cell volume in stomatal movements. *Current Opinion Plant Biol.*, 3, 196–204.

Blatt, M.R. and Thiel, G. (1993) Hormonal control of ion channel gating. *Annu. Rev. Plant Physiol. Plant Mol. Biol.*, 44, 543–567.

Büchsenschütz, K., Marten, I., Becker, D., Philippar, K., Ache, P. and Hedrich, R. (2005) Differential expression of K^+ channels betweem guard cells and subsidiary cells within the maize stomatal complex. *Planta*, 222, 968–976.

Butt, A.D., Blatt, M.R. and Ainsworth, C.C. (1997) Expression, evolution and genomic complexity of potassium ion channel genes of *Arabidopsis thaliana*. *J. Plant Physiol.*, 150, 652–660.

Canny, M.J. (1987) Locating active proton extrusion in leaves. *Plant Cell Environ.*, 10, 271–274.

Canny, M.J. (1990a) What becomes of the transpiration stream? *New Phytol.*, 114, 341–368.

Canny, M.J. (1990b) Rates of apoplastic diffusion in wheat leaves. *New Phytol.*, 116, 263–268.

Canny, M.J. (1990c) Fine veins of dicotyledon leaves as sites for enrichment of solutes of the xylem sap. *New Phytol.*, 115, 511–516.

Canny, M.J. (1993) The transpiration stream in the leaf apoplast: water and solutes. *Phil. Trans. Roy. Soc. Lond. B. Biol. Sci.*, 341, 87–100.

Canny, M.J. (1995a) Apoplastic water and solute movement: new rules for an old space. *Ann. Rev. Plant Physiol. Plant Mol. Biol.*, 46, 215–236.

Canny, M.J. (1995b) Potassium cycling in *Helianthus*: ions of the xylem sap and secondary vessel formation. *Trans. Roy. Soc. Lond. B. Biol. Sci.*, 348, 457–469.

Czempinski, K., Gaedeke, N., Zimmermann, S. and Mueller-Roeber, B. (1999) Molecular mechanisms and regulation of plant ion channels. *J. Exp. Bot.*, 50, 955–966.

Deeken, R., Geiger, D., Fromm, J., Koroleva, O., Ache, P., Langenfeld-Heyser, R., Sauer, N., May, S.T. and Hedrich, R. (2002) Loss of the AKT2/3 potassium channel affects sugar loading into the phloem of *Arabidopsis*. *Planta*, 216, 334–344.

Dietrich, P., Sanders., D. and Hedrich, R. (2001) The role of ion channels in light-dependent stomatal opening. *J. Exp. Bot.*, 52, 1959–1967.

Engels, C. and Kirkby, E.A. (2001) Cycling of nitrogen and potassium between shoots and roots in maize as affected by shoot and root growth. *J. Plant Nutr. Soil Sci.*, 164, 183–191.

Evert, R.F., Botha, C.E.J. and Mierzwa, R.J. (1985) Free-space marker studies on the leaf of *Zea mays* L. *Protoplasma*, 126, 62–73.

Evert, R.F., Eschrich, W. and Heyser, W, (1977) Distribution and structure of the plasmodesmata in mesophyll and bundle-sheath cells of *Zea mays* L. *Planta*, 136, 77–89.

Evert, R.F., Russin, W.A. and Bosabalidis, A.M. (1996) Anatomical and ultrastructural changes associated with sink-to-source transition in developing maize leaves. *Internat. J. Plant Sci.*, 157, 247–261.

Fricke, M.D. and Willmer, C.M. (1990) Nitrate-sensitive ATPase activity and proton pumping in guard cell protoplasts of *Commelina*. *J. Exp. Bot.*, 41, 193–198.

Fritz, E., Evert, R.F. and Heyser, W. (1983) Microautoradiographic studies of phloem loading and transport in the leaf of *Zea mays* L. *Planta*, 159, 193–206.

Furbank, R.T., Scofield, G.N., Hirose, T., Wang, X., Patrick, J.W., Offler, C.E. and Wang, X.D. (2001) Cellular localisation and function of a sucrose transporter OsSUT1 in developing rice grains. *Aust. J. Plant Physiol.*, 28, 1187–1196.

Gaur, S., Yamaguchi, H. and Goodman, H.M. (2000) Activation of the sodium pump blocks the growth hormone-induced increase in cytosolic free calcium in rat adipocytes. *Endocrinology*, 141, 513–519.

Gradmann, D. (2001) Impact of apoplast volume on ionic relations in plant cells. *J. Membrane Biol.*, 184, 61–69.

Gradmann, D. and Blatt, M.R. (1997) K^+-sensitive gating of the K^+ outward rectifier in *Vicia* guard cells. *J. Membrane Biol.*, 158, 241–256.

Gradmann, D. and Hoffstadt, J. (1998) Electrocoupling of ion transporters in plants: interaction with internal ion concentration. *J. Membrane Biol.*, 166, 51–59.

Gaur, S., Yamaguchi, H. and Goodman, H.M. (2000) Activation of the sodium pump blocks the growth hormone-induced increase in cytosolic free calcium in rat adipocytes. *Endocrinology*, 141, 513–519.

Hattersley, P.W. and Browing, A.J. (1981) Occurrence of the suberized lamella in leaves of grasses of different photosynthetic types. I. In parenchymatous bundle sheaths and PCR ("Kranz") steaths. *Protoplasma*, 109, 371–401.

Hedrich, R., Bregante, M., Dreyer, I. and Gambale, F. (1995) The voltage-dependent potassium-uptake channel of corn coleoptiles has permeation properties different from other K^+ channels. *Planta*, 197, 193–199.

Heldt, H.W. (1999) Pflanzenbiochemie Spektrum. Akademischer Verlag, Heidelberg, Berlin.

Hirsch, R.E., Lewis, B.D., Spalding, E.P. and Sussman, M.R. (1998) A role for the AKT1 potassium channel in plant nutrition. *Science*, 280, 918–921.

Hugouvieux, V., Murata, Y., Young, J.J., Kwak, J.M., Mackesy, D.Z. and Schroeder, J.I. (2002) Localization, ion channel regulation, and genetic interactions during abscisic acid signaling of the nuclear mRNA cap-binding protein, ABH1. *Plant Physiol.*, 130, 1276–1287.

Hoth, S., Dreyer, I., Dietrich, P., Becker, D., Müller-Röber, B. and Hedrich, R. (1997) Molecular basis of plant-specific acid activation of K^+ uptake channels. *Proc. Natl. Acad. Sci. USA*, 94, 4806–4810.

Hoth, S., Geiger, D., Becker, D. and Hedrich, R. (2001) The pore of plant K^+ channels is involved in voltage and pH sensing: domain swapping between different K^+ channel α-subunits. *Plant Cell*, 13, 943–952.

Jahn, T., Baluska, F., Michalke, W., Harper, J.F. and Volkmann, D. (1998) Plasma membrane H^+-ATPase in the root apex: evidence for strong expression in xylem parenchyma and asymmetric localization within cortical and epidermal cells. *Physiol. Plant*, 104, 311–316.

Kelly, W.B., Esser, J.E. and Schroeder, J.I. (1995) Effects of cytosolic calcium and limited possible dual effects of G protein modulators on guard cell inward potassium channels. *Plant J.*, 1995, 479–489

Keunecke, M. and Hansen, U.P. (2000) Different pH dependences of K^+ channel activity in bundle sheath and mesophyll cells of maize leaves. *Planta*, 210, 792–800.

Keunecke, M. and Hansen, U.P. (2001) Potassium transport in bundle sheath cells of maize and its role for plant nutrition. In: *Plant Nutrition. Food Security and Sustainability of agro-ecosystems through basic and applied research.* W. J. Horst et al., (eds), Kluwer Academic Publishers. Dordrecht, Boston, London, pp. 276–277

Keunecke, M., Lindner, B., Seydel, U., Schulz, A. and Hansen, U.P. (2001) Bundle sheath cells of small veins in maize are the location of uptake from the xylem. *J. Exp. Bot.*, 52, 709–714.

Keunecke, M., Sutter, J.U., Sattelmacher, B. and Hansen, U.P. (1997) Isolation and patch clamp measurements of xylem contact cells for the study of their role in the exchange between apoplast and symplast of leaves. *Plant Soil*, 196, 239–244.

Kinoshita, T. and Shimazaki, K. 2002 Biochemical evidence for the requirement of 14-3-3 protein binding in activation of the guard-cell plasma membrane H^+-ATPase by blue light. *Plant Cell Physiol.*, 43, 1359–1365.

Köhler, B. and Raschke, K. (2000) The delivery of salts to the xylem. Three types of anion conductance in the plasmalemma of the xylem parenchyma of roots of barley. *Plant Physiol.*, 122, 243–254.

Leonhardt, N., Marin, E., Vavasseur, A. and Forestier, C. (1997) Evidence for the existence of a sulfonylurea-receptor-like protein in plants: Modulation of stomatal movements and guard cell potassium channels by sulfonylureas and potassium channel openers. *Proc. Nat. Acad. Sci. USA*, 94, 14156–14161.

Leonhardt, N., Vavasseur, A. and Forestier, C. (1999) ATP-binding casette modulators control abscisic acid-regulated slow anion channels in guard cells. *Plant Cell*, 11, 1141–1151.

Lohaus, G., Hussmann, M., Pennewiss, K., Schneider, H., Zhu, J.J. and Sattelmacher, B. (2000) Solute balance of a maize (*Zea mays* L.) source leaf as affected by salt treatment with special emphasis on phloem retranslocation and ion leaching. *J. Exp. Bot.*, 51, 1721–1732.

Maathuis, F.J.M., Filatov, V., Herzyk, P., Krijger, G.C., Axelsemn, K.B., Chen, S., Green, B.J., Li, Y., Madagan, K.L., Sánchez-Fernandez, R., Forde, B.G., Palmgren, M.G., Rea, P.A., Williams, L.E., Sanders, D. and Amtmann, A. (2003) Transcriptionell analysis of root transporters reveals participation of multiple gene families in the response to cation stress. *Plant J.*, 35, 675–692.

MacRobbie, E.A.C. (1995) Effects of ABA on $^{86}Rb^+$ fluxes at plasmalemma and tonoplast of stomatal guard cells. *Plant J.*, 7, 835–843.

Marschner H., Kirkby, E.A. and Engels, C. (1997) Importance of cycling and recycling of mineral nutrients within plants for growth and development. *Bot. Acta*, 110, 265–273.

Marten, I., Hoth, S., Deeken, R., Ache, P., Ketchum, K.A., Hoshi, T. and Hedrich, R. (1999) AKT3, a phloem-localized K^+ channel, is blocked by protons. *Proc. Natl. Acad. Sci. USA*, 96, 7581–7586.

McAinsh, M.R., Brownlee, C. and Hetherington, A.M. (1997) Calcium ions as second messengers in guard cell signal transduction. *Physiol. Plant.*, 100, 16–29.

McClure, P.R., Kochian, L.V., Spanswick, R.M. and Shaff, J.E. (1990) Evidence for cotransport of nitrate and protons in maize roots. *Plant Physiol.*, 93, 290–294.

Mühling, K.H. and Läuchli, A. (2000) Light-induced pH and K^+ changes in the apoplast of intact leaves. *Planta,* 212, 9–15.

Mühling, K.H. and Sattelmacher, B. (1995) Apoplastic ion concentration of intact leaves of field bean (*Vicia faba*) as influenced by ammonium and nitrate nutrition. *J. Plant Physiol.*, 147, 81–86.

Nelson, T. and Dengler, N. (1997) Leaf vascular pattern formation. *Plant Cell*, 9, 1121–1135.

Philippar, K., Fuchs, I., Luthen, H., Hoth, S., Bauer, C.S., Haga, K., Thiel, G., Ljung, K., Sandberg, G., Böttger, M., Becker, D. and Hedrich, R. (1999) Auxin-induced K^+ channel expression represents an essential step in coleoptile growth and gravitropism. *Proc. Natl. Acad. Sci. USA*, 96, 12186–12191.

Philippar, K., Büchsenschütz, K., Abshagen, M., Fuchs, I., Geiger, D., Lacombe, D. and Hedrich, R. (2003) The K^+ channel KZM1 mediates potassium uptake into the phloem and guard cells of the C4 grass *Zea mays*. *J. Biol. Chem.*, 278, 16973–16981.

Pilot, G., Lacombe, B., Gaymard, F., Cherel, I., Boucherez, J., Thibaud, J.B. and Sentenac, H. (2001) Guard cell inward K^+ channel activity in *Arabidopsis* involves expression of the twin channel subunits KAT1 and KAT2. *J. Biol. Chem.*, 276, 3215–3221.

Pratelli, R., Lacombe, B., Torregrosa, L., Gaymard, F., Romieu, C., Thibaud, J.B. and Sentenac, H. (2002) A grapevine gene encoding a guard cell K^+ channel displays developmental regulation in the grapevine berry. *Plant Physiol.*, 128, 564–577.

Prima, P. D. and Botton, B. (1998) Organic and inorganic compounds of xylem exudates from five woody plants at the stage of bud breaking. *J. Plant Physiol.*, 153, 670–676.

Riessner, T., Woelk, F., Abshagen-Keunecke, M., Caliebe, A. and Hansen, U.P. (2002) A new level detector for ion channel analysis. *J. Membrane Biol.*, 189, 105–118.

Roelfsema, M.R.G., Hanstein, S., Felle, H.H. and Hedrich, R. (2002) CO_2 provides an intermediate link in the red light response of guard cells. *Plant J.*, 32, 65–75.

Russell, S.H. and Evert, R.F. (1985) Leaf vasculature in *Zea mays* L. *Planta*, 164, 448–458.

Russin, W.A., Evert, R.F., Vanderveer, P.J., Sharkey, T.D. and Briggs, S.P. (1996) Modification of a specific class of plasmodesmata and loss of sucrose export ability in the sucrose export defective maize mutant. *Plant Cell*, 8, 645–658.

Salva, S. and Matteson, D.R. (1991) Voltage-dependent slowing of K^+ channel closing kinetics by Rb^+. *J. Gen. Physiol.*, 98, 535–554.

Sanders, D. and Hansen, U.P. (1981) Mechanism of Cl^- transport at the plasma membrane of *Chara corallina*. II. Transinhibition and the determination of H^+/Cl^- binding order from a reaction kinetic model. *J. Membrane Biol.*, 58, 139–153.

Sarda, K., Tousch, D., Ferrare, K., Legrand, E., Dupuis, J.M., Casse-Delbart, F. and Lamaze, T. (1997) Two TIP-like genes encoding aquaporins are expressed in sunflower guard cells. *Plant J.*, 12, 1103 – 1111.

Sattelmacher, B. (2001) The apoplast and its significance for plant mineral nutrition. *New Phytol.*, 149, 167–192.

Schneider, H., Zhu, J.J. and Zimmermann, U. (1997). Xylem and cell turgor pressure probe measurements in intact roots of glycophytes: transpiration induces a change in the radial and cellular reflection coefficients. *Plant Cell Environ.*, 20, 221–229.

Schroeder, J.I. (1995) Anion channels as central mechanism for signal transduction in guard cells and putative functions in roots for plant–soil interaction. *Plant Mol. Biol.*, 28, 353–361.

Shen, M.R., Chou, C.Y. and Ellory, J.C. (2000) Volume-sensitive KCl cotransport associated with human cervical carcinogenesis. *Pfluegers Archiv Eur. J. Physiol.*, 440, 751–760 .

Shi, H.Z., Quintero, F.J., Pardo, J.M. and Zhu, J.K. (2002) The putative plasma membrane Na^+/H^+ antiporter SOS1 controls long-distance Na^+ transport in plants. *Plant Cell*, 14, 465–477.

Steudle, E. (2002) Transport of water in plants. *Environ. Contr. Biol.*, 40, 29–37.

Su, Y.-H., North, H., Grignon, C., Thibaud, J.-B., Sentenac, H. and Véry, A.-A. (2005) Regulation by external K^+ in a maize inward *Shaker* channel targets transport activity in the high concentration range. *Plant Cell*, 17, 1532–1548.

Thompson, J. and Begenisich, T. (2001) Affinity and location of an internal K^+ ion binding site in *Shaker* channels. *J. Gen. Physiol.*, 117, 373–384.

Tyree, M.T. (1997) The cohesion-tension theory of sap ascent: Current controversies. *J. Exp. Bot.*, 48, 1753–1765.

Wang, X.Q., Ullah, H., Jones, A.M. and Assmann, S. (2001) G-protein regulation of ion channels and abscidic acid signaling in *Aradidopsis* guard cells. *Science*, 292, 2070–2072.

Wegner, L.H., De Boer, A.H. and Raschke, K. (1994) Properties of the K^+ inward rectifier in the plasma membrane of xylem parenchyma cells from barley roots: Effects of TEA^+, Ca^{2+}, Ba^{2+} and La^{3+}. *J. Membrane Biol.*, 142, 363–379.

Weiner, H., Burnell, J.N., Woodrow, J.E., Heldt, H.W. and Hatch, M.D. (1988) Metabolite diffusion into bundle-sheath cells from C4 plants. *Plant Physiol.*, 88, 815–822.

Wilkinson, S. and Davies, W.J. (2002) ABA-based chemical signalling: The co-ordination of responses to stress in plants. *Plant Cell Environ.*, 25, 195–210.

Yamada, S. and Bohnert, H.J. (2000) Expression of the PIP aquaporin promoter-MipA from the common ice plant in tobacco. *Plant Cell Physiol.*, 41, 719–725.

Zimmermann, U., Schneider, H., Wegner, L., Wagner, H.J., Szimtenings, M., Haase, A. and Bentrup, F.W. (2002) What are the driving forces for water lifting in the xylem conduit? *Physiol. Plant*, 114, 327–335.

LOADING OF IONS INTO THE XYLEM OF THE ROOT

B. KÖHLER[1] and K. RASCHKE
Albrecht-von-Haller-Institut für Pflanzenwissenschaften, Georg-August-Universität Göttingen, Germany, kraschk at gwdg.de; [1]now at Universität Potsdam, Institut für Biochemie und Biologie, Lehrstuhl für Molekularbiologie, Golm, Germany, bakoehl@uni-potsdam.de

Abstract. Anion conductances and an electrogenic pump participating in the loading of ions into the xylem of the root were investigated on protoplasts isolated from the xylem parenchyma of barley roots, using the patch-clamp technique. Previous studies had mainly focused on K^+ channels; they showed that loading of the stelar apoplast was thermodynamically downhill, not requiring a second pump at the exit of the symplast of the root to the apoplast. Our work confirmed this view after concentrating on anion conductances and an electrogenic pump. Three major anion conductances were found residing in the plasmalemma of the xylem parenchyma: an inwardly rectifying anion channel (X-IRAC) with open times of up to several seconds, generating the largest currents at hyperpolarization, a quickly activating anion conductance (X-QUAC), important for anion loading at voltages between –50 mV and the equilibrium voltage for the permeating anion, and a slowly activating anion conductance (X-SLAC), activating above –100 mV. Anion currents through X-SLAC and X-QUAC, in combination with K^+ currents through the outwardly directed K^+ channel, KORC, were estimated to be large enough to account for reported rates of xylem loading. In the presence of nitrate in the xylem, the current-voltage relationship of X-QUAC shifted towards hyperpolarization, exerting positive feedback on loading of nitrate, so that nitrate efflux into the xylem would be maintained even at high concentrations of nitrate in the xylem, which occurs for instance during the night. Current-voltage relationships of the protoplasts showed also the existence of an electrogenic pump. It was stimulated by fusicoccin and inhibited by dicyclohexylcarbodiimide (DCCD): it exhibited features of an H^+-ATPase. The pump was short-circuited by other conductances, mainly for anions. Simultaneous activity of pump and anion conductances provided a condition for acid release into the xylem. Beyond participating in controlling the membrane voltage, the pump appears to be involved in energizing the absorption of ions from the xylem during the circulation of nutrients within the plant. Ion conductances in the xylem parenchyma of roots of barley and maize are similar. A brief review is given on what is known about their control.

B. Sattelmacher† and W.J. Horst (eds.), The Apoplast of Higher Plants: Compartment of Storage, Transport and Reactions, 181–200.

Key words: xylem loading, salt transport, anion conductance, H^+-pump activity, patch clamp, xylem parenchyma, barley roots

1. THE IMPORTANCE OF THE XYLEM PARENCHYMA

Nutrient traffic between root and shoot occurs simultaneously on a divided highway. The upward long distance transport from the root to the shoot uses the xylem conduits of the apoplast, whereas the phloem serves long distance downward transport. The sieve tubes and companion cells of the phloem are alive and possess molecular and supramolecular structures that absorb, transmit and release their cargo. The xylem strands are dead and must be charged with nutrients through the activity of the living parenchyma cells surrounding them. Xylem-parenchyma cells also serve the task of unloading nutrients from the xylem. The anatomy of the root is complex with different, specialized, cell types in close proximity. The metaxylem is presumed to be the most effective longitudinal transport pathway for ions to the shoot (Läuchli et al., 1974). Xylem and phloem are not isolated from each other. The cell walls of the vascular tissue provide apoplastic paths between the upward and downward traffic lanes, allowing nutrient circulation at a smaller scale than between root and shoot and nutrient transport even at low mass flow through the xylem or in the absence of a transpiration stream (Tanner and Beevers, 2001). Although xylem loading appears to be the main function of the xylem parenchyma in the root, mechanisms for ion uptake must exist side by side with mechanisms for ion release to make ion circulation possible. Both types of mechanisms are also required to supply balancing charges when more cationic nutrients (like K^+) are taken out of the stream than anions and vice versa.

Turning to ion uptake and transport into the shoot, we recognize that, during the evolution of the root, the function acquisition of nutrients from the soil was separated from that of putting them into the forwarding system. A buffering storage system was inserted between them, primarily represented by the vacuoles in the root cortex. Dunlop and Bowling (1971a, b, c) on the basis of their experiments with maize roots, concluded that ion movement to the xylem vessels is mainly symplasmic, where the outer boundary of the symplasm is the plasmalemma of the epidermal cells and that the inner boundary is the plasmalemma of the xylem parenchyma, as was originally proposed by Crafts and Broyer (1938). Uptake of ions was thought to be uphill and release into the xylem to be passive; but Pitman (1972) found that the uncoupler CCCP inhibited not only absorption of Cl^-

into roots of barley seedlings but also the export of Cl^- from them. He concluded on the presence of "a second active transport prior to entry into the xylem"

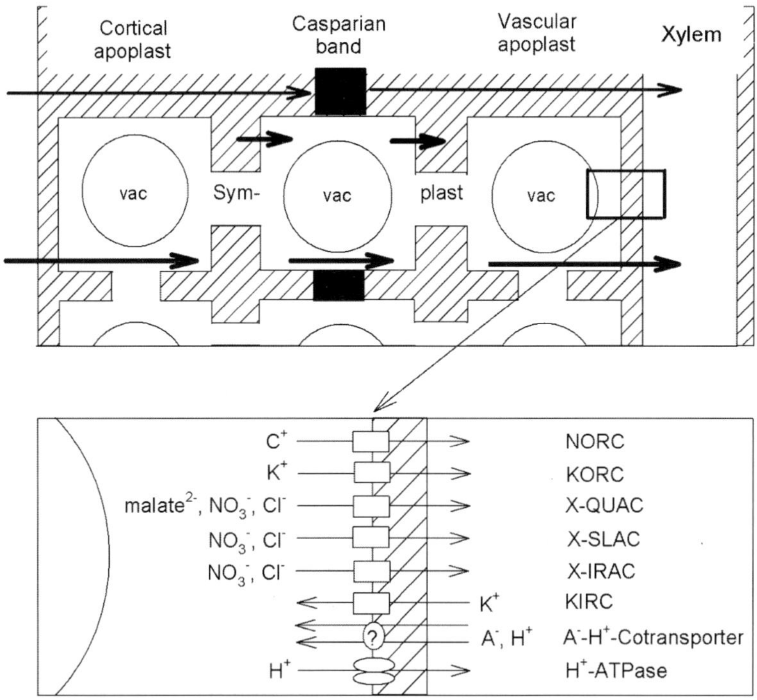

Fig. 1. The Casparian band in the cell walls of the endodermis separates the apoplast of the cortex from that of the stele, making an apoplastic transport of water and ions into the vascular apoplast impossible (upper arrows); also the electrical connection is interrupted. Nutrients taken up by the cortical cells can move within the symplast of the root from the cortex to the stele and are released there into the vascular apoplast for transport to the shoot (lower heavy arrows). Substances moving in the symplast may be transferred into the vacuoles of the various cells of the root for various times. Disregarding this mechanism of sequestration and buffering, at least two membranes must be passed by a nutrient, one at its entry into the root, the other during its loading into the vascular apoplast for transport into the plant body. In our contribution we are concerned with mechanisms controlling the passage of ions into the vascular apoplast and the xylem (and out of it). The detail in the lower part of this figure lists ion transporters involved in xylem loading and unloading; NORC, nonselective cation conductance; KORC, K^+-selective outwardly rectifying conductance (at the molecular level identified as SKOR, stelar K^+ outward rectifier); KIRC, K^+-selective inwardly rectifying conductance; X-QUAC, quickly activating anion conductance; X-SLAC, slowly activating anion conductance; X-IRAC, inwardly rectifying anion channel: A^-/H^+cotransport, several presumed anion/H^+- symporters (evidence exists for H^+ - anion cotransport during anion uptake into the root: Dunlop, 1989; McClure et al., 1990; Ullrich and Novacky, 1990), and such NO_3^- transporters have been cloned (Forde, 2000).

By extrapolation, it was thought that, in general, the tasks of ion uptake from the soil and transfer to the xylem were accomplished by two separate pumps. We know now that none of the major nutrients (except Ca^{2+}) can be moved directly by specific ion pumps. Rather, electrogenic mechanisms, in general proton pumps, produce electrical potential differences and differences in pH across the plasmalemma of the cells which absorb ions from the soil solution, and this electrochemical potential difference is large enough to extract ions like K^+, NO_3^- or Cl^- from the soil and raise their electrochemical potentials high enough to allow a downhill flux into the xylem. Hanson (1978) proposed a hypothesis in which not only ion influx into the "symplasm" but also efflux are "energy-linked" through electrogenic H^+ efflux and that control of ion currents was exerted by variation of resistances at entry into the cortex and release into the stele.

In contrast to this hypothesis, the activities of the major nutrient ions in the cortex appear to suffice for a passive release into the vascular apoplast and thus into the xylem, although Dunlop and Bowling (1971a) did not discover a radial gradient in vacuolar K^+ activity. In this context, one must remember that the processes of ion uptake and ion release are spatially and temporally separated; there must be ion storage capacities. With respect to the flow of water and ions, the apoplast of the stele is disconnected from that of the cortex by the Casparian band in the wall of the endodermis, and the parenchymas of the cortex and the stele provide the ion capacitances which allow a dissociation of phases of ion release from those of ion uptake (Fig. 1). The contributions of Gilmer and Schurr (this volume, pp 225-234) and Siebrecht et al. (this volume, pp 269-288) provide examples for a lack of synchronization between absorption and release. Although energetically a second pumping stage appears not to be required for xylem loading, evidence points to xylem loading as a second control point in the transfer of nutrients to the shoot (Glass et al., 2001; Herdel et al., 2001). At the level of ion transfer through the endodermis to the pericycle and further on to the xylem parenchyma an electrogenic pump could have a function; De Boer and Volkov (2003) summarized the evidence. Activity of an electrogenic pump was indeed discovered in protoplasts of the xylem parenchyma of barley roots (Köhler and Raschke, 1998), and De Boer and Volkov (2003) found by xylem perfusion experiments, that fusicoccin, an activator of the plasma membrane H^+-ATPase, strongly acidified the xylem sap in roots of *Plantago maritima* and barley.

The recent review by De Boer and Volkov (2003) on structure and functioning of the xylem focused on water transport, pump activity, and cation loading, in particular on loading with K^+. Here we complement their review by reporting on the transfer of anions into the xylem and on characteristics of an electrogenic pump. Our results were obtained with protoplasts from the xylem parenchyma of roots of barley.

2. ISOLATION OF PROTOPLASTS FROM THE XYLEM PARENCHYMA AND RECORDING ION CURRENTS THROUGH THEIR PLASMA MEMBRANES

L.H. Wegner developed a method for the selective isolation of protoplasts from the parenchyma bordering the metaxylem vessels of barley roots, taking advantage of the lignification pattern in the stele of nodal root cells (Wegner and Raschke, 1994). In brief, after discarding the apical 1–2 cm of the roots, sections of 4–6 cm length were selected. The cortex was pulled off and the distal 1–2 cm of the resulting steles were chopped and incubated in a solution containing 2% (w/v) cellulase, 0.02% (w/v) pectolyase, 2% (w/v) bovine serum albumin, 10 mM Na ascorbate, and 1 mM $CaCl_2$, adjusted to pH 5.5 and to an osmolality of 500 mosmol kg^{-1} with mannitol. Protoplasts were separated from the incubation mixture by filtering through a 100 μm nylon mesh and two centrifugations in 1 mM $CaCl_2$ containing mannitol at 500 mosmol kg^{-1}. The electrical properties of ion channels and conductances and of the electrogenic pump in the plasma membrane were investigated with the patch-clamp technique (Hamill et al., 1981). Provided a plant species can be transformed easily, a promising approach to isolate cell-type specific protoplasts is the use of transgenic plants expressing green fluorescent protein under the control of cell-specific promoters. This approach has been successfully applied to *Arabidopsis* roots where protoplasts from epidermal, cortical and pericycle cells were isolated and subjected to patch-clamp experiments (Kiegle et al., 2000).

3. BRIEF OVERVIEW ON CATION CONDUCTANCES IN THE XYLEM PARENCHYMA

The first ion channels identified in the plasma membrane of xylem-parenchyma protoplasts were two outwardly directed conductances, the K^+ outward rectifying conductance, KORC, and the non-specific outward rectifying conductance, NORC, and one inward conductance, the K^+ inward rectifying conductance, KIRC (Wegner and Raschke, 1994; Fig. 1). They differed in their voltage dependence and their ion specificity. KORC and NORC activated at membrane depolarization, namely positive of –50 and above 20 mV, and KIRC activated below a membrane potential of -100 mV (Wegner and Raschke, 1994). The activation potential of KORC depended on the extracellular K^+ concentration (Wegner and De Boer, 1997a). KORC and KIRC were highly selective for K^+, whereas NORC selected poorly among monovalent cations and even anions (Wegner and De Boer, 1997a). Gaymard et al. (1998) succeeded in identifying the K^+ outward rectifier on the molecular level; they called it SKOR. Like the K^+ inward rectifier it belongs to the *Shaker* family, but it was the first one to show outward rectification. Selectivity and activation kinetics of SKOR and KORC were identical (Gaymard et al., 1998; Wegner and De Boer, 1997a). SKOR is expressed exclusively in the pericycle and the xylem parenchyma of *Arabidopsis* roots (Gaymard et al., 1998). The K^+ content of the shoot was reduced significantly in knock-out mutants of *Arabidopsis* lacking SKOR (Gaymard et al., 1998), as was the K^+ concentration in the xylem sap after inhibition of KORC in a xylem-perfusion experiment by Wegner and De Boer (1997a). These results indicate strongly that K^+ channels serve as major pathways for the release of K^+ into the xylem. Inward and outward K^+ channels, similar to those in barley roots, were also found in roots of maize (Roberts and Tester, 1995).

4. THREE MAJOR ANION CONDUCTANCES CONTRIBUTE TO XYLEM LOADING

Since salt export is electroneutral and K^+ and NO_3^- fluxes are closely coupled (as confirmed by Herdel et al., 2001) the presence of anion channels had to be postulated. Indeed they were found, and three different types of major anion channels have been identified in the plasmalemma of xylem-parenchyma cells of barley. They are: the quickly activating anion conductance (X-QUAC), the slowly activating anion conductance (X-SLAC), and the inward rectifying anion channel (X-IRAC) (Köhler and Raschke, 2000). These anion conductances differed widely in their kinetics and voltage dependence (Fig. 2) and also in their sensitivity to the intracellular Ca^{2+} concentration. X-QUAC activated completely within milliseconds. It was active over the whole range of membrane voltages occurring in the cells of the xylem parenchyma. Current was minimal around –40 mV and increased both with more positive and more negative membrane potentials. In contrast, deactivation and activation of X-SLAC occurred within several seconds. X-SLAC activated at voltages positive of –100 mV. So far, it was not possible to record single-channel events of X-QUAC and X-SLAC.

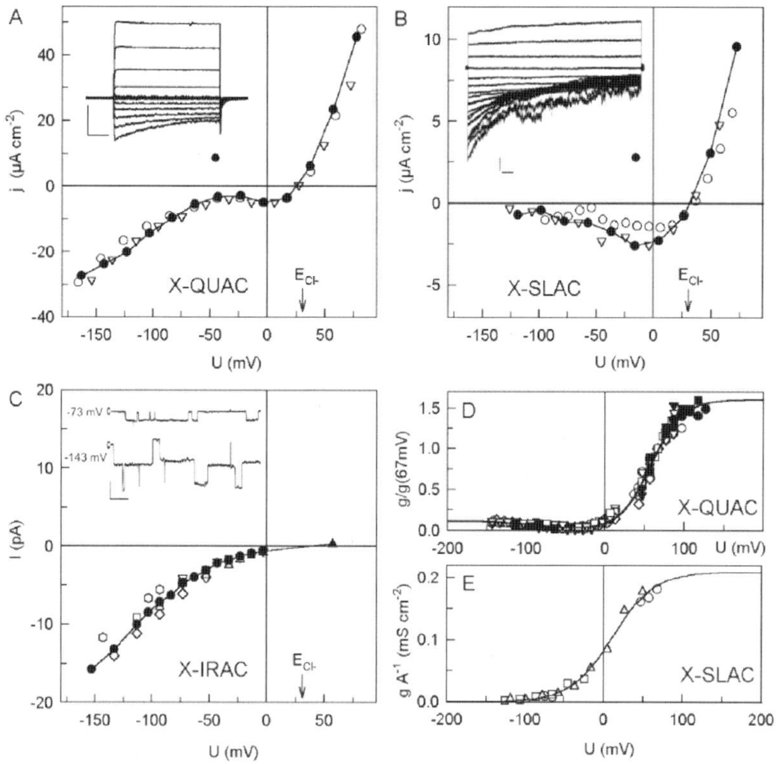

Fig. 2. Properties of the three anion conductances in the plasmalemma of xylem-parenchyma cells in barley roots. Current-voltage relationships (A, B, C) derived from current responses to voltage pulses (insets in A, B show examples). In all cases currents reversed at the Nernst potential of Cl^- (E_{Cl^-}). A, X-QUAC (quickly activating anion conductance), current-voltage relationships of three cells, normalized with respect to the sample plotted in the inset and represented by the filled circles in the current-voltage plot; vertical bar, 400 pA; horizontal bar, 100 ms. Holding potential between pulses was –43 mV. B, X-SLAC, current-voltage curves of three cells. Filled circles correspond to the example shown in the inset. Note that time scale differs from that of panel A (vertical bar, 40 pA, horizontal bar, 4 s). Holding potential was 37 mV. C, X-IRAC, current-voltage relationship of six cells. Here single channel amplitudes at different voltages are plotted, as resolved in the whole-cell configuration. Examples of current traces at –73 and –143 mV are shown in the inset, Downward changes correspond to channel openings; c, channel closed; vertical bar, 10 pA, horizontal bar, 1 s. A-C, modified from Köhler and Raschke (2000). D, E, Boltzmann equations (lines) fitted to chord conductances of X-QUAC and X-SLAC, respectively. Chord conductances of X-QUAC were obtained after subtraction of background currents, which were determined after the addition of the anion channel inhibitors DIDS (solid symbols) or IAA-94 (open symbols); data from 6 cells, conductances normalized with respect to values measured at 67 mV (Köhler et al., 2002); data for X-SLAC are from three protoplasts. D, The concavity in the current-voltage relationship of X-QUAC can be explained by the assumption that this conductance possessed two gates with opposite voltage dependences and considerably different conductivities.

In contrast, single channel events of X-IRAC could be resolved in whole-cell measurements (Fig. 2 C). X-IRAC was characterized by long open times, lasting up to several seconds, and its strong inward rectification. Currents through this channel were largest at negative voltages. X-IRAC was particularly active at high cytoplasmatic Ca^{2+} concentrations, whereas the activity of X-QUAC increased when the cytoplasmatic Ca^{2+} concentration was lowered (Köhler and Raschke, 2000). X-IRAC and X-QUAC-like activities have also been reported in maize-root stelar cells (Gilliham and Tester, 2005). Both K^+ and anion conductances from maize and barley display remarkably similar properties, indicating that these conductances belong to the common equipment of stelar cells and that they have general functions. The molecular identity of the anion conductances is still unknown. All of the three conductances can serve anion efflux and participate in xylem loading. However, passive salt release is restricted to a limited range of membrane potentials in which conductances for anions and cations are active simultaneously. KORC is active at voltages positive of E_K^+, and both X-QUAC and X-SLAC open with depolarization (Fig. 3). Permeating anions were Cl^- and NO_3^-. Nitrate is quantitatively the most important inorganic anion that is transported from the root to the shoot. X-QUAC was three times more permeable for NO_3^- than for Cl^-. In addition, X-QUAC was permeable for malate (Köhler and Raschke, 2000). A saturable, low-affinity transport system with an unusual high K_M-value has been postulated for xylem loading with Cl^- from Cl^- flux measurements on barley, and NO_3^- had a suppressive effect on Cl^- shoot flux (Britto et al., 2004). Absolute and relative concentrations of NO_3^- and Cl^- vary with growth conditions, e.g. with the nitrogen source or under salinity. Possibly, the diversity of anion conductances found in xylem-parenchyma cells of roots is a result of the adaptation to various growth conditions. Barley is known to be a salt includer whereas maize is a salt excluder. It would be interesting to compare how the anion conductances are controlled in these two species.

Furthermore, anion conductances provide putative pathways for sulfate and phosphate efflux into the xylem. So far, experiments on anion conductances were done with Cl^-, NO_3^- and malate. The permeabilities of X-QUAC, X-IRAC and X-SLAC for other physiologically relevant anions and for anions of different sizes need to be investigated. The low-affinity transporters identified so far mediated sulfate influx (Takahashi et al., 2000; Kataoka et al., 2004). The *Arabidopsis pho1* mutant is characterized by a severe deficiency in shoot phosphate content. *PHO1* is active in stelar cells of the root and encodes a membrane protein with six putative membrane spanning domains with no homology to any characterized solute transporter identified in plants. Although PHO1 is likely a component of a phosphate

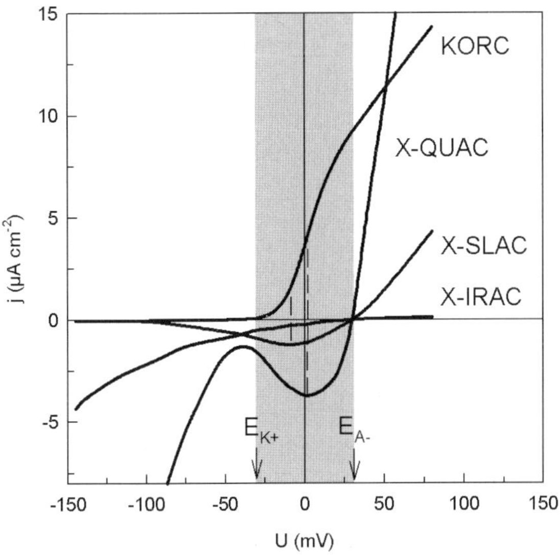

Fig. 3. Representative current-voltage relationships of the three anion conductances, X-QUAC, X-SLAC, X-IRAC and of the outwardly rectifying K^+ conductance, KORC. $E_K{}^+$ and $E_A{}^-$, equilibrium potentials of K^+ and anions, respectively. Shaded area, voltage range in which salt release is possible; vertical dashed lines, examples for simultaneous anion and cation currents of equal magnitude but opposite sign through X-SLAC and KORC in one case, and through X-QUAC and KORC in the other (Köhler and Raschke, 2000).

transport system, no transport activity could be shown for PHO1 until now (Hamburger et al., 2002).

It was possible to describe the voltage dependence of X-QUAC on the assumption of two gates, one opening at depolarization and the other with hyperpolarization. The conductance of the positive gate was much larger than that of the negative gate (Köhler et al., 2002). There appears to be some similarity to the slow and the quick anion conductances in guard cells. It was suggested that they represented two gating modes of the same protein (Dietrich and Hedrich, 1994; Raschke, 2003). However, alternations between X-QUAC and X-SLAC, as they occurred between QUAC and SLAC, did not appear in whole-cell records of the xylem-parenchyma protoplasts. On the basis of our present knowledge we suggest that X-QUAC is unique and the most important conductance for loading anions into the apoplast of barley roots. Its large transport capacity matches the Cl^- transport rates reported by Pitman (1971). X-QUAC is inhibited by the anion-channel inhibitor 4,4'-diisothiocyanatostilbene-2,2'-disulfonic acid (DIDS) (Köhler et al., 2002). In barley, application of DIDS or anthracene-9-carboxylic acid

(A-9-C) reduced the NO_3^- concentration in the xylem sap significantly without affecting NO_3^- uptake (Kawachi et al., 2002). This supports a fundamental role of X-QUAC in xylem loading with NO_3^-, although DIDS sensitivities of X-SLAC and X-IRAC have not been investigated so far. The effect of DIDS on the translocation of Cl^- was less pronounced than the effect of A-9-C (Kawachi et al., 2002). Further inhibitor studies might provide means to distinguish the functions of the three anion conductances and assess their contributions to xylem loading.

5. CONTROL OF X-QUAC BY NITRATE

Extracellular, but not cytoplasmic, NO_3^- caused a shift of the midpoint potential of the depolarization-activated gate towards more negative potentials, increasing the transport capacity of the membrane in the voltage range between –60 and 100 mV not only for nitrate, but for all permeating anions (Fig. 4A). The composition of the xylem sap generally changes during the course of a day, depending on nutrient supply and various stresses (Marschner, 1995; Herdel et al., 2001; Siebrecht et al., 2003), and the apoplastic ion concentration is considered to be an important factor in the circulation of ions (see also Drew et al., 1990). Half maximum shift of the depolarization activated gate of X-QUAC occurred at a NO_3^- concentration of 3.4 mM (Fig. 4A, inset). This lies within the reported physiological concentration range of NO_3^- (Mattsson et al., 1988; Herdel et al., 2001). In *Ricinus*, the NO_3^- concentration in the xylem sap reached values up to 10 mM during the night (Herdel et al., 2001); we measured 20 mM in barley root-exudates. At such high concentrations, the driving force for NO_3^- would be reduced, yet NO_3^- efflux into the xylem could be maintained because of the NO_3^- -induced shift in the voltage dependence of X-QUAC. In the model shown in Fig. 4B, this shift enhanced the flux by about one-third. The effect of NO_3^- on X-QUAC gating can therefore be described in terms of positive feedback during NO_3^- loading into the xylem, counteracting the negative feedback produced by the concentration increase. A shallow nitrate gradient could thus be compensated and further nitrate efflux into the xylem be ensured. If such a mechanism applies not only to barley but also to castor bean, we could explain, at least in part, the finding of Herdel et al. (2001), who reported that both nitrate concentration in the xylem sap and nitrate efflux increased in the afternoon and partly in the night. To a large extent, these changes could not be ascribed to variations in transpiration rate. Changes in anion and cation conductances alter the membrane potential (Fig. 4B, inset).

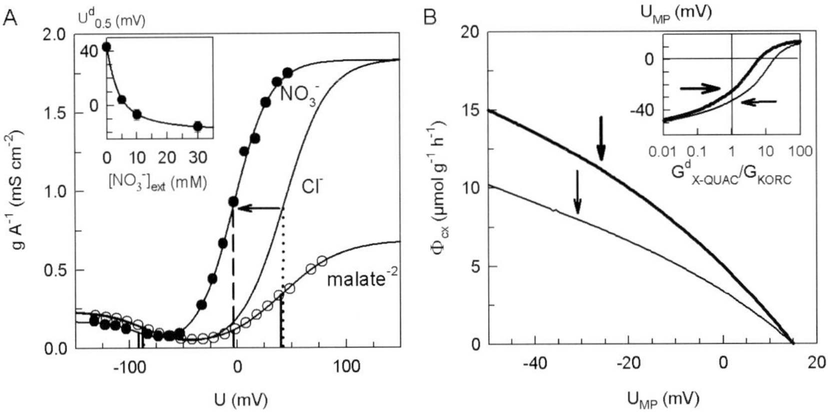

Fig. 4. A, Voltage dependence of X-QUAC is modulated by apoplastic NO_3^-. Replacement of external Cl^- by NO_3^- shifted the voltage dependence of the depolarization-activated gate towards more hyperpolarized voltages (arrow), whereas a change to external malate^{2-} remained without effect on the midpoint potential. The perpendicular broken or dotted lines indicate midpoint potentials as derived from Boltzmann curves. Inset, concentration dependence of the nitrate induced shift (midpoint potentials plotted against external NO_3^- concentration); a Hill equation was fitted. The parameters were Km = 3.4 mM, Hill coefficient = 1.25. The maximum shift was by −63 mV. B, nitrate flux (cx) into the xylem. cx, was computed as a function of the membrane potential at cytosolic and extracellular nitrate concentrations of 10 and 5 mM, respectively. Bold lines were calculated for a midpoint potential of 5 mV (in the presence of extracellular nitrate), thin lines for a midpoint potential of 43 mV (in the absence of extracellular nitrate). The inset shows the membrane potential as function of the ratio of the maximum conductances of X-QUAC and KORC for the two midpoint potentials. Arrows indicate nitrate efflux rates at membrane potentials taken from panel A, bold arrows apply to a midpoint potential of 5 mV, thin arrows to a midpoint potential of 43 mV (Köhler et al., 2002).

Apoplastic K^+ did not affect X-QUAC, indicating that coupling of K^+ and NO_3^- fluxes was rather indirect, via the membrane potential (and through messengers?). The response of X-QUAC to its substrate in the xylem differed from that of KORC. While nitrate exerted positive feedback on its loading into the xylem, an increase in K^+ in the xylem produced negative feedback in that K^+ efflux decreased through a change in the activation potential of KORC (Wegner and De Boer, 1999).

6. AN ELECTROGENIC PUMP

The activities of the cation and anion conductances from the xylem parenchyma depend on the membrane potential, which, in turn, may be determined, in part, by the action of an electrogenic pump in the plasmalemma.

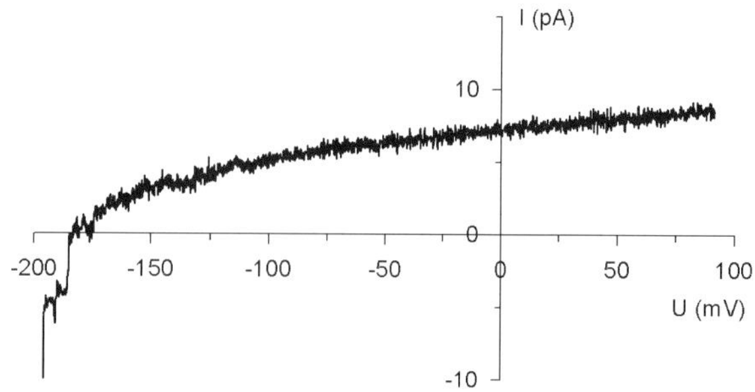

Fig. 5. Activity of an electrogenic pump in xylem-parenchyma protoplasts. Current response to voltage ramps displayed a negative reversal potential and an asymptotic approach to a maximum current (resembling saturation). For the investigation of the electrogenic pump, solutions were designed which aimed at the repression of ion conductances which would disturb the detection of the pump current. The bath solution representing the apoplastic space was (mM): 5 $Ca(Glc)_2$, 8 $MgCl_2$, 10 Mes, pH 5.8 (adjusted with Tris). The intracellular solution representing the cytoplasm was (mM): 10 Mg^{2+}-ATP, 5 μM free Ca^{2+} (added as $Ca(Glc)_2$, 10 mM $MgCl_2$, 10 N-hydroxyethyl-ethylenediamine-tri-acetic acid (HEEDTA), 10 Tris, pH = 7.2 (adjusted with Mes).

Xylem-parenchyma cells of barley roots were strongly labeled by antibodies against the plasma membrane H^+-ATPase (Samuels et al., 1992) which supports the presence of an H^+-ATPase in this cell type. However, a functional assay was lacking. In patch-clamp experiments, we discovered manifestations of such a pump (Köhler and Raschke, 1998; Fig. 5). It was identified as an H^+-pump, because the fungal toxin fusicoccin stimulated pump activity, and application of the pump inhibitor dicyclohexylcarbodiimide (DCCD) resulted in the disappearance of the hyperpolarization as well as of the transmembrane current at zero membrane potential. Furthermore, as expected for an H^+-pump, pump activity in xylem parenchyma was considerably enhanced by an enlarged transmembrane pH gradient (J. Zhu et al., in press). In general, H^+-pump activity is about 100 to 1,000 times higher than Ca^{2+}-pump activity. The Ca^{2+}-pump inhibitor, erythrosin B, did not produce an effect on the transmembrane current at zero membrane potential, or on the polarization of the membrane (J. Zhu et al., 2007), indicating that Ca^{2+}-pumps did not contribute significantly to the detected electrogenic activity. Pump current measured in root-xylem parenchyma of barley was 0.6 μA cm^{-2} on average. This is in line with currents of electrogenic pumps in the root cortex of wheat (Findlay et al., 1994; Tyerman et al., 2001), or in mesophyll cells and guard cells from *Vicia faba* (Lohse and Hedrich, 1992). H^+-pumps are ubiquitous for the provision of energy for active transport processes and thought to be important for the adjustment of the membrane potential and pH regulation (Sze et al., 1999;

Serrano, 1990, 1993; Palmgren, 2001; Sondergaard et al., 2004). Plasma membrane H^+-ATPases are encoded by a multigene family, whose members are expressed or controlled differently in specialized cell types (Morsomme and Boutry, 2000; Palmgren, 2001). By heterologous expression in yeast it was shown that isoforms have distinct enzymatic properties, for instance, with respect to the dependence on intracellular pH (Luo et al., 1999). Most H^+-ATPases show maximum activity at a pH slightly below 7.0, but they differ with respect to their profile on the alkaline side. The activity of the H^+-pump in the xylem parenchyma was not affected by changes in the intracellular pH in the range between 6.9 and 7.7 (J. Zhu et al., in press). This indicates that this H^+-pump is not involved in cytoplasmatic pH regulation. It differed from the ATPase from guard cells and mesophyll cells also in its insensitivity to changes in the cytosolic Ca^{2+} concentration from 0.15 to 5 μM (J. Zhu et al., in press). The guard-cell enzyme is inhibited by intracellular Ca^{2+} with a K_i of 0.3 μM (Kinoshita et al., 1993). The Ca^{2+} concentration required for half inhibition of ATP hydrolysis was with 150 μM much higher in plasma membranes of maize roots than in guard cells (Leonard and Hotchkiss, 1976). Apparently, isoforms of H^+-ATPases with low sensitivity to Ca^{2+} are expressed in roots.

7. ION CONDUCTANCES SHORT-CIRCUIT THE PUMP – CONSEQUENCES FOR MEMBRANE VOLTAGE AND XYLEM PH

Current-voltage recordings of the pump frequently varied in appearance; they indicated modulation by short-circuiting conductances. Application of La^{3+} reduced part of the modification. Further addition of DIDS brought out the typical features of the pump (Fig. 6A), showing that anion conductances constituted the major transmembrane shunt. La^{3+} is frequently used as inhibitor to test the involvement of calcium channels. And indeed, its effect observed in xylem-parenchyma cells and the effects of removing and reapplying extracellular Ca^{2+} (Fig. 6B) could be interpreted as Ca^{2+} permeable channels contributing to the short-circuiting of the electrogenic pump. In stelar protoplasts form maize roots, a hyperpolarization activated cation conductance (HACC) with high selectivity for Ca^{2+} and Ba^{2+} over K^+ and Na^+ was identified (Gilliham and Tester, 2005). However, the clear evidence of HACC in xylem parenchyma from barley roots still has to be provided. La^{3+} might affect other ion conductances including the electrogenic pump, and the Ca^{2+} effect shown in Fig. 6 B might be due to Ca^{2+}-dependent ion conductances rather than representing a Ca^{2+} conductance itself. A further interaction is

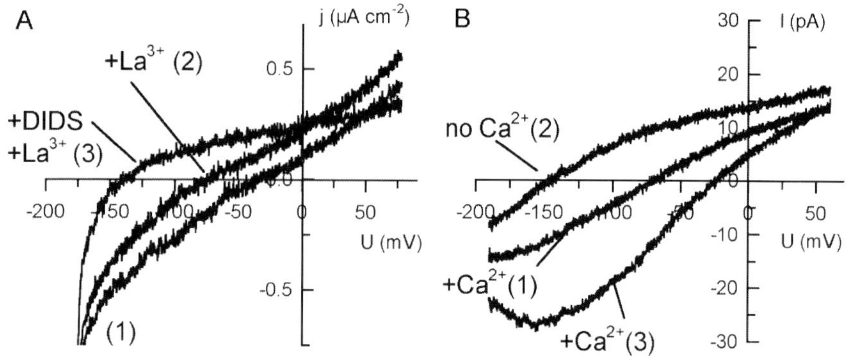

Fig. 6. Conductances for anions and possibly also for calcium short-circuit the electrogenic pump. A, application of the Ca^{2+} channel blocker La^{3+} (1 mM) enhanced apparent pump activity, indicated by an increase in the current at 0 mV and a hyperpolarization of the membrane. The following addition of the anion channel inhibitor DIDS (100 µM) led to a further hyperpolarization without a change in the current at 0 mV. B, removal of external Ca^{2+} brought out a current-voltage relationship indicative of the pump (compare 1 and 2). This activity was concealed after renewed addition of Ca^{2+} (5 mM) to the bath (3). For the composition of solution see Fig. 5. In Fig. 6 A $Ca(OH)_2$ was used instead of $Ca(Glc)_2$.

expected between the activities of pump and anion- (and possibly Ca^{2+}-) conductances with those of KORC. This remains to be investigated.

X-QUAC, X-IRAC and X-SLAC activity occurred simultaneously with pump activity resulting in intermediate reversal potentials. Repeatedly, current-voltage relationships showed triple zero crossings (Fig. 7). Application of DIDS led to the disappearance of two of the three reversal potentials; only the one at hyperpolarization remained; the one that was produced by the pump (J. Zhu et al., 2007). The simultaneous activity of the H^+-pump and of X-IRAC or X-QUAC, provide mechanisms for electroneutral acid release into the xylem also at very negative membrane potentials (Köhler and Raschke, 2000). De Boer and Volkov (2003) measured an acidification of the xylem sap down to a pH of 4.2 after stimulation of the pump by addition of fusicoccin to the medium with which they perfused the xylem. L. H. Wegner, in a personal communication, reported that the xylem pH in complete roots of maize can be equally low, even in the absence of fusicoccin. It is conceivable that malic acid (pK_1 = 3.5) enters the xylem through pump and anion conductances and functions there as a buffer.

Proton pumping, together with chloride transport through X-QUAC, was suggested to function in the loading of borate into the xylem through the boron transporter, BOR1, located in the pericycle (Takano et al., 2002; Frommer

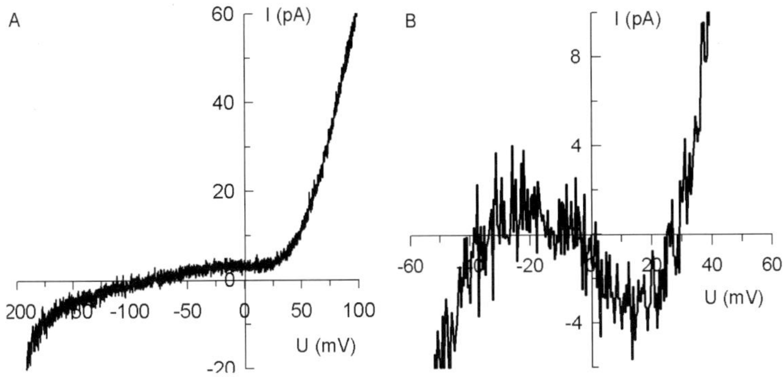

Fig. 7. Simultaneous activities of the electrogenic pump and X-QUAC with (A) one and (B) three zero crossings. Two whole-cell experiments on two xylem-parenchyma protoplasts are shown (note different scales). Solutions were (mM): 30 TEACl, 5 (A) and 40 (B) $Ca(Glc)_2$, 2 $MgCl_2$, 10 Mes, pH 5.8 (adjusted with Tris) (bath solution) and 120 TEACl, 0.15 µM free Ca^{2+} (added as $Ca(Glc)_2$), 10 EGTA, 2 ATP, 2 $MgCl_2$, 10 Tris, pH 7.2 (adjusted with Mes) (pipette solution).

and Wirén, 2002). Admittedly, the presence of X-QUAC in the pericycle was not proven so far, but it can be expected by extrapolation: Activity of KORC was discovered in protoplasts derived from the xylem parenchyma (Wegner and Raschke, 1994) but its molecular correspondence, SKOR, appeared in abundance also in the pericycle (Gaymard et al., 1998). Reduction in pH lowered the activity of SKOR (Lacombe et al., 2000), and, indeed, De Boer and Volkov (2003) recorded a diminution of K^+ efflux into an acidified xylem. In contrast, the efflux of Na^+ was stimulated. This effect could also be explained by the decrease in pH, if Na^+ is released into the xylem through an Na^+/H^+ antiporter, one like the product of the gene SOS1 (De Boer and Volkov, 2003).

The pump polarizes the plasmalemma, short-circuiting by ion conductances leads to depolarization. Correspondingly, the xylem parenchyma will switch from a state of ion uptake to one of ion loss. Negative membrane voltages will be required for re-absorption of NO_3^- by the putative NO_3^-/H^+-symporter (Fig. 1) and also for the uptake of amino acids (Okumoto et al., 2002). Absorption of K^+ from the xylem during nutrient cycling (Marschner et al., 1997) requires a potential below −100 mV (Fig. 1, KIRC, Wegner and Raschke, 1994). Depolarization by increased activity of anion conductances would establish the condition for xylem loading (Fig. 3). Thus, the activity of the electrogenic pump in the xylem parenchyma does not energize the transfer of salts to the xylem but, in combination with the short-circuiting conductances, plays a crucial role in controlling xylem loading and unloading through modulation of the voltage difference across the membrane.

8. CONTROL OF STELAR ION CONDUCTANCES BY ABA

The relationships among the activities of pumps, channels, symporters and antiporters are not only controlled by voltage, but also by phytohormones, Ca^{2+}, pH, metabolites, and other messengers (as for example GTP, Wegner and De Boer, 1997b). The application of the phytohormone abscisic acid (ABA) prior to protoplast isolation reduced K^+ and anion outward currents in stelar cells of maize possibly due to a lower expression level (Roberts and Snowman, 2000; Gilliham and Tester, 2005). Indeed, SKOR transcript levels strongly decreased 3 hours after ABA application (Pilot et al., 2003). The reduction of anion outward currents is attributed to the inhibition of *Zm*-X-QUAC (Gilliham and Tester, 2005). ABA is known to increase solute accumulation within the root by having little effect on initial ion uptake but by significantly inhibiting release of ions into the xylem (Roberts and Snowman, 2000 and references therein). The observed ABA action on K^+ and anion conductances in maize and *Arabidopsis* roots indicates that ABA effects on KCl/KNO_3 transport in roots are, at least in part, ion channel-mediated. Consistent with these observations, the inwardly-rectifying K^+ channel (KIRC) of the maize root cortex was insensitive to ABA (Roberts, 1998). In contrast, ABA pre-treatment activated KIRC and X-IRAC from the stele of maize (Roberts and Snowman, 2000; Gilliham and Tester, 2005). The differing effects of ABA on *Zm*-X-QUAC and *Zm*-X-IRAC may indicate different roles of the anion conductances under different (stress) conditions. However, the increase in *Zm*-X-IRAC activity could not compensate the observed overall reduction of anion currents (Gilliham and Tester, 2005). In addition to the hypothesized effect on transcript level, ABA affected the anion conductances immediately. Addition of ABA to protoplasts inhibited *Zm*-X-QUAC within minutes. This response has been proposed to be mediated via an increase in the cytosolic Ca^{2+} concentration (Gilliham and Tester, 2005). Nothing is known about the effect of ABA on the H^+-pump in xylem-parenchyma cells. The membrane potential of stelar cells from ABA-treated maize roots was significantly more negative than those from control roots (Roberts and Snowman, 2000). However, the mechanism by which ABA hyperpolarizes the maize root is still unknown. A possible explanation would be a stimulation of the H^+-ATPase activity, which would be in contrast to other cell types like guard cells and suspension cells where ABA inhibited H^+-ATPase activity (Brault et al., 2004; Zhang et al., 2004). A possible ABA effect on H^+-pump activity in xylem-parenchyma cells is unlikely to be mediated via the cytosolic Ca^{2+} concentration. Other than with K^+ and anion conductances, H^+-pump activity was independent of the intracellular Ca^{2+} concentration in this cell type (see above, Wegner and De Boer, 1997a; Köhler and Raschke, 2000).

Further regulatory factors of major interest are amino acids and pH. The pH sensitivity of the K^+ outward rectifier SKOR indicates that internal and external pH plays a role in the control of K^+ secretion into the xylem sap (Lacombe et al., 2000). Both, an internal pH decrease from 7.4 to 7.2 and external acidification from 7.4 to 6.4 induced a strong voltage-independent decrease of the macroscopic current. Nothing is known about the effect of the intra- and extracellular pH on X-QUAC, X-IRAC and X-SLAC. Also, the effect of amino acids on ion conductances was not investigated yet. Amino acids might be important signals in nutrient cycling. Their content in xylem sap varies dependent on the nitrogen source and if NO_3^- reduction takes place in roots or shoots. Major amino acids such as glutamine, asparagine, and glutamate and γ-aminobutyric-acid (GABA) should be included in further investigations.

9. OUTLOOK

Xylem loading is a control point in the transfer of nutrients from the root to the shoot. Control is exerted through several ion conductances. The molecular identity of some of them needs to be determined, this need applies in particular to the anion conductances. A number of other transporters were discovered in stelar cells: BOR1, a boron transporter, involved in boron transport into the xylem (Takano et al., 2002; Frommer and Wirén, 2002), SOS1, a Na^+ transporter, which might function in retrieving Na^+ from the xylem (Shi et al., 2002), SULTR2;1 a low affinity sulphate transporter, most likely re-absorbing sulphate from intercellular spaces in conjunction with SULTR3;5 (Takahashi et al., 2000; Kataoka et al., 2004), and AAP6, a high-affinity amino acid symporter, which may function in uptake of amino acids from the xylem sap (Okumoto et al., 2002). The example of AAP6 underlines the importance of cell-type specific characterization of transporters. AAP6 from the xylem parenchyma is the only member of a multigene family with an affinity for aspartate in the physiologically relevant range. Increasing proton concentrations strongly activate transport of amino acids. Thus, the actual apoplastic concentration of amino acids and the pH will determine what is transported *in vivo* (Fischer et al., 2002; Okumoto et al., 2002).

There is need for modelling electroneutral salt loading through anion- and cation conductances of differing voltage dependences and sensitivities to second messengers, including Ca^{2+} and phytohormones, in the presence of high or low activities of electrogenic pumps. We require knowledge about how the activities of pumps, channels, symporters and antiporters in the stele are adjusted among each other in order to understand the loading of the apoplast in the root.

REFERENCES

Brault, M., Amiar, Z., Pennarun, A.M., Monestiez, M., Zhang, Z., Cornel, D., Dellis, O., Knight, H., Bouteau, F. and Rona, J.P. (2004) Plasma membrane depolarization induced by abscisic acid in *Arabidopsis* suspension cells involves reduction of proton pumping in addition to anion channel activation, which are both Ca^{2+} dependent. *Plant Physiol.*, 135, 231-43.

Britto, D.T., Ruth, T.J., Lapi, S. and Kronzucker, H.J. (2004) Cellular and whole-plant chloride dynamics in barley: insughts into chloride-nitrogen interactions and salinity responses. *Planta*, 218, 615-622.

Crafts, A.S. and Broyer, T.C. (1938) Migration of salts and water into xylem of the roots of higher plants. *Am. J. Bot.*, 25, 529-535.

De Boer, A.H. and Volkov, V. (2003) Logistics of water and salt transport through the plant: structure and functioning of the xylem. *Plant Cell Environ.*, 26, 87-101.

Dietrich, P. and Hedrich, R. (1994) Interconversion of fast and slow gating modes of GCAC1, a guard cell anion channel. *Planta*, 195, 301-304.

Drew, M.C., Webb, J. and Saker, L.R. (1990) Regulation of K^+ uptake and transport to the xylem in barley roots: K^+ distribution determined by electron probe x-ray microanalysis of frozen-hydrated cells. *J. Exp. Bot.*, 41, 815-825.

Dunlop, J. (1989) Phosphate and membrane electropotentials in *Trifolium repens* L. *J. Exp. Bot.*, 40, 803–807.

Dunlop, J. and Bowling, D.J.F. (1971a) The movement of ions to the xylem exudate of maize roots I. Profiles of membrane potential and vacuolar potassium activity across the root. *J. Exp. Bot.*, 22, 434–444.

Dunlop, J. and Bowling, D.J.F. (1971b) The movement of ions to the xylem exudate of maize roots II.A comparison of the electrical potential and the electrochemical potentials of ions in the exudate and in the root cells. *J. Exp. Bot.*, 22, 445-452.

Dunlop, J. and Bowling, D.J.F. (1971c) The movement of ions to the xylem exudate of maize roots III.The location of the electrical and electrochemical potential differences between the excudate and the medium. *J. Exp. Bot.*, 22, 453-464.

Findlay, G.P., Tyerman, S.D., Garrill, A. and Skerrett, H. (1994) Pump and K^+ inward rectifiers in the plasmalemma of wheat root protoplasts. *J. Memb. Biol.*, 139, 103-116.

Fischer, W.N., Loo, D.D., Koch, W., Ludewig, U., Boorer, K.J., Tegeder, M., Rentsch, D., Wright, E.M. and Frommer, W.B. (2002) Low and high affinity amino acid H^+-cotransporters for cellular import of neutral and charged amino acids. *Plant J.*, 29, 717-731.

Forde, B.G. (2000) Nitrate transporters in plants: structure, function and regulation. *Biochim. Biophys. Acta.*, 1465, 219-235.

Frommer, W.B. and von Wirén, N. (2002) Ping-pong with boron. *Nature*, 420, 282-283.

Gaymard, F., Pilot, G,. Lacombe, B., Thibaud, J.P. and Sentenac, H. (1998) Identification and disruption of a plant Shaker-like outward channel involved in K+ release into the xylem sap. *Cell*, 94, 647-655.

Gilliham M, Tester M (2005) The regulation of anion loading to the maize root xylem. *Plant Physiol.*, 137, 819-828.

Glass, A.D.M., Brito, D.T., Kaiser, B.N., Kronzucker, H.J., Kumar, A., Okamoto, M., Rawat, S.R., Siddiqi, M.Y., Silim, S.M., Vidmar, J.J. and Zhuo, D. (2001) Nitrogen transport in plants, with an emphasis on the regulation of fluxes to match plant demand. *J. Plant Nutr. Soil Sci.*, 164, 199-207.

Hamburger D, Rezzonico E, MacDonald-Comber Petetot J, Sommerville C, Poirier Y (2002) Identification and characterization of the *Arabidopsis* PHO1 gene involved in phosphate loading to the xylem. *Plant Cell*, 14, 889-902.

Hamill, O.P., Marty, A., Neher, E., Sakmann, B. and Sigworth, F.J. (1981) Improved path-clamp techniques for high-resolution current recording from cells and cell-free membrane patches. *Pflügers Arch.*, 391, 85-100.

Hanson, J.B. (1978) Application of the chemiosmotic hypothesis to ion transport across the root. *Plant Physiol.*, 62, 402-405.

Herdel, K., Schmidt, P., Feil, R., Mohr, A. and Schurr, U. (2001) Dynamics of concentrations and nutrient fluxes in the xylem of *Ricinus communis* – diurnal course, impact of nutrient availability and nutrient uptake. *Plant Cell Environ.*, 24, 41-52.

Kataoka, T., Hayashi, N., Yamaya, T. and Takahashi, H. (2004) Root-to-shoot transport of sulfate in *Arabidopsis*. Evidence for the role of SULTR3;5 as a component of low-affinity sulfate transport system in the root vasculature. *Plant Physiol.*, 136, 4198-4204.

Kawachi, T., Nishijo, C., Fujikake, H., Abdel-Latif, S., Ohtake, N., Sueyoshi, K., Ohyama, T., Shigeta-Ishioka, N., Watanabe, S., Osa, A., Sekine, T., Matsuhashi, S., Ito, T., Mizuniwa, C., Kume, T., Hashimoto, S., Uchida, H. and Tsuji, A. (2002) Effects of anion channel blockers on xylem nitrate transport in Barley seedlings. *Soil Sci. Plant Nutr.*, 48, 271-277.

Kiegle, E., Gilliham, M., Haseloff, J. and Tester, M. (2000) Hyperpolarisation-activated calcium currents found only in cells from the elongation zone of *Arabidopsis thaliana* roots. *Plant J.*, 21, 225-229.

Kinoshita, T., Shimazaki, K.I. and Nishimura, M. (1993) Phosphorylation and dephosphorylation of guard-cell proteins from *Vicia faba* L in response to light and dark. *Plant Physiol.*, 102, 917-923.

Köhler, B. and Raschke, K. (1998) An electrogenic pump in cells of the xylem parenchyma of barley roots. 11th International Workshop on Plant Membrane Biology, Cambridge, U.K. Experimental Biology Online, www.link.springer.de/link/service/journals/00898/toc.htm:

Köhler, B. and Raschke, K. (2000) The delivery of salts to the xylem. three types of anion conductance in the plasmalemma of the xylem parenchyma of roots of barley. *Plant Physiol.*, 122, 243-254.

Köhler, B., Wegner, L.H., Osipov, V. and Raschke, K. (2002) Loading of nitrate into the xylem: apoplastic nitrate controls the voltage dependence of X-QUAC, the main anion conductance in xylem-parenchyma cells of barley roots. *Plant J.*, 30, 133-142.

Lacombe, B., Pilot, G., Gaymard, F., Sentenac, H. and Thibaud, J.B. (2000) pH Control of the Plant Outwardly-Rectifying Potassium Channel Skor. *FEBS LETT*, 466, 351-354.

Läuchli, A., Kramer, D., Pitman, M.G. and Lüttge, U. (1974) Ultrastructure of barley roots in relation to ion transport to the xylem. *Planta*, 110, 85-99.

Leonard, R.T. and Hotchkiss, C.W. (1976) Cation-stimulated adenosine triphosphatase activity and cation transport in corn roots. *Plant Physiol.*, 58, 331-335.

Lohse, G. and Hedrich, R. (1992) Characterization of the plasma-membrane H^+-ATPase from *Vicia faba* guard cells. Modulation by extracellular factors and seasonal changes. *Planta*, 166, 206-214.

Luo, H., Morsomme, P. and Boutry, M. (1999) The two major types of plant plasma membrane H^+-ATPases show different enzymatic properties and confer differential pH sensitivity of yeast growth. *Plant Physiol.*, 119, 627-634.

Okumoto, S., Schmidt, R., Tegeder, M., Fischer, W.N., Rentsch, D., Frommer, W.B. and Koch, W. (2002) High affinity amino acid transporters specifically expressed in xylem parenchyma and developing seeds of *Arabidopsis*. *J. Biol. Chem.*, 277, 45338-46.

Marschner, H. (1995) *Mineral Nutrition of Higher Plants*. 2nd edn. Academic Press, London

Marschner, H., Kirkby, E.A. and Engels, C. (1997) Importance of cycling and recycling of minreal nutrients within plants for growth and development. *Bot. Acta*, 110, 265-273.

Mattsson, M., Lundborg, T. and Larsson, C. (1988) Nitrate utilization in barley: relations to nitrate supply and light/dark cycles. *Physiol. Plant.*, 73, 380-386.

McClure, P.R., Kochian, L.V., Spanswick, R.M. and Shaff, J.E. (1990) Evidence for cotransport of nitrate and protons in maize roots. I. Effects of nitrate on the membrane potential. *Plant Physiol.*, 93, 281–289.

Morsomme, P. and Boutry, M. (2000) The plasma membrane H^+-ATPase: structure, function and regulation. *Biochem. Biophys. Acta*, 1465, 1-16 .

Okumoto, S., Schmidt, R., Tegeder, M., Fischer, W.N., Rentsch, D., Frommer, W.B. and Koch, W. (2002) High affinity amino acid transporters specifically expressed in xylem parenchyma and developing seeds of *Arabidopsis*. *J. Biol. Chem.*, 277, 45338-45346.

Palmgren, M.G. (2001) Plant plasma membrane H^+-atpases: powerhouses for nutrient uptake. *Annu. Rev. Plant Hysiol.*, 52, 817-845.

Pilot, G., Gaymard, F., Mouline, K., Cherel, I. and Sentenac, H. (2003) Regulated expression of *Arabidopsis* shaker K^+ channel genes involved in K^+ uptake and distribution in the plant. *Plant Mol. Biol.*, 51, 773-87.

Pitman, M.G. (1971) Uptake and transport of ions in barley seedlings I. Estimation of chloride fluxes in cells of excised roots. *Austr. J. Biol. Sci.*, 24, 407-421.

Pitman, M.G. (1972) Uptake and transport of ions in barley seedlings II. Evidence for two active stages in transport to the shoot. *Austr. J. Biol. Sci.*, 25, 243-257.

Raschke, K. (2003) Alternation of the slow with the quick anion conductance in whole guard cells effected by external malate. *Planta*, 217, 651-657.

Roberts, S.K. (1998) Regulation of K^+ channels in maize roots by water stress and abscisic acid. *Plant Physiol.*, 116, 145-153.

Roberts, S.K. and Snowman, B.N. (2000) The effects of aba on channel-mediated K^+ transport across higher plant roots. *J. Exp. Bot.*, 51(350) Special Iss. SI, 1585-1594.

Roberts, S.K. and Tester, M. (1995) Inward and outward K^+-selective currents in the plasma membrane of protoplasts from maize root cortex and stele. *Plant J.*, 8, 811-825.

Roberts, S.K. (1998) Regulation of K^+ channels in maize roots by water stress and abscisic acid. *Plant Physiol.*, 116, 145-153.

Samuels, A.L., Fernando, M., Glass, A.D.M. (1992) Immunofluorescent localization of plasma membrane H^+-ATPase in barley roots and effects on K^+ nutrition. *Plant Physiol.*, 99, 1509-1514.

Serrano, R. (1990) Plasma membrane ATPase. In: C. Larsson, I.M. Mřller (eds), *The Plant Plasma Membrane. Sructure, Function and Molecular Biology*. Springer-Verlag, Berlin etc., pp. 127-153

Serrano, R. (1993) f Plasma Membrane H^+-ATPase. *FEBS Lett.*, 325, 108-111.

Shi, H., Quintero, F.J., Pardo, J.M. and Zhu, J.K. (2002) The putative plasma membrane Na^+/H^+ antiporter SOS1 controls long-distance Na^+ transport in plants. *Plant Cell*, 14, 465–477.

Shimazaki, K. and Kondo, N. (1987) Plasma membrane H^+-ATPase in guard-cell protoplasts from *Vicia faba* L. *Plant Cell Physiol.*, 28, 893-900.

Siebrecht, S., Herdel, K., Schurr, U. and Tischner, R. (2003) Nutrient translocation in the xylem of poplar – diurnal variations and spatial distribution along the shoot axis. *Planta*, 217, 783-793.

Sondergaard, T.E., Schulz, A.and Palmgren, M.G. (2004) Energization of transport processes in plants. Roles of the plasma membrane H^+-ATPase. *Plant Physiol.*, 136, 2475-2482.

Sze, H., Li, X. and Palmgren, M.G. (1999) Energization of plant cell membranes by H^+-pumping: regulation and biosynthesis. *Plant Cell*, 11, 677-689.

Takahashi, H., Watanabe-Takahashi, A., Smith, F.W., Blake-Kalff, M., Hawkesford, M.J. and Saito, K. (2000) The roles of three functional sulphate transporters involved in uptake and translocation of sulphate in *Arabidopsis thaliana*. *Plant J.*, 23, 171-82.

Takano, J., Noguchi, K., Yasumori, M., Kobayashi, M., Gajdos, Z., Miwa, K., Hayashi, H., Yoneyama, T. and Fujiwara, T. (2002) *Arabidopsis* boron transporter for xylem loading. *Nature*, 420, 337-340.

Tanner, W. and Beevers, H. (2001) Transpiration, a prerequisite for long-distance transport of minerals in plants? *Proc. Nat. Acad. Sci. USA*, 98, 9443-9447.

Tyerman, S.D., Beilby, M., Whittington, J., Juswono, U., Newman, I. and Shabala, S. (2001) Oscillations in proton transport revealed from simultaneous measurements of net current and net proton fluxes from isolated root protoplasts: mife meets patch-clamp. *Austr. J. Plant Physiol.*, 28, 591-604.

Ullrich, C.I. and Novacky, A.J. (1990) Extra- and intracellular pH and membrane potential changes Induced by K^+, Cl^-, $H_2PO_4^-$, and NO_3^- uptake and fusicoccin in root hairs of *Limnobium stoloniferum*. *Plant Physiol.*, 94, 1561-1567.

Wegner, L.H. and De Boer, A.H. (1997a) Properties of two outward-rectifying channels in root xylem parenchyma cells suggest a role in K^+ homeostasis and long-distance signaling. *Plant Physiol.*, 115, 1707-1719.

Wegner, L.H. and De Boer, A.H. (1997b) Two inward K^+ channels in the xylem parenchyma cells of barley roots are regulated by G-protein modulators through a membrane-delimited pathway. *Planta*, 203, 506-516.

Wegner, L.H. and De Boer, A.H. (1999) Activation kinetics of the K^+ outward rectifying conductance (KORC) in xylem parenchyma cells from barley roots. *J. Membr. Biol.*, 170, 103-119.

Wegner, L.H. and Raschke, K. (1994) Ion channels in the xylem parenchyma of barley roots. A procedure to isolate protoplasts from this tissue and a patch-clamp exploration of salt passageways into xylem vessels. *Plant Physiol.*, 105, 799-813.

Zhang, X., Wang, H., Takemiya, A., Song, C.P., Kinoshita, T. and Shimazaki, K. (2004) Inhibition of blue light-dependent H_+ pumping by abscisic acid through hydrogen peroxide-induced dephosphorylation of the plasma membrane H^+-ATPase in guard cell protoplasts. *Plant Physiol.*, 136, 4150-4158.

Zhu, J., Raschke, K. and Köhler, B. (2007). An electrogenic pump in the xylem parenchyma of barley roots. *Physiol. Plant.* 129 (2), pp. 397-406.

Section 4

The Significance of the Apoplast as a Compartment for Long-Distance Transport

NEW TOOLS TO EXPLORE THE APOPLAST

F.W. BENTRUP
Division of Plant Physiology, Department of Cell Biology, University of Salzburg, Austria, Friedrich.Bentrup@sbg.ac.at

In the cormophyte the long-distance transport route through xylem and phloem is established in terms of anatomy, ultrastructure and basic composition of the fluids passing through these conduits. As for the involved driving forces, the source-sink model of osmotically driven mass flow satisfactorily explains phloem transport, whereas the question of which forces drive the flow of water and solutes through the xylem still is a matter of vived debate triggered by new tools and concepts (cf. Zimmermann et al., 2004). This apoplastic route of water and minerals from the rhizosphere to the tissues and organs deserves detailed biophysical and biochemical elucidation in order to facilitate a sound understanding of the water relations and nutrient fluxes within a given plant in a given environmental context.

The contributions to this section, to my opinion, are paradigmatic for biological research "at the cutting edge", because they illustrate that scientific progress especially depends upon progress in pioneering new tools which, likewise typically, have emerged from interdisciplinary cooperation. Within a volume dedicated to the achievements of such cooperation, it seems adequate to recall briefly pertinent pioneering work of previous decades.

In 1967 André Läuchli published his first X-ray study of element-specific nutrient distribution and long-distance translocation, namely of K, Ca, Sr and P, in plant tissues and organs (see Läuchli and Boursier, 1989). In 1969 Ulrich Zimmermann and coworkers described the development of an intracellular turgor pressure probe, and 20 years later the likewise minimal-invasive xylem pressure probe (see Schneider et al., this volume, pp. 251-264). In 1984 Hubert Felle introduced a turgor pressure-tight ion sensitive microelectrode to monitor pH and pCa in plant cells (see Felle, 1993).

B. Sattelmacher† and W.J. Horst (eds.), The Apoplast of Higher Plants: Compartment of Storage, Transport and Reactions, 203–205.

Recently, he accomplished non-invasive recording from, and infusion of agents into, the substomatal cavity of leaves by inserting the probe through the stomatal aperture (Hanstein and Felle, 2004; Felle and Hanstein, this volume, pp 299-310). Operation of a multifunctional probe for the simultaneous recording of xylem pressure, voltage and ion activities in the xylem sap, again from Zimmermann´s laboratory, will be described by Wegner et al. (this volume, pp 211-224). Insertion of this probe also does not affect the hydraulic equilibrium between the vessel and surrounding parenchyma cells, so the composition of the xylem sap remains basically undisturbed.

These minimal-invasive tools have been complemented by non-invasive in planta imaging of the distribution and flow of water by ^{1}H NMR spectroscopy (Kuchenbrod et al., 1995). This spectroscopy relies upon the fact that proton spin relaxation signals from the water molecule differ between bound water (e.g. in the plant cell wall) and mobile water flowing, for instance, in a xylem vessel. The article of Schneider et al. (this volume, pp 255-268) illustrates that the tuned application of the non-invasive NMR and the above outlined probes renders possible to record both the water flow in the xylem as well as the hydrostatic pressure gradients causing it; clearly, recording flow-force relationships will be indispensable for the understanding of the long-distance transport between root and shoot. Particularly, the meanwhile remarkable spatial resolution of the NMR imaging approach renders possible to sever phloem and xylem transport activity in planta.

Eventually, monitoring the intact transpiring plant in the field is the ultimate experimental scenario; a significant step toward this goal is outlined by Gilmer and Schurr (this volume, pp 225-234) comprising, for instance, continuous in planta recordings of the bulk flow parameter in the xylem conduit. Special attention is given to collect the xylem sap. Siebrecht et al. (this volume, pp 269-288) have already analyzed a wide range of nutrient ions, that is, nutrient content and nutrient flow, respectively, under a diurnal light regime.

In the future increased attention will include in planta monitoring of regulatory molecules, for instance, of the abscisic acid (ABA), or of the apoplast-bound γ–aminobutyric acid (GABA); the concentration of GABA has turned out to increase in the leaf apoplast after invasion of phytopathogenic fungi (Salomon and Oliver, 2001; Oliver and Salomon, 2004).

Finally, as any botanist is well aware, the microscopic intercellular network of the apoplast space in the plant body outside the conducting bundles consists of two phases divided by a gaseous-liquid meniscus. This labile meniscus network will immediately respond to any experimental shift of the ambient pressure, for instance, upon pressurization of a plant organ, and thus will markedly affect size and architecture of the apoplastic route. Therefore, the promising progress outlined in this Section largely resides with the fact that minimal or non-invasive tools now are available which rendered accessible the apoplast in situ without affecting the hydraulic equilibrium which is poised essentially by transpiration and the hydraulic coupling of the apoplast and symplast.

REFERENCES

Felle, H.H. (1993) Ion-selective microlectrodes: their use and importance in modern plant cell biology. *Botanica Acta*, 106, 5–12.

Hanstein, S.M. and Felle, H.H. (2004) Nanoinfusion: an integrating tool to study elicitor. *New Phytol.*, 161, 595–606.

Kuchenbrod, E., Haase, A., Benkert, R., Schneider, H. and Zimmermann, U. (1995) Quantitative NMR Microscopy on Intact Plants. *Magn. Reson. Imag.*, 13, 447–455.

Läuchli, A. and Boursier, P.J. (1989) Compartmentation in root cells and tissues – X-ray microanalysis of specific ions. *Methods in Enzymology*, 174, 267–277.

Oliver, R.P. and Salomon, P.S. (2004) Does the oxidative stress used by plants for defence provide a source of nutrients for pathogenic fungi? *Trends in Plant Sci.*, 9, 472–473.

Salomon, P.S. and Oliver, R.P. (2001) The nitrogen content of the tomato leaf apoplast increases during infection by *Cladosporium fulvum*. *Planta*, 213, 241–249.

Zimmermann, U., Schneider, H., Wegner and L.H., Haase, A. (2004) Tansley review: water ascent in tall trees: does evolution of land plants rely on a highly metastable state? *New Phytol.*, 162, 575–615.

ON-LINE MEASUREMENTS OF ION RELATIONS IN THE XYLEM SAP OF INTACT PLANTS

L.H. WEGNER, H. SCHNEIDER and U. ZIMMERMANN
Lehrstuhl für Biotechnologie, Biozentrum, Würzburg, Germany,
lars.wegner@biozentrum.uni-wuerzburg.de

Abstract. The ion-selective xylem probe is a new tool that allows for on-line recording of ion activities in the xylem sap when water is under tension. Ion-selective electrodes are combined with a xylem pressure probe additionally measuring the electrical potential in a vessel. Using a K^+-selective probe, xylem K^+ in maize roots was recorded simultaneously with the xylem pressure and the trans-root potential (the electrical potential in the root xylem with respect to the ambient medium) under conditions of varying light irradiation and K^+ supply. Furthermore, a new pH-selective xylem probe is presented. The impact of the new xylem probes for studies on plant nutrition is discussed.

Key words: ion-selective electrodes, long-distance transport of K^+, trans-root potential, xylem pH, xylem pressure, xylem probe

1. INTRODUCTION

In higher plants, the xylem conduits are the main pathway for long-distance transport of water and nutrients from the root to the shoot (Marschner, 1995). Accurate concentration measurements of the micro- and macronutrients in the xylem sap are needed to quantify nutrient supply to the shoot. Unfortunately, there are obstacles to the analysis of the composition of the xylem sap. Firstly, xylem vessels are small in diameter and not easily accessible, because they are often deeply located inside the tissue. Secondly,

B. Sattelmacher† and W.J. Horst (eds.), The Apoplast of Higher Plants: Compartment of Storage, Transport and Reactions, 207–220.

the sap is usually under tension, i.e. pressure is below atmospheric (Balling and Zimmermann, 1990; Zimmermann et al., 2004). At metastable pressures below vacuum, water may undergo cavitation, i.e. a spontaneous formation of gas bubbles. The sensitivity to cavitation at negative pressures precludes extraction of the sap using microcapillaries (but see Lohaus et al., 2000). Thus, most xylem sap analyses are based on gross, highly invasive techniques (Schurr, 1998). When the shoot (or a plant organ) is decapitated, tension in the xylem is released (i.e. pressure becomes atmospheric), and the xylem vessels become accessible from the cut surface. Detached root systems tend to exudate spontaneously at the cut surface ("root-pressure exudation"). This exudate has frequently been analysed to get an estimate of the composition of the xylem sap in the intact, transpiring plant (e.g. Rowan et al., 2000 and literature cited therein). However, Rowan et al. (2000), using cryo scanning electron microscopy, presented evidence that fluids collected that way do not originate from the xylem alone. But even if the exudate represents the composition of the xylem sap as is generally believed, it must be kept in mind that the process of exudation itself results from an accumulation of solutes in the root xylem vessels and, compared to the intact transpiring plant, the flow velocity is dramatically reduced (Schurr and Schulze, 1995). An improvement is to pressurise the detached root system to induce flow rates comparable to those in transpiring plants (White, 1997; Liang and Zhang, 1997). More sophisticated versions of this approach include sap collection from single veins of leaves and coupling of the applied pressure to the sap flow by feedback control (Schurr and Schulze, 1995; Schurr, 1998). However, all these techniques suffer from several shortcomings that cast doubt on the physiological relevance of the data obtained by this approach (see also Canny, 1993): (i) Cutting of the entire shoot leads to a breakdown of radial turgor pressure and osmotic gradients known to exist in the roots of intact plants (Zimmermann et al., 1992; Rygol et al., 1993), and similar effects are likely to occur in the leaf upon cutting of veins. (ii) "Xylem sap" which is pressed out of the cut surface may be contaminated by solutes released from damaged cells. (iii) Most importantly, due to cutting of the tissue and root pressurization, tension in the vessels is released and above-atmospheric pressures are established (Zimmermann et al., 1995). This will affect the turgor pressure of the adjacent xylem parenchyma cells, and, in turn, may influence ion channel activity in these cells and alter ion concentrations in the vessels.

An alternative approach that circumvents these problems is provided by the use of insects that are known to feed on xylem sap (so-called "xylem suckers"). Malone et al. (1999), using the spittlebug *Philaenus spumarius*, found that the insects did not affect xylem tension when feeding on the sap despite earlier speculations that beak insertion may induce cavitation inside a vessel under tension (Sperry et al., 1996). Ponder et al. (2002) perfused stem segments of tomato with solutions of known composition. When insects were feeding on these segments, good agreement was found between inorganic ion concentrations in the perfusate and excreta produced by the insects. However, the applicability of the technique is limited by the seasonal availability of the insects and the limited range of plants that can be investigated with a certain species.

The limitations of the avialable methods and the importance of obtaining precise, reliable values for the concentrations of the main nutrients under various environmental conditions called for a new, generally applicable technique that allows continuous monitoring of ion concentrations in individual xylem vessels of intact, transpiring plants. This method should interfere as little as possible with the physiology of the plant, i.e. it should be minimal- or non-invasive and should operate at physiological pressures (i.e. under tension). The above criteria are met by the ion selective microelectrode technique for in situ monitoring of ion concentrations.

2. ION-SELECTIVE XYLEM PROBES: IN SITU RECORDING OF K^+ IN THE XYLEM SAP OF INTACT PLANTS

Ion-selective microelectrodes have successfully been used for both intra- and extracellular recordings of ion concentrations in plants before (e.g. Maathuis and Felle, 1993; Sanders, 1993; Walker et al., 1996; Felle et al., 2000). The basics of this technique have been described in detail elsewhere (e.g. Ammann, 1986; for use in plant physiology see Felle, 1993; Miller, 1995). Briefly, the very tip of a microcapillary is filled with a matrix separating the bath solution from the medium inside the capillary. Theoretically, if this matrix is permeable to a single type of ion only, the electrical potential difference generated across this "membrane" corresponds to the Nernst potential of the particular ion. From the measured electrical potential difference and the concentration inside the capillary, unknown ion concentrations (or, more correctly, ion activities) in the solution surrounding the pipette tip can be calculated, provided that the capillary has been pre-calibrated against bath solutions of known composition (to account for non-ideal selectivity of the electrode). Most common nowadays are the so-called liquid-membrane electrodes (Ammann, 1986) which basically consist of a highly selective ionophore (e.g. valinomycine for measuring K^+) dissolved in a polar, hydrophobic solvent (e.g. 1,2-dimethyl-3-nitrobenzene). Ionophores that are selective for various ions are commercially available, including those for the most important nutrients like K^+, Ca^{2+}, Mg^{2+} and NO_3^- as well as Na^+ and H^+ (Ammann, 1986). We decided to start with measuring K^+ activities in roots of hydroponics-grown maize seedlings.

For placement of a K^+-selective microelectrode (and ion-selective microelectrodes in general) into xylem vessels, several technical obstacles had to be overcome. Most important was the verification of the electrode tip position. This problem could be solved by making use of the xylem pressure probe technique previously designed by Balling and Zimmermann (1990). Hydrostatic pressure in a xylem vessel is measured by inserting a microcapillary that is attached to a small perspex chamber with integrated pressure transducer (see also Schneider et al., this volume, pp 255–268). Here, the xylem pressure probe technique was combined with a K^+-selective electrode for simultaneous monitoring of the K^+ activity and the pressure in a single vessel using single-tipped, double-barrelled microcapillaries (Fig. 1). One barrel is designed as a K^+-selective electrode, whereas the other barrel is connected with the perspex chamber of the pressure probe. The "pressure" barrel also serves as the reference for the ion-selective electrode. To this end, an Ag/AgCl electrode is integrated into the perspex chamber of the probe (so-called xylem pressure-potential probe; Wegner and Zimmermann, 1998) and the capillary is filled with an electrolyte solution instead of destilled water. For technical details, the reader is referred to Wegner and Zimmermann (2002).

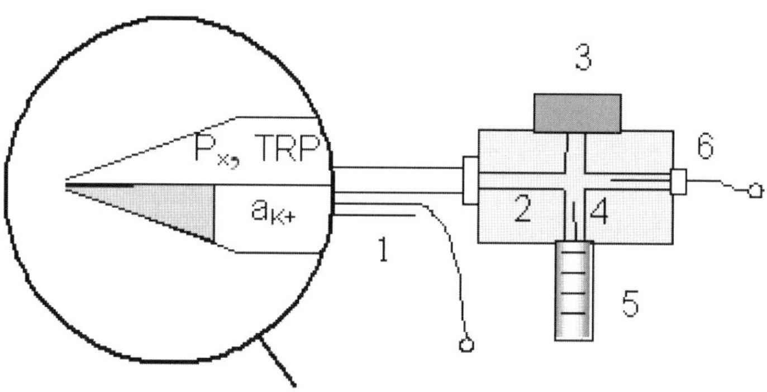

Fig. 1. Multifunctional xylem probe for measuring xylem K^+ activity (a_{K+}), xylem pressure (P_x) and trans-root potential (TRP). 1, double-barrelled capillary; 2, four-channel perspex chamber; 3, pressure transducer; 4, metal rod; 5, micrometer screw; 6, potential-measuring Ag/AgCl electrode. Both barrels form a single tip as shown under the "magnification glass". Reproduced from Wegner and Zimmermann, 2002, with permission.

When the double-barrelled capillary is advanced through the tissue, a rapid drop of the probe pressure to a sub-atmospheric value serves as a clear indication that a vessel has been hit, since stable underpressures are confined to the xylem (Zimmermann et al., 2004). 16 years of experience with the xylem pressure probe have established clear criteria for identifying proper location of the capillary tip in the lumen of a vessel and detecting tip clogging or leaks induced by impalement (for details, see Zimmermann, 2003; Zimmermann et al., 2004). Pressures above vacuum may result from tiny air bubbles in the probe or from leaks, but under these conditions negative pressures cannot be established, e.g. by increasing light irradiation.

Fig. 2 shows a typical example for the time course of the xylem pressure, the electrical potential in the xylem with respect to the ambient medium (termed "trans-root potential", TRP) and the K^+ activity recorded with the multifunctional xylem K^+ probe. The probe was inserted into a vessel close to the root base. Note that the pressure drops rapidly to a new stable value below vacuum (here –0.018 MPa) once a vessel has been hit (marked by an asterisk), confirming that the new method operates at negative pressures. This has not yet been proven for other methods. The electrical potential attains negative values during penetration of the cortex and stelar tissues due

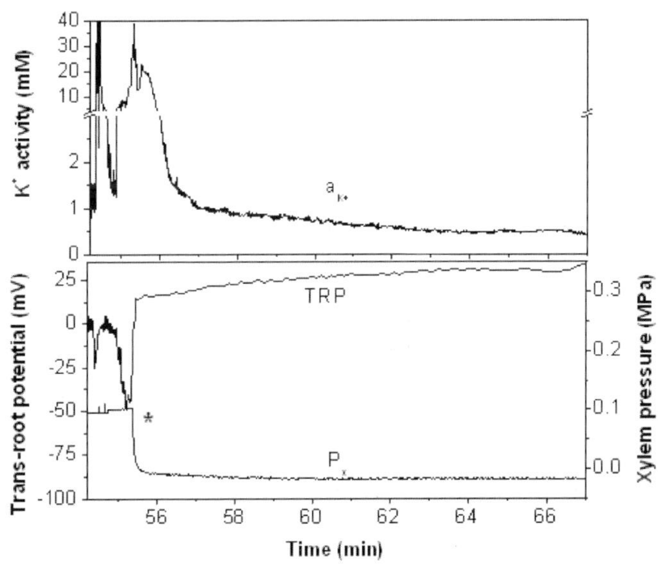

Fig. 2. Xylem K^+ activity (a_{K+}), xylem pressure (P_x) and trans-root potential (TRP) after a root xylem vessel of a maize seedling was hit by the K^+ selective xylem probe (indicated by asterisk). The root was impaled 22.8 cm above the tip (total root length 24.0 cm). Light irradiation: 300 μmol $m^{-2}s^{-1}$.

to the negative membrane potential of these cells. It relaxes towards a stable TRP once the tip is located in a xylem vessel; in this case the value was +33 mV. TRP recordings are not significantly affected by leaks or damage of the tissue at the site of impalement, as was shown by extensive testing of the xylem pressure-potential probe (Wegner and Zimmermann, 1998; Wegner et al., 1999). Before the tip contacts the root surface, the probe reads the K^+ activity in the ambient medium (1 mM). The value increases when the tip is advancing through the tissue, since K^+ concentration is much higher in the vacuolar sap of the cells that are penetrated. When the pressure trace indicates placement of the tip in a vessel, the K^+ activity relaxes rapidly to a stable value; here, it was 0.49 mM. The low value definitely excludes the possibility that the probe tip is located in immature xylem vessels as proposed previously (Milburn, 1996). In these vessels, xylem K^+ concentrations are extremely high (McCully, 1994). Moreover, the rapid attainment of this low value suggests that K^+ released from root cells that are inevitably destroyed by advancing the tip through the tissue is not a major source of error.

This is also supported by the experiment shown in Fig. 3 designed to test for proper functioning of the K^+ probe in situ. After impaling a root xylem vessel of a transpiring maize seedling, the root was cut about 1 cm away from the site of impalement in apical direction (horizontal arrow). As expected, this led to a rapid increase of the pressure (half-time 21 sec) in the open vessel from a value below vacuum (about –0.04 MPa) to atmospheric. Correspondingly, the TRP transiently became more negative and attained a final value of +15 mV. The lumen of the open vessels equilibrated rapidly with the bath upon cutting; this was reflected by an increase in the K^+ signal from 0.34 to about 1 mM, confirming that the probe was reading the correct K^+ values. Apparently, leakage of K^+ from the cells that were destroyed by impalement, or dilution of the sap by the probe contents were negligible. Correspondingly, when K^+-free pipette medium was injected into the xylem using a volume pulse (see upwardly-directed arrow in Fig. 3), this led only to a transient decrease in the measured K^+ activity, apparently due to a rapid exchange of the vessel content at the site of impalement by flow. It should also be noted that the rapid responses of xylem pressure and K^+ activity to the cutting of the root renders it impossible that the probe tip is located other than in the lumen of a vessel. This was claimed by several authors, e.g. by Sperry et al. (1996). Simultaneous recording of xylem pressure, TRP and K^+ activity is not only a technical prerequisite for localizing the probe tip in a vessel and for calculating the electrical potential drop across the tip of the ion-selective barrel in the tissue. It also offers a unique opportunity to relate changes of xylem K^+ (e.g. when plants are subjected to environmental stress)

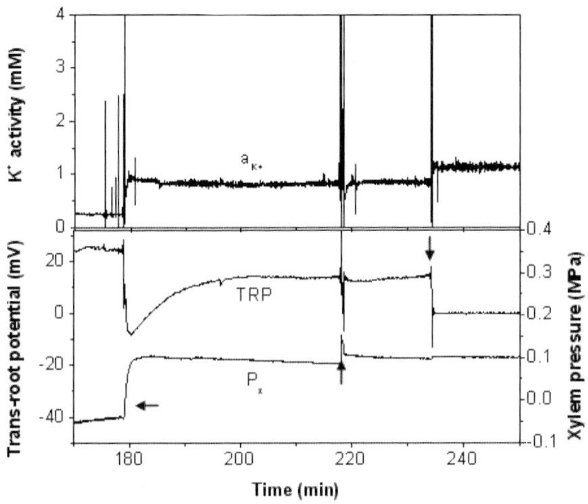

Fig. 3. Response of K^+ activity (a_{K+}), xylem pressure (P_x) and trans-root potential (TRP) to root excision. About 2 h after insertion of the probe the root was cut (horizontal arrow) 1 cm below the insertion point that was 28.3 cm above the root tip. The other arrows indicate:↑, injection of a volume pulse; ↓, removal of the probe. Reproduced from Wegner and Zimmermann, 2002, with permission; see there for further details.

to hydraulic and electrical responses of the plant. For the significance of the xylem pressure as an indicator of the plant water status, see e.g. Zimmermann et al. (2002, 2004) and Schneider et al. (this volume, pp 255–268). The TRP reflects changes in membrane potentials of both cortex cells and xylem parenchyma cells, as shown previously for excised roots (De Boer et al., 1983) as well as roots of intact transpiring plants (Wegner et al., 1999; see also De Boer, 1999). Hence the TRP, together with xylem K^+ activity, can be interpreted with respect to membrane transport processes that are related to xylem loading of K^+.

3. RESPONSE OF XYLEM K^+ TO CHANGES IN LIGHT IRRADIATION

A typical example of the response of root xylem-pressure, K^+ activity and the TRP to varying light intensities is given in Fig. 4. In this experiment, the multifunctional xylem K^+ probe was inserted into the root of a maize seedling 28.6 cm above the root tip, close to the root base. Impalement was performed at laboratory light (about 10 µmol m^{-2} s^{-1}). Once a xylem vessel was hit, stable values for the xylem pressure (0.069 MPa), the K^+ activity (about 3 mM) and the TRP (+13 mV) could be recorded after a few minutes.

Subsequently, the light irradiation was increased to 300 µmol m^{-2} s^{-1}. After a lag time of about 3 min, xylem pressure started to decrease; it passed through a minimum of –0.12 MPa and levelled off at a new stationary value of 0.03 MPa. It was previously shown that this undershoot upon an increase in light irradiation is due to a delay in the regulation of stomatal aperture (Wegner and Zimmermann, 1998; Schneider et al., this volume, pp. 251–264). Xylem K^+ activity responded to the increase in transpiration by a rapid decrease to about 0.6 mM.

Xylem pressure and K^+ activity increased again when light intensity was reduced to the initial level. Similar patterns were evoked by two further periods of elevated light intensity in the course of this experiment (Fig. 4) as well as in comparable experiments performed on other individuals (see Wegner and Zimmermann, 2002). Reduction of xylem K^+ after the onset of transpiration can be interpreted in terms of a dilution of the xylem sap when radial water flow in the root increased but K^+ release into the xylem was not affected to the same extent. Similar observations were made when conventional techniques of xylem sap sampling were applied (Schurr and Schulze 1995), but Wegner and Zimmermann (2002) also showed that xylem K^+ activities are considerably lower (by a factor of two to three) in intact

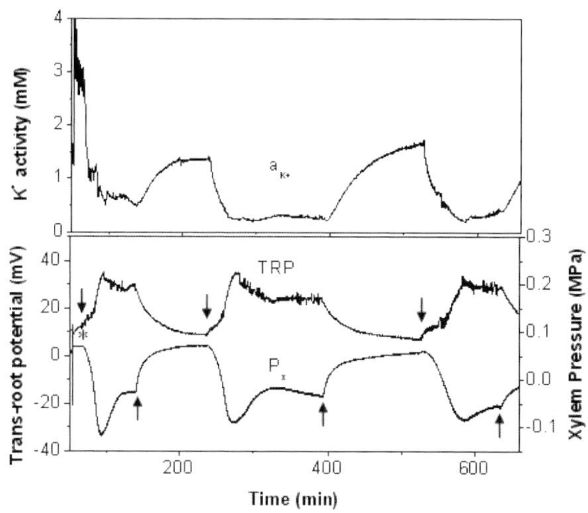

Fig. 4. Effect of light on the xylem K^+ activity (a_{K+}), xylem pressure (P_x) and trans-root potential (TRP). A 29.7-cm-long root was impaled 28.6 cm above the tip (*) at laboratory light (10 µmol m^{-2} s^{-1}). Subsequently, the plant was subjected alternately to elevated light irradiation (300 µmol m^{-2} s^{-1},↓) and low light irradiation (↑). Reproduced from Wegner and Zimmermann, 2002, with permission; see there for further details.

plants than reported previously (e.g. Miller, 1985, for maize). Xylem K^+ activities at 300 μmol m^{-2} s^{-1} light irradiation were often below the value of 1 mM that prevailed in the ambient medium. In the experiment shown in Fig. 4, the TRP became more positive when the light irradiation was increased (by about +21mV), passed through a maximum and returned to lower positive values. This effect on the TRP was also reversible.

4. ROOT XYLEM K^+ VERSUS K^+ ACTIVITY IN THE NUTRIENT SOLUTION

Another obvious application of the multifunctional K^+ probe to physiological studies on plant nutrition is to vary the external K^+ activity ($a_{K+,ext}$) and to monitor the subsequent effect on root xylem K^+. A typical example of such an experiment is shown in Fig. 5. A xylem vessel was impaled at the elevated light irradiation (300 μmol m^{-2} s^{-1}) at a concentration of 1 mM KCl in the bath.

Subsequently, $a_{K+,ext}$ was lowered to 0.1 mM and then increased stepwise to 42 mM. The bath solution was exchanged for the higher concentration

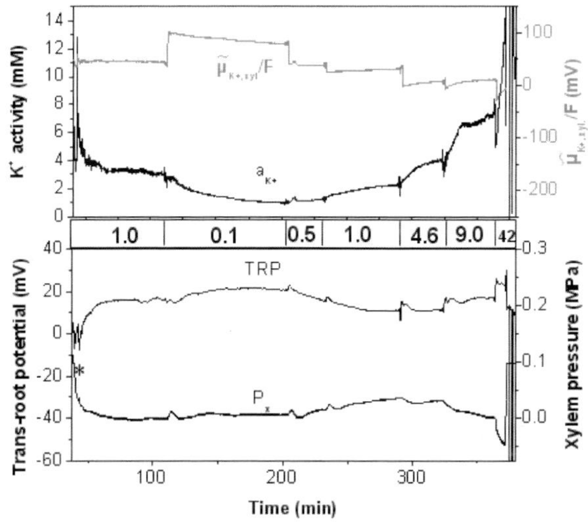

Fig. 5. Xylem K^+ activity (a_{K+}), K^+ electrochemical potential in the xylem with respect to the bath ($\mu_{K+,xyl}/F$), xylem pressure (P_x) and trans-root potential (TRP) upon stepwise changes of ambient K^+ as indicated in the figure (acitivities in mM). The 29.8-cm-long root of a 18-day-old plant was impaled 27.3 cm above the tip. Light irradiation: 300 μmol m^{-2} s^{-1}; rel. humidity 34%.

when the K^+-selective barrel returned to stable potentials. In this experiment the whole root system, including the site of impalement, was subjected to this K^+ regime whereas in other experiments the K^+ activity at the root base was kept constant at 1 mM throughout the experiment (Wegner and Zimmermann, 2002).

Qualitatively similar results were obtained, indicating that leakage of K^+ at the site of impalement was negligible. In the experiment shown in Fig. 5, a K^+ activity of 3.1 mM in the xylem was recorded after the impalement at 1 mM K^+ in the bath. When $a_{K+,ext}$ was reduced to 0.1 mM, K^+ in the xylem sap decreased slowly to 0.98 mM. With a progressive increase of $a_{K+,ext}$ to 0.5 and 1 mM, xylem K^+ increased slightly to 1.26 and 2.23 mM, respectively. Note that the latter value is somewhat lower than the K^+ activity recorded directly after the impalement in an identical bath medium. The difference may be due to the intermittent incubation in a diluted medium. A further increase of the external K^+ activity to 4.6 and 9 mM led to more pronounced changes in the K^+ activity of the xylem sap. It increased to 4.5 and 7.0 mM, respectively. At 42 mM external K^+, cavitation occurred before a new steady value was established.

Xylem pressure fluctuated around vacuum at the begin of the experiment. At 9 mM KCl and, more pronounced, at 42 mM, pressure responded by a rapid decrease due to the increase in the osmotic pressure of the bath until the vessel cavitated. The TRP ranged from +10 to +20 mV throughout the experiment. From the TRP and the ratio of xylem K^+ ($a_{K+,xyl}$) to external K^+ ($a_{K+,ext}$), the electrochemical potential of K^+ in the xylem ($\mu_{K+,xyl}$) with respect to the bath can be calculated (expressed as a voltage by division with the Faraday constant, F; R = gas constant; T = absolute Temperature):

$$\mu_{K+,xyl}/F = TRP + RT/F \ln(a_{K+,xyl}/a_{K+,ext}) \qquad (1)$$

This parameter was also plotted for the experiment shown in Fig. 5. At $a_{K+,ext}$ values up to 1 mM, the electrochemical potential difference would allow for K^+ efflux from the xylem. Above this value, the driving force was about zero. However, in this case the electrochemical potential gradient is of limited use for predicting the direction of net K^+ transport, because K^+ transport in the root is predominantly symplastic and governed by the mechanism of salt uptake at the cortex and K^+ release from the xylem parenchyma cells in the stele. Moreover, coupling of K^+ transport to radial water flow (i.e. solvent drag) in the transpiring plant is not taken into account.

Despite these limitations, it can be concluded that the K^+ activity in the xylem is well buffered against short-term changes in the external K^+ supply, especially when $a_{K+,ext}$ is below 5 mM. K^+ uptake by the root strongly increases with increasing $a_{K+,ext}$ in this concentration range (Kochian and Lucas, 1982), indicating that K^+ uptake and K^+ release into the xylem are independent of each other (Wegner and Zimmermann, 2002). Further work will focus on the effects of long-term K^+ deficiency on xylem K^+. Modification and partial revision of long-distance K^+ transport models (e.g. White, 1997; Peuke et al., 2002) may be required when minimal-invasive techniques like the multifunctional K^+ probe become available. This is also true for data on the xylem sap pH that plays a key role in long-distance signalling, e.g. under conditions of drought stress (Wilkinson, 1999), and in Fe^{2+}/Fe^{3+} nutrition (Mengel et al., 1994).

5. THE XYLEM pH PROBE

The ion-selective xylem probe technique could also be applied to pH measurements. To this end, a pH-selective fluid membrane was designed to meet all the requirements of this kind of measurement in the xylem sap of transpiring plants. 4-nonadecylpyridine was used as H^+-ionophore (for technical details, see Wegner and Zimmermann, 2004). A typical example for a synchronous recording of xylem sap pH, xylem pressure and the TRP is depicted in Fig. 6. The experiment was started at laboratory illumination (10 $\mu mol\ m^{-2}\ s^{-1}$). After calibration (Fig. 6, inset), a xylem vessel was impaled as indicated by a pressure drop from atmospheric to 0.02 MPa. The TRP relaxed more slowly from an initial value of about –25 mV to a less negative steady value (–5 mV). The delay (about 5 min) presumably reflected the healing process of stelar cells at the insertion site (Wegner and Zimmermann, 1998).

By contrast, the pH changed rapidly to a value of about 4.8 and, apart from minor fluctuations, remained stable. When the light intensity was increased to 300 $\mu mol\ m^{-2}\ s^{-1}$, the TRP responded first with a hyperpolarization to a final value of 11 mV. This result was in contrast to the experiment shown in Fig. 4, but in agreement with a previous publication (Wegner and Zimmermann, 1998). With a delay of 2–3 min, the xylem pressure started to decrease and entered the negative pressure range. The experiment was terminated by a cavitation, as indicated by a rapid pressure increase to values slightly above vacuum (see arrow). The pH appeared to be independent of short-term fluctuations of the light irradiation.

The pH values recorded by the multifunctional pH probe in this and 24 other experiments ranged between 4.2 and 4.9. It appears that the xylem sap pH, at least in maize, is more acidic than inferred from measurements on root-pressure exudate (pH 5.5–5.7, Davis and Higinbotham, 1969) or on sap

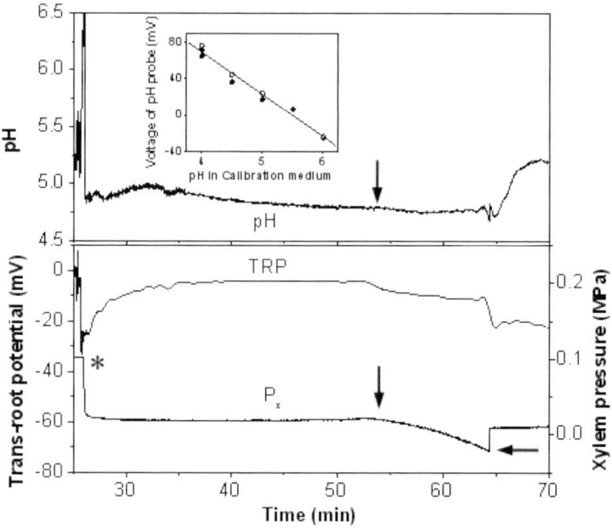

Fig. 6. Recording of xylem sap pH, xylem pressure (P_x) and trans-root potential (TRP) in a 16.5-cm-long maize root. The root was impaled 15.0 cm above the tip at laboratory light (10 $\mu mol\ m^{-2} s^{-1}$). 25 min later, the light irradiation was increased to 300 $\mu mol\ m^{-2} s^{-1}$ (↓). Xylem pressure decreased until cavitation occurred (←). Inset: Calibration of the pH probe before (●) and after(○) the experiment. The slope was 46 mV (continuous line).

extracted by the root-pressure bomb (pH 5.01; Miller, 1985). This may have far-reaching consequences, e.g. for modelling mobilization of abscisic acid from vacuolar stores (Daeter et al., 1993). Interestingly, addition of bicarbonate to the bath at concentrations of 5 and 20 mM led to an alkalinization of the xylem sap; xylem pH increased on average by 0.59 (n = 3) and 1.07 degrees (n = 5), respectively (Wegner and Zimmermann, 2004). Bicarbonate was supplied at a constant pH and at constant osmolality of the bath.

6. CONCLUSION AND PERSPECTIVE

It was the intention of this review to give an overview of the data that have been obtained so far with multifunctional xylem probes for K^+ and pH. These probes have proved to be reliable tools for continuously measuring K^+ activity and pH, respectively, in the xylem sap of intact transpiring plants. Errors inherent to the conventional techniques of xylem sap sampling can be avoided.

In future work, the K^+ and pH xylem probes will be combined with simultaneous recordings of water uptake by the root or leaf gas exchange for an additional monitoring of the flows. This is required for quantitative modelling, e.g. of K^+ fluxes. We are also currently developing probes that are selective for NO_3^- and Ca^{2+}.

REFERENCES

Ammann, D. (1986) Ion-selective microelectrodes. Springer , Berlin etc.

Balling, A. and Zimmermann, U. (1990) Comparative measurements of the xylem pressure of *Nicotiana* plants by means of the pressure bomb and pressure probe. *Planta*, 182, 325–338.

Canny, M.J. (1993) The transpiration stream in the leaf apoplast: water and solutes. *Phil. Trans. R. Soc. Lond. B*, 341, 87–100.

Daeter, W., Slovik, S. and Hartung, W. (1993) The pH gradients in the root system and the abscisic acid concentration in xylem and apoplastic saps. *Philos. Trans. R. Soc. Lond. B*, 341, 49–56.

Davis, R.F. and Higinbotham, N. (1969) Effects of external cations and respiratory inhibitors on electrical potential of the xylem exudate of excised corn roots. *Plant Physiol.*, 44, 1383–1392.

De Boer, A.H. (1999) Potassium translocation into the root xylem. *Plant Biology*, 1, 36–45.

De Boer, A.H., Prins, H.B.A. and Zanstra, P.E. (1983) Biphasic composition of transport electrical potential in roots of *Plantago* species: involvement of spatially separated electrogenic pumps. *Planta*, 157, 259–266.

Felle, H.H. (1993) Ion-selective microelectrodes: their use and importance in modern plant cell biology. *Bot. Acta*, 106, 5–12.

Felle, H.H., Hanstein, S., Steinmeyer, R. and Hedrich, R. (2000) Dynamics of ion-activities in the apoplast of the substomatal cavity of intact *Vicia faba* leaves during stomatal closure evoked by ABA and darkness. *Plant J.*, 24, 287–304.

Kochian, L.V. and Lucas, W.J. (1982) Potassium transport in corn roots. I. Resolution of kinetics into a saturable and linear component. *Plant Physiol.*, 70, 1723–1731.

Lohaus, G., Hussmann, M., Pennewiss, K., Schneider, H., Zhu, J.-J. and Sattelmacher, B. (2000) Solute balance of a maize (*Zea mays* L.) source leaf as affected by salt treatment with special emphasis on phloem re-translocation and ion leaching. *J. Exp. Bot.*, 51, 1721–1732.

Liang, J. and Zhang, J. (1997) Collection of xylem sap at flow rate similar to *in vivo* transpiration flux. *Plant Cell Physiol.*, 38, 1375–1381.

Maathuis F.J.M. and Sanders, D. (1993) Energization of potassium uptake in *Arabidopsis thaliana*. *Planta*, 191, 302–307.

Malone, M., Pritchard, J. and Watson, R.J. (1999) The spittlebug *Philaenus spumarius* feeds from mature xylem at the full hydraulic tension of the transpiration stream. *New Phytol.*, 123, 261–271.

Marschner, H. (1995) Mineral nutrition of higher plants. Academic Press, London etc., 2nd ed.

McCully, M.E. (1994) Accumulation of high levels of potassium in the developing xylem elements in roots of soybean and some other dicotyledons. *Protoplasma*, 183, 116–125.

Mengel, K., Plänker, R. and Hoffmann, B. (1994) Relationship between leaf apoplast pH and iron chlorosis of sunflower (*Helianthus annuus* L). *J. Plant Nutr.*, 17, 1053–1065.

Milburn, J.A. (1996) Sap ascent in vascular plants: challengers to the cohesion theory ignore the significance of immature xylem and the recycling of Münch water. *Annals Bot.*, 78, 399–407.

Miller, A.J. (1995) Ion-selective microelectrodes for measurement of intracellular ion concentrations. In: Methods in Cell Biology, Vol. 49, Methods in Plant Cell Biology, Part A (Galbraith, D.W., Bohnert, H.J. and Bourque, D.P., eds.). San Diego etc.: Academic Press, pp. 275–291.

Miller, D.M. (1985) Studies of root function in *Zea mays*. III. Xylem sap composition at maximum root pressure provides evidence of active transport into the xylem and a measurement of the reflection coefficient of the root. *Plant Physiol.*, 77, 162–167.

Peuke, A.D., Jeschke, W.D. and Hartung, W. (2002) Flows of elements, ions and abscisic acid in *Ricinus communis* and site of nitrate reduction under potassium limitation. *J. Exp. Bot.*, 53, 241–250.

Ponder, K.L., Watson, R.J., Malone, M. and Pritchard, J. (2002) Mineral content of excreta from the spittlebug *Philaenus spumarius* closely matches that of xylem sap. *New Phytol.*, 153, 237–242.

Rowan, A., McCully, M.E. and Canny, M.J. (2000) The origin of the exudate from cut maize roots. *Plant Physiol. Biochem.*, 38, 957–967.

Rygol, J., Pritchard, J., Zhu, J.J., Tomos, A.D. and Zimmermann, U. (1993) Transpiration induces radial turgor pressure gradients in wheat and maize roots. *Plant Physiol.*, 103, 493–500.

Schurr, U. (1998) Xylem sap sampling – new approaches to an old topic. *Trends Plant Sci.*, 3, 293–298.

Schurr, U. and Schulze, E.-D. (1995) The concentrations of xylem sap constituents in root exudate, and in sap from intact transpiring castor bean plants (*Ricinus communis* L.). *Plant Cell Environ.*, 18, 409–420.

Sperry, J.S., Saliendra, N.Z., Pockman, W.T., Cochard, H., Cruiziat, P., Davis, S.D., Ewers, F.W. and Tyree, M.T. (1996) New evidence for large negative xylem pressures and their measurement by the pressure chamber method. *Plant Cell Environ.*, 19, 427–436.

Walker, D.J., Leigh, R.A. and Miller, A.J. (1996) Potassium homeostasis in vacuolate plant cells. *Proc. Natl. Acad. Sci. USA*, 93, 10510–10514.

Wegner, L.H., Sattelmacher, B., Läuchli, A. and Zimmermann, U. (1999) Trans-root potential, xylem pressure, and root cortical membrane potential as influenced by nitrate and ammonium. *Plant Cell Environ.*, 22, 1549–1559.

Wegner, L.H. and Zimmermann, U. (1998) Simultaneous recording of xylem pressure and trans-root potential in roots of intact glycophytes using a novel xylem pressure probe technique. *Plant Cell Environ.*, 21, 849–865.

Wegner, L.H. and Zimmermann, U. (2002) On-line measurements of K^+ activity in the tensile water of the xylem conduit of higher plants. *Plant J.*, 32, 409–417.

Wegner, L.H. and Zimmermann, U. (2004) Bicarbonate-induced alkalinization of the xylem sap in intact corn seedlings as measured in situ with a novel xylem pH probe. *Plant Physiol.*, 136, 3469–3477.

White, P.J. (1997) The regulation of K^+ influx into roots of rye (*Secale cereale* L.) seedlings by negative feedback via the K^+ flux from shoot to root in the phloem. *J. Exp. Bot.*, 48, 2063–2073.

Wilkinson, S. (1999) PH as a stress signal. *Plant Growth Reg.*, 29, 87–99.

Zimmermann, U. (2003) in www.biozentrum.uni-wuerzburg.de/physikomedica/aktuelles/streitgespräch .html.

Zimmermann, U., Rygol, J., Balling, A., Klöck, G., Metzler, A. and Haase, A. (1992) Radial turgor and osmotic pressure profiles in intact and excised roots of *Aster tripolium*. *Plant Physiol.*, 99, 186–196.

Zimmermann, U., Schneider, H., Thürmer, F. and Wegner, L.H. (2002) Pressure probe measurements of the driving forces for water transport in intact higher plants: effects of transpiration and salinity. In: Salinity: Environment – Plants – Molecules, (Läuchli, A. and Lüttge, U., eds.). New York: Kluwer Academic Publishers, pp. 249–270.

Zimmermann, U., Schneider, H., Wegner, L.H. and Haase, A. (2004) Tansley review: water ascent in tall trees: does evolution of land plants rely on a highly metastable state? *New Phytol.*, 162, 575–615.

Zimmermann, G., Zhu, J.J., Benkert, R., Schneider, H., Thürmer, F. and Zimmermann, U. (1995) Xylem pressure measurements in intact laboratory plants and excised organs: a critical evaluation of methods in the literature and the xylem pressure probe. In: Tree Sap, (Terazawa, M., McLeod, C.A. and Tamai, Y., eds.). Sapporo: Hokkaido University Press, pp. 59–70.

DYNAMIC AND NUTRIENT FLUXES IN THE XYLEM

F. GILMER and U. SCHURR
Institute for Chemistry and Dynamics of the Geosphere: ICG-III Phytosphere, Forschungszentrum Jülich GmbH, Germany, u.schurr@fz-juelich.de

Abstract. Fluxes of nutrients in the xylem sap vary dynamically. They are indicative of the status of the plant – similar to the blood in medical applications – and can be used to address basic as well as applied questions. Innovative methods recently have been developed and used to study the impact of internal mechanisms in the plant as well as the effect of external conditions on xylem-sap composition, nutrient fluxes and allocation in intact, transpiring plants. A brief comparison of these methods describes their application potential and drawbacks. The relevance of the dynamic variation of nutrient fluxes is discussed by using key studies of the dynamic impact of external and internal conditions. The need for additional research on the dynamics of solute fluxes in the xylem and its combination with phloem flux analysis to understand the allocation of substances within plants mechanistically is put forward.

Key words: flow velocity, online xylem analysis, root pressure chamber, xylem

1. INTRODUCTION

The xylem mediates water and nutrient fluxes as well as signals responsible for the balance between roots and shoots in plants. Water and nutrient availability vary with time and environmental conditions, from which the signals in the xylem originate, change dynamically. The dynamics of these processes and the active response of plant activity are mirrored by xylem sap composition and fluxes of various substances in the xylem. In this

B. Sattelmacher† and W.J. Horst (eds.), The Apoplast of Higher Plants: Compartment of Storage, Transport and Reactions, 221–229.

respect xylem sap can be regarded as a compartment with similar diagnostic potentials as blood, used for its general diagnostic options in medicine.

It is an important prerequisite for such purposes to use xylem sap sampled from the intact, working transport system without impairing its function. A very limited number of methods are capable to fulfil this task. Thus – even though methods are not the primary scope of this paper - the potential and limitations of methods available need to be described and will then be discussed in some applications, in which the dynamic of xylem sap composition and nutrient fluxes in the xylem has been evaluated. It will be shown that the dynamics of nutrient transport in the xylem sap is not only characterising the transport in the xylem, but also that this dynamic is important for the understanding of nutrient relations and active responses of plants to external factors. The role of internal anatomical and physiological features for the dynamics of nutrient fluxes in the xylem will also be addressed, as they provide the framework for the dynamics of water and matter fluxes in the xylem.

2. METHODS FOR *IN VIVO* SAMPLING OF XYLEM SAP

2.1 A large variety of methods – options and drawbacks

A large number of different physical principles have been used to sample xylem sap (Schurr, 1998). The major problem that all of these approaches try to overcome is the tension in the xylem, which causes air entering the xylem rather than sap bleeding out of the vessels. Some of the classical techniques circumvent this tension by cutting parts form the plant and sample by suction or positive pressure (e.g. Scholander methods or centrifugation techniques) and root exudation methods use the natural positive pressure from root stumps to extract root exudates. The most striking drawback of such methods is that the sampling method may cause alterations of xylem sap composition e. g. by disturbance of the equilibria between xylem sap and the adjacent tissue or more specific effects of sampling (Schurr and Schulze, 1995). More modern techniques allow sampling of xylem sap *in vivo* and can thus open the study of dynamics of xylem sap composition in relation to plant internal conditions and in response to external stimuli. These techniques overcome the tension in the xylem by application of positive pressure in a root-pressure chamber or – in even less-invasive techniques – sample xylem sap under tension (see Wegner et al., this volume, pp. 207–220). This paper will thus concentrate on the root-pressure chamber, its benefits and drawbacks in the following paragraph.

2.2 Root-pressure chamber

In the root-pressure chamber (shown in fig. 1) the tension in the xylem is overcome by application of positive pressure in a pressure device that houses the root system and the medium in which it is placed. In the original version, build to study hydraulic limitations in the xylem transport-pathway (Passioura and Tanner, 1985), root systems were placed in the soil. This allowed using the root-pressure chamber to analyse the dynamics of xylem-sap composition in response to soil drought – an important step in identifying non-hydraulic root-shoot signals (Gollan et al., 1992). Subsequent developments enhanced the performance of the system to be used at different soil compaction status to study the effect on plant growth and water relations (Passioura, 2002).

Recently, further developments of the root-pressure chamber were made that allow the analysis of the dynamic responses of xylem sap composition to dynamic changes in the root environment (nutrients, gas composition, temperature, etc.). Besides soil cultures, a system was developed that allowed aeroponic delivery of nutrients inside the root-pressure chamber (Herdel et al., 2001). This system could be used to study the diurnal variation at constant nutrient availability as well as the responses to short-term changes in nutrient availability.

While the root-pressure chamber still is the only method to provide bulk samples for a full analysis of xylem sap from intact plants (Gerendas and Schurr, 1999) and has proved its potential in analysing dynamic situations in xylem-sap composition (Herdel et al., 2001), it has limitations that also need to be considered: the plants must be grown in specialised containers in order to guarantee the required sealing at the hypocotyl, the size of the root system is limited and thus field-grown plants are difficult to analyse.

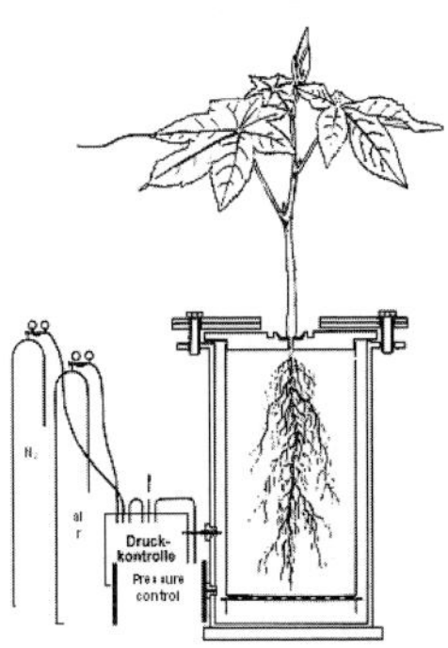

Fig. 1. Scheme of the root-pressure chamber for non-destructive sampling of xylem sap from intact transpiring plants (e.g. *Ricinus communis*). This device can be used with plants grown in soil or in aeroponic culture.

2.3 Combined analysis of xylem-sap composition and nutrient fluxes in intact plants

Analysing dynamic changes in xylem transport cannot be limited to variations in xylem-sap composition, but needs to include variation in fluxes as well as temporal and spatial heterogeneities of lateral transport along the transport pathway. Thus in addition to the devices described above new equipment has been made available in the past decade that allows minimal-invasive approaches to study xylem transport. These range from standard systems like heat pulse (Smith and Allen, 1996) and gas exchange techniques (Jahnke and Krewitt, 2002) to nuclear magnetic resonance imaging (Schneider et al., this volume, pp 255–268). While NMR imaging allows detailed analysis of the internal water fluxes (Köckenberger et al., 1997) and even combined water and substance fluxes (Peuke et al., 2001) the former techniques are best suited for quantitative balances of matter fluxes in the xylem (Herdel et al., 2001).

3. IMPACT OF AVAILABILITY OF NUTRIENTS ON XYLEM-SAP COMPOSITION AND NUTRIENT FLUXES

3.1 Heterogeneity of nutrient availability in the soil and nutrient uptake of plants

Nutrient availability in the soil is subject to significant fluctuations that are highly dependent on the nutrient species. Concentrations in the rhizosphere, that are critical for the uptake into plants, can differ significantly from bulk soil solution due to the activity of roots (Wang and Göttlein, 2001). In addition, small scale fluctuations of nutrient concentrations are characteristic for specific soils (Göttlein et al., 2001).

Plants have highly efficient ion transport systems in the root to acquire nutrients from its soil environment. However transport into the root is by no means constant, even when the external supply is continuous (e.g. MacDuff and Bakken, 2003). This is at least partly due to a diurnal regulation of the uptake systems (Delhon et al., 1995a). Xylem loading is an additional control point for xylem loading (see Koehler and Raschke, this volume, pp 183–204). Accumulation of ^{15}N in the roots during the night, that is transported to the shoot later during the day indicates active control processes in the cooperation between xylem loading and root storage (Delhon et al., 1995b).

3.2 Xylem sap dynamics in response to external conditions

Diurnal variations of xylem-sap composition have been observed in root exudates (e.g. MacDuff and Bakken, 2003) as well as in intact plants (Herdel et al., 2001). When plants are grown at constant light conditions the amplitude of the diurnal variations of xylem sap composition are significantly diminished (Gilmer et al., 2001). Short-term changes of nitrogen availability to the root result in a significant increase of the xylem sap concentration, but this is not maintained. Characteristic oscillations with declining amplitudes have been observed even when the nutrient concentration in the root medium was kept constant (Gilmer et al., 2001). These results strongly suggest active control of nutrient transport to the shoot and of root development, by shoot-born signals indicating the nutrient demand of the shoot (Marschner et al., 1996 Forde and Lorenzo, 2001).

Xylem sap concentrations are commonly higher during the night (Herdel et al., 2001, Gilmer et al., 2001). The higher concentrations of nutrients during the night cause an almost constant delivery of nutrients to the shoot despite the lower transpiration rates in darkness (Herdel et al., 2001). However, after an initial increase, concentration declines even at constantly low transpiration rates (Fig. 2).

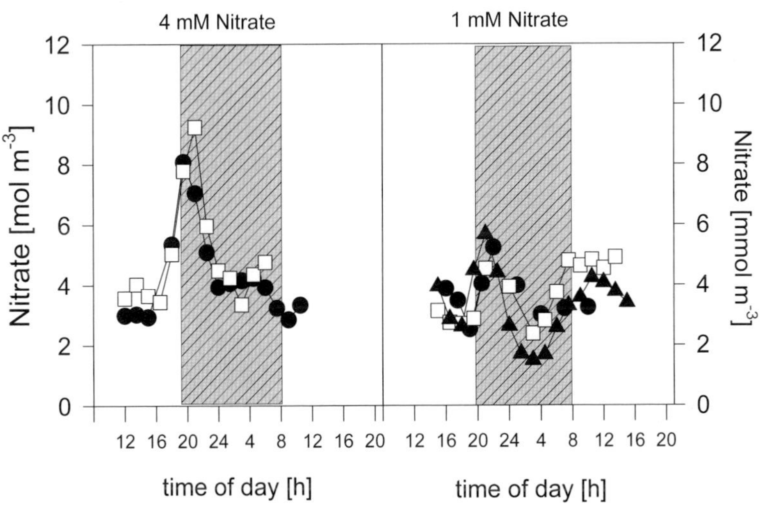

Fig. 2. Diurnal course of nitrate concentrations in the xylem sap of *Ricinus communis* plants growth at high (4 mM) and low (1 mM) nitrate in the rooting medium. The symbols indicate two plants in each treatment. Xylem-sap samples were taken by the root-pressure chamber technique from individual plants.

Reduced nutrient availability does not directly lead to reduced concentrations of the xylem sap (Fig. 2): nitrate concentrations in the xylem sap of plants with different nitrate supply were almost identical during the day, but the decline in concentration in the second half of the night started earlier. This causes a significant difference in the timing of delivery of N to the shoot. While shoots of high-N plants obtained N at an almost constant rate day and night, shoots of low-N plants received the same amount of N during daytime, but significantly less during the dark period. As the regulation of nitrate reductase responds quickly to changes in nitrate availability (Kaiser and Huber, 2001) the diurnally changing delivery of nitrate to the leaves may be a significant link between whole-plant physiology and the molecular regulation of nitrate assimilation.

4. IMPACT OF THE TRANSPORT PATHWAY ON FLUXES OF NUTRIENTS IN THE XYLEM

4.1 Interaction with the xylem walls and the parenchyma

After entering the xylem, nutrients are transported along the xylem towards the leaf. The major medium for this transport obviously is the transpiration stream. Additional routes could involve ion-exchange sites along the xylem walls.

Lateral release and retrieval is provided by transport systems located in the xylem-paraenchyma cells (see Ache and Deeken, this volume, pp. 151–164). Lateral transport is of major importance in the balancing of substance allocation within the whole plant according to the models of Jeschke, Pate and collegues (e.g. Jeschke and Pate, 1991, Peuke, Jeschke and Hartung, 2002). Mechanistic analysis however requires much smaller time scales, as biochemical and molecular regulation acts in minutes and hours rather than weeks. Lateral exchange of substances can be determined by anatomical features like in sectorial plants (Orians et al., 2002), but besides this large scale heterogeneity local exchange between the xylem and the surrounding paranchyma can be observed, e.g. by perfusion experiments (Clarkson and Hanson, 1986). Strong impact of lateral exchange on the concentrations in the perfusion solution have been interpreted as the result of mechanisms to support ion homeostasis in the xylem (Lacan and Durand, 1996).

The impact of lateral transport is dependent on the nutrient. In the perfusion experiments (e.g. Fig. 3) the nitrate concentration in the outflow from the isolated stem segment reached the input concentration at very high

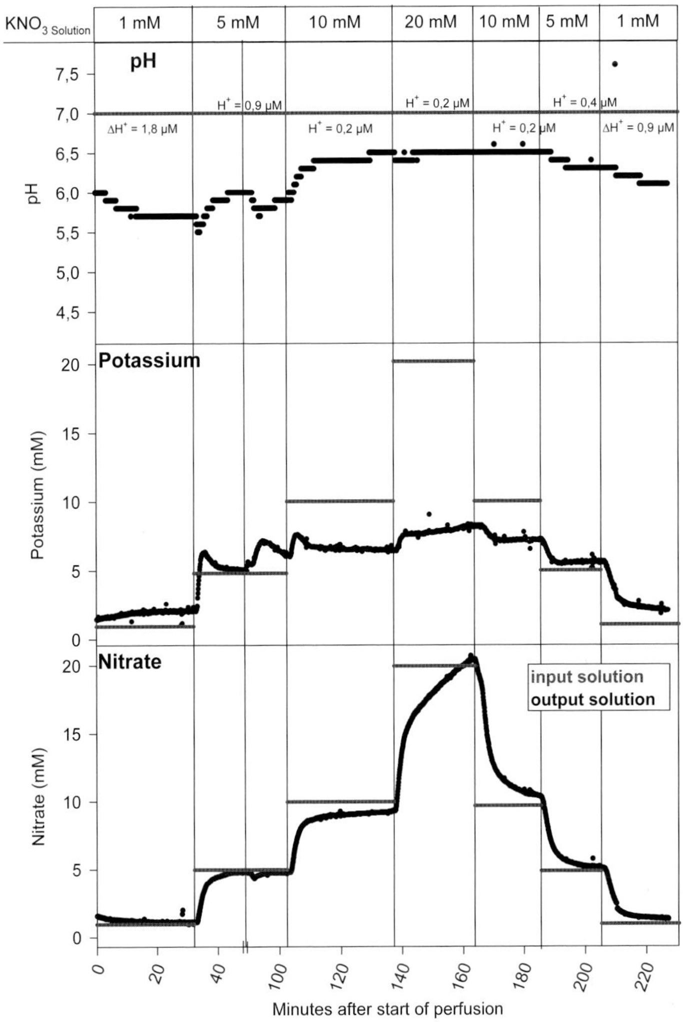

Fig. 3. Perfusion of *Ricinus communis* stem segments with potassium nitrate of various concentrations. Concentrations of potassium, nitrate and protons in the outflow were determined by online ion-electrode analysis (dark traces) and compared to the input solution (gray traces).

and very low concentrations. In contrast, the concentration of potassium was maintained in a much smaller band (2–7 mM) due to lateral exchange along the stem segment. Changing the through-flow did only affect the half-times necessary to reach the new concentration. For reasons of counterbalance other cations must have been released from the stem tissue. Changes in pH were not sufficient to counterbalance the cation exchange. Such exchange processes haven even been shown to interact with signal transport in the xylem (Sauter and Hartung, 2002). Exchange processes may however not been distributed homogenously along the xylem pathway. The nodes intercalating between the transport segments and special anatomical features at these sites may provide additional exchange sites between xylem and parenchyma as well as between xylem and phloem tissues (e.g. Jeschke and Pate, 1991). However, very little is known about the actual sites and quantities of substances that are exchanged in short-term between the long-distance transport pathways of plants.

5. CONCLUSIONS

Xylem transport is highly dynamic and responds to internal features of plants as well as to environmental conditions. The methods available now provide new insights into these dynamic properties and we are just at the beginning of understanding consequences of nutrient-flux dynamics in plants. However, such analyses on the whole-plant level must be linked to molecular approaches to understand the plant-internal mechanisms of nutrient allocation.

A crucial step for this will be to combine analysis and understanding of *in vivo* transport in both the xylem and the phloem. It will be necessary to continue the development of non-invasive techniques allowing integrative studies of both transport pathways. Nuclear magnetic resonance imaging is a powerful technology in that respect, because it presently provides the only option to analyse water fluxes in xylem and phloem simultaneously (Schneider et al., this volume, pp 255–268). Recent developments additionally allow the analysis of lateral water transport into and out of the tissue (Scheenen et al., 2001). However additional techniques will be needed to study the transport of substances other than water. Besides transport in the xylem, phloem contributes to the distribution of substances in plants. Therefore, studying dynamics of nutrient fluxes in the xylem need to be combined with studies on the dynamics of fluxes in the phloem resulting in a much more appropriate picture of nutrient allocation in plants.

REFERENCES

Clarkson, D.T. and Hanson, J.B. (1986). Proton fluxes and the activity of a stelar proton pump in onion roots. *J. Exp. Bot.*, 37, 1136–1150.

Delhon, P., Gojon, A., Tillard, P. and Passama, L. (1995a). Diurnal regulation of NO_3^- uptake in soybean plants. I. Changes in NO_3^- influx, efflux, and N utilisation in the plant during the day/night cycle. *J. Exp. Bot.*, 46, 1585–1594.

Delhon, P., Gojon, A., Tillard, P. and Passama, L. (1995b). Diurnal regulation of NO_3^- uptake in soybean plants. II. Relationship with accumulation of NO_3^- and asparagine in the roots. *J. Exp. Bot.*, 46, 1595–1602.

Forde, B.R. and Lorenzo, H. (2001). The nutritional control of root development. *Plant Soil*, 232, 51–68.

Gerendas, J. and Schurr, U. (1999). Physicochemical aspects of ion relations and pH regulation in plants – a quantitative approach. *J. Exp. Bot.*, 50, 1101–1014.

Gollan, T., Schurr, U. and Schulze, E.D. (1992). Stomatal response to drying soil in relation to changes in the xylem sap composition of *Helianthus annuus*. I. The concentration of cations, anions, amino acids and the pH of the xylem sap. *Plant, Cell Environ.*, 15, 551–559.

Göttlein, A., Lindenmair, J. and Kuhn, A.J. (2001). Experimental approaches for studying rhizosphere chemistry and nutrient uptake of tree roots by in-situ methods. In: Horst, W.J. et al. (Eds.) Plant nutrition - food security and sustainability of agro-ecosystems, pp 536–537, Kluwer Academic Publishers, Dordrecht, Netherland.

Herdel, K. Schmidt, P., Feil, R., Mohr, A. and Schurr, U. (2001). Dynamics of concentrations and nutrient fluxes in the xylem of Ricinus communis – diurnal course, impact of nutrient availability and nutrient uptake. *Plant, Cell Environ.*, 24, 41–52.

Jahnke, S. and Krewitt, M. (2002). Atmospheric CO_2 concentrations may directly affect leaf respiration measurement in tobacco, but not respiration itself. *Plant, Cell Environ.*, 25, 641–651.

Jeschke, W.D. and Pate, J.S. (1991). Modelling the uptake, flow and utilization of C, N and H_20 within whole plants of *Ricinus communis* L. based on empirical data. *J. Exp. Bot.*, 137, 488–498.

Kaiser, W. and Huber, S.C. (2001). Post-translational regulation of nitrate reductase: mechanism, physiological relevance and environmental triggers. *J. Exp. Bot.*, 52, 1981–1989.

Köckenberger, W., Pope, J.M., Xia, Y., Jeffrey, K.R., Komor, E. and Callaghan, P.T. (1997). A non-invasive measurement of phloem and xylem water flow in castor bean seedlings by nuclear magnetic resonance microimaging. *Planta*, 201, 53–63.

Lacan, D., Durand, M. (1996). Na^+-K^+-exchange at the xylem/ symplast boundary. *Plant Physiol.*, 110, 705–711.

MacDuff, J.H. and Bakken, A.K. (2003). Diurnal variation in uptake and xylem contents of ionorganic and assimilated N under continuous and interrupted N supply to *Phleum pratense* and *Festuca pratensis*. *J. Exp. Bot.*, 54, 431–444.

Marschner, H., Kirkby, E.A. and Cakmak, I. (1996). Effect of mineral nutritional status on shoot-root partitioning of photoassimilates and cycling of mineral nutrients. *J. Exp. Bot.*, 47, 1255–1263.

Orians, C.M., Ardon, A. and Mohammad, B.A. (2002). Vascular architecture and patchy nutrient availability generate within-plant resource heterogeneity in plant traits important for herbivores. *Am. J. Bot.*, 89, 270–278.

Passioura, J.B. and Tanner, C.B. (1985). Oscillations in apparent hydraulic conductance of cotton plants. *Austr. J. Plant Physiol.*, 11, 341–350.

Passioura, J.B. (2002). Soil conditions and plant growth. *Plant, Cell Environ.*, 25, 311–318.

Peuke, A.D., Rokitta, M., Zimmermann, U., Schreiber, L. and Haase, A. (2001). Simultaneous measurement of water flow velopcity and solute transport in xylem and phloem of adult plants of *Ricinus communis* over a daily time course by nuclear magnetic resonance spectrometry. *Plant, Cell Environ.*, 24, 491–503.

Peuke, A.D., Jeschke, W.D. and Hartung, W. (2002) Flows of elements, ions and abscisic acid in *Ricinus communis* and site of nitrate reduction under potassium limitation. *J. Exp. Bot.*, 53, 241–250.

Sauter, A. and Hartung, W. (2002). The contribution of internode and mesocotyl tissues to root-to-shoot signalling of abscisic acid. *J. Exp. Bot.*, 53, 297–302.

Schurr, U. (1998). Xylem sap sampling – new approaches to an old topic. *Trends in Plant Sciences*, 3, 293–298.

Schurr, U. and Schulze, E.D. (1995). The concentrations of xylem sap constituents in root exudates, and in sap from intact, transpiring castor bean plants (*Ricinus communis* L.). *Plant, Cell Environ.*, 18, 409–420.

Scheenen, T.W.J., Vergeldt, F.J., Windt, C.W., de Jager, P.A. and Van As, H. (2001). Microscopic imaging of slow flow and diffusion: a pulsed field gradient stimulated echo sequence combined with turbo spin echo imaging. *J. Magnetic Resonance*, 151, 94–100.

Smith, D.M. and Allen, S.J. (1996) Measurement of sap flow in plant stems. *J. Exp. Bot.*, 47, 1833–1844.

Wang, Z. and Göttlein, A. (2001). Rhizosphere chemistry and soil exploration by seedlings of beech and spruce in a rhizotron experiment. In: Horst, W.J. et al. (Eds.) Plant nutrition - food security and sustainability of agro-ecosystems, pp. 596–597, Kluwer Academic Publishers, Dordrecht, Netherland.

RELATIONSHIP BETWEEN APOPLASTIC NUTRIENT CONCENTRATIONS AND THE LONG-DISTANCE TRANSPORT OF NUTRIENTS IN THE *RICINUS COMMUNIS* L. SEEDLING

E. KOMOR, G. ORLICH and H. BAUER-RUCKDESCHEL
Pflanzenphysiologie, Universität Bayreuth, Germany, ewald.komor@uni-bayreuth.de

Abstract. The seedling of *Ricinus communis* was used to study the fluxes of nutrients from the apoplastic space into the symplast of the cotyledons, from the apoplast and the symplast of the cotyledons into the phloem, and the recirculation of nutrients into the cotyledons via the xylem. Cotyledons of the intact seedling were incubated in media of defined nutrient concentrations till a steady-state of fluxes was reached, then a small cut of two to three bundles at the hypocotyl hook was made and the exudates from phloem and xylem were sampled and analysed. Together with the information obtained for nutrient concentrations in the medium (apoplast) and in the leaf (symplast) a quantitative scheme of nutrient fluxes at different apoplastic concentrations was designed. All tested nutrients (sucrose, glutamine, potassium ion) were found in higher concentrations in the symplast and in the phloem sap than in the apoplast, but sucrose was the only substrate which was clearly enriched in the phloem sap compared to the symplast. Recirculation (i.e. export via the phloem and re-import via the xylem) of sucrose was negligible at all tested sucrose concentrations, whereas recirculation of glutamine and potassium ions was up to 80% in case of high supply at the cotyledon apoplast. The recirculation of these nutrients enables the apoplastic space of the cotyledons to function as a storage space for homeostatic nutrient supply to the phloem for up to 1h. The concentrations of all tested nutrients in phloem and in xylem were determined by the high accumulation capacity of the cotyledons′ symplast and the high phloem loading activity. Therefore, the concentrations of nitrogenous compounds and of potassium reached in the xylem by providing these nutrients to the roots were not higher than obtained through recirculation of phloem-derived nutrients. This situation may be typical for the seedling stage with low transpirational water flux and high nutrient resources in the cotyledons.

Key words: glutamine, nutrient circulation, phloem transport, potassium, sucrose, xylem transport

B. Sattelmacher[†] *and W.J. Horst (eds.), The Apoplast of Higher Plants: Compartment of Storage, Transport and Reactions*, 231–249.

1. INTRODUCTION

Assimilates are loaded into the phloem of source leaves and transported to roots, growing organs and seeds. Minerals which are taken up by the roots are transported to transpiring leaves where they are either deposited or reloaded into the phloem and transferred to growing tissues. Pate elaborated the net fluxes of nitrogen, carbon, and water in phloem and xylem and showed that there is a substantial cycling of these nutrients in some plant organs by phloem-xylem and xylem-phloem exchange (Pate, 1973, Pate et al., 1979, Ho et al., 1987, Jeschke and Pate, 1991). Marschner emphasized the importance of mineral-ion recirculation in the plant for the supply of plant organs with the needed compounds (Marschner et al., 1997, Peuke et al., 2002). The power of this internal nutrient and water cycling through phloem and xylem was demonstrated in several aspects. The cycling of mineral ions via the phloem may determine the induction of mineral-ion transporters in the root thus tuning the uptake of minerals to the need of the entire plant (e.g. White, 1997). The water cycling through phloem and xylem may also be the sufficient transport flow to ascertain the nutrient requirements and distribution of growing plants in the virtual absence of transpiration (Tanner and Beevers, 2001). Special anatomical structures of the vascularization in the corms of grasses which ease exchange of nutrients between phloem and xylem may be even the decisive factor for the competitive advantage for growth in harsh, nutrient-poor ecological environments (Cholewa and Griffith, 2004). Xylem transport proceeds in the highly conductive, wide tracheary cells which are devoid of a plasma membrane so that their solute is in unhindered exchange to the surrounding sclerenchyma and cell-wall space, thus being a part of the apoplast of for instance a source leaf. This apoplastic space stands in exchange with the cell-wall space and air space of the mesophyll. Depending on the plant species and the developmental state of the leaf, suberine-containing diffusion barriers around the bundle sheath may separate the xylem apoplast from the mesophyll apoplast. In young leaves these barriers are usually less developed. The mesophyll apoplast of source leaves is bordered by water impermeable cuticles on the surface of the epidermis layers to avoid nutrient loss during wetting, although some substantial leaching of minerals by rain has been considered for leaves (Tukey and Mecklenburg, 1964, Lohaus et al., 2000).

In previous studies, either the net fluxes of nutrients were determined because the quantitative determination of phloem transport especially the chemical composition of the transported substances in the sieve tube sap was not possible with the used plant material, or the transfer of labelled nutrients from a fed source leaf to other leaves or twigs was studied with some difficulties of quantification from the label distribution (Gessler et al., 2003). To get quantitative data on the relationship between defined apoplastic concentrations of particular nutrients and the resulting export through the phloem and, possibly, its re-import through the xylem the *Ricinus* seedling system was used. The cotyledons of the *Ricinus* seedling are embedded in the endosperm, in which during germination nutrients are mobilized e.g. phosphate, potassium and amino acids or synthesized such as sucrose (Kriedemann and Beevers, 1967). The nutrients are taken up from the external space by the lower epidermis of the cotyledons, which is devoid of a cuticle. The apoplast (cell-wall space) of the cotyledons is in direct contact with the external space so that it can be considered as an "open apoplast". This feature was used in the past to measure the uptake and phloem loading characteristics of the cotyledons, its dependence on different sugar concentrations and its mechanism of active transport (Komor, 1977). The castor bean also offers the unique chance to sample phloem sap from a cut of conducting bundles at the hypocotyl hook, because in contrast to nearly all other plants it apparently lacks an efficient sieve-tube plugging-mechanism (Kallarackal et al., 1989). In addition, the opened xylem strands of the conducting bundle exude xylem sap at the basal part of the wounded hypocotyl which can be sampled (Schobert and Komor, 1990). Thus the *Ricinus* seedling allows the simultaneous measurement of nutrient fluxes from a defined apoplastic concentration into the symplast of the cotyledon cells, the export of nutrients via the phloem, and the import of nutrients from the xylem. The quantitative determination of circulation of nutrients between phloem and xylem at different apoplastic nutrient concentrations was the aim of this study.

2. MATERIAL AND METHODS

2.1 Plant material and growth conditions

Ricinus seeds (*Ricinus communis* L. var. *sanguineus*) were obtained from Jelitto Staudensamen (Schwarmstedt, Germany). The seeds were soaked in water overnight, then surface-sterilized with 8-hydroxy-chinoline-sulfate (0.3% in water) and germinated on sterile agar at 27°C for 2 days in the dark. Then the sterile seedlings were transferred to growth vessels with liquid medium (0.5 mM $CaCl_2$) at the roots and the endosperm-covered cotyledons in humid air, and grown under continuous aeration of the rooting medium at 27°C in the dark for further 3–4 days. The seedlings were used for experiments when 6 days-old when the hypocotyls had a length of ca. 50 mm and when the endosperm was still intact (not cracked).

2.2 Incubation conditions for loading via the cotyledons

The cotyledons of the seedlings were carefully exposed by removal of the endosperm and then incubated still attached to the hypocotyls in medium containing 5 mM Na-MES buffer pH 5.5, 0.2 mM $CaCl_2$, and the nutrient of the desired concentration (e.g. sucrose, glutamine, potassium chloride). The incubation solution was stirred to avoid local anaerobiosis. The roots were immersed all the time in 0.5 mM $CaCl_2$. The incubation was performed under a hood to achieve high air humidity at 27°C for up to 8 h till a steady state of fluxes had been reached. Since sucrose loading is the major driving force of phloem transport, 100 mM sucrose had been always present in the incubation medium at the cotyledons except in those cases where sucrose loading at different concentrations was tested.

2.3 Incubation conditions for loading via the roots

The endosperm was removed from the cotyledons and the cotyledons were incubated as described above in a medium of 100 mM sucrose, 0.2 mM $CaCl_2$, and 5 mM MES-KOH of pH 5.5. The medium at the roots contained 1 mM MES-KOH pH 5.5, 0.5 mM $CaCl_2$ and the nutrient of interest. The incubation was at 27°C.

2.4 Phloem-sap and xylem-sap collection

When the steady state of fluxes was reached (which had been determined in parallel experiments) the hypocotyl hook was cut with a razor blade and the exuding phloem sap was collected with glass capillaries for up to 15 min. The xylem sap was sampled from a small vertical cut executed with a scalpell point which separated and opened two to three of the conducting bundles. The cut ends of the bundles were dipped into 1 M $CaCl_2$ for 1 s to stop phloem sap exudation, washed and then the exuding xylem sap was collected with glass capillaries. For calculation of the xylem flux the measured exudation rate from the opened bundles was extrapolated to the value for all eight bundles of the hypocotyls. The phloem sap and the xylem sap were stored in the capillaries at –20°C until they were analysed.

2.5 Flux measurements

The unidirectional rate of influx into the cotyledons was determined when a steady state of concentrations of the nutrient of interest (e.g. sucrose) was reached during incubation. Then a labelled marker of the nutrient was added to the medium and seedlings were harvested in 30 min intervals to obtain an influx kinetic. The cotyledons of the harvested seedlings were carefully blotted dry and rinsed in non-labelled medium three times for 2 min. Then the seedling was chopped in pieces and extracted in water at 95°C for 15 min. The extract (including the tissue remains) was used for isotope analysis. For measurement of sucrose flux, glutamine flux, phosphate flux and sulphate flux the following radioactive isotopes were used: ^{14}C-sucrose, ^{14}C-glutamine, ^{33}P-phosphate and ^{35}S-sulfate. The radioactivity was measured by a scintillation counter (Packard 2500 TR, Canberra-Packard, Dreieich, Germany) after addition of Lumasafe Plus (Luma-LSC, Groningen,Netherlands) or Rotiszint (Roth, Karlsruhe, Germany) to the extract. For flux measurements of K, Mg, and Ca the stable isotopes ^{41}K, ^{25}Mg and ^{44}Ca were added to the incubation medium and samples from the medium were taken in 30 min intervals. The isotope content of the samples was determined by LAMMA (laser microprobe mass analyser) and SIMS (secondary ion mass spectroscopy), performed by Dr. Schröder and Dr. Kuhn, Forschungszentrum Jülich.

2.6 Analytical methods

Phloem-sap and xylem-sap samples were used for analysis directly without prior extraction procedures. Tissue samples were immediately after sampling ground in 80% ethanol buffered with 5 mM HEPES-KOH pH 7.0 and incubated at 70°C for 1 h for complete extraction of water-soluble compounds. After centrifugation the clear supernantant was used for analysis, or stored in liquid nitrogen till analysis.

The sugars glucose, fructose and sucrose were determined by coupled enzymatic tests. Glucose was first converted to glucose-6-phosphate by hexokinase which was then determined by $NADP^+$-coupled glucose-6-phosphate dehydrogenase. The same enzymatic reaction was used for determination of fructose via fructose-6-phosphate, which after phosphorylation by hexokinase was converted to glucose-6-phosphate by phosphoglucose isomerase. Sucrose was hydrolysed to glucose and fructose by acid invertase (30 µl sample in 140 µl 15 mM MES-KOH ph 5.5 plus 5 U acid invertase from yeast) for 30 min at room temperature. The resulting amounts of glucose and fructose were determined as described above. All measurements were conducted with an Eppendorf photometer (Hamburg, Germany).

Glutamine and other amino acids were measured by HPLC after derivatization according Schurr and Gebauer (1989). Phloem-sap and xylem-sap samples were directly subjected to derivatization,. Tissue samples were ground in liquid nitrogen and then in 80% acetone, the powder was then incubated at 4°C for 30 min to precipitate the proteins and then centrifuged.

After centrifugation the supernatant was brought to dryness and then taken up in water. The amino acids in this sample were coupled to o-phtalic acid dialdehyde and the product was separated on HPLC (sperisorb ODS II) and quantified by fluorometer (Kontron, Eching, Germany).

C and N contents were determined from 2–4 mg dry samples in a C/N analyser (Carlo Erba istrumentazioni, Milano, Italy).

Total contents of K, Na, S and P were determined after wet ashing according Schramel et al. (1980) under pressure. 50–100 mg of fresh plant material were emersed in 1 ml 65% HNO_3 and then ashed at 170°C for 6–10 h at 100–120 kPa in a pressure oven (Aufschlusstechnik Seif, Unterschleißheim, Germany). The resulting product was diluted in water and analysed for K and P in an atomic emission spectrometer coupled with inductive plasma mass-spectrometry (ICP-MS, GBC-Integra, Australia).

Water soluble phosphate was determined in extracts of tissue, obtained after grinding of the plant material in liquid nitrogen and incubating the powder in water at 95°C for 10 min. After centrifugation the supernatant was determined according Ames (1966) as molybdate complex in a LKB Biochrom Ultraspec II spectrophotometer (Pharmacia, Freiburg, Germany). Water-soluble phosphate in phloem and xylem sap was determined in the same way without prior extraction procedures.

Potassium in phloem and xylem sap was measured by atomic absorption spectroscopy without prior sample preparation (AAS, Perkin-Elmer, Norwalk, USA).

2.7 Statistical treatments

All experiments were repeated at least three times and for each experimental set three seedlings were used. The standard deviation from the mid value was usually in the range of 20%. For clarity of the graphs only the mid values are depicted.

3. RESULTS

3.1 The *Ricinus* seedling as a model system for analysis of phloem loading and xylem transport at different apoplastic nutrient concentrations at the cotyledons and at the root system

The cotyledons of 6 days old *Ricinus* seedlings are tightly packed with cells nearly without air-filled intercellular spaces (Fig. 1C). The cells appear like meristematic cells with well developed organelles but without a large central vacuole. The water space of the cotyledons except of the cell-wall space can, therefore, fairly be considered as the symplastic space. The apoplast of the cotyledons consists of the cell-wall space which is ca. 14% of the cotyledon volume and of the xylem tracheal space which is ca. 1% of the cotyledon volume (Köhler et al., 1991). It is unknown how well these two components of the apoplastic space are in unhindered apoplastic nutrient exchange.

The veins in the cotyledons appear well developed and differentiated into phloem and xylem. They unite in the hypocotyl to eight bundles, which remain separated along the hypocotyl without any apparent horizontal anastomoses. The bundles unite at the root crown and form a ring from where they continue into a root-typical radial bundle morphology (Fig. 1B).

The phloem sap exudation proceeds after complete cutting of the hypocotyl hook, so that the entire flux of nutrients from the phloem out of the cotyledons is obtained (Fig. 1A). The volume flux-rates depend strongly on the nutrient availability in the cotyledons. In well-fed cotyledons such as in case of endosperm containing intact seedlings the exudation rate is in the range of 20–40 µl/h. This is indeed the natural flow rate as was shown by non-invasive NMR-microimaging (Köckenberger et al., 1997).

Xylem-sap exudate was obtained by cutting two to three bundles at the hypocotyl hook. Phloem-sap exudation from the cut bundle was stopped by dipping a drop of $CaCl_2$ on the cut surface for a second, thus initiating callose formation at the sieve plates. The absence of phloem sap in the sampled xylem exudates was monitored as absence of ATP in the exudates, since phloem sap in contrast to xylem sap contains 1 mM ATP. Only two to three bundles were cut for xylem sap sampling so that the remaining five to six bundles stayed unhurt. There had been evidence before that long term and complete interruption of phloem or xylem transport may result in changes of the not-interrupted fluxes and giving therefore misleading results (Fishman et al., 2001). Phloem transport could proceed within the unhurt bundles and nutrient exchange between phloem and xylem was possible in the bundle ring at the root crown and further down in the root central cylinder.

The nutrients of interest were delivered either to the medium bathing the cotyledons or to the root medium. Usually a concentration was chosen which was thought to be close to the "natural" concentration at the cotyledon or the root surface and then one concentration 3-fold above and another 3-fold below this "natural" one was applied to cover the interesting range.

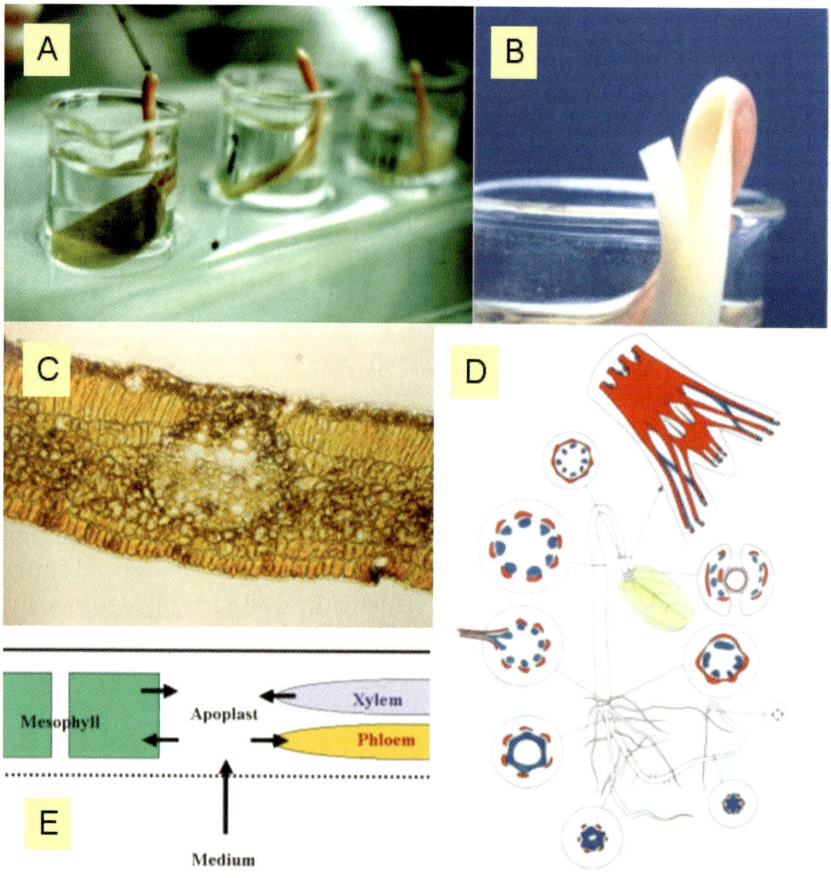

Fig. 1. *The Ricinus seedling as experimental system for studies on phloem and xylem transport. A: Excised cotyledons which incubate in a medium containing nutrients. The phloem sap exudes from the cut hypocotyl hook and is collected with a microcapillary (left beaker). B: Hypocotyl hook which is partially scliced so that two to three bundles are severed. The cotyledons are incubated in a beaker with medium (visible in the background) and the seedling axis in the front extends to the root medium. From the horizontally cut plain surface the xylem sap exudes and is collected with microcapillaries. C: Cross section (30 µm) through the cotyledon showing the densely packed mesophyll cells and a major vein. The cotyledon's thickness is 100 µm. D: Vasculature of the Ricinus seedling axis with phloem (red) and xylem (blue) strands. At the top, the apical zone with the two petioles extending from the apical node. Further down the hypocotyl with the eight parallel bundles which unite to a phloem ring at the root crown. The roots have the typical radial bundle vasculature with a central xylem. E: Scheme of the nutrient fluxes from the medium to the cotyledon apoplast and from there to mesophyll cells or phloem and nutrient import into the apoplast from the xylem.*

3.2 Sucrose loading from the cotyledon apoplast, sucrose concentrations, and sucrose fluxes

Sucrose is the major nutrient in phloem sap of *Ricinus* seedlings and adult plants. During germination sucrose is synthesized in the endosperm and transferred to the apoplast of the cotyledons. Cells of the lower epidermis, of palisade parenchyma and of the phloem (Bick et al., 1998, Yan, unpublished) are equipped with sucrose transporters. Experiments with position-labelled sucrose showed that at least half of the sucrose found in phloem sap was not randomized along the uptake path from the medium to the sieve tube cytosol, despite high invertase activity in the mesophyll (Orlich and Komor, 1992). It had, therefore, been concluded that at least that part of sucrose was loaded into the phloem by direct apoplastic loading. The other half which had been randomized during phloem loading was probably loaded from the apoplast by passage through a symplastic pathway (so-called indirect apoplastic loading). When low, medium and high sucrose concentrations were offered to the cotyledons a stable steady state was established after a few hours with constant sucrose concentrations and fluxes. At all concentrations of sucrose in the medium (i.e. the apoplast) sucrose was accumulated in the cotyledon symplast (although only weakly in case of 300 mM in the medium) and even more so in the phloem sap. Under all conditions the sucrose concentration in the phloem sap was 120–150 mM higher than in the cotyledons (Fig. 2A). The sucrose concentration in the

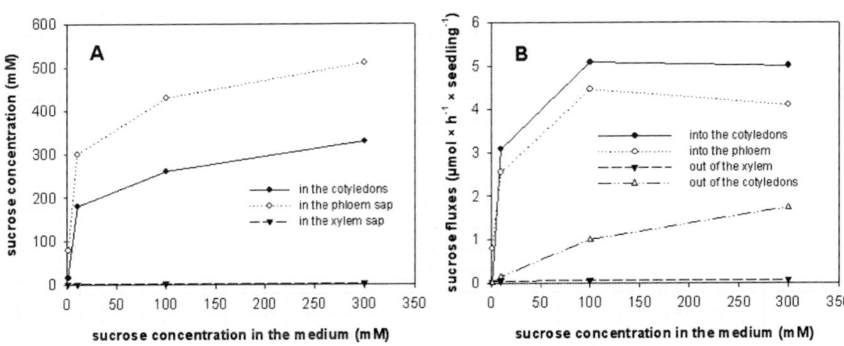

Fig. 2. Sucrose concentrations (A) and fluxes (B) in the cotyledons at different apoplastic (= medium) concentrations of sucrose. The cotyledons attached to the seedling axis were incubated for 5 h in 5 mM Na-MES buffer pH 5.5, 0.2 mM $CaCl_2$, and sucrose as indicated. Then the concentrations in phloem sap and xylem sap were determined after cutting the hypocotyl. The cotyledons were then harvested and extracted to measure their content of sucrose; the sucrose concentration was calculated at the basis of 55% water content in the cotyledons. For xylem sap the sucrose equivalents were calculated from the amount of hexoses found.

xylem sap was under all conditions very low, in fact usually only glucose and fructose at approximately equal concentrations were found, which increased slightly from 1.4 mM to 3 mM with the increasing sucrose concentration in the phloem sap. The sucrose concentration in xylem sap was extremely low (>0.1 mM) even when sucrose was fed to the root system of the seedlings (data not shown). Obviously there is no efficient sucrose loading in the xylem, and the small amounts which may get into the root central cylinder are split by extracellular invertase. These data are in agreement with other reports where the tests for sucrose in xylem sap were unsuccessful (e.g. Peuke et al., 2002).

The sucrose fluxes into the cotyledons and into the phloem showed saturation behaviour and were not increased when the apoplastic sucrose concentration had been increased from 100 mM to 300 mM (Fig. 2 B), although there had been a slight but significant increase in sucrose concentrations in cotyledons and in phloem sap. The reason is that high apoplastic sucrose concentration of 300 mM decreased the volume flow from the apoplast to the phloem so that the mass flow (concentration × volume flow) did not change (Orlich, 1998). The recirculation fluxes via the xylem were negligibly small (0.07 – 0.15 μmol h^{-1} seedling^{-1}). There was an efflux of sucrose from the cotyledon apoplast into the medium increasing with the external sucrose concentrations. This efflux had been measured by preloading the cotyledons with labelled sucrose followed by exchange of the labelled sucrose of the medium with unlabelled sucrose and measurement of appearance of label in the medium. The efflux may derive from either exchange of sucrose through the sucrose transporter (so-called exchange flux) or from leakage from the cotyledon apoplast. The "loss" of sucrose from the cotyledon into the phloem and into the medium is sometimes not fully compensated by uptake into the cotyledons. It is supposed that the high intracellular starch reserves contribute slightly (5–10%) to sucrose synthesis and to the sucrose fluxes.

3.3 Glutamine loading, glutamine concentrations, and fluxes and correlation with N transport

Amino acids are the second largest group of phloem compounds in *Ricinus* and glutamine makes up half of the concentration of amino acids. Therefore, the loading of glutamine as major amino acid was studied. Similar to sucrose loading a saturation-like curve was obtained when increasing apoplastic concentrations of glutamine were supplied (Fig. 3A). The cotyledons accumulated glutamine 5 – 10-fold above the concentration in the medium and also the concentration in the phloem sap was higher than in the medium. But in contrast to the situation of sucrose, glutamine in the phloem sap was clearly lower than in the cotyledon (= symplast). There was also a substantial concentration of glutamine in the xylem sap depending clearly on the glutamine concentration offered at the cotyledons, indicating a phloem to xylem transfer of glutamine and its recirculation up into the cotyledons (Fig. 3A). The flux rates of glutamine into the cotyledons were twice of the fluxes into the phloem. Surprisingly the recirculation of glutamine via the xylem was as high as the export of glutamine via the phloem, except only at the lowest apoplastic glutamine concentration (Fig. 3B). Similar to the situation of sucrose there was also some efflux of glutamine into the medium at high external glutamine concentrations.

The virtually complete recirculation of glutamine is suggesting that the sink organs do not consume glutamine when it comes at high concentrations from the phloem. However, glutamine is subject to conversion to other amino acids which are transported by a multitude of transporters with overlapping specificity. The interaction with the biosynthetic and transport routes between amino acids seems complex since the relative content of the other amino acids (besides glutamine) in the cotyledons was increasingly lowered by high levels of glutamine, but their concentration in the phloem sap increased nevertheless (data not shown). The same was true for the xylem sap.

In addition to the amino acids there were other soluble reduced N-containing compounds in cotyledons, phloem sap and xylem sap, which

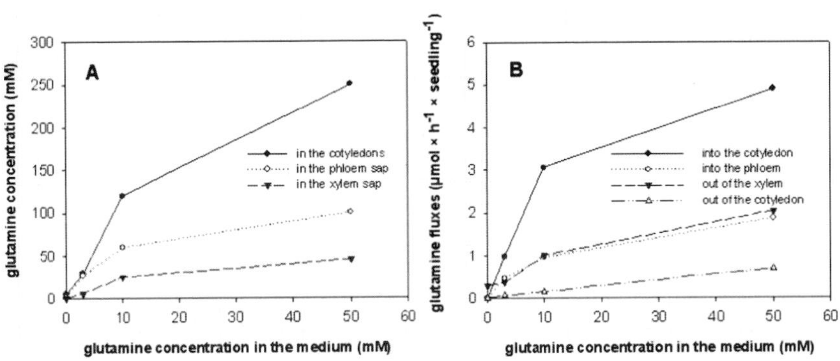

Fig. 3. Glutamine concentrations (A) and fluxes (B) in the cotyledons at different apoplastic (= medium) concentrations of glutamine. The cotyledons attached to the seedling axis were incubated for 5 h in 5 mM Na-MES buffer pH 5.5, 0.2 mM $CaCl_2$, 100 mM sucrose and glutamine as indicated. Then the concentrations in phloem sap and xylem sap were determined after cutting the hypocotyl. The cotyledons were then harvested and extracted to measure their content of glutamine; the concentration was calculated at the basis of 55% water content in the cotyledons.

added up to the N economy and N transport of the seedling. These "other N sources" were constant in the cotyledons at various apoplastic glutamine levels, but increased strongly in the phloem sap and slightly in the xylem sap (Fig. 4A). The fluxes of N at the different external glutamine concentrations were so that there was always a higher export rate of N via the phloem than the recirculation rate of N via the xylem (Fig. 4B). Obviously, there is a preferential consumption of "other N compounds" by the sink tissue especially at high glutamine levels with the consequence that the "surplus" glutamine is preferentially transferred to the xylem and recirculated to the cotyledons. The chemical nature of these "other N compounds" is not completely known, but a large part is added up by ricinine, nucleotides, proteins and peptides, and glutathione.

Adult *Ricinus* plants derive nitrogen from nitrate via the root system. The *Ricinus* seedling can also feed on nitrate which is given to the roots (Schobert and Komor, 1990). Nitrate is partly assimilated to glutamine in the roots and then transported in the xylem upward. The glutamine flux-rates reached in the xylem are similar to those reached by feeding glutamine to the cotyledons and they seem to be saturated at low nitrate levels (Fig. 5). Also the glutamine flux in the phloem is increased by feeding nitrate to the roots. But again low nitrate levels seem to be saturating already. The result is further evidence for the circulatory flux of glutamine from xylem to phloem and from phloem to xylem.

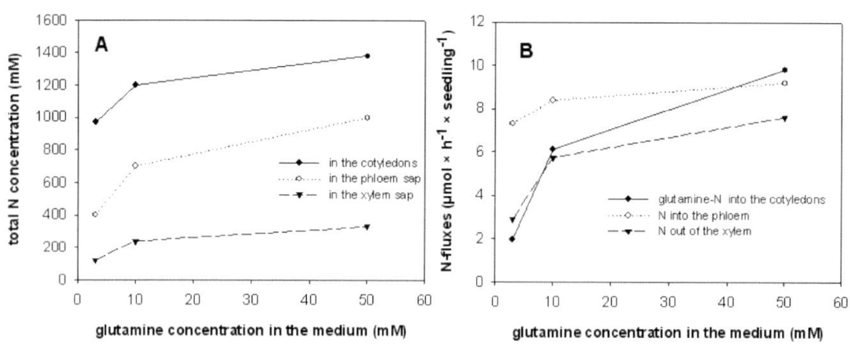

Fig. 4. Total nitrogen concentrations (A) and fluxes (B) in the cotyledons at different apoplastic (= medium) concentrations of glutamine. The cotyledons attached to the seedling axis were incubated for 5 h in 5 mM Na-MES buffer pH 5.5, 0.2 mM $CaCl_2$, 100 mM sucrose and glutamine as indicated. Then the concentrations of N in phloem sap and xylem sap were determined after cutting the hypocotyl. The cotyledons were then harvested and extracted to measure their content of N; the concentration was calculated at the basis of 55% water content in the cotyledons.

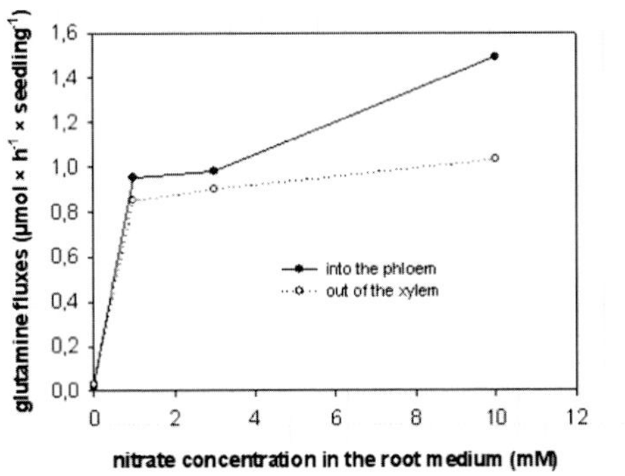

Fig. 5. Glutamine fluxes in the cotyledons at different concentrations of nitrate at the root medium. The root system of the seedlings were incubated for 5 h in 1 mM K-MES buffer pH 5.5, 0.5 mM $CaCl_2$ and K-nitrate as indicated. The same time the cotyledons were incubated in 5 mM Na-MES pH 5.5 and 100 mM sucrose. Then the concentrations of glutamine in phloem sap and xylem sap were determined after cutting the hypocotyl.

3.4 Potassium loading, potassium concentrations, and potassium fluxes

Potassium (K) is the most abundant mineral nutrient in the phloem sap and its loading had been studied together with phloem loading of other mineral ions (Zhong et al., 1998). Feeding of potassium chloride to the cotyledons of seedlings resulted in an increase of K content in the cotyledons without increasing the K concentration in the phloem sap, which stayed at high level even when no K ions had been added to the medium (Fig. 6A). Similarly the K concentration in the xylem sap was the same at all incubation conditions. The K fluxes in the phloem also were rather constant under all conditions, the recirculation fluxes in the xylem showed a small increase when K in the medium had been increased (Fig. 6B).

In all cases the recirculating xylem fluxes were slightly smaller than the phloem fluxes. When potassium chloride was fed to the roots the xylem fluxes became larger whereas the phloem fluxes remained constant (Fig. 7). Obviously, the phloem loading system is under saturation at all applied K concentrations and cannot increase anymore by further addition of external K ions, whereas the xylem loading system still had the capacity for a small increase. In general it appears that a large fraction of K ions is circulated between phloem and xylem and only a small part is withdrawn from the circulating fluxes by the sink tissues. A similar situation was found for the nutrients phosphate, magnesium ions and sulphate (data not shown).

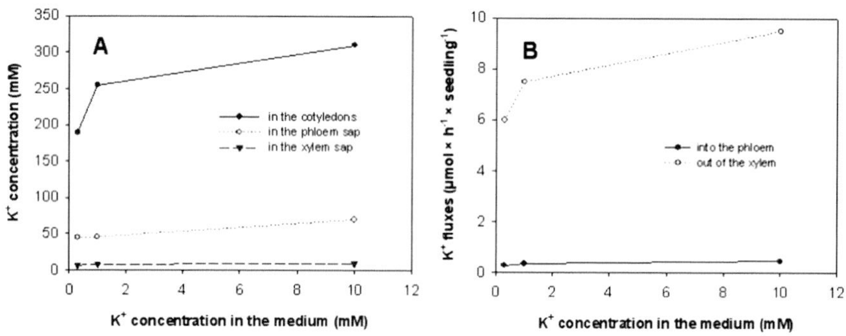

Fig. 6. *Potassium ion concentrations (A) and fluxes (B) in the cotyledons at different apoplastic (=medium) concentrations of KCl. The cotyledons attached to the seedling axis were incubated for 5 h in 5 mM Na-MES buffer pH 5.5, 0.2 mM $CaCl_2$, 100 mM sucrose and KCl as indicated. Then the concentrations in phloem sap and xylem sap were determined after cutting the hypocotyl. The cotyledons were then harvested and extracted to measure their content of potassium ions; the concentration was calculated at the basis of 55% water content in the cotyledons.*

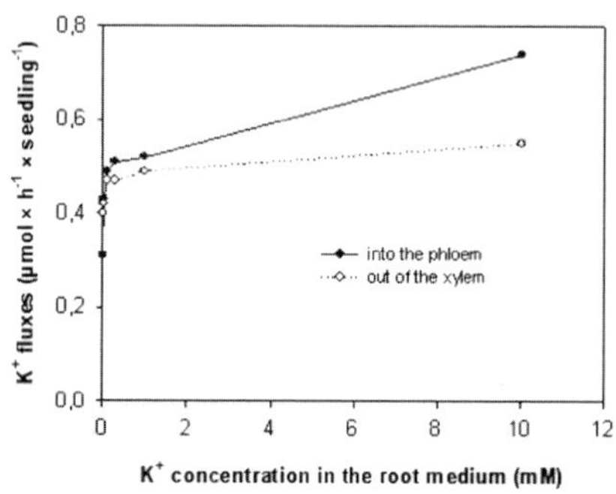

Fig. 7. *Potassium ion fluxes in the cotyledons at different concentrations of KCl at the medium of the root system. The seedlings were incubated at their root system for 5 h in 1 mM Na-MES buffer pH 5.5, 0.5 mM $CaCl_2$ and KCl as indicated. The same time the cotyledons were incubated in 5 mM Na-MES pH 5.5 and 100 mM sucrose. Then the concentrations in phloem sap and xylem sap were determined after cutting the hypocotyl.*

3.5 Functional nutrient buffer capacity of the cotyledon apoplast

The apoplast of cotyledons consists of the cell-wall space in the mesophyll and the xylem vessels which make up ca. 14% and 1% of the

Table 1. Estimated nutrient buffer capacity of the cotyledon apoplast to maintain the phloem loading activity. The apoplast of the cotyledons comprises 14.4% of the cotyledon's volume (10.7% cell wall space, 2.7% intercellular space and 1.0% tracheary space). The fresh weight of the cotyledon pair of the seedling is ca. 50 mg. It was assumed that under steady state conditions the apoplastic concentration of the nutrients is equal to that in the medium. The phloem loading rate was taken from the previous figs 2, 4 and 6. In case of glutamine the phloem loading of all N-compounds was considered. The time period where the apoplastic nutrient could feed phloem loading was calculated from the amount of nutrient in the apoplast and the phloem loading rate under the particular condition. In case where recirculation of nutrients via the xylem occurs the difference between export by the phloem and import by the xylem was taken as effective export rate.

Nutrient in the apoplast	concentration cotyledon	Time period allowing phloem loading without nutrient recirculation (min)	Time period allowing phloem loading in presence of nutrient recirculation (min)
Sucrose	10 mM	1.8	1.8
	100 mM	10.4	10.4
	300 mM	33.9	33.9
Glutamine	3 mM	0.19	0.32
	10 mM	0.55	1.74
	50 mM	2.51	14.4
Potassium	0.3 mM	0.29	0.66
	1 mM	0.94	2.9
	10 mM	8.7	58

cotyledons′ volume. Under the assumption that the nutrient concentration is equal in the medium and the apoplast under steady-state conditions the amount of particular nutrients in the apoplastic spaces can be estimated including the time how long this buffer would last to supply the particular nutrient for phloem loading. Without import of recirculated nutrients the buffer capacity of the apoplastic space is rather low for sucrose, glutamine and potassium. In all these cases it would last only for 1 min at low apoplastic concentration and up to 1 h at high concentrations (Table 1). With the import of nutrients by recirculating fluxes via the xylem the functional buffer capacity increases 5 – 10 fold or more in case of N and K (and also for P and Mg, data not shown), not for sucrose, which shows negligible recirculation (Table 1). In summary, the buffer capacity of the apoplast for nutrients is small and insufficient to support the nutrient demand of the phloem loading system for more than a few minutes, however in case of "non-consumable" nutrients such as mineral salts, the recirculating fluxes can provide a homeostasis-like situation in the cotyledons for a time range of 30 min to 1 h just by using the minerals of the apoplastic space. In cases where mineral nutrients were taken up by the roots and delivered to the apical regions via the xylem (K and nitrate) the import into the cotyledon is larger than the export via the phloem (see Figs. 5 and 7).

4. DISCUSSION AND CONCLUSIONS

Nutrients given to the medium in which the cotyledons are incubated are taken up by the cotyledons and loaded into the phloem. The amounts taken up and transferred to the phloem strongly depend on the type of nutrient. Sucrose uptake is highest and 90% of it is loaded into the phloem. Amino acid uptake and loading of amino acids and other nitrogenous compounds into the phloem is 10–20% of the rate for sucrose and potassium loading is 5–10% of the sucrose rate. All these nutrients occur at higher concentration in the phloem sap than in the medium, but only sucrose is accumulated relative to the concentration in the cotyledon symplast. Previous studies had shown that at least half of the sucrose is loaded apoplastically without passage through the symplast and the accumulation of sucrose in the sieve-tube sap can only be explained by a dominant apoplastic sucrose transport route for phloem loading, since symplastic transport would not be able to achieve accumulation in the sieve tube sap. The finding of sucrose carrier expression in the phloem of cotyledons gives further indication of active accumulative phloem loading (Bick et al., 1998). This does not imply that the decisive accumulative step is only at the sieve tube-companion cell complex. Indeed Orlich et al. (1998) had shown that a small but significant amount of sucrose was transiently loaded directly after complete blockage of the apoplastic loading path by a mercury compound. This observation was interpreted that this amount of sucrose had been accumulated previously in phloem cells outside the sieve tubes and was transferred to the sieve tubes by a symplastic route. Besides the effect of poisoning the apoplastic sucrose carriers there are further possibilities of modulation of apoplastic and symplastic phloem loading routes, for example the presence of high concentrations of mineral salts can favour the apoplastic route over the symplastic route and vice versa (Schobert et al., 1998).

The example of sucrose concentrations in symplast and sieve tube sap is of great importance for the interpretation of phloem loading of other nutrients (glutamine, K), which are at lower concentration in the sieve tube sap than in the symplast. Although a symplastic loading path might be concluded for these nutrients at first sight, the fact that the same phloem strands transport sucrose and the other nutrients implies that there are no significant functional symplastic connections to the mesophyll and the transport of amino acids, K, etc., therefore, also has to be through an apoplastic route.

The apoplast of the cotyledons is open to the medium and nutrients can exchange between medium and the mesophyll apoplast. There is indirect evidence for an open apoplast of the cotyledon. For example there is a sucrose-specific depolarization occurring in mesophyll cells after addition of sucrose to the medium (Köhler et al., 1991) and the presence of sucrose carriers in the mesophyll and the phloem (Bick et al., 1998, Yan, unpublished) points to uptake of sucrose from the cell-wall space. In addition, there is efflux of labelled sucrose and of other nutrients out of preloaded cotyledons. It is not known, however, how conductive the cell-wall space at the lower epidermis and in the mesophyll really is. There may be gradients of nutrients in the apoplast from the mesophyll to the phloem. On one side the lower epidermis is especially rich in sucrose carrier expression (Bick et al., 1998) pointing to an intermediate accumulation of sucrose in this and close to this tissue creating possibly a gradient from the epidermis to the bundles; on the other side there may be a strong water flow from the medium to the phloem because of phloem loading. The amount of water exported via the sieve tubes in one hour corresponds to the entire water content of the cotyledon pair. Whether that water is derived from the medium in our experimental condition or whether it is derived from the xylem sap as shown for the intact, endosperm-surrounded seedling (Köckenberger et al., 1997), is unknown. In case that a significant part really passes through the mesophyll then a flow-derived nutrient gradient may exist. Despite all these uncertainties we think that the nutrient concentrations in the medium are not far off from the concentrations in the apoplast because the cotyledons were brought into a steady state condition by incubation for 5-7 hours and considering the relatively small dimensions of the cotyledon with a distance of only 30 μm between the outer surface of the cotyledon and the phloem.

The nutrient recirculation from phloem to xylem and back from xylem to phloem was nil in case of sucrose, indicating that the *Ricinus* seedling grows under source limitation: all sucrose is taken up by the sink tissues (see also Orlich, 1998). Amino acids (and other nitrogenous compounds), however, are well recirculated and re-imported via the xylem and the extent of recirculation depends on the availability of amino acids for phloem loading. Obviously, the phloem loading-capacity for amino acids is higher than their maximal consumption by the growing seedling. *Ricinus* is a plant which grows readily on nitrogen-poor and on nitrogen-rich soils. The surplus uptake of amino compounds may be an adaptation of the plant to the fact that in nature availability of nitrogen is sometimes at shortage and the accumulation of amino acids in case of a transient N availability may be a means of assuring sufficient nitrogen for growth also for times when the nitrogen source is depleted. The cotyledons of the seedling store reduced nitrogen to support initial seedling growth. The storage of mineral nutrients such as potassium, phosphorus, magnesium and sulphur is also very high and since these minerals are not really "consumed" by metabolism but only incorporated into the newly grown cells, they are at ample supply in the cotyledon and in the phloem. This fact limited strongly the possibilities to manipulate their content in cotyledons and sieve tube sap. There was just always enough for phloem loading close to saturation and no significant reduction of mineral content in the cotyledons could be achieved by preincubation.

Despite this limitation our data show that the cycling of the minerals and N compound fluxes may be so high that there is a nearly complete compensation of phloem export by xylem import. Relatively high recycling rates for mineral ions and N compounds had been also found in adult maize and *Ricinus* plants (Lohaus et al., 2000, Peuke et al., 2002), thus it is not a specialty of the seedling stage. The "engine" driving the cyclic water and nutrient flow is the active proton-coupled loading of the phloem with sucrose together with the complete consumption of sucrose in the sink tissues. The apoplast as an intermediate source of nutrients is in case of the *Ricinus* cotyledons rather small, which means that the efflux of nutrients from storage cells or cells involved in nutrient assimilation into the apoplast is of great importance for nutrient supply to the long distance transport systems. Whether that is only the case for the cotyledons with their small apoplastic space is unknown. But also the relatively large apoplast of green source leaves consists mainly of air-filled space, so that the space for apoplastic nutrients as an intermediate storage compartment en route to the phloem is rather limited, too.

In conclusion our data support quantitatively the previous observations that the cycling of water and nutrients in phloem and xylem (Köckenberger et al., 1997, Tanner and Beevers, 2001) are a successful mechanism to supply nutrients at sufficient amounts to all plant parts even in the absence of a transpirational water flow and that the cycling fluxes may provide a homeostasis-like supply of nutrients to all plant parts.

REFERENCES

Bick, J.A., Neelam, A., Smith, E., Nelson, S.J., Hall, J.L. and Williams, L.E. (1998) Expression analysis of a sucrose carrier in the germinating seedling of *Ricinus communis*. *Plant Mol. Biol.*, 38, 425–435.

Cholewa, E. and Griffith, M. (2004) The unusual vascular structure of the corm of *Eriophorum vaginatum*: implications for efficient retranslocation of nutrients. *J. Exp. Bot.*, 55, 731–741.

Fishman, S., Genard, M. and Huguet, J.G. (2001) Theoretical analysis of systematic errors introduced by a pedicel-girdling technique used to estimate separately the xylem and phloem flows. *J. Theor. Biol.*, 213, 435–446.

Gessler, A., Weber, P., Schneider, S. and Rennenberg, H. (2003) Bidirectional exchange of amino compounds between phloem and xylem during long-distance transport in Norway spruce trees (*Picea abies* (L.) *Karst*). *J. Exp. Bot.*, 54, 1389–1397.

Ho, L.C., Grange, R. and Picken, A.J. (1987) An analysis of the accumulation of water and dry matter in tomato fruit. *Plant Cell Envir.*, 10, 157–162.

Jeschke, W.D. and Pate, J.S. (1991) Modelling of the uptake, flow and utilization of C, N, and H_2O within whole plants of *Ricinus communis* L. based on empirical data. *J. Plant Physiol.*, 137, 488–498.

Kallarackal, J., Orlich, G., Schobert, C. and Komor, E. (1989) Sucrose transport into the phloem of *Ricinus communis* L. seedlings as measured by the analysis of sieve tube sap. *Planta*, 177, 327–335.

Köckenberger, W., Pope, J.M., Xia, Y., Jeffrey, K.R., Komor, E. and Callaghan, P.T. (1997) A non-invasive measurement of phloem and xylem water flow in castor bean seedling by nuclear magnetic resonance microimaging. *Planta*, 201, 53–63.

Köhler, J., Fritz, E., Orlich, G. and Komor, E. (1991) Microautoradiographic studies of the role of mesophyll and bundle tissues of the *Ricinus* cotyledon in sucrose uptake. *Planta*, 183, 251–257.

Komor, E. (1977) Sucrose uptake by cotyledons of *Ricinus communis* L.: characteristics, mechanism and regulation. *Planta*, 137, 119–131.

Kriedemann., P. and Beevers, H. (1967) Sugar uptake and translocation in the castor bean seedling. I. Characteristics of transfer in intact and excised seedlings. *Plant Physiol.*, 42, 161–173.

Lohaus, G., Hussmann, M., Pennewiss, K., Schneider, H., Zhu, J.J. and Sattelmacher, B. (2000) Solute balance of a maize (*Zea mays* L.) source leaf as affected by salt treatment with special emphasis on phloem retranslocation and ion leaching. *J. Exp. Bot.*, 51, 1721–1732.

Marschner, H., Kirkby, E.A. and Engels, C. (1997) Importance of cycling and recycling of mineral nutrients within plants for growth and development. *Bot. Acta*, 110, 265–273.

Orlich, G. (1998) Analysis of the driving forces of phloem transport in *Ricinus* seedlings: sucrose export and volume flow are determined by the source. *Planta*, 206, 266–271.

Orlich, G., Hofbrückl, M. and Schulz, A. (1998) A symplasmic flow of sucrose contributes to phloem loading in *Ricinus* cotyledons. *Planta*, 206, 108–116.

Orlich, G. and Komor, E. (1992) Phloem loading in *Ricinus* cotyledons: sucrose pathways via the mesophyll and the apoplasm. *Planta*, 187, 460–474.

Pate, J.S. (1973) Uptake, assimilation and transport of nitrogen compounds by plants. *Soil Biol. Biochem.*, 5, 109–119.

Pate, J.S., Layzell, D.B. and McNeil, D.E. (1979) Modelling the transport and utilization of carbon and nitrogen in a nodulated legume. *Plant Physiol.*, 63, 730–737.

Peuke, A.D., Jescke, W.D. and Hartung, W. (2002) Flows of elements, ions and abscisic acid in *Ricinus communis* and site of nitrate reduction under potassium limitation. *J. Exp. Bot.*, 53, 241–250.

Schobert, C. and Komor, E. (1990) Transfer of amino acids and nitrate from the roots into the xylem of *Ricinus communis* seedlings. *Planta*, 181, 85–90.

Schobert, C., Zhong, W.J. and Komor, E. (1998) Inorganic ions modulate sucrose export and phloem loading in *Ricinus communis* seedlings. *Plant Cell Envir.*, 21, 1047–1054.

Tanner, W. and Beevers, H. (2001) Transpiration, a prerequisite for long-distance transport of minerals in plants? *Proc. Natl. Acad. Sci. USA*, 98, 9443–9447

Tukey, H.B. and Mecklenburg, R.A. (1964) Leaching of metabolites from foliage and subsequent reabsorption and redistribution of the leachate in plants. *Amer. J. Bot.*, 51, 737–742.

White, P.J. (1997) The regulation of K^+ influx into roots of rye (*Secale cereale* L.) seedlings by negative feedback via the K^+ flux from shoot to root in the phloem. *J. Exp. Bot.*, 48, 2063–2073.

Zhong, W.J., Kaiser, W., Köhler, J., Bauer-Ruckdeschel, H. and Komor, E. (1998) Phloem loading of inorganic cations and anions by the seedling of *Ricinus communis* L. *J. Plant Physiol.*, 152, 328–335.

LONG-DISTANCE WATER TRANSPORT UNDER CONTROLLED TRANSPIRATIONAL CONDITIONS: MINIMAL-INVASIVE INVESTIGATIONS BY MEANS OF PRESSURE PROBES AND NMR IMAGING

H. SCHNEIDER[1], L.H. WEGNER[1], A. HAASE[2] and U. ZIMMERMANN[1]
[1]*Lehrstuhl für Biotechnologie, Biozentrum, Würzburg, Germany, he.schneider@fz-juelich.de*
[2]*Lehrstuhl für Experimentelle Physik V, Würzburg, Germany.*

Abstract. Herbaceous plants exhibit complex reaction patterns as a response to changes in environmental factors such as light intensity, relative humidity, temperature and root water-supply. The reactions of volume flows and hydrostatic pressures within the intact plant can only be described in an authentic way if minimal- and non-invasive methods are used for investigation. This review article highlights some aspects on the correlation of xylem volume flows and corresponding hydrostatic pressures, with special emphasis put on the role of the water supply to the roots. From these studies it can be concluded that the tissue cells play an important role in determining and maintaining xylem pressures.

Key words: cavitation, water flow, ^{1}H NMR imaging, pressure probe, turgor, xylem

1. INTRODUCTION

It is generally believed that the xylem conduits of higher plants form tightly hydraulically coupled entities from the roots to the leaves. But, there is no general agreement on the degree of the radial hydraulic connection of the xylem vessels to the tissue cells. Supporters of the well-known, more than 100-year-old, Cohesion Theory (CT) believe that water is driven through the plant within impervious tubes. However, increasing evidence

B. Sattelmacher† and W.J. Horst (eds.), The Apoplast of Higher Plants: Compartment of Storage, Transport and Reactions, 251–264.

indicates that the xylem conduit is substantially leaky and, in turn, tightly hydraulically connected to the symplast (Canny, 1993; Zimmermann et al., 1994, 2002a, b, 2004; Stahlberg and Cosgrove, 1995; Schneider et al., 1997b, 1999, 2000a, b, 2003; Wagner et al., 2000). As a consequence, one of the main tenets of the CT, namely that the xylem pressure is solely affected by the balance between water uptake and transpirational water loss has been called into question.

A substantial portion of this new experimental evidence has been obtained by applying non- and minimal-invasive ^{1}H NMR imaging and pressure-probe techniques for the investigation of the distribution and movement of water as well as the corresponding driving forces. With these approaches, the complex coupling of flows and forces within the intact plant remains completely undisturbed. Thus, artefacts arising from cutting of plant organs (Rygol et al., 1993) are avoided.

In this review article, we present a survey of data on the correlation of xylem volume-flows and pressures in higher plants as obtained by minimal-invasive measurements, with special emphasis put on the role of the water supply to the root.

2. BRIEF DESCRIPTION OF THE TECHNIQUES USED

2.1 Pressure-probe techniques

At present, hydrostatic pressures in single xylem vessels or individual tissue cells are routinely measured on intact, transpiring plants by means of the minimal-invasive pressure-probe technique, which was originally introduced by Zimmermann and co-workers (Zimmermann et al., 1969, 1994, 2002a, b, 2004; Pritchard et al., 1989; Steudle, 1989; Balling and Zimmermann, 1990; Benkert et al., 1991, 1995; Rygol et al., 1993; Fricke, 1997; Schneider et al., 1997a, b, 1999, 2000a; Franks et al., 1998; Melcher et al., 1998; Wegner and Zimmermann, 1998, 2002; Thürmer et al., 1999; Wei et al., 1999, 2001; Wistuba et al., 2000). Through combination of this technique with microanalytical methods, osmotic pressures of cell and xylem saps can additionally be determined in a minimal-invasive manner (Rygol et al., 1993; Tomos and Leigh, 1999; Schneider et al., 1999; Thürmer et al., 1999; Lohaus et al., 2000). Recently, a new generation of pressure probes has been developed which allow direct measurements of K^+ concentrations and pH in the xylem of the intact plant (Wegner and Zimmermann, 2002, 2004; see also Wegner et al., this volume, pp 211–224).

The principle of the pressure probe techniques is physically sound: A single cell or a xylem vessel is impaled by a microcapillary which is connected to a pressure sensor via a perspex chamber. The entire system is filled with oil in the case of turgor-pressure measurements and with degassed, deionised water in the case of xylem-pressure measurements. In this way, a hydraulic continuum between the cell or xylem sap and the pressure sensor is created upon impalement, allowing continuous pressure measurements lasting minutes up to a few days.

Rigorous testing has confirmed the reliability of the xylem probe as a tool for measuring xylem pressures. Namely, it was clearly demonstrated that the probe reads absolute negative pressures down to about -1 MPa (Balling and Zimmermann, 1990; Thürmer et al., 1999). Pressures that were imposed artificially by means of a Hepp-type osmometer attached to a leaf were accurately recorded (Balling et al., 1988). This result as well as the fact that consecutive probe insertions into the same or different vessels always yielded identical xylem pressure values (see e.g. Benkert et al., 1995; Schneider et al., 1997a; Zimmermann et al., 2004) refuted the argument of Milburn (1996) and Sperry et al. (1996) that the insertion of the probe capillary might shift the xylem pressure to more positive values. Likewise, the assumption that the probe capillary might not be positioned properly within a xylem vessel (Milburn, 1996; Steudle, 2001) was disproved by staining of the probed vessel during the pressure measurement (Balling and Zimmermann, 1990; Benkert et al., 1991; Zimmermann et al., 1994, 2002a, 2004).

2.2 NMR imaging

Since it is completely non-invasive, high-resolution ^{1}H NMR imaging is to date the most suitable technique for investigations on the distribution and flow of water within the intact higher plant (Kuchenbrod et al., 1995, 1996; Rokitta et al., 1999a, b; Schneider et al., 2000a, 2003; Wagner et al., 2000; Wistuba et al., 2000; Zimmermann et al., 2000, 2002c; Peuke et al., 2001). The principle of the method is roughly described as follows. Positioning of the plant in a static homogeneous magnetic field leads to the alignment of all proton spins (which are predominantly contained in water molecules) within the specimen. Application of a short high-energy radio frequency pulse forces the spins to change their orientation. The subsequent relaxation of the spins to the original orientation is accompanied by the emission of a radio frequency signal which is detected and processed by a computer. Since this signal is strongly affected by the local chemical and physical environment of the protons, information about anatomical and chemical features of a specimen can be obtained by appropriate choice of the experimental parameters (Kuchenbrod et al., 1995; Schneider et al., 2003). Furthermore, differentiation between stationary and mobile water within a virtual cross-section of an intact plant makes it possible to monitor water diffusion

(Kuchenbrod et al., 1995; Schneider et al., 2000a, 2003) as well as flow velocities in the xylem and phloem with high spatial and temporal resolution (Kuchenbrod et al., 1996; Köckenberger et al., 1997; Rokitta et al., 1999a, b; Wistuba et al., 2000).

3. CORRELATION OF HYDROSTATIC PRESSURES AND FLOWS

For a given plant species, entire flow-force patterns can be obtained when the same plant is investigated simultaneously by means of pressure probes and NMR imaging. However, due to disturbance of the pressure-probe measurements by the high magnetic field strengths needed for NMR imaging on plants, this is at present only possible for lianas which allow spatial separation of the measuring sites (Wistuba et al., 2000). In smaller plants, xylem-pressure values can be related to flow data measured separately when transpiration rates are essentially identical in both experiments. Transpiration rates can also roughly be taken as a measure of volume flows since comparison of ^{1}H NMR flow imaging and gas-exchange measurements revealed that both parameters are closely correlated (Kuchenbrod et al., 1996).

3.1 Flow-force patterns under conditions of sufficient water supply to the roots

3.1.1 Conditions leading to elevated xylem volume-flow

In well-watered plants, the xylem volume-flow can generally be increased by an elevation of the light intensity, as proven by ^{1}H NMR flow imaging (Kuchenbrod et al., 1996; Rokitta et al., 1999a, b). Correspondingly, transpiration rates increase and xylem pressures decrease as demonstrated in Fig. 1A for a maize plant grown in hydroculture. At constant laboratory illumination (ca. 6 μmol m^{-2} s^{-1}) a root xylem-pressure value of about +0.09 MPa was registered, along with a transpiration rate close to zero. A sudden increase of the light intensity to ca. 300 μmol m^{-2} s^{-1} (dotted lines in Fig. 1A) by means of an additional light source resulted in an almost immediate increase of the transpiration rate. A corresponding xylem pressure decrease was observed only after a delay of about 5 min. During the following 30 min, xylem pressure and transpiration rate underwent a typical "overshoot" reaction (compare Wegner and Zimmermann, 1998; Zimmermann et al., 2002a, 2004; Wegner et al., this volume, pp. 207–220)

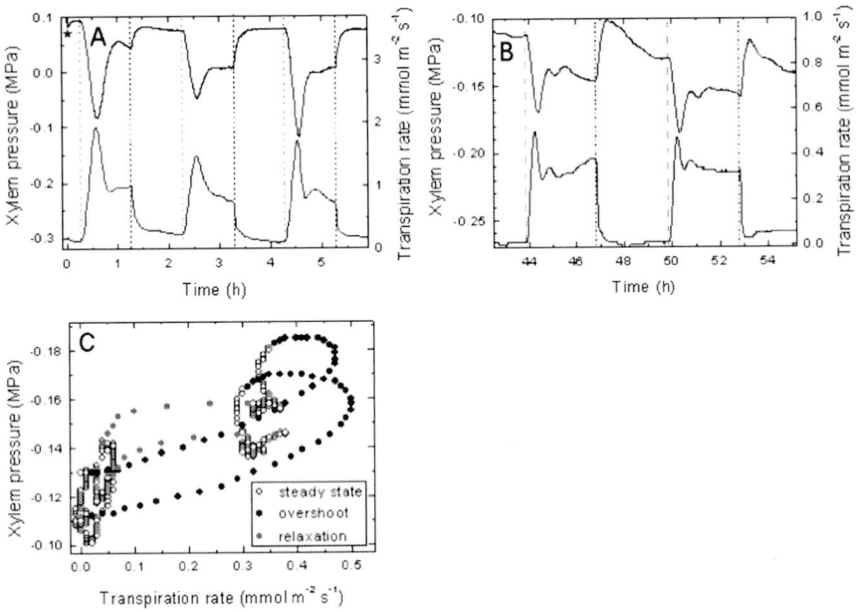

Fig. 1. Parts of simultaneous recordings of light-induced changes in transpiration rate (thin traces) and root and shoot xylem pressure, respectively (thick traces), of a maize plant grown in hydroculture (A; probe insertion marked by asterisk) and a tomato plant grown in soil culture (B) at constant laboratory conditions (ca. 24°C and 45% relative humidity (RH)). Light intensity (LI) was increased (dotted lines) from 6 µmol $m^{-2}s^{-1}$ to 300 µmol $m^{-2}s^{-1}$ (A) or to 150 µmol $m^{-2}s^{-1}$ (B) and later reduced to the initial value (dashed lines). (C) is a plot of the xylem pressures from (B) against corresponding transpiration rates.

whereby xylem pressure reactions usually lagged slightly behind stomatal reactions. Namely, after passing a minimum xylem pressure value of –0.08 MPa in combination with a maximum transpiration rate of ca. 2 mmol m^{-2} s^{-1}, both parameters assumed new quasi-stationary values of +0.05 MPa and ca. 1 mmol m^{-2} s^{-1}, respectively, at elevated light intensity.

When the additional lamp was again switched off (dashed lines in Fig. 1A), xylem pressure and transpiration rate responded instantaneously by mirrored relaxations back to the original values at laboratory illumination. As also indicated in Fig. 1A, comparable light-dependent reaction patterns of the xylem pressure and the transpiration rate could be repeatedly measured on the same plant.

Tobacco plants exhibited no clear-cut overshoot reactions as long as the increase in light intensity occurred in steps of less than ca. 40 µmol m^{-2} s^{-1} (Fig. 2). This was found irrespective whether the increase occurred from laboratory illumination (Fig. 2) or from already elevated light intensities

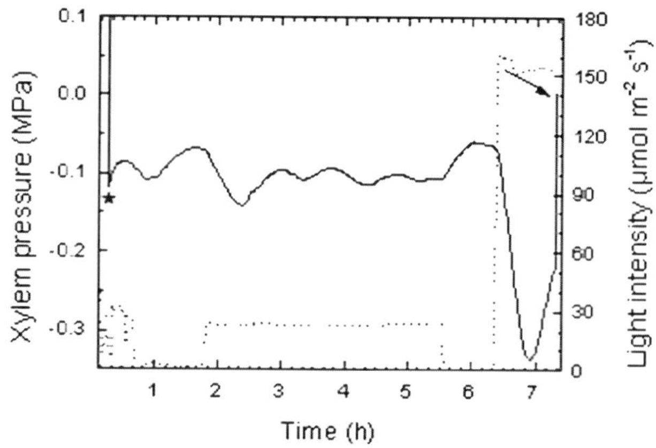

Fig. 2. Shoot xylem-pressure responses (solid trace) of a well-watered tobacco plant grown in soil culture to sudden changes in LI (dotted trace) under laboratory conditions (23–25°C, ca. 50% RH). The instant of xylem puncturing is marked by an asterisk. Whereas slight changes in LI between 4 μmol m^{-2} s^{-1} and ca. 30 μmol m^{-2} s^{-1} did not change the average xylem pressure significantly, a rapid pressure decrease was observed upon application of 150 μmol $m^{-2}s^{-1}$ until cavitation (i.e. formation of a water-vapor bubble) occurred (arrow).

(data not shown). Similar observations have also been made on maize for light intensity increases by 50–60 μmol m^{-2} s^{-1} (Wei et al., 1999). Interestingly, however, unlike other plant species (e.g. tropical lianas; Thürmer et al., 1999; Wistuba et al., 2000) tobacco exhibited no significant change in the quasi-stationary xylem pressures upon light intensity increases from laboratory illumination to less than ca. 40 μmol m^{-2} s^{-1} (Fig. 2).

About 10 years ago, Balling and Zimmermann (1990) and Benkert et al. (1991) presented similar results for slowly transpiring tobacco plants treated with a conventional desk lamp, which provoked at that time a fierce debate on the reliability of xylem pressure-probe data. Today, however, we understand that different plant species obviously react differently and that some species may even be completely insensitive to changes in light intensity (Schneider et al., 1999).

From all concurrent pressure-probe and transpiration measurements on well-watered crop plants, we can state that changes in xylem pressures were generally correlated with changes in xylem volume-flows (Fig. 1). This finding is consistent with pressure probe and ^1H NMR flow-imaging data obtained previously for a well-watered tropical liana (Wistuba et al., 2000), and does not contradict the assumptions of the CT. However, a closer inspection of concurrently measured response patterns of xylem pressures and transpiration rates revealed that the correlation was by no means perfect (Fig. 1C). Theoretically, such a lack of agreement could arise from local tailbacks of water at sites forming bottlenecks for water flow or from pronounced local differences in transpiration rates (Sperry et al., 1996; Thürmer et al., 1999). However, this does not apply to the rather small (up to ca. 60 cm) plants used in this study.

Another, more straightforward explanation for temporal differences between transpiration and xylem pressure patterns is that a xylem pressure decrease induced by changes in the transpiration rate is buffered either by the tissue cells feeding water into the xylem or by changes in root hydraulic conductance. Water feeding from the tissue into the xylem is very likely because of the generally very rapid water-exchange times of tissue cells (some 10 s; Zimmermann, 1989; Steudle, 1989; Malone, 1993) and because of the hydraulic coupling between both compartments which can be visualised by simultaneous xylem pressure and cell turgor-pressure measurements (Schneider et al., 1997b, 1999; Thürmer et al., 1999; Wistuba et al., 2000; Zimmermann et al., 2002a, b, 2004). Hydraulic coupling is demonstrated in Fig. 3, which shows concurrent xylem and cortical-cell turgor-pressure changes upon changes in light intensity in the shoot of a well-watered potato plant grown in soil culture. Upon a light intensity increase from 1 μmol m^{-2} s^{-1} to ca. 110 μmol m^{-2} s^{-1} (upward-directed arrow) the pressures of both compartments exhibited simultaneous decreases. The light-induced changes of the xylem pressure were, however, more

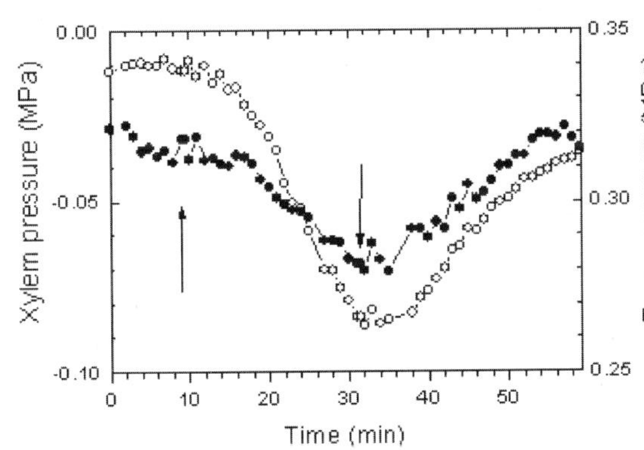

Fig. 3. Simultaneous recording of xylem (○) and cortical-cell turgor pressure responses (●) in the shoot of a well-watered potato plant grown in soil culture to changes in LI. Both probes were inserted under laboratory conditions at an LI of 1 μmol m^{-2} s^{-1}, yielding stable xylem and turgor pressure values of 0.01 and +0.32 MPa, respectively. Upon an increase in LI to 110 μmol m^{-2} s^{-1} (↑) both values dropped in parallel by 0.09 and 0.04 MPa, respectively, returning to the initial pressures at low LI (↓).

pronounced than those of the turgor pressure as demonstrated previously for other plant species and organs (Schneider et al., 1997b, 1999; Zimmermann et al., 2002a, b, 2004), leading to a ratio between both parameters of ca. 1:0.5 (Fig. 3). Return to the low light intensity (downward-directed arrow) resulted in a simultaneous relaxation of both pressures. Interestingly, the original xylem pressure value was reached with a delay compared to the turgor pressure value (Fig. 3).

3.1.2 Conditions leading to a decrease in xylem volume flow

More insight into the coupling of transpiration and xylem pressure in well-watered plants was obtained by choosing experimental conditions which cause a substantial decrease in flow, namely by increasing ambient relative humidity and by submerging the leaves in water. Under these conditions, the volume flow is usually proportional to the flow velocity since the water-conducting area remains constant.

3.1.2.1 Elevation of the ambient relative humidity

Elevation of the ambient relative humidity to values close to 100% generally leads to an immediate decrease in the transpiration rate and increase of the xylem pressure to more positive values, again demonstrating a clear-cut correlation between xylem volume flow and pressure. Typical xylem pressure changes of a well-watered tomato plant grown in soil culture to changes in the relative humidity are depicted in Fig. 4. Successful probing of a shoot vessel at a relative humidity of ca. 85% resulted in a xylem-pressure value of +0.01 MPa. Negative xylem-pressure values could be established when the relative humidity was gradually reduced, thus excluding the presence of air in the xylem-probe continuum. Subsequently, the relative humidity was increased again in a stepwise manner, leading to a corresponding incremental increase in the xylem pressure until slightly above-atmospheric values were reached at a relative humidity of ca. 100%. After some time, root pressure manifested itself by visible guttation.

3.1.2.2 Leaf submersion in water

As opposed to the case of elevated relative humidity, where xylem volume flow is not reduced to zero (compare Tanner and Beevers, 2001), submersion of the leaves in water should lead to complete cessation of transpiration.

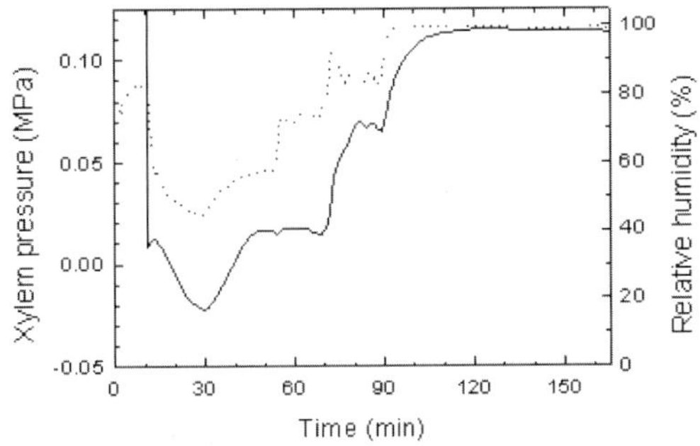

Fig. 4. Response pattern of the shoot xylem pressure (solid line) of a tomato plant grown in soil culture to changes in RH (dotted line). Measurements were performed in a climate chamber at almost constant illumination (10 µmol m^{-2} s^{-1}) and temperature (ca. 23–25°C). RH was increased by temporarily blowing humidified air into the chamber. Xylem pressure changed immediately in response to changes in RH, exceeding the atmospheric value at ca. 100% RH.

As shown in Fig. 5 for a tomato plant grown in soil culture that had been turned into an upside-down position (inversion of the plants to the upside-down orientation in air had no effect on the xylem pressure; data not shown), complete submersion of the entire foliage in water resulted in an immediate xylem pressure increase to more positive steady-state values. Similar results had been obtained previously for intact wheat plants grown in hydroculture

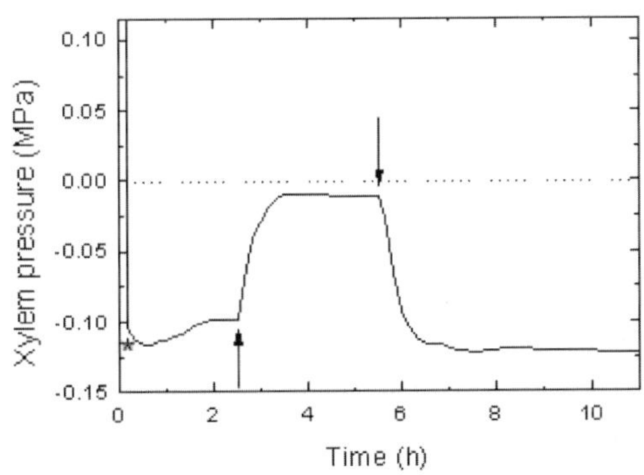

Fig. 5. Response of the shoot xylem pressure of an upside down tomato plant grown in soil culture under laboratory conditions (21°C, 30% RH, LI ca. 10 µmol m^{-2}s^{-1}), to submersion of the leaves in water (upward-directed arrow) and to the subsequent removal of the water (downward-directed arrow). The instant of xylem vessel puncturing by the pressure probe capillary is marked by an asterisk. The dotted line marks vacuum.

(Schneider et al., 1997b). However, while above-atmospheric pressures were established in submerged wheat (Schneider et al., 1997b), the xylem pressure of tomato mostly remained in the slightly negative range (Fig. 5). This demonstrated that zero transpiration (i.e. xylem volume-flow) does not necessarily correspond to (above-) atmospheric xylem-pressure values.

The new steady-state values remained constant as long as the plants were kept in the submerged state (i.e. for hours or even days; Fig. 5 and Schneider et al., 1997b). The "submersion effect" was completely reversible (Fig. 5; Schneider et al., 1997b). Thus, the submersion experiments, as well as experiments in which the leaf blades were entirely covered with oil (Balling and Zimmermann, 1990), confirm the previously formulated statement (Schneider et al., 1999; Thürmer et al., 1999; Wistuba et al., 2000; Zimmermann et al., 2004) that the xylem pressure is determined by the "water potential" of the tissue cells, but not a priori by the transpiration rate, and that xylem pressure will not change as long as the turgor pressure remains constant (compare the respective equations in the above references).

3.2 Flow-force patterns under conditions of restricted water supply to the roots

Severe restriction of the water supply to the roots due to drought, osmotic stress or severing part of the root system (Fig. 6) also leads to decreased or completely ceased volume flow as proven by ^{1}H NMR imaging (data not shown). In heavily stressed plants xylem volume flow is no longer proportional to the flow velocity, since the fraction of conducting xylem elements may unpredictably change due to cavitation (compare Rokitta, 1999a). This aspect comes into play since under conditions of restricted water supply the xylem pressure decreases to minimum stable values of ca. –0.6 MPa (Fig. 6; see also Zimmermann et al., 1994, 2004) before finally the majority of the xylem vessels become cavitated. This can also be demonstrated by ^{1}H NMR imaging (Fig. 7). A detailed outline of the correspondence of flows and driving forces under conditions of restricted root water-supply will be published separately (Rokitta, Westhoff et al., manuscript in preparation). Briefly, it can be stated that upon restriction of the root water-supply, negative or no correlation between xylem volume-flow and pressures was found (compare Wistuba et al., 2000).

The experiments conducted under conditions of restricted root water-supply again provided evidence for the feeding of water by the tissue cells into the adjacent xylem elements. As demonstrated by the ^{1}H NMR image of the shoot of a drought-stressed tomato plant grown in soil culture in Fig. 7, pronounced shrinkage of the tissue cells occurred during drought, until the

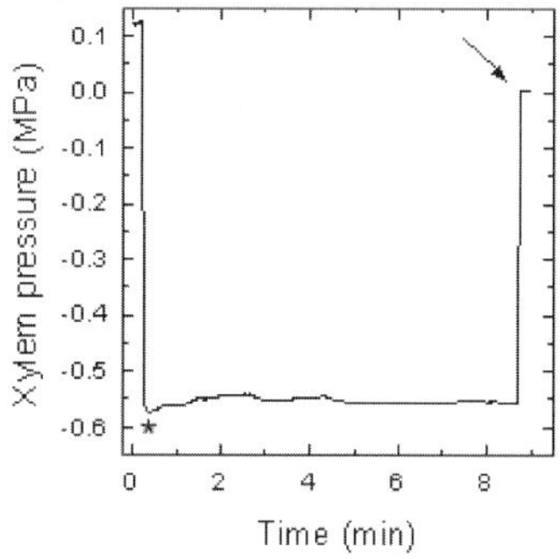

Fig. 6. Xylem pressure recording in the shoot of a tomato plant grown in soil culture which was heavily stressed due to cutting of part of the fine roots. The measurement was performed under constant laboratory conditions (24°C, 45% RH), at a light intensity of ca. 6 μmol m^{-2}s^{-1}. The successful probe insertion is indicated by an instantaneous decrease of the xylem pressure from above-atmospheric values to a negative quasi-stationary value of –0.55 MPa within a few seconds (marked by an asterisk). The measurement is ended after about 9 min by a cavitation (arrow).

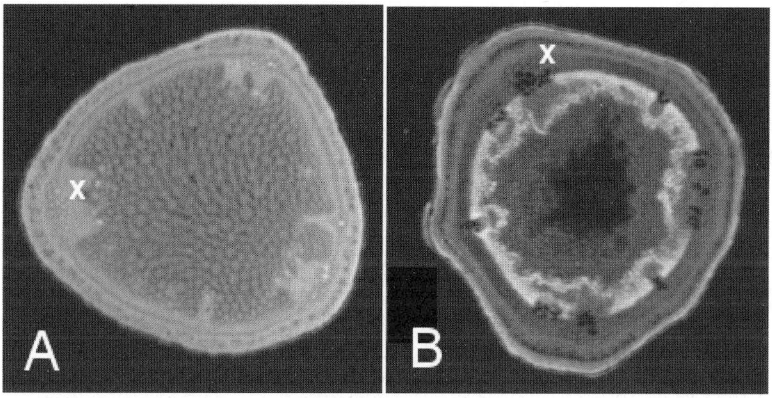

Fig. 7. ^{1}H NMR spin-echo images of the shoots of a well-watered (A) and a severely drought-stressed tomato plant (B). The brighter the image pixels the higher the amount of protons (i.e. water). Single cells and xylem vessels can be distinguished in both images. Whereas the bright xylem vessels ("x") in (A) indicate water filling, those in (B) are completely dark, confirming that they are cavitated. Comparison of both images suggests a shift of water between tissue regions during drought, as indicated in (B) by a high-intensity ring-shaped region within the pith and relatively dark regions nearby the xylem.

entire xylem was cavitated. Accordingly, cell saps seemed to become highly concentrated as indicated by ^{1}H NMR T_1 measurements (Rokitta, Westhoff et al., manuscript in preparation).

Qualitatively similar results were obtained for tobacco and tomato plants. However, in contrast to tobacco, comparison of NMR images of the shoots of well-watered (Fig. 7A) and drought-stressed tomato plants (Fig. 7B) revealed significant local changes in the signal intensity, indicating a substantial shift of water between tissue regions. Most interestingly, portions of tissue cells directly adjacent to the xylem exhibited extraordinarily low signal intensities in drought-stressed plants (Fig. 7B), as expected when water is shifted from the cells into the xylem.

4. CONCLUSIONS AND PERSPECTIVES

Our results have shown that the combination of pressure-probe measurements with NMR imaging and transpiration measurements can be used for the evaluation of water-relations parameters of higher plants in a highly accurate, minimal-invasive manner. We could demonstrate that the plant has to be considered as a highly hydraulically coupled system which reacts very sensitively to changes in the environmental conditions. Because of this tight hydraulic coupling, reaction patterns of the xylem pressure (and flow) cannot be evaluated separately from the corresponding reaction patterns of the surrounding tissue. Thus, future projects have to deal with the evaluation of radial water flows and diffusion, as well as with the investigation of the mechanisms of water feeding from the tissue cells into adjacent xylem vessels, in order to obtain a more detailed image of the mechanisms of water transport in the intact, higher plant.

ACKNOWLEDGMENTS

The authors are grateful to M. Schubert, K. Schwuchow, F. Thürmer and M. Westhoff for their assistance in performing some of the experiments and to an anonymous reviewer for her/his excellent suggestions.

REFERENCES

Balling, A. and Zimmermann, U. (1990) Comparative measurements of the xylem pressure of *Nicotiana* plants by means of the pressure bomb and pressure probe. *Planta,* 182, 325–338.

Balling, A., Zimmermann, U., Büchner, K.-H. and Lange, O.L. (1988) Direct measurement of negative pressure in artificial-biological systems. *Nat. wiss.,* 75, 409–411.

Benkert, R., Balling, A. and Zimmermann, U. (1991) Direct measurements of the pressure and flow in the xylem vessels of *Nicotiana tabacum* and their dependence on flow resistance and transpiration rate. *Bot. Acta,* 104, 423–432.

Benkert, R., Zhu, J.-J., Zimmermann, G., Türk, R., Bentrup, F.-W. and Zimmermann, U. (1995) Long-term xylem pressure measurements in the liana *Tetrastigma voinierianum* by means of the xylem pressure probe. *Planta,* 196, 804–813.

Canny, M.J. (1993) The transpiration stream in the leaf apoplast: water and solutes. *Phil. Trans. R. Soc. Lond.,* B 341, 87–100.

Franks, P.J., Cowan, I.R. and Farquhar, G.D. (1998) A study of stomatal mechanics using the cell pressure probe. *Plant Cell Environ.*, 21, 94–100.

Fricke, W. (1997) Cell turgor, osmotic pressure and water potential in the upper epidermis of barley leaves in relation to cell location and in response to NaCl and air humidity. *J. Exp. Bot.*, 48, 45–58.

Köckenberger, W., Pope, J.M., Xia, Y., Jeffrey, K.R., Komor, E. and Callaghan, P.T. (1997) A non-invasive measurement of phloem and xylem water flow in castor bean seedlings by nuclear magnetic resonance microimaging. *Planta*, 201, 53–63.

Kuchenbrod, E., Haase, A., Benkert, R., Schneider, H. and Zimmermann, U. (1995) Quantitative NMR microscopy on intact plants. *Magn. Res. Imag*., 13, 447–455.

Kuchenbrod, E., Landeck, M., Thürmer, F., Haase, A. and Zimmermann, U. (1996) Measurement of water flow in the xylem vessels of intact maize plants using flow-sensitive NMR imaging. *Bot. Acta,* 109, 184–186.

Lohaus, G., Hussmann, M., Pennewiss, K., Schneider, H., Zhu, J.J. and Sattelmacher, B. (2000) Solute balance of a maize (*Zea mays* L.) source leaf as affected by salt treatment with special emphasis on phloem re-translocation and ion leaching. *J. Exp. Bot.*, 51(351), 1721–1732.

Malone, M. (1993) Hydraulic signals. *Phil. Trans. R. Soc. Lond.*, B 341, 33–39.

Melcher, P.J., Meinzer, F.C., Yount, D.E., Goldstein, G. and Zimmermann, U. (1998) Comparative measurements of xylem pressure in transpiring and non-transpiring leaves by means of the pressure chamber and the xylem pressure probe. *J. Exp. Bot.*, 49, 1757–1760.

Milburn, J.A. (1996) Sap ascent in vascular plants: Challengers to the Cohesion Theory ignore the significance of immature xylem and the recycling of Münch water. *Ann. Bot*., 78, 399–407.

Peuke, A.D., Rokitta, M., Zimmermann, U., Schreiber, L. and Haase, A. (2001) Simultaneous measurement of water flow velocity and solute transport in xylem and phloem of adult plants of *Ricinus communis* over a daily time course by nuclear magnetic resonance spectroscopy. *Plant Cell Environ.,* 24, 491–503.

Pritchard, J., Williams, G., Wyn Jones, R.G. and Tomos, A.D. (1989) Radial turgor pressure profiles in growing and mature zones of wheat roots - a modification of the pressure probe. *J. Exp. Bot*., 40, 567–571.

Rokitta, M., Peuke, A.D., Zimmermann, U. and Haase, A. (1999a) Dynamic studies of phloem and xylem flow in fully differentiated plants by fast nuclear-magnetic-resonance microimaging. *Protoplasma,* 209, 126–131.

Rokitta, M., Zimmermann, U. and Haase, A. (1999b) Fast NMR flow measurements in plants using FLASH imaging. *J. Magn. Reson*., 137, 29–32.

Rygol, J., Pritchard, J., Zhu, J.J., Tomos, A.D. and Zimmermann, U. (1993) Transpiration induces radial turgor pressure gradients in wheat and maize roots. *Plant Physiol*., 103, 493–500.

Schneider, H., Manz, B., Westhoff, M., Mimietz, S., Szimtenings, M., Neuberger, T., Faber, C., Krohne, G., Haase, A., Volke, F. and Zimmermann, U. (2003) The impact of lipid distribution, composition and mobility on xylem water refilling of the resurrection plant *Myrothamnus flabellifolia. New Phytol*., 159, 487–505.

Schneider, H., Thürmer, F., Zhu, J.J., Wistuba, N., Geßner, P., Herrmann, B., Zimmermann, G., Hartung, W., Bentrup, F.-W. and Zimmermann, U. (1999) Diurnal changes in xylem pressure of the hydrated resurrection plant *Myrothamnus flabellifolius*: evidence for lipid bodies in conducting xylem vessels. *New Phytol*., 143, 471–484.

Schneider, H., Wistuba, N., Miller, B., Geßner, P., Thürmer, F., Melcher, P., Meinzer, F., Zimmermann, U. (1997a) Diurnal variation in the radial reflection coefficient of intact maize roots determined with the xylem pressure probe. *J. Exp. Bot*., 48, 2045–2053.

Schneider, H., Wistuba, N., Reich, R., Wagner, H.-J., Wegner, L.H. and Zimmermann, U. (2000a) Minimal- and noninvasive characterization of the flow-force pattern of higher plants. In M. Terazawa (ed.), Tree Sap II. *Hokkaido University Press*, Sapporo, pp. 77–91.

Schneider, H., Wistuba, N., Wagner, H.-J., Thürmer, F. and Zimmermann, U. (2000b) Water rise kinetics in refilling xylem after desiccation in a resurrection plant. *New Phytol*., 148, 221–238.

Schneider, H., Zhu, J.J. and Zimmermann, U. (1997b) Xylem and cell turgor pressure measurements in intact roots of glycophytes: transpiration induces a change in the radial and cellular reflection coefficients. *Plant Cell Environ*., 20, 221–229.

Sperry, J.S., Saliendra, N.Z., Pockman, W.T., Cochard, H., Cruiziat, P., Davis, S.D., Ewers, F.W. and Tyree, M.T. (1996) New evidence for large negative xylem pressures and their measurement by the pressure chamber method. *Plant Cell Environ.*, 19, 427–436.

Stahlberg, R. and Cosgrove, D.J. (1995) Comparison of electric and growth responses to excision in cucumber and pea seedlings. II. Long-distance effects are caused by the release of xylem pressure. *Plant Cell Environ.*, 18, 33–41.

Steudle, E. (1989) Water transport in roots. In C. Loughman (ed.), Structural and Functional Aspects of Transport in Roots, *Kluwer Academic Publishers*, pp. 139–145, Dordrecht, Netherland.

Steudle, E. (2001). The cohesion-tension mechanism and the acquisition of water by plant roots. Annual Review of Plant Biology 52, pp. 847–875

Tanner, W. and Beevers, H. (2001) Transpiration, a prerequisite for long-distance transport of minerals in plants? *PNAS,* 98, 9443–9447.

Thürmer, F., Zhu, J.J., Gierlinger, N., Schneider, H., Benkert, R., Geßner, P., Herrmann, B., Bentrup, F. W. and Zimmermann, U. (1999) Diurnal changes in xylem pressure and mesophyll cell turgor pressure of the liana *Tetrastigma voinierianum*: the role of cell turgor in long-distance water transport. *Protoplasma*, 206, 152–162.

Tomos, A.D. and Leigh, R.A. (1999) The pressure probe: a versatile tool in plant cell physiology. *Ann. Rev. Plant Physiol. Plant Mol. Biol.*, 50, 447–472.

Wagner, H.-J., Schneider, H., Mimietz, S., Wistuba, N., Rokitta, M., Krohne, G., Haase, A. and Zimmermann, U. (2000) Xylem conduits of a resurrection plant contain a unique lipid lining and refill following a distinct pattern after desiccation. *New Phytol.*, 148, 239–255.

Wegner, L.H. and Zimmermann, U. (1998) Simultaneous recording of xylem pressure and trans-root potential in roots of intact glycophytes using a novel xylem pressure probe technique. *Plant Cell Environ.*, 21, 849–865.

Wegner, L.H. and Zimmermann, U. (2002) On-line measurements of K^+ activity in the tensile water of the xylem conduit of higher plants. *Plant J.*, 32, 1–9.

Wegner, L.H. and Zimmermann, U. (2004) Bicarbonate-induced alkalinization of the xylem sap in intact maize seedlings as measured in situ with a novel xylem pH probe. *Plant Physiol.*, 136, 3469–3477.

Wei, C., Steudle, E., Tyree, M.T. and Lintilhac, P.M. (2001) The essentials of direct xylem pressure measurement. *Plant Cell Environ.*, 24, 549–555.

Wei, C., Tyree, M.T. and Steudle, E. (1999) Direct measurement of xylem pressure in leaves of intact maize plants. A test of the cohesion-tension theory taking hydraulic architecture into consideration. *Plant Physiol.*, 121, 1191–1205.

Wistuba, N., Reich, R., Wagner, H.-J., Zhu, J.J., Schneider, H., Bentrup, F.-W., Haase, A. and Zimmermann, U. (2000) Xylem flow and its driving forces in a tropical liana: concomitant flow-sensitive NMR imaging and pressure probe measurements. *Plant Biol.*, 2, 1–4.

Zimmermann, U. (1989) Water relations of plant cells: pressure probe technique. *Meth. Enzymol.*, 174, 338–366.

Zimmermann, U., Meinzer, F.C., Benkert, R., Zhu, J.J., Schneider, H., Goldstein, G., Kuchenbrod, E. and Haase, A. (1994) Xylem water transport: is the available evidence consistent with the cohesion theory? *Plant Cell Environ.*, 17, 1169–1181.

Zimmermann, U., Räde, H., Steudle, E. (1969) Kontinuierliche Druckmessung in Pflanzenzellen. *Nat.wiss.*, 56, 634.

Zimmermann, U., Schneider, H., Thürmer, F. and Wegner, L.H. (2002a) Pressure probe measurements of the driving forces for water transport in intact higher plants: effects of transpiration and salinity. In A. Läuchli, U. Lüttge (eds) Salinity: Environment – Plants – Molecules, *Kluwer Academic Publishers*, pp. 249–270, Dordrecht, Netherland.

Zimmermann, U., Schneider, H., Wegner, L.H. and Haase, A. (2004) Tansley Review: Water ascent in tall trees: does evolution of land plants rely on a highly metastable state? *New Phytol.*, 162, 575–615.

Zimmermann, U., Schneider, H., Wegner, L.H., Wagner, H.-J., Szimtenings, M., Haase, A. and Bentrup, F.-W. (2002b) What are the driving forces for water lifting in the xylem conduit? *Physiol. Plant.*, 114, 327–335.

Zimmermann, U., Wagner, H.-J., Heidecker, M., Mimietz, S., Schneider, H., Szimtenings, M., Haase, A., Mitlöhner, R., Kruck, W., Hoffmann, R. and König, W. (2002c) Implications of mucilage on pressure bomb measurements and water lifting in trees rooting in high-salinity water. *Trees*, 16, 100–111.

Zimmermann, U., Wagner, H.-J., Rokitta, M., Schneider, H., Haase, A. and Bentrup, F.-W. (2000) Water ascent in plants: the ongoing debate. *Trends Plant Sci.*, 5(4), 145–146.

CHANGES IN COMPOSITION OF THE XYLEM SAP AS WELL AS IN ION FLUXES IN *POPULUS TREMULA X ALBA* L. XYLEM IN DEPENDENCE ON EXOGENOUS FACTORS

S. SIEBRECHT, G. FIEBELKORN and R. TISCHNER
Albrecht von Haller Institut für Pflanzenwissenschaften, Universität Göttingen, rtischn@gwdg.de

Abstract. This investigation shows diurnal variations in the xylem-sap composition of poplar. All major macronutrients reached a maximum concentration in the first half of the light period and decreased to the middle of the night. The relative abundance of the nutrients did not change during the day. The sap flow which responded very fast to the environmental changes (2.2 fold increase within 10–20 min of illumination) reached a maximum value in the second half of the light period. Transpiration (and photosynthesis) was constant throughout the light phase. The calculated translocation rates displayed a maximum in the first half of the light period and, therefore, did not fit the time course of sap flow. During the night, translocation rates were 63–69% lower than the maximum. The regulation of nutrient translocation is discussed taking the active xylem loading into account. The axial distribution located the nitrate assimilation in younger and storage of nitrate (and other macro nutrients) in older leaves. However, the sap flow was greater in younger shoot sections compared to older sections. We assume that the greater demand for nitrate in the younger shoot section was satisfied via an increased volume flow rather an increased nitrate concentration. Salt treatment mainly affected young leaves increasing their Na^+ and Cl^- content on the expense of Mg^{2+} and NO_3^- content respectively. A threshold (pH 3.25) was observed concerning the effect of low pH on *in vivo* sap flow.

Key words: axial distribution, ion flux, nitrogen metabolism, xylem sap composition

B. Sattelmacher† and W.J. Horst (eds.), The Apoplast of Higher Plants: Compartment of Storage, Transport and Reactions, 265–283.

1. INTRODUCTION

Xylem vessels are responsible for the long distance transport of water and nutrients from the roots to the shoot. Root pressure and transpiration stream are the driving forces of this transport (Marschner, 1995) the former being important under environmental conditions of low transpiration (e.g. high humidity at night). The latter, however, is the major component forcing xylem transport and is reliable for the negative pressure (compared to atmospheric pressure) in the xylem vessels (Zimmermann et al., 1994). Loading of nutrients into the xylem for transport is still under debate concerning the energy dependence of this process or the dependence from passive diffusion along the electrochemical gradient. Anion- and K^+-channel activity measured by patch-clamp techniques (Wegner and Raschke, 1994; Köhler et al., 2002) may be responsible for passive xylem loading. It has been accepted that xylem loading with nutrients can occur independent of water transport (Smith, 1991). A decrease in sap flow rate by 50% (at high humidity) did not affect the nitrate translocation in maize (Shaner and Boyer, 1976) and soybean (Delhon et al., 1995a). Translocation was also independent of nitrate uptake (Delhon et al., 1995b; Rufty et al., 1989). Therefore, the plant can regulate nutrient translocation dependent on the shoots demand (Marschner, 1995). Long-distance transport is also modified by the interaction between xylem and phloem. Nutrients can be transferred from the phloem into the xylem which results in nutrient circulation. This includes most of the nutrients. Exceptions are Ca^{2+} and NO_3^- (Peuke and Jeschke, 1998) which are almost not phloem-mobile. In wheat and rice more than 60% of the amino-N and 26% and 36% of K^+; respectively, in the xylem originates from the phloem (Grignon and Sentenac, 1991). The circulation of K^+ between root and shoot tissue has the advantage for the plant that the anion/cation balance can be easily maintained. In many plants NO_3^- is the dominant anion in the xylem and, therefore, changes in nitrate availability will affect the translocation of other nutrients.

Nitrate and sulphate assimilation (Kopriva et al., 1999), uptake of NO_3^- and K^+ vary in a diurnal manner indicating changes in the demand for nutrients with time. The demand for nutrients varies also along the shoot axis and, therefore, is correlated to leaf age. Young leaves have a higher demand for nitrogen compared to older leaves (Gonzales-Real and Baille, 2000).

This investigation aimed to analyse diurnal changes in nutrient translocation with respect to xylem-flow rate and xylem-sap composition with high time resolution. We also report on the nutrient distribution along the plant axis and the interaction between younger and older shoot sections. The effect of salt and pH-stress on xylem flow is evaluated additionally.

2. ANALYSIS OF XYLEM SAP COMPOSITION AND SAP FLOW

2.1 Diurnal investigations

The concentration of macronutrients measured in the root-pressure exudate changed during the day (Fig. 1) exhibiting a maximum in the first half of the light period (between 10.00 and 14.00 h).

In the second half of the light period a drop in the concentrations of K^+, SO_4^{2-} and NO_3^- was observed. At late night the concentrations increased again to values close to the maximum. The concentrations of ions usually present in low amounts (Cl^- 0.17 ± 0.06, Na^+ 0.09 ± 0.05; NH_4^+ 0.03 ± 0.02 mM) did not show significant diurnal variations. The relative abundance for all inorganic ions was nearly constant during the day (data not shown), therefore, the average abundance during the day is shown (Table 1).

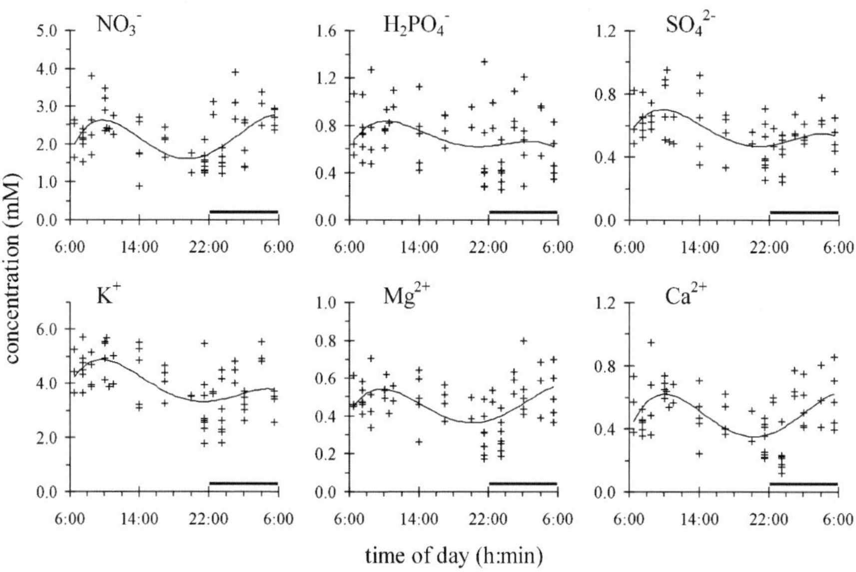

Fig. 1. Diurnal changes of the ion concentration in the root-pressure exudates of poplar. Root-pressure exudate was sampled 5–10 min after decapitation. Plants were cultivated hydroponically with a day/night cylce of 16/8 h and a temperature of 22/18°C. The dark phase is indicated by horizontal lines along the x-axis. Each data point represents the sap composition of a single plant, up to seven plants per time point.

Table 1. Ionic composition of the root-pressure exudates of poplar. Data were calculated based on the concentration of macronutrients (shown in Fig. 1) and the concentration of Cl^- Na^+ and NH_4^+ in xylem exudate collected at different times of day. Because the relative abundance of the ions in the xylem exudate was constant during the light/dark cycle, mean values SE are given (n = 64).

Xylem-exudate composition (% of total inorganic ions concentration)								
K^+	NO_3^-	$H_2PO_4^-$	SO_4^{2-}	Ca^{2+}	Mg^{2+}	Cl^-	Na^+	NH_4^+
46.1±5.5	25.5±5.2	7.9±1.9	6.6±1.1	5.6±1.4	5.3±1	1.9±1	1.1±0.8	0.4±0.3

K^+ was the major cation and NO_3^- the major anion in xylem exudate of poplar (46% and 26% of total inorganic ions, respectively). The relative abundance of the other inorganic ions was less than 8% each.

The sap flow-rate responded to changes in the light regime very fast (Fig. 2).

Within 10–20 min a 2.2 fold increase was observed after illumination started. The sap-flow rate increased further (by 23%) and reached a maximum in the second half of the light period (15.00–16.00 h). The flow rate dropped immediately when the dark period began and stayed constant during the night. This night-rate was not zero as it was not for transpiration. Also transpiration responded immediately (within 15 min) on illumination (Fig. 3).

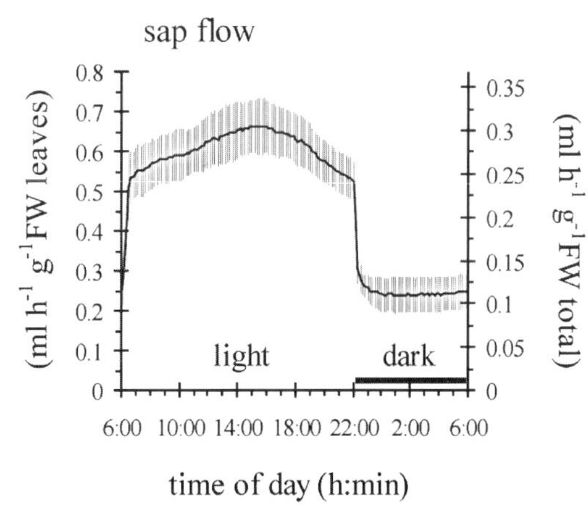

Fig. 2. Sap flow in the xylem of poplar during the day. Sap flow was measured at the stem base in intervals of 10 min continuously during the whole light/dark cycle. The dark phase is indicated by a horizontal line along the x axis. Sap flow was linearly correlated to the fresh weight of the leaves and to the total fresh weight of the plant. Therefore, flow rates are expressed in relation to the fresh weight. Values are means ± SD of 16 independent measurements.

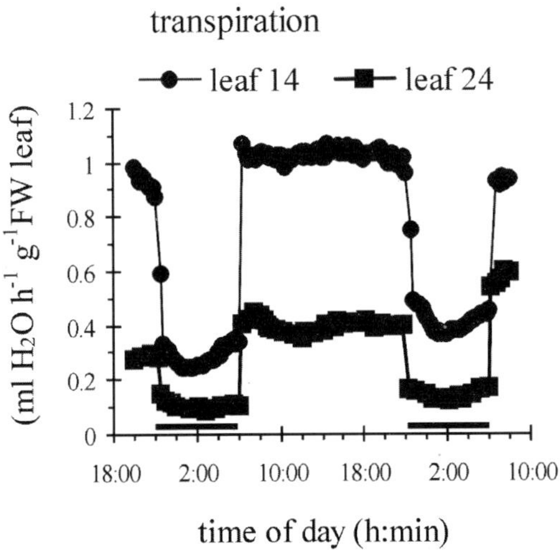

Fig. 3. Diurnal time course of transpiration of poplar plants. The transpiration was determined for two attached leaves of different age: leaf 14 = mid-aged leaf, leaf 24 = old leaf (leaf 1 = youngest expanded leaf, total leaves = 33) with supplemental illumination (170 µmol photo m^{-2} s^{-1}). Transpiration rates were measured in intervals of 15 min continuously during almost two days. The plants were cultivated hydroponically with a day/night cycle of 16/8 h (light on at 6.00 h, 170 µmol photon m^{-2} s^{-1}) and a temperature of 22/18°C. The dark phase is indicated by horizontal lines along the x axis. The curves represent averages from three independent repetitions.

However, during the light period the rates of net photosynthesis (data not shown) and transpiration were constant. Data are presented for two leaves of different age (Fig. 3). The transpiration rate of the younger leaf was higher (1 ml H_2O $g^{-1}FW$ h^{-1}) than that of the older leaf (0.4 ml H_2O $g^{-1}FW$ h^{-1}) but the diurnal time course was identical., The *in vivo* sap flow rate, measured at the shoot basis, was 0.6 ml $g^{-1}FW$ h^{-1} (Fig. 2) and, therefore, corresponded to the average transpiration rate, because the sap flow in the xylem was driven by the combined transpiration of all leaves.

Based on the sap flow rates and the concentrations measured in the root-pressure exudate the translocation rate was calculated. It is obvious that the diurnal variations of the nutrient translocation did not correspond to that of the sap flow (Fig. 4). The maximum translocation rate for all macronutrients was found in the first half of the illumination period (10.00 h) and decreased by 40–53% (dependent on the ion species) to the end of the light period. Contrary, maximum sap-flow rate was observed several hours later (16.00 h) and dropped by 22% to the end of the light period. This indicates that the translocation of ions into the shoot is regulated not exclusively by the sap

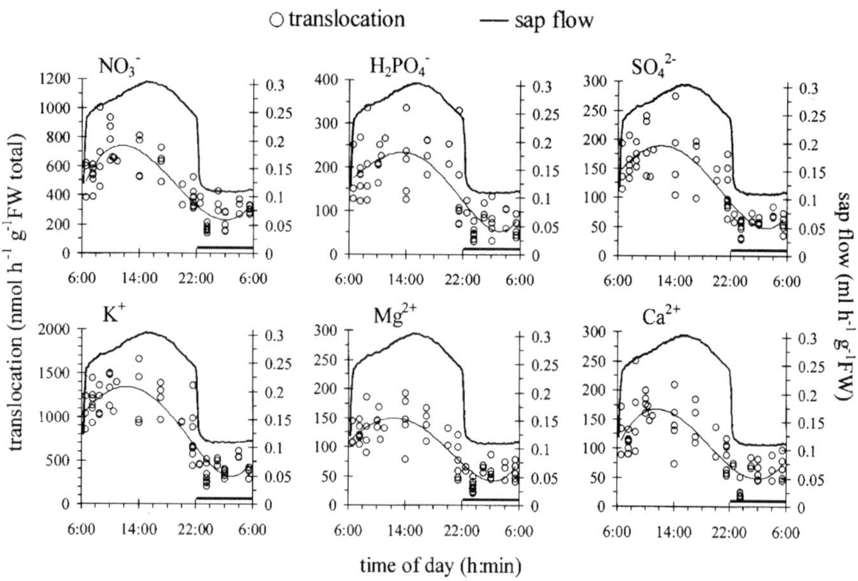

Fig. 4. Nutrient translocation from root to shoot of poplar plants dependent on the time of day. Translocation rates were calculated by multiplying the ion concentration in the root pressure exudate sampled at the stem base (see Fig. 1) with the in vivo *sap-flow rate measured at the stem base (see Fig. 2) obtained during the day. For comparison sap-flow rates are also shown in the figures. The dark phase is indicated by horizontal lines along the x axis. Each data point represents the translocation rates from an individual plant, up to seven plants per time point.*

flow, which depends on the transpiration stream, but also by the process of xylem loading which could be regulated by nutrient demand of the shoot. During the night, nutrient translocation into the shoot continued, but translocation rates were 63–69% (dependent on the ion species) lower compared to the maximum rate in the light.

Opposite to NO_3^- translocation, NO_3^- net uptake rate was constant during the light period and decreased by only 20% during the dark period (Fig. 5). NO_3^- net uptake rate in the light fitted the maximal NO_3^- translocation rate during the day (0.55 ± 0.06 µmol h^{-1} g^{-1}FW, Fig. 5 compared to 0.74 ± 1.2 µmol h^{-1} g^{-1}FW, Fig. 4), whereas at night NO_3^- uptake was 1.7 fold higher than NO_3^- translocation to the shoot (0.43 ± 0.05 µmol h^{-1} g^{-1}FW, Fig. 5 compared to 0.26 ± 0.09 µmol h^{-1} g^{-1}FW, Fig. 4).

2.2 Investigations along the axis of the shoot

The leaves along the axis, the youngest leaf being leaf 1, differed in fresh weight (Fig. 6). This parameter increased from leaf 1 to leaf 9 and reflected

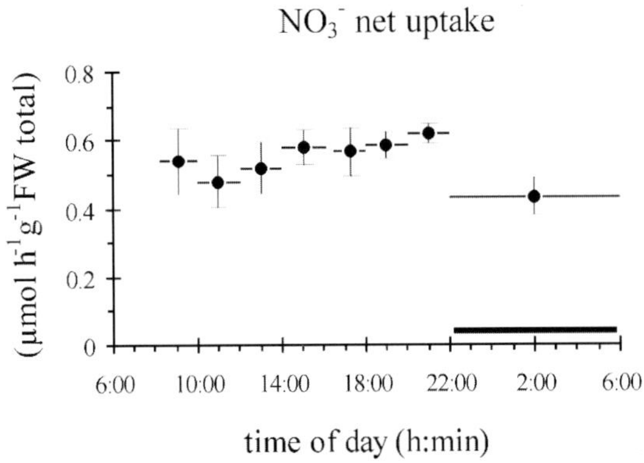

Fig. 5. NO_3^- net uptake of poplar plants dependent on the time of day. Net uptake rates were measured as NO_3^- depletion of the medium in 2 h intervals during the light phase. The dark phase is indicated by a horizontal line along the x-axis. During the dark phase the average net uptake rate was determined. The period of uptake measurement is indicated by horizontal bars. Values are means ± SD (N=5). Vertical bars indicate SD.

the shoot section of young and still growing leaves. The shoot section with the leaves 10–20 contained fully expanded leaves with almost identical fresh weights. The leaves 21–35 of the lower shoot section showed decreasing fresh weights with increasing leaf age. To compare metabolite contents and NRA between leaves of different age, these parameters were expressed on fresh weight basis.

Soluble protein contents were highest in young growing leaves and decreased continuously with increasing leaf age (Fig. 6). Chlorophyll contents of the leaves increased as long the leaves were still expanding and did not change for older leaves, while the rate of net photosynthesis was maximum in the fully expanded leaves 13–15 and decreased in older leaves (Fig. 6).

The in vitro NRA and the amino-N contents were highest in the young still growing leaves 5-7 and decreased with increasing leaf age (Fig. 7), indicating the greatest capacity for NO_3^- assimilation in those leaves. Contrary to the NRA and the amino-N content, the NO_3^- content of the leaves increased with increasing age (Fig. 7). Therefore, NO_3^- storage occurred mainly in older leaves (>leaf 15). This holds true not only for NO_3^-; also the contents of Mg^{2+}, Ca^{2+} and SO_4^{2-} increased with increasing leaf age (Fig. 8). However the K^+ content was the same in all leaves and the $H_2PO_4^-$ content was maximum in mid aged leaves (Fig. 8).

The *in vivo* sap flow (stem heat balance method), estimated for two different shoot sections simultaneously, was significantly greater in the upper part of the shoot (Fig. 9a) As soon as the section under investigation

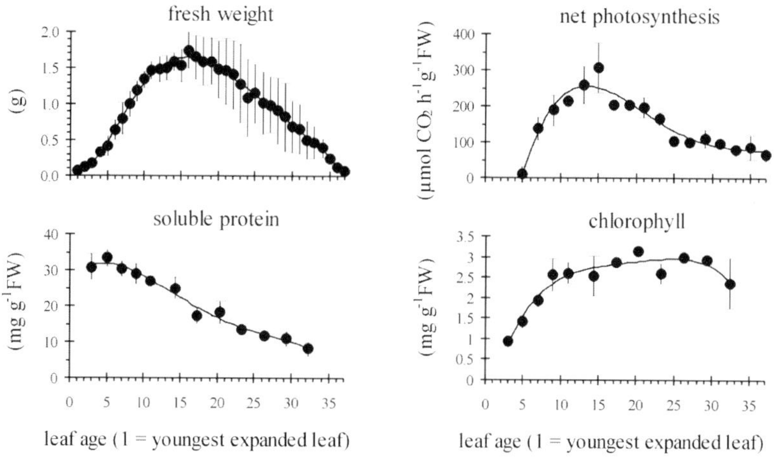

Fig. 6. *Six Leaf fresh weight, net photosynthesis, and contents of soluble protein and chlorophyll dependent on the age of the leaves. Poplar plants were grown hydroponically with a light/dark cycle of 16/8 h. Photosynthesis was measured at light saturation (600 µmol photon m^{-2} s^{-1}). Values are means ± SD of six plants (fresh weight) or three plants (net photosynthesis, soluble protein, chlorophyll) with 32–37 leaves. All data were obtained in the first half of the light period. Data are expressed based on the fresh weight.*

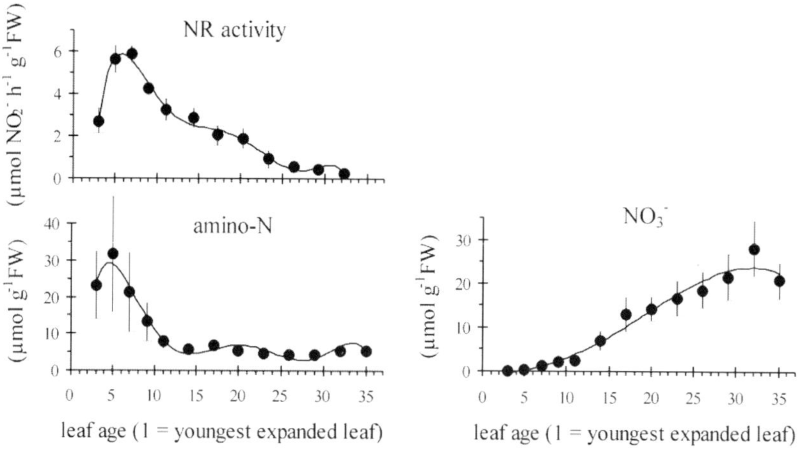

Fig. 7. Nitrate reductase activity (NRA) and contents of soluble amino-N and NO_3^- dependent on the age of the leaves of poplar plants. NRA was measured in vitro *with additional NO_3^- and NADH. Values are means ± SD of three plants with 32–35 leaves. All data were obtained in the first half of the light period.*

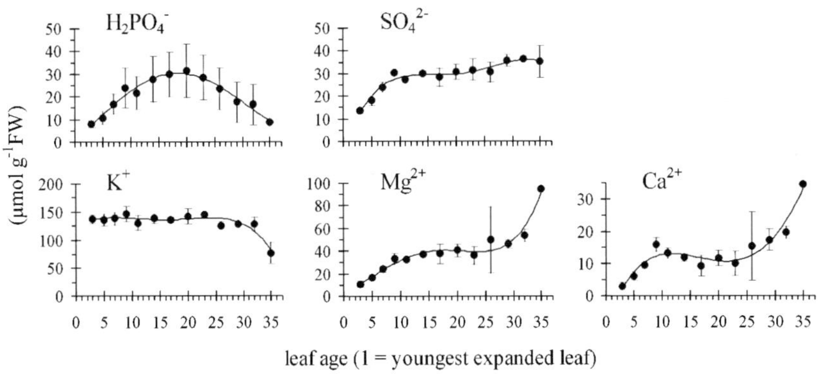

Fig. 8. Nutrients contents dependent on the age of the leaves. Values are means ± SD of three poplar plants with 37 leaves. All data were obtained in the first half of the light period. Data are expressed based on fresh weight.

included more older leaves, the flow rate decreased (Fig. 9a). This clearly demonstrated a greater volume flow into younger leaves. This was confirmed by transpiration measurements using leaf cuvettes (Fig. 9b), which gained a 2.5 fold higher transpiration rate for young leaves (leaf 7–13) compared to old leaves (leaf 23–37).

Using leaf cuvettes for transpiration measurements can lead to an overestimation of the *in vivo* sap-flow rate due to changes in the microclimate caused by the high gas flow. This in turn can increase the transpiration rate. However, the calculated sap-flow rate based on transpiration measurements fitted that measured *in vivo* using the stem heat-balance method with a maximum deviation of ± 20% (Table 2). Therefore the transpiration rate reflects the *in vivo* sap-flow rate into the single leaf.

Table 2. Comparison of sap flow measured in vivo *(stem heat-balance method) to that calculated on the basis of transpiration measurements (leaf cuvette). The sap flow in two sections of the stemof poplar plants was measured* in vivo *(stem hea- balance method). The average sap-flow rate during the light period is shown for three independent experiments (plant 1-3, compare Fig. 10a). In a second set of experiments the transpiration rate was measured dependent on the age of the leaves (compare Fig. 10b). Based on the transpiration rates and the fresh weights of the different old leaves in each shoot section of plant 1-3 the potential sap flow rate in the two shoot sections was calculated.*

	Upper shoot section			Lower shoot section		
	Sap flow (ml h^{-1} g^{-1} FW)					
	Measured	Calculated	Deviation[1]	Measured	Calculated	Deviation[1]
Plant 1	0.652	0.705	+ 8%	0.398	0.319	–20%
Plant 2	1.070	0.851	–20%	0.434	0.411	– 5%
Plant 3	0.716	0.842	+18%	0.359	0.359	0%

[1] *Deviation between measured and calculated sap flow; measured sap flow = 100 %*

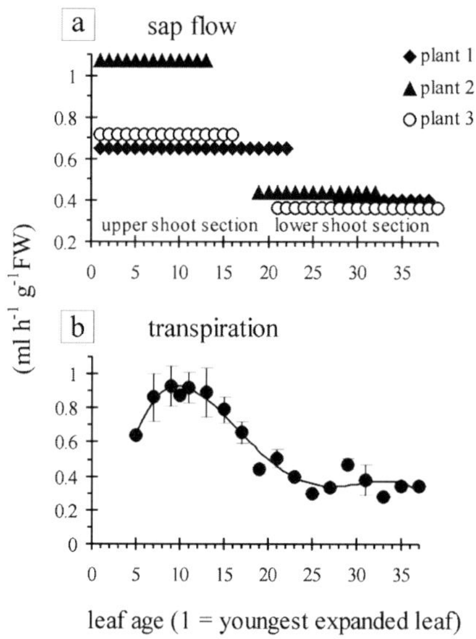

Fig. 9. Sap flow in xylem and transpiration of leaves along the axis of the shoot of poplar plants. (a) For calculation of sap-flow rates in two different shoot sections, sap flow was measured in vivo *(stem hea- balance method) at two different positions of the stem (stem base, upper stem position). The rate obtained at the upper position gives the flow rate into the upper shoot section. The flow rate into the lower shoot section is given by the difference between the flow rate measured at the stem base and the upper stem position. Sap-flow rates of three plant individuals with 32–39 leaves are presented. Data are expressed based on the fresh weight of the supplied leaves. (b) Transpiration rates were measured at leaves of different age using a leaf cuvette and rates are expressed based on the leaf fresh weigh. Values are means ± SD of three plants with 37 leaves. For sap flow the average rate during the whole light period is shown. Transpiration rates were measured in the first half of the light period, but the rate was constant during the whole light period (Fig. 3).*

NO_3^- translocation rates into single leaves were calculated based on the specific transpiration rates for each leaf and the NO_3^- concentration of the xylem exudate collected at the stem base, as this concentration did not change along the shoot axis (Fig. 10). In this case the translocation rates were calculated per leaf to get information about the distribution pattern along the shoot axis (taking the leaf size into account). The data for NRA and NO_3^- content were recalculated on the same basis to compare the distribution pattern of these three parameters along the poplar axis (Fig. 10).

In the upper half of the shoot maximum rates for NO_3^- influx via the xylem and for NRA (which is a measure for NO_3^- assimilation) were found, but the NO_3^- content was low (Fig. 10). In the lower half of the shoot the opposite holds true. Here, the NO_3^- content was high, but the NO_3^- influx and the NRA reached minimal rates (Fig. 10). This distribution pattern

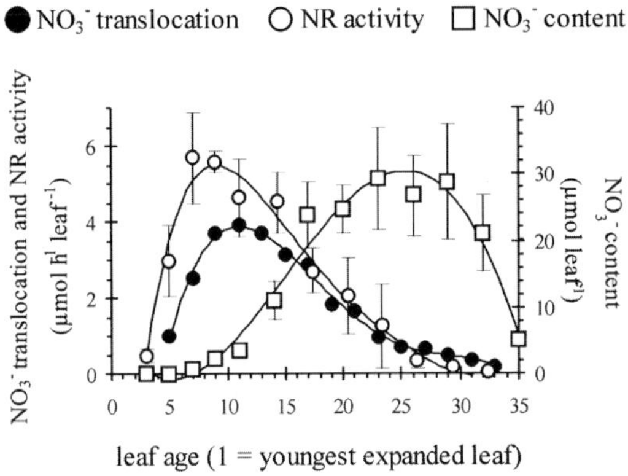

Fig. 10. NO_3^- translocation, nitrate reductase activity (NRA) and NO_3^- storage along the axis of the shoot of poplar plants. Translocation rates into leaves were calculated by multiplying the average NO_3^- concentration in the xylem exudate at 10.00 h (Fig. 1; 3 mM), which could be used because the NO_3^- concentration was constant along the shoot axis (Fig. 9), and the transpiration rate of the different old leaves measured in the first half of the light period (Fig. 10b). Data (means ± SD, N= 3) for NRA (measured in vitro*) and NO_3^- content were obtained in the first half of the light period. All data are expressed per leaf. For direct comparison with NRA, translocation rates are calculated based on the fresh weight of the leaves used for analysis of NRA.*

clearly demonstrates that NO_3^- assimilation occurred mainly in the upper half of the shoot, whereas NO_3^- storage takes place predominantly in the leaves at the lower shoot part.

The rates of NO_3^- influx and NRA were identical in the middle section of the shoot (leaves 11–23), but differed in the upper and lower shoot sections (Fig. 10). In the upper part containing the still growing leaves (leaves 3–9), the rate of NRA was 52% (leaves 9) to 200% (leaf 5) higher than that of NO_3^- influx. Contrary, in the lower part of the shoot (leaves 26–33) NRA was 48% to 64% lower than NO_3^- influx via the xylem.

2.3 Salt stress and pH stress

The addition of different NaCl concentrations into the medium affected both, the tissue and root-pressure exudates ion compositions. For root-pressure exudates only the data after 3 d of salt treatment are presented (Table 3) as after 10 d almost no exudate could be collected.

The Na^+ content increased slightly after 3 d on the expense of Mg^{2+} and Ca^{2+} while K^+ content was almost not changed. This effect is supported by the data on xylem flow (Fig. 11) in plants treated with increasing NaCl concentrations.

Table 3. Ion composition of root-pressure exudates obtained from poplarplants exposed to 10, 25 or 50 mM NaCl for 3 d. Growth of poplar and collection of the root pressure exudates were as described in the material and method section. Data are given as concentrations (mM), and significant differences (SD) to the control plants are marked by an asterisk.

	Na^+	NH_4^+	K^+	Mg^{2+}	Ca^{2+}	Cl^-	NO_3^-	PO_4^{3-}	SO_4^{2-}
Control	0.09	0.05	6.2	0.8	1.07	0.003	4.3	1.2	0.63
10 mM NaCl	0.44*	0.14	6.7	0.8	0.88	0.24*	3.9	2.2*	0.88
25 mM NaCl	1.44*	0.1	8.9*	0.83	0.81	0.12*	3.5*	1.5	1.0
50 mM NaCl	1.84*	0.06	5.8	0.31*	0.27*	0.18*	1.2*	0.6*	0.7

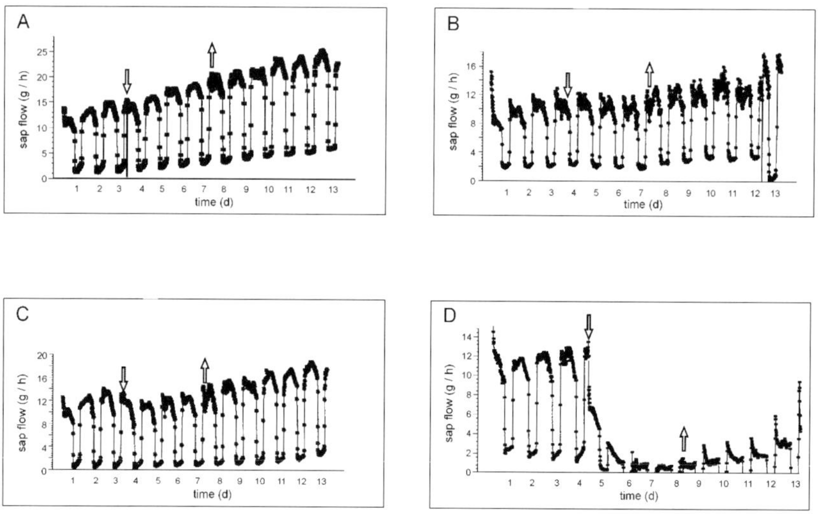

Fig. 11. The effect of different NaCl concentrations on sap flow in the xylem of poplar plants. Plants were grown with nitrate as sole nitrogen source to the 35 leaf stage. Recording of the flow rate started 3 d before the salt-treatment. The first arrow indicates the start of the treatment (A: control; B:10 mM NaCl; C: 25 mM NaCl; D: 50 mM NaCl), the second arrow indicates the change back into salt free medium. Recording of the flow rate continued to d 13.

Xylem flow reacted shortly after NaCl supply and decreased immediately. There was a recovery to observe for 10 mM and 25 mM but not for 50 mM in the time scale of our observation. The ion composition is shown in Table 4 for a young leaf (leaf 4), an old leaf (leaf 20), and for fine roots after 10 d of the salt treatment.

Table 4. Ion composition of a young leaf (leaf 4), an old leaf (leaf 20), and fine roots of poplar plants dependent on treatment with different NaCl concentrations for 10 d. Data are given as μmol (g fresh wt)$^{-1}$. SD was 10%.

Leaf No (NaCl treatment)	Na^+	NH_4^+	K^+	Mg^{2+}	Ca^{2+}	Cl^-	NO_3^-	PO_4^{3-}	SO_4^{2-}
Leaf 4 (control)	1.26	1.87	107.9	15.7	12.4	0	4.9	18.3	12.0
Leaf 4 (10 mM)	0.7	1.64	134.8	16.5	–	0	5.0	23.6	14.0
Leaf 4 (25 mM)	34	1.1	105.8	8.9	8.6	3.2	0	17.7	11.3
Leaf 4 (50 mM)	79.2	0	107.7	5.3	3.0	22.7	0	22.3	8.7
Leaf 20 (control)	0.61	0.51	102.6	13.9	13.0	7.5	8.9	45.4	25.8
Leaf 20 (10 mM)	0.25	0.17	100.2	8.7	12.6	9.0	7.34	57.6	18.8
Leaf 20 (25 mM)	3.0	0.16	40.03	1.5	4.8	7.5	3.8	52.1	16.5
Leaf 20 (50 mM)	25.7	0	26.5	0	0.5	29.0	0		16.3
Fine roots (control)	4.1	0.35	77.2	6.7	0.14	2.2	17.0	3.6	17.7
Fine roots (10 mM)	14.0	0.08	55.5	3.4	0	3.4	15.2	2.8	11.7
Fine roots (25 mM)	51.8	0	34.5	6.5	0	3.9	11.0	3.0	15.7
Fine roots (50 mM)	68.3	0	11.2	2.5	0	6.4	16.1	5.5	18.6

It became obvious that the sodium concentration increased with NaCl concentration in the medium. The young leaf and the root were more affected than the old leaf. The accumulation of Na^+ decreased the concentrations of the other cations measured. Also the Cl^- content increased after salt treatment mainly in the young leaf and that was negatively correlated to the NO_3^- content. The effects on other anions was less pronounced, slight decreases were found for SO_4^{2-}. These effects were even more distinct with NH_4^+ as sole N source (data not presented).

The changes in pH of the medium also affected sap flow, ion composition of plant tissue and root-pressure exudates. The most obvious effect was observed for the sap flow (Fig. 12) which decreased strongly within a small variation of the outer pH. A change from pH 3.5 (which itself did not affect sap flow very much compared to the control pH 5.8) to pH 3.25 reduced sap flow to half of the control rate (we observed recovery within 4 d). Further decrease to pH 3.0 decreased sap flow to zero within 2 d, no recovery was detected within 4 d after pH was changed back to pH 5.8.

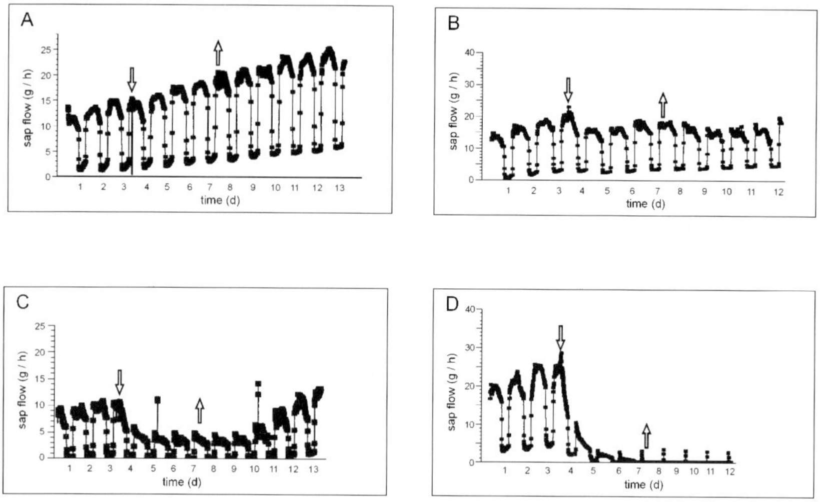

Fig. 12. The effect of low pH in the medium on the sap flow of poplar plants. Plants were grown with nitrate as sole nitrogen source to the 35 leaf stage. Recording of the flow rate started 3 d before the pH treatment. The first arrow indicates the start of the treatment (A: control pH 5.8; B: pH 3.5; C: pH 3.25; D: pH 3.0), the second arrow indicates the change back into control medium. Recording of the flow rate continued to d 13.

3. DISCUSSION

The isolation of root-pressure exudate after shoot decapitation can create major problems. Cutting the shoot can cause contamination of the exudate with components from destroyed cells (Else et al., 1994); removing the shoot interrupts phloem flow which may affect the concentration of nutrients which cycle between root and shoot (Jeschke and Pate, 1991); removing the shoot also interrupts the transpiration stream and this will affect the exudate composition (Schurr and Schulze, 1995). We measured a decrease in the flow rate after cutting and this decrease corresponded to the increase in ion concentration of the exudate. These changes became obvious about 15 min after cutting. Therefore, we performed the collection of the exudate according to the method published previously (Siebrecht and Tischner, 1999). This guaranteed that the nutrient composition of the root pressure exudates closely reflected the xylem-sap composition of intact plants.

The nutrient concentrations reached a maximum in the late morning and decreased until the end of the light period (Fig. 1) similar as reported for barley (Mattson et al., 1988). The same time course was found for nitrate and potassium but not for the other macronutrients in *Ricinus* (Schurr and Schulze, 1995). Obviously, the diurnal rhythm of xylem-sap components is variable, presumably dependent on the plant species and/or growth conditions as poplar and barley have been grown in hydroponic culture whereas *Ricinus* was grown in soil.

Xylem-sap flow was estimated using the „stem heat-balance method". This measured the net flow i.e. the difference between xylem flow and the phloem flow. However, the latter contributes only 2–3% (Jeschke and Pate, 1991) up to 9% (Tanner and Beevers, 1990) to total water flow. During the light phase, transpiration is the major component driving xylem flow which reacts immediately on illumination (Schneider et al., this volume, pp 255–268). Also diurnal stem radius changes (Zweifel et al., 2001) and xylem pressure (Wei et al., 1999) react on changes in illumination rapidly.

After the end of illumination the flow rate dropped rapidly due to the change from transpiration to osmotically driven water flow (Crosett, 1968). This decrease was only 63% in poplar (Fig. 2) indicating a high transpiration also at night. This is supported by the close fit between leaf transpiration and xylem flow at night (Figs. 2 and 3). The water loss at night may occur via the cuticula, which is not as strong in plants grown hydroponically under constant humidity. A decrease in transpiration at night by only 80% has been shown for soybean (Delhon et al., 1995b) and *Lolium perenne* (Ourry et al., 1996) both cultivated in hydroponic culture. The highest flow rate (in the middle of the day) fits the time of maximum leaf expansion (Ourry et al., 1996) and, therefore, the high demand for water, while leaf transpiration was constant during the light period.

The nutrient translocation rates were calculated based on the nutrient concentration in the root-pressure exudate (Fig. 1) and the xylem-sap flow-rates (Fig. 2). Translocation rates from the root to the shoot were reduced during the night by 65% for NO_3^- and by 63–69% for the other nutrients compared to the maximum rate observed (Fig. 4). This is similar in barley (Mattson et al., 1988; Oji et al., 1989), soybean (Delhon et al., 1995a) and tobacco (Rufty et al., 1989). However, these authors measured an average translocation during the day or the night, while we estimated a precise time course over the whole day. From the greatest translocation rate (4 h after illumination started) this rate decreased continuously to the end of the light period. Nutrient translocation obviously took place in the first half of the light period, later xylem loading was reduced. At high humidity transpiration stream was reduced by 50%, but nitrate translocation to the shoot was not reduced (Shaner and Boyer, 1976; Delhon et al., 1995b). Therefore, nutrient translocation to the shoot can be independent of the sap flow rate and might be regulated at the level of xylem loading (Smith, 1991). This is supported by our data, where changes in the nutrient translocation rate were based exclusively on changes in the transpiration-independent xylem loading rather than on changes in sap flow. The maximum rates for both were measured at different time points (Fig. 4). At identical sap flow-rates (0.27 mL g^{-1} FW at 10 a.m. and at 8 p.m.) the NO_3^- translocation rate was reduced by 55% at 8 p.m. The reduced demand of the shoot for nutrients in the dark period regulates xylem loading. A demand-driven regulation has been reported for K^+ and loading into the xylem is mediated by SKOR, a K^+ channel identified in *Arabidopsis* (Gaymard et al., 1998).

Diurnal changes have been reported for both nitrate and sulphate assimilation. The highest activities of nitrate reductase and adenosine 5′-phosphosulphate reductase have been reported in the first half of the light period. A decrease towards the end of illumination (Huber et al., 1992c; Kopriva et al., 1999) continued into the night period (Huber et al., 1992b; De Cires et al., 1993). The diurnal changes of these both pathways correspond closely to the diurnal changes in the nutrient translocation rates reported here supporting the idea that xylem loading is regulated according to the nutrient demand of the shoot.

However, also the regulation of nutrient uptake can adjust nutrient translocation to meet the shoots demand. Regulatory substances transported from the shoot to the root (Marschner et al., 1997) including carbohydrates (Rideout et al., 1994), organic nitrogen compounds (Muller and Touraine, 1992) and malate (Touraine et al., 1992) can affect nutrient uptake. For poplar, however, we found no changes in the nitrate uptake-rate during the light period. Therefore, the nitrate translocation-rate during the light period is independent of nitrate uptake. This is similar in spinach (Scaife and Schloemer, 1994), barley (Peuke and Jeschke, 1998), and holds true for the night period (Delhon et al., 1995b) and under different growth conditions (Poirier et al., 1991).

To summarize: Nutrient translocation in poplar changes in a circadian manner independently of xylem-sap flow and nutrient uptake (demonstrated for nitrate). The shoots demand for nutrients is regulated at the level of xylem loading.

The spatial distribution of NO_3^- translocation and NO_3^- reduction along the shoot axis suggest that the demand for NO_3^- is highest in growing not completely expanded leaves. This is reflected in the distribution of NO_3^--content, NRA, amino-N content and NO_3^- translocation rate. A measure for NO_3^- assimilation capacity is total NRA. The plants under investigation had up to 37 leaves and we distinguished three sections:

i) Leaf numbers 1–10: The highest NR capacity and amino-N content and the lowest NO_3^- content were found in leaves 7–9. These leaves displayed a relative high photosynthetic capacity and depend on a continuous NO_3^- supply via the xylem.

ii) Leaf numbers 11–15: These leaves showed the highest rate of photosynthesis and a reduced NRA.

iii) Leaf numbers 17–35: Here, NO_3^- content was highest, indicating a nitrate storage tissue. NRA decreased with leaf age as also photosynthesis did.

The decreasing N assimilation capacity with leaf age reported here fits data for soybean (Santora and Magalhaes, 1983), spinach (Huber et al., 1992a) and Ricinus by Jeschke and Pate (1991). The youngest leaves depend on a continuous NO_3^- supply in soybean (Delhon et al., 1995a), spinach (Scaife and Schloemer, 1994). The high demand for nitrate of the younger leaves can be satisfied either by an increased sap flow or an increasing NO_3^- concentration in the upper shoot section. Such increasing nutrient concentrations (NO_3^-, K^+) along the shoot axis have been demonstrated for beach trees in spring (Else et al., 1995; Schill et al., 1996) and for Ca^{2+} and Mg^{2+} in *Acer platanoides* (Jeschke and Pate, 1991). In poplar, the NO_3^-, K^+ and Ca^{2+} concentrations did not change along the shoot axis, while Mg^{2+} and SO_4^{2-} were highest in young leaves. Amino-N and $H_2PO_4^-$ reached maximum contents in older leaves similar as in *Ricinus* (Jeschke and Pate, 1991) and barley (Wolf et al., 1990). However, here the in situ measured flow rate and the simultaneously estimated leaf transpiration are highest in the upper shoot section. Both parameters decreased when the shoot sections under investigation included more older leaves indicating a greater volume flow into younger leaves. Therefore, we present evidence, that the demand for nutrients of the young leaves is satisfied by an increased sap flow rate potentially from older leaves which function as storage pools rather than an increase in NO_3^- concentration.

Salt stress up to 10 mM NaCl was managed by poplar by retaining Na^+ ions in the root system, similar as reported for barley by Wolf et al. (1990). However, at higher concentrations a transport via the xylem into the shoot can no longer be avoided. Na^+ content increased in younger leaves (where the sap flow was more rapid; see above) and less in older leaves. This accumulation caused Mg^{2+} contents to decrease most considerably compared to other cations measured. From the anions estimated Cl^- increased obviously to the expense of NO_3^- while other anions did not change significantly.

Lowering the pH of the medium affected sap flow most rapidly and these changes were reversible after changing back to pH 5.8 (control). The threshold for pH stress was found to be very close, as a pH of 3.25 could be survived (complete recovery within 6 days), while a treatment with pH 3.0 killed the plants. One may speculate if the high capacity of poplar to compensate for pH changes from 5.8 to 3.25 is based on the recently reported rapid alkalisation factors (Haruta and Constable, 2003).

REFERENCES

Crossett, R.N. (1968). Effect of light upon the translocation of phosphorus by seedlings of Hordeum vulgare (L.). *Aust. J. Biol. Sci.* 21, 225–233.

De Cires, A., De la Torre, A., Delgado, B. and Lara, C. (1993) Role of light and CO_2 fixation in the control of nitrate reductase activity in barley leaves. *Planta*, 190, 227–283.

Delhon, P., Gojon, A., Tillard, P. and Passama, L. (1995a) Diurnal regulation of NO_3^- uptake in soybean plants I. Changes in NO_3^- influx, efflux, and N utilization in the plant during the day/night cycle. *J. Exp. Bot.*, 46, 1585–1594.

Delhon, P., Gojon, A., Tillard, P. and Passama, L. (1995b) Diurnal regulation of NO_3^- uptake in soybean plants II. Relationship with accumulation of NO_3^- and asparagine in the roots. *J. Exp. Bot.*, 46, 1595–1602.

Else, M.A., Davis, W.J., Whitford, P.N., Hall, K.C. and Jackson, M.B. (1994) Concentrations of abscisic acid and other solutes in xylem sap from root systems of tomato and castor-oil plants are distorted by wounding and variable sap flow rates. *J. Exp. Bot.*, 45, 317–323.

Else, M.A., Hall, K.C., Arnold, G.M., Davis, W.J. and Jackson, M.B. (1995) Export of abscisic acid, 1-aminocyclo-propane-1carboxylic acid, phosphate, and nitrate from roots to shoots of flooded tomato plants. Accounting for effects of xylem sap flow rate on concentration and delivery. *Plant Physiol.*, 107, 377–384.

Gaymard, F., Pilot, G., Lacombe, B., Bouchez, D., Bruneau, D., Boucherez, J., Michaux-Ferriere, N., Thibaud, J.B. and Sentenac, H. (1998) Identification and disruption of a plant shaker-like outward channel involved in K^+ release into the xylem sap. *Cell*, 94, 647–655.

Gonzales-Real, M.M. and Baille, A. (2000) Changes in leaf photosynthetic parameterswith leaf position and nitrogen content within a rose plant canopy (*Rosa hybrida*). *Plant Cell Environ.*, 23, 351–357.

Grignon, C. and Sentenac, H. (1991) pH and ionic conditions in the apoplast. *Ann. Review Plant Physiol. Plant Mol. Biol.*, 42, 103–128.

Haruta, M. and Constable, C.P. (2003) Rapid alkalization factors in poplar cultures. Peptide isolation, cDNA cloning, and differential expression in leaves and methyl jasmonate treated cells. *Plant Physiol.*, 131, 814–823.

Huber, J.L., Huber, S.C., Campell, W.H. and Redinbaugh, M.G. (1992a) Reversible light/dark modulation of spinach leaf nitrate reductase activity involves protein phosphorylation. *Arch. Biochem. Biophys.*, 296, 58–65.

Huber, J.L., Huber, S.C., Campell, W.H. and Redinbaugh, M.G. (1992b) Apparent dependence of the light acitvation of nitrate reductase and sucrose-phosphate synthase activities in spinach leaves on protein synthesis. *Plant Cell. Physiol.*, 33, 639–646.

Huber, S.C., Huber, J.L., Campell, W.H. and Redinbaugh, M.G. (1992c) Comparative studies of the light modulation of nitrate reductase and sucrose-phosphate synthase activities in spinach leaves. *Plant Physiol.*, 100, 706–712.

Jeschke, W.D. and Pate, J.S. (1991) Modelling of water uptake, flow and utilization of C, N and H_2O within whole plants of *Ricinus communis* L. Based on empirical data. *J. Plant Physiol.*, 137, 488–498.

Kopriva, S., Muheim, R., Koprivova, A., Trachsel, N., Catalano, C., Suter, M. and Brunold, C. (1999) Light regulation of assimilatory sulphate reduction in *Arabidopsis thaliana*. *Plant J.*, 20, 37–44.

Köhler, B., Wegner, L., Osipov, V. and Raschke, K. (2002) Loading of nitrate into the xylem: apoplastic nitrate controls the voltage dependence of X-QUAC, the main conductance in xylem-parenchyma cells of barley roots. *Plant J.*, 30, 133–142.

Marschner, H. (1995) Mineral nutrition of higher plants. Second edition, Academic Press, London.

Marschner, H., Kirkby, E.A. and Engels, C. (1997) Importance of cycling and recycling of mineral nutrients within plants for growth and development. *Bot. Acta*, 110, 265–273.

Mattson, M., Lundborg, T. and Larsson, C. (1988) Nitrate utilization in barley: relations to nitrate supply and light/dark cycles. *Physiol. Plant*, 73, 380–386.

Muller, B. and Touraine, B. (1992) Inhibition of NO_3^- uptake by various phloem-translocated amino acids in soybean seedlings. *J. Exp. Bot.*, 43, 617–623.

Oji, Y., Otani, Y., Hosomi, Y., Wakiuchi, N. and Shiga, H. (1989). Nitrate reduction in in root and shoot and exchange of reduced nitrogen between organs in two-row barley seedlings under light-dark cycles. *Planta*. 179, 359–366.

Ourry, A., Macduff, J.H., Prudhomme, M.P. and Boucaud, J. (1996) Diurnal variation in the simultaneous uptake and ‚sink' allocation of NH_4^+ and NO_3^- by Lolium perenne in flowing solution culture. *J. Exp. Bot.*, 47, 1853–1863.

Peuke, A.D. and Jeschke, W.D. (1998) The effects of light on induction, time courses, and kinetic patterns of net nitrate uptake in barley. *Plant Cell Environ.*, 21, 765–774.

Poirier, Y., Thoma, S., Sommerville, C. and Schiefelbein, J. (1991) A mutant of *Arabidopsis* deficient in xylem loading of phosphate. *Plant Physiol.*, 97, 1087–1093.

Rideout, J.W., Chaillou, S., Raper, C.D. and Morot-Gaudry, J.F. (1994) Ammonium and nitrate uptake by soybean during recovery from nitrogen deprivation. *J. Exp. Bot.*, 45, 23–33.

Rufty, T.W., Mackown, C.T. and Volk, R.J. (1989). Effects of altered carbohydrate availability on whole-plant assimilation of 15NO-3. *Plant Physiol.*, 89, 457–463.

Santora, L.G. and Magalhaes, A.C.N. (1983) Changes in nitrate reductase activity during development of soybean leaf. *J. Plant Physiol.*, 112, 113–121.

Scaife, A. and Schloemer, S. (1994). The diurnal pattern of nitrate uptake and reduction by spinach (*Spinacia oleracea* L.). *Ann. Bot.*, 73, 337–343.

Schill, V., Hartung, W., Orthen, B. and Weiselseel, M.H. (1996) The xylem sap of maple (*Acer platanoides* L.) trees-sap obtained by a novel method shows changes with season and height. *J. Exp. Bot.*, 47, 123–133.

Schurr, U. and Schulze, E.D. (1995) The concentration of xylem sap constituents in root exudate, and in sap from intact, transpiring castor bean plants (*Ricinus communis* L.). *Plant Cell Environ.*, 18, 409–420.

Shaner, D.L. and Boyer, J.S. (1976) Nitrate reductase activity in maize (*Zea mays* L.) leaves. II Reguation by nitrate flux at low leaf water potential. *Plant Physiol.*, 58, 505–509.

Siebrecht, S. and Tischner, R. (1999) Changes in the xylem exudate composition of poplar (*Populus tremula* x *P. alba*) - dependent on the nitrogen and potassium supply. *J. Exp. Bot.*, 50, 1797–1806.

Smith, J.A.C. (1991) Ion transport and the transpiration stream. *Bot. Acta*, 104, 416–421.

Tanner, W. and Beevers, H. (1990) Does transpiration have an essential function in long-distance ion transport in plants? *Plant Cell Environ.*, 13, 745–750.

Touraine, B., Muller, B. and Grignon, C. (1992) Effect of phloem-translocated malate on NO_3^- uptake by roots of intact soybean plants. *Plant Physiol.*, 99, 1118–1123.

Wei, C., Tyree, M.T. and Steudle, E. (1999) Direct measurement of xylem pressure in leaves of intact maize plants. A test of the cohesion-tension theory taking hydraulic architecture into consideration. *Plant Physiol.*, 121, 1191–1206.

Wegner, L.H. and Raschke, K. (1994) Ion channels in the xylem parenchyma of barley roots: a procedure to isolate protoplasts from this tissue and a patch-clamp exploration of salt passageways into xylem vessels. *Plant Physiol.*, 105, 799–813.

Wolf, D., Jeschke, W.D. and Hartung, W. (1990) Concentrations and transport of solutes in xylem and phloem along the leaf axis of NaCl-treated *Hordeum vulgare*. *J. Exp. Bot.*, 41, 1133–1141.

Zimmermann, U., Meinzer, F.C., Benkert, R., Zhu, J.J., Schneider, H., Goldstein, G., Kuchenbrod, E. and Haase, A. (1994) Xylem water transport: is the available evidence consistent with the cohesion theory? *Plant Cell Environ.*, 17, 1169–1181.

Zweifel, R., Item, H. and Hasler, R. (2001) Link between dirunal stem radius changes and tree water relations. *Tree Physiol.*, 12–13, 869–877.

Section 5

Ion Relations in the Apoplast of Leaves

ION DYNAMICS IN THE APOPLAST OF LEAF CELLS

Z. RENGEL
Soil Science and Plant Nutrition, School of Earth and Geographical Sciences, The University of Western Australia, Australia, Zed.Rengel@uwa.edu.au

The apoplast is a complex compartment on the outside face of the plasma membrane, comprising water- and gas-filled spaces within the cell wall fibrillar network. The leaf apoplast is involved in delivering water and ions to the mesophyll cells as well exporting sugars and re-translocating water and ions back to roots via phloem. Hence, ion dynamics in the leaf apoplast are important for many cellular processes and interactions.

The papers in this section of the volume have covered a wide range of issues relevant in characterisation and functioning of the leaf apoplast. Two papers deal with technical and scientific challenges of quantifying ion dynamics in the leaf apoplast by using ion-sensitive microelectrodes (Felle and Hanstein, this volume, pp. 295–306) or fluorescent and luminescent ion-selective indicators (Plieth et al., this volume, pp. 373–392). Two papers discuss the role of the leaf apoplast in phloem loading (Lohaus, this volume, pp. 323–336) and mechanisms of long-distance transport (Merbach et al., this volume, pp. 337–352). Finally, two papers explore specific physiological phenomena in which the apoplast plays an important role: manganese toxicity (Fecht-Christoffers et al., this volume, pp. 307–324) and iron deficiency (Nikolic and Römheld, this volume, pp. 353–372).

Measuring the ion milieu in the apopast of intact leaf cells is a difficult task because of the small volume of solution involved and the rapid changes occurring due to metabolic activity. Yet, such measurements *in vivo* are indispensable in gaining the knowledge of the processes and interactions occurring at the outside face of the plasma membrane.

B. Sattelmacher† *and W.J. Horst (eds.), The Apoplast of Higher Plants: Compartment of Storage, Transport and Reactions*, 287–293.

Since the mid 1980s Felle and collaborators have been working on perfecting a CO_2 micro-sensor comprising a H^+-selective microelectrode combined with a microelectrode with carbonate buffer containing carbonate anhydrase and thus suitable for measuring gas exchange in the leaf apoplast (see Hanstein and Felle, 2002). The CO_2 micro-sensor as well as a reference electrode are inserted into the stomatal cavity and positioned so that the electrode tip is in contact with the apoplastic fluid. Other relevant microelectrodes can be made by using appropriate ion-selective sensors. The measured concentration ranges of various ions in leaf apoplast are quite wide (eg. 50–150 μM Ca^{2+} and 2–5 mM K^+) (Felle and Hanstein, this volume, pp. 295–306).

The values for apoplastic pH reported in the literature range between 4 and 7 (ie. spanning three orders of magnitude) possibly because of technical problems associated with the measurements. The apoplastic pH is expected to be slightly acidic due to H^+-ATPase activity in the plasma membrane; indeed, Felle and Hanstein (this volume, pp. 295–306) found a relatively narrow range of pH 4.7–5.1 in the leaf apoplast of five plant species.

The pH in the leaf apoplast is buffered to an order of magnitude smaller extent than the pH in the cytosol (Felle and Hanstein, this volume, pp. 295–306); thus, ion dynamics in the apoplastic liquid has a potential to substantially influence pH and thus affect photosynthesis and other processes (eg. Cleland, 2002). Stomatal closing is accompanied by increased activity of K^+ and Cl^- and decreased activity of Ca^{2+} in the apoplastic fluid (consistent with guard cells loosing K^+ and Cl^- and taking up Ca^{2+} to bring about the stomatal closure) (Felle and Hanstein, this volume, pp. 295–306). In contrast, low concentration of CO_2 opens and high concentration closes the stomata; however, there are fine points in this regulation that future research will need to address, namely CO_2 causes substantial stomata closure only after prolonged exposure.

Both H^+ and Ca^{2+} have a signalling function in the cellular metabolism. Voluminous literature exists on changes in the cytosolic activity of these ions serving as a secondary messenger signalling system allowing the cells to receive and respond to signals from the environment. In addition, H^+ and Ca^{2+} activities in the apoplast, as the first cellular compartment in contact with the environment, may also play an important role in signal perception and transduction. Studying signalling functions in the apoplast will benefit from the new, the least invasive techniques for measuring ion dynamics in the leaf apoplast based on creating plants that synthesise ion indicators and target them specifically to the apoplast. Plieth et al. (this volume, pp. 373–392) have transformed *Arabidopsis thaliana* to express fused H^+ and Ca^{2+} indicators in the apoplast or the cytosol, thus opening up exciting possibilities of monitoring effects of environmental signals on the dynamics of H^+ and Ca^{2+} ions acting as secondary messengers in the apoplast and symplast.

The technique developed by Plieth et al. (this volume, pp 379–398) is based on fluorescent and luminescent protein indicators, eg. aequorin for Ca^{2+} (Campbell et al., 1989) and Green Fluorescent Protein (GFP) altered to enhance pH dependency and create "pHluorins" sensitive to H^+ (Miesenböck et al., 1998), with good spatial and temporal resolution. The difficulties usually associated with loading these indicators into intact plant cells (e.g. see Zhang et al., 1998) were overcome by expressing relevant genes and allowing for compartment-specific, *in vivo* reconstitution (eg. Kiegle et al., 2000). Aequorin and pHluorins were expressed in *Arabidopsis* in either the cytosol or, by creating fusion proteins using an *Arabidopsis* chitinase signal sequence, the apoplast (Plieth et al., this volume, pp. 373–392). Aequorin luminescence and pHluorin fluorescence imaging was then performed on roots of transgenic plants.

The advantage of the method in terms of *in vivo* measurements on intact plants should be balanced against disadvantage arising from a weak luminescence signal requiring the whole roots to be used (or relatively large root segments for ratiometric fluorescence measurements), thus masking any potential differences among various cell and tissue types. Further developments of the technique will be watched with immense interest in expectations that the above disadvantage will be overcome. Measurements of the ionic milieu in the leaf as opposed to the root apoplast are also eagerly awaited. The next major step should be to specifically target ion-specific indicators into the cytosol and the apoplast of a single plant (or even better a cell) to allow studies of potential interactions between the two compartments.

The total concentration of anions and cations in the apoplastic fluid is about 10–20 mM (Lohaus, this volume, pp. 323–336). Substantial differences in concentration exist among different compounds as well as among different plant species. Nitrate and chloride are main anions, whereas K^+ represents up to ¾ of all cations. Interestingly, apoplastic concentrations of toxic ions (like Na^+), even though being increased upon exposure to saline solutions, are regulated tightly to minimise such an increase (Lohaus, this volume, pp. 323–336).

Ionic environment in the leaf apoplast can be finely regulated, eg. for phosphate (Mimura et al., 1992) and NH_4^+ (Nielson and Schjoerring, 1998) maintaining near-homeostatic concentrations (Lohaus, this volume, pp. 323–336), and is strongly influenced by inflow of the xylem sap. Transport of solutes in the xylem sap is enhanced by transpiration, directing most of the flow into vigorously transpiring leaves from which re-distribution into other organs takes place via phloem (Merbach et al., this volume, pp. 337–352).

Regulation of the ionic environment in the leaf apoplast is dependent on the cell types. For example, bundle-sheath cells in C4 plants have their cytoplasm connected to the cytoplasm of mesophyll cells by numerous plasmodesmata, but apoplastic spaces of these two cell types are separated by suberinised lamellae, allowing for different ionic environments (Mühling and Läuchli, 2002). These differences should be taken into account when assessing the appropriateness of the infiltration-centrifugation technique for the task of extracting apoplastic fluid, because it does so indiscriminately from the whole leaf or leaf part subjected to extraction.

Solute concentrations in the leaf apoplast influence phloem loading because most common plant species unload solutes from eg. bundle-sheath cells into the apoplast and then load the phloem sieve-element companion-cell complex from the apoplast (Lohaus, this volume, pp. 323–336). There is also strong relationship between the solute concentration in the xylem (part of the apoplast, but with quite different sap composition compared with the apoplastic fluid) and phloem sap, with substantial re-distribution of solutes imported into leaf by xylem and exported by phloem. Such solute re-distribution may have an important role in the signalling network linking shoots and roots (Marschner et al., 1996).

The role of the apoplast in the long-distance transport of nutrients varies for different nutrients. Concentrations of K^+ and phosphate are kept relatively constant in the apoplast, with K^+ in the apoplast serving as a dynamic reservoir. A similar claim can be made for Fe which can be precipitated in and remobilised from the apoplast (Welch, 1995), even though there are different opinions in the literature as to whether Fe precipitated in the apoplast is trapped rather than stored there, especially under alkaline conditions (eg. Kosegarten and Koyro, 2001), or whether Fe uptake into cells is independent of pH (Nikolic and Römheld, 2002).

Reduction of Fe^{3+} chelates in the leaf apoplast is required prior to Fe^{2+} uptake into leaf-cell symplast. The pH dependency of the leaf Fe^{3+} reductase, reported for several plant species (eg. Moog and Bruggemann, 1994), has been questioned because such dependency was not found in sunflower (Nikolic and Römheld, this volume, pp. 353–372). *In vivo* photoreduction of Fe(III)-citrate in the leaf apoplast was also not pH-dependent. Moreover, the pH of the apoplastic fluid did not appear to be influenced by Fe deficiency. However, given the shortcomings of the infiltration-centrifugation technique, any potential spatial differences in the pH gradients would be lost in the process of extracting apoplastic fluid.

Studying Fe uptake from the leaf apoplast across the plasma membrane hinges on a reliable method for differentiating apoplastic and symplastic Fe in leaves (Nikolic and Römheld, 2002, 2003). Most Fe in the leaf apoplast (>95%) is bound to the cell walls (Nikolic and Römheld, 2003).

A dynamic equilibrium between the cell-wall pool and soluble Fe^{3+} complexes in the apoplastic fluid governs availability of leaf apoplastic Fe for uptake across the plasma membrane (Nikolic and Römheld, this volume, pp. 353–372).

The role of high supply of bicarbonate and nitrate in influencing Fe uptake into mesophyll cells has attracted considerable attention in the past decade. One school of thought (Mengel, 1994) was that increased uptake of bicarbonate and/or nitrate increases pH in the leaf apoplast, thus inhibiting Fe^{3+} reductase activity and minimising Fe^{2+} uptake into mesophyll cells. However, the other thinking (Nikolic and Römheld, this volume, pp 359–378) is based on the findings that the inhibiting effect on Fe^{3+} reductase and Fe uptake by high pH due to bicarbonate happens primarily in roots (Römheld, 2000), thus making it unlikely that large amounts of Fe would be taken up by roots, transported to leaves and immobilised there by bicarbonate action on the leaf apoplastic pH and leaf-cell Fe^{3+} reductase. Indeed, supplying high concentrations of bicarbonate in the root-growing medium did not result in an increase in leaf apoplastic pH regardless of the measurement techniques used (Kosegarten et al., 1999; Nikolic and Römheld, 2002). Similarly, supplying plants with high concentrations of nitrate in the root medium resulted only in small and localised increases in the leaf apoplastic pH (Kosegarten et al., 1999; Nikolic and Römheld, 2003), making them an unlikely direct cause of Fe deficiency.

The apoplastic compartment in the leaf is important in governing ion toxicity (eg. sodium, Mühling and Lauchli, 2001); hence, Fecht-Christoffers et al. (this volume, pp. 307–322) have tested the hypothesis that the leaf apoplast is important in tolerance to Mn toxicity in cowpea. Oxidation of Mn^{2+} to Mn^{3+} in the leaf apoplast causing oxidative stress was proposed as a mechanism of Mn toxicity (Horst, 1988) and was corroborated by increased H_2O_2 production (Horst et al., 1999) and peroxidase activity in the apoplastic fluid (Fecht-Christoffers et al., 2003) coinciding with the formation of Mn toxicity-related brown spots in leaves. Increased formation of H_2O_2 may lead to secretion of a range of wound-induced proteins, organic acids and phenols into the apoplast (Fecht-Christoffers et al., this volume, pp. 307–322), with the significance of these responses in the overall Mn toxicity syndrome yet to be fully established. However, apoplastic peroxidases producing and consuming H_2O_2 have been suggested to play a central role in controlling the severity of Mn toxicity (Fecht-Christoffers et al., 2006).

Increased Si supply alleviated Mn toxicity in a range of plant species (Wiese et al., this volume, pp. 33–48). In cowpea, Si decreased oxidative stress and lowered peroxidase activity (Fecht-Christoffers et al., this volume, pp. 307–322). However, the mechanisms of Si-related alleviation of Mn toxicity remain obscure.

The papers discussed above attest to intensity and quality of the research on dynamics of ionic milieu in the leaf apoplast. Availability of powerful ion-selective microelectrodes as well as transformation of plants to specifically target ion-sensitive indicators into either the cytosol or the apoplast offer exciting opportunities of precise *in vivo* measurements of ion activities in the leaf apoplast, which will enhance our knowledge of the signaling sequences and other cellular processes (eg. xylem and phloem transport, ion homeostasis, uptake of Fe, tolerance to Mn toxicity).

REFERENCES

Campbell, A.K., Patel, A., Houston, W.A.J., Scolding, N.J., Frith, S., Morgan, B.P. and Compston, D.A.S. (1989). Photoproteins as indicators of intracellular Ca^{2+}. *J. of Bioluminescence and Chemiluminescence*, 4, 463–474.

Cleland, R.E. (2002). The role of the apoplastic pH in cell wal extension and cell enlargement. In: Handbook of Plant Growth. pH as a master variable (Z. Rengel, ed.), *Marcel Dekker, Inc,* New York. pp. 131–148,

Fecht-Christoffers., M.M., Führs H., Braun, H.-P., Horst, W.J. (2006). The role of H_2O_2-producing and H_2O_2-consuming peroxidases in the leaf apoplast of *Vigna unguiculata* L. in manganese tolerance. *Plant Physiol.*, 140, 1451–1463.

Fecht-Christoffers, M.M., Maier, P. and Horst, W.J. (2003). Apoplastic peroxidase and ascorbate are involved in manganese toxicity and tolerance of *Vigna unguiculata. Physiol. Plant.*, 117, 237–244.

Hanstein, S. and Felle, H.H. (2002). CO_2 triggered anion release from guard cells in intact faba bean leaves: kinetics of the onset of stomatal closure. *Plant Physiol.*, 130, 940–950.

Horst, W.J. (1988). The physiology of manganese toxicity. In: Manganese in Soils and Plants (R.D.Graham, Hannam, R.J. and Uren, N.C., eds.), *Kluwer Academic Publishers*, Dordrecht. pp. 175–188.

Horst, W.J., Fecht, M., Naumann, A., Wissemeier, A.H. and Maier, P. (1999). Physiology of manganese toxicity and tolerance in *Vigna unguiculata* (L.) Walp. *J. Plant Nutr. Soil Sci.*, 162, 263–274.

Kiegle, E., Moore, C.A., Haseloff, J., Tester, M.A. and Knight, M.R. (2000). Cell-type-specific calcium responses to drought, salt and cold in the *Arabidopsis* root. *Plant J.*, 23, 267–278.

Kosegarten, H. and Koyro, H.W. (2001). Apoplastic accumulation of iron in the epidermis of maize (*Zea mays*) roots grown in calcareous soil. *Physiologia Plantarum*, 113, 515–522.

Kosegarten, H.U., Hoffmann, B. and Mengel, K. (1999). Apoplastic pH and Fe^{3+} reduction in intact sunflower leaves. *Plant Physiol.*, 121, 1069–1079.

Marschner, H., Kirkby, E.A. and Cakmak, I. (1996). Effect of mineral nutritional status on shoot-root partitioning of photoassimilates and cycling of mineral nutrients. *J. Exp. Bot.*, 47, 1255–1263.

Mengel, K. (1994). Iron availability to plant tissues - iron chlorosis on calcareous soils. *Plant Soil*, 165, 275–283.

Miesenböck, G., De Angelis, D.A. and Rothman, J.E. (1998). Visualizing secretion and synaptic transmission with pH-sensitive green fluorescent protein. *Nature*, 394, 192–195.

Mimura, T., Dietz, K.-J., Kaiser, W.M., Schram, M.J., Kaiser, G. and Heber, U. (1992). Phosphate transport across biomembranes and cytosolic phosphate hoemostasis in barley leaves. *Planta*, 180, 139–146.

Moog, P.R. and Bruggemann, W. (1994). Iron reductase systems on the plant plasma membrane – A review. *Plant Soil*, 165, 241–260.

Mühling, K.H. and Läuchli, A. (2002). Determination of apoplastic Na^+ in intact leaves of cotton by *in vivo* fluorescence ratio-imaging. *Funct. Plant Biol.*, 29, 1491–1499.

Mühling, K.H. and Lauchli, E. (2001). Physiological traits of sodium toxicity and salt tolerance. In Plant Nutrition: Food Security and Sustainability of Agro-Ecosystems through Basic and Applied Research (W.J. Horst et al., eds.), *Kluwer Academic Publishers*, Dordrecht, Netherlands, pp. 378–379.

Nielson, K.H. and Schjoerring, J.K. (1998). Regulation of apoplastic NH_4^+ concentration in leaves of oilseed rape. *Plant Physiol.*, 109, 735–742.

Nikolic, M. and Römheld, V. (2002). Does high bicarbonate supply to roots change availability of iron in the leaf apoplast? *Plant Soil*, 241, 67–74.

Nikolic, M. and Römheld, V. (2003). Nitrate does not result in iron inactivation in the apoplast of sunflower leaves. *Plant Physiol.*, 132, 1303–1314.

Römheld, V. (2000). The chlorosis paradox: Fe inactivation as a secondary event in chlorotic leaves of grapevine. *J. Plant Nutr.*, 23, 1629–1643.

Welch, R.M. (1995). Micronutrient nutrition of plants. *Critical Rev. Plant Sci.*, 14, 49–82.

Zhang, W.-H., Rengel, Z. and Kuo, J. (1998). Determination of intracellular Ca^{2+} in cells of intact wheat roots: Loading of acetoxymethyl ester of Fluo-3 under low temperature. *Plant J.*, 15, 147–151.

PROBING APOPLASTIC ION RELATIONS IN *VICIA FABA* AS INFLUENCED BY NUTRITION AND GAS EXCHANGE

H.H. FELLE and S. HANSTEIN
Botanisches Institut I, Justus-Liebig Universität Gießen, Germany,
hubert.felle@bio.uni-giessen.de

Abstract. A novel and non-invasive method to probe the apoplastic space is presented. Ion-selective microprobes are inserted through open stomata until contact with the apoplastic fluid is achieved. It is demonstrated how absolute values of H^+-, K^+-, Ca^{2+}-, NH_4^+-, Cl^-- activities, and CO_2 partial pressures, as well as continuous changes thereof can be reliably measured while superimposing physiological challenge. The technique also permits the use of intact plants, thus guaranteeing the acquisition of data from an undisturbed biological system. Details of microprobe fabrication are given.

Key words: apoplastic buffering; CO_2-sensor; gas exchange ion-selective microprobes; stomata

1. INTRODUCTION

The extracellular space within a plant, the apoplast, is highly important for transport of matter, storage, enzymatic reactions or during plant-microbe interactions. Since the apoplastic fluid layer is mostly very thin, already small transmembrane fluxes rapidly change the ionic activities which subsequently will influence other parameters. Therefore, the apoplastic ion milieu has to be well regulated. This holds especially for the leaf apoplast, because it is directly influenced from different sides: gas exchange, transpiration stream, light conditions and microbial attack. Although extracellular, the determination of apoplastic ionic composition has always been difficult and was prone to errors.

B. Sattelmacher† and W.J. Horst (eds.), The Apoplast of Higher Plants: Compartment of Storage, Transport and Reactions, 295–306.

Due to the development of the non-invasive ion-probe technique (Hanstein and Felle, 1999), the sources of error could be reduced considerably. The insertion of specific probes through open stomata and the direct and continuous monitoring of ion activities as well as CO_2 is an important step forward in non-invasive cell biology.

2. TECHNIQUES

2.1 Ion-selective microprobes

The basic principle of these probes are glass micro-pipettes with an ion-selective liquid sensor in the tip (e.g. Ammann, 1986; Felle and Bertl, 1986). The voltage, generated at the interface on either side of the sensor with the aqueous phases, is picked up by a high-impedance differential amplifier (10^{15} Ohm). Wherever required, two or more microprobes, each harbouring different sensors, may be bundeled. Such multibarreled arrangement permits the continuous measurement of several ion activities at the same spot and at the same time. Since interfering voltages may also be registered by the selective probe, an additional non-selective reference electrode is mandatory. After placing both probes in the substomatal cavities of a leaf, the voltage difference from these electrodes represents the net signal of interest. Since these probes have response times in the lower second range (i.e. H^+: 90% response ≈ 5 s) faster ion dynamics will not be picked up with the real kinetics.

The fabrication and use of these electrodes has been described in detail (Felle and Bertl, 1986; Felle et al., 1998; Hanstein and Felle, 1999).

2.2 The CO_2 sensor

The fabrication and use of this specific micro-sensor has been described recently (Hanstein et al., 2001; Hanstein and Felle, 2002). The sensor assembly consists of a pH-microelectrode concentrically arranged within a sheathing micropipette. The tip of the latter is plugged with a silicone membrane through which the CO_2 diffuses. The space behind is filled with carbonate buffer, the pH of which quickly responds to CO_2 concentration changes due to the incorporated carbonic anhydrase (Fig. 1).

2.3 Cuvette, positioning of the probes, and recording the signals

The leaf is fixed within a two-chambered cuvette (Fig. 2). One chamber in which the probe positioning takes place remains dry, the other chamber

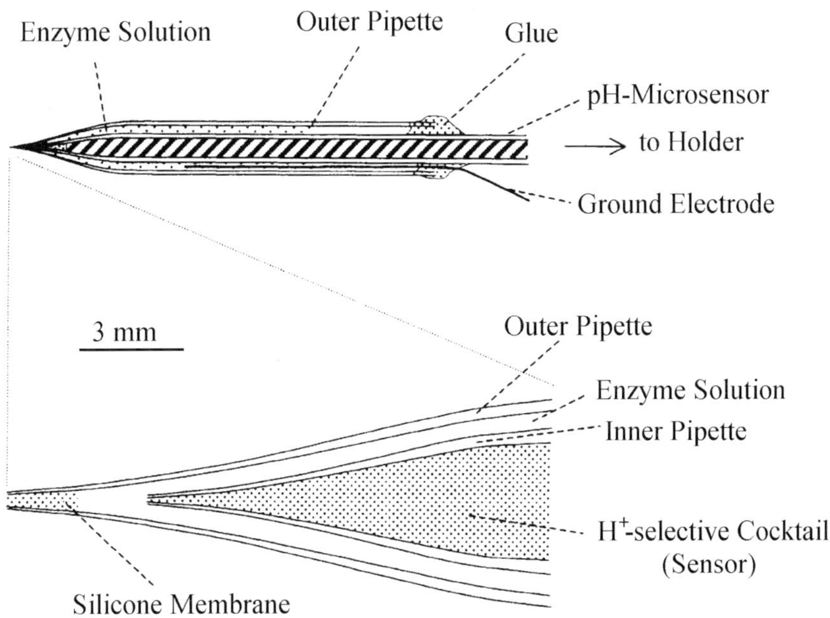

Fig. 1. Basic design of the CO_2 microsensor. It consists of two concentrically arranged glass pipettes, glued together. The inner pipette is a pH-sensitive microelectrode which measures CO_2 –dependent pH changes in a carbonic anhydrase solution, placed in the tip of the outer pipette. The pH-electrode sits some 20 μm behind the tip of the outer pipette, which is plugged with a silicone membrane. An Ag/AgCl-reference electrode connects the enzyme solution with ground. Reprinted from, Hanstein et al., 2001. With permission of Elsevier.

holds the cut petiole and the reference electrode placed within a solution the composition of which was chosen as close as possible to the apoplastic ion activities (e.g. 1 mM KCl, 0.1 mM $CaCl_2$, 5 mM buffer). In this chamber solutions can be manipulated during the experiment. The transpiration stream carries these changes within the xylem and responses are measured within the substomatal cavity by the probes. The probes are placed inside the substomatal cavity at an angle of about 45 degree. As soon as the probe tip is in contact with the apoplastic fluid, the electrical circuit is closed. Since the thin apoplastic layer within the leaf represents a relatively high electrical resistance which prevents the electrical charges to flow off rapidly to ground, a second reference electrode had to be placed in a stoma near the selective probe to pick up interfering voltage changes. The electrical difference obtained from the ion-selective probe and from the non-selective electrode is the ion-related net signal of interest.

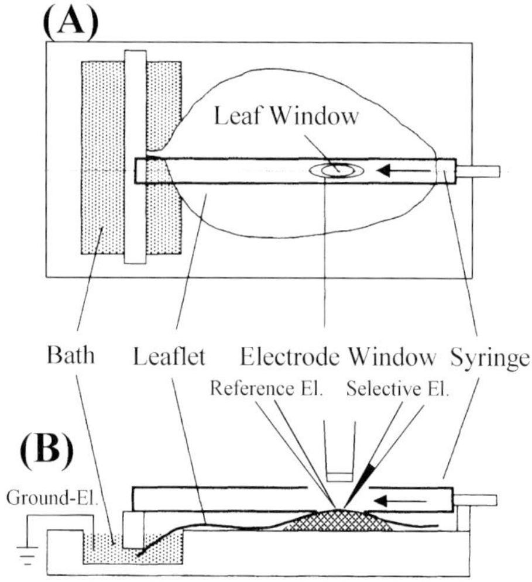

Fig. 2. Cuvette to hold the leaflet (dry chamber) and the cut petiole (bath). Along the dry chamber a minicuvette is placed to allow for controlled gas flow. A lower opening in the minicuvette (leaf window) exposes a small part of the leaf to the gas flow, whereas the upper (electrode) window permits the approach of the two electrodes (ion-selective or CO_2 probes and leaf reference). Probes are placed opposite each other approx. at an angle of 45 degrees to the leaf surface. (A) View from top; (B) view from the side. From, Hanstein et al. 2001, altered. With permission of Elsevier.

3. PHYSIOLOGY OF THE LEAF

3.1 The apoplastic ion milieu in leaves of *Vicia faba* and other plant species

In light-adapted *Vicia faba* leaves mean activities (from 20 to 50 measurements, each) were 2.5 mM K^+ (pK 2.6), 1.3 mM Cl^- (pCl 2.9), 64 µM Ca^{2+} (pCa 4.2) and 16 µM H^+ (pH 4.8). These values may change considerably depending on the bath reference solution, the transpiration stream, light- or the nutritional conditions in general (see below). Table 1 shows that, wherever determined, the ionic activities of other plant species are in the same range. The apoplastic activities of K^+, Cl^- or Ca^{2+} are not given frequently in the literature. However, the apoplastic pH has been recorded and discussed more frequently, probably because of its multifunctional role. Often, the range of apoplastic pH is given in a wide range (e.g. five to seven).

Table 1. Ionic activities, measured with ion-selective microprobes within the substomatal cavity of leaves from different plant species, as indicated. All values refer to light-adapted leaves, ambient 350 ppm CO_2, and a reference bath solution of pH 5 containing 5 mM KCl and 0.1 mM $CaCl_2$.

Plant	K^+ (mM)	H^+ (pH)	Cl^- (mM)	Ca^{2+} (µM)	References
Arabidopsis thaliana	3 – 5	n.d.	n.d.	n.d.	Szyroki et al. 2001
Bromus erect.	8	4.7– 4.9	n.d.	n.d.	Hanstein and Felle 1999
Brassica spec.	n.d.	5.0	n.d.	n.d.	Hanstein et al. unpublished
Hordeum vulg.	5 – 8	4.8 – 5.1	n.d.	80 –150	Felle et al. 2003
Vicia faba	2 – 5	4.7 – 5.1	0.7 – 2.4	50 – 80	Felle et al. 2000

Although in part this may have methodological causes, recent investigations on barley during infection with mildew show that apoplastic pH may change from 4.8 to near 7 or even above (Felle et al., 2004). Since these changes are long lasting, it may well be that high apoplastic pH values – wherever found – may have been recorded in a state of physiological imbalance.

3.2 Apoplastic pH buffering

Apoplastic pH is not as well buffered as the cytosol. Whereas in the latter 30–50 mM/pH is a frequently reported buffer capacity (Guern et al., 1991) the apoplast of leaves has roughly one tenth of this value. The apoplastic buffer capacity can be determined *in vivo* by fumigating the leaves with NH_3, which dissolves in the apoplastic fluid to form NH_4^+. This will change the pH which is picked up by the pH-electrode. The ratio of NH_4^+ formed (measured with a NH_4^+-electrode) and pH changed yields the buffer capacity (see 4.1). Using this method, Hanstein and Felle (1999) found 6–8 mM/pH in *Bromus erectus* and 2–3 mM/pH in *Vicia faba* (Felle and Hanstein, 2002); in potato, Oja et al. (1999) report 4 mM/pH. These values indicate that upon transport-activity changes apoplastic pH may change more rapidly and more substantial than the cytoplasm, as shown below.

3.3 Light and darkness

Upon "light-off" ion activities rapidly change, as shown in Fig. 3. The nearness of the electrodes to the guard cells (specific stoma responses to light) clearly influenced the individual kinetics, especially that of K^+ which does not return to basic levels but remains elevated, in accordance to guard cells loosing K^+ while closing. As such the response to "light-off" or "light-on" will somewhat look different when measured in the mesophyll (not shown). Figs. 3a and 3b show that ion activities do not change into the same direction. Whereas K^+ and Cl^- increase, H^+ and Ca^{2+} decrease following "light-off". Similar light/dark-induced kinetics were shown in leaves of bean by Shabala and Newman (1999), in pea by Stahlberg and van Volkenburgh (1999), and in sunflower by Mühling and Läuchli (2000).

3.4 The influence of the transpiration stream on the apoplastic ionic milieu

The xylem is an integral part of the apoplast that is interconnected with the apoplast of the leaf mesophyll and substomatal cavity. Since there is no membrane barrier, substances that are added to the cut end of the leaf are transported within the xylem elements and diffuse into the leaf apoplast to

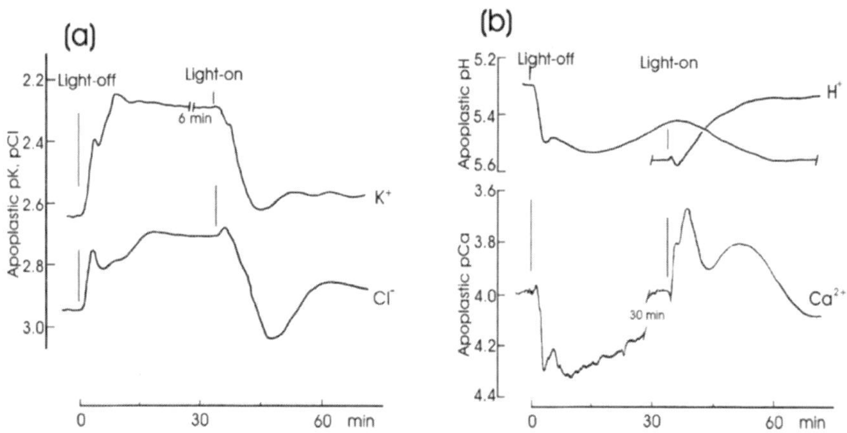

Fig. 3. Apoplastic activities (K^+, Cl^-, H^+, Ca^{2+}) of Vicia faba leaves, measured within the sub-stomatal cavity using ion-selective microelectrodes. Traces show the effect of ‚light-off' first and then of ‚light-on". K^+- and Cl^- kinetics are measurements from the same leaf and were measured with a double-barreled K^+/Cl^- micro-electrode ; H^+ and Ca^{2+} kinetics are kinetics from different leaves, each.

reach the stomatal area within minutes. Depending on their chemical nature or electrical charge the thus translocated matter will change apoplastic conditions and hence the apoplastic ionic milieu.

a) ABA: as shown in Fig. 4, this phytohormone is rather active with respect to changing apoplastic ion activities in the vicinity of stomata. ABA, fed through the cut petiole, within minutes causes increases in K^+- and Cl^- activity but decreases in H^+- and Ca^{2+}-activity. This behaviour is similar to that observed after 'light-off and is very likely related to stomatal closure (Felle et al., 2000).

b) Changes in the ionic composition: a change in the ion activity within the xylem (reference bath) may be carried fully to the leaf apoplast, as can be shown for instance for K^+ (Fig. 5), but also may alter apoplastic pH, a response that apparently is due to the osmotic change. The latter notion is supported by the observation that sugar alcohols like mannitol or sorbitol, added at the same osmolarity as K^+, will cause similar pH effects. On the other hand, pH changes within the xylem (by addition to the cut petiole) will only weakly be transferred (Fig. 5b). More stress-related data are presented in Felle and Hanstein (2002) and Felle et al. (2005).

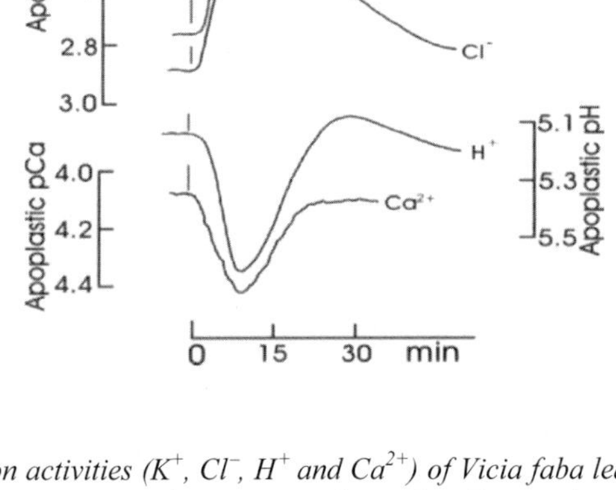

Fig. 4. Apoplastic ion activities (K^+, Cl^-, H^+ and Ca^{2+}) of Vicia faba leaves, measured within sub-stomatal cavities using ion-selective microelectrodes. Traces show the influence of ABA added to the bath reference and taken up through the cut petiole. K^+- and Cl^- measurements from the same leaf, H^+ and Ca^{2+} measurements from separate leaves, each. From Felle et al. (2000, Fig. 1), reprinted with permission of Blackwell Publ. Ltd.

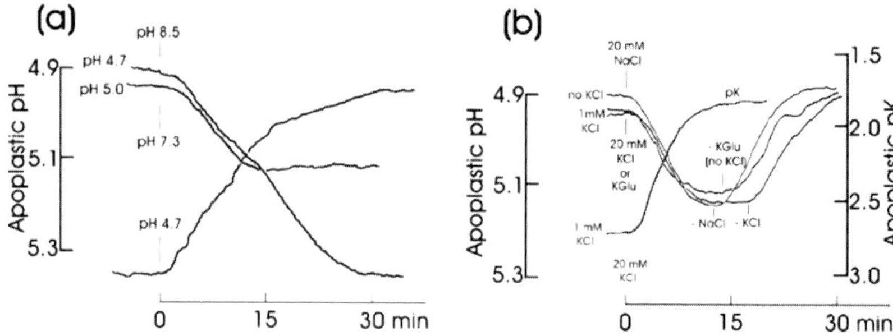

Fig. 5. Influence of xylem-carried factors (pH, K^+, Na^+), added to the reference bath-solution, on apoplastic pH and K^+. (a) Effect of different pH, (b) effect of different ions. To exclude anion effects, Cl^- was exchanged for gluconate. From Felle and Hanstein (2002, Fig. 1), reprinted with permission of Oxford University Press.

4. STOMATAL GAS EXCHANGE

CO_2 is taken up to be fixed in photosynthesis and NH_3 as reduced N is used to supplement the nitrogen requirements of a plant. As will be shown below, both gases have a marked effect on the apoplastic ionic milieu and thus on the stomatal opening.

4.1 NH_3

Due to the acidic apoplastic pH and its high pK-value, NH_3 upon contact with the apoplastic fluid will readily form NH_4^+. This process removes H^+ from the fluid and tends to alkalize it. An example thereof is given in Fig. 6a. The kinetics show a rapid increase in apoplastic pH which is accompanied by a just as rapid increase (formation) of NH_4^+ (measured as NH_4^+ with a NH_4^+-selective microelectrode). The rapid pH increase is followed by a slower one, also accompanied by a further increase in NH_4^+. Removal of NH_3 reverses the effects. The formed NH_4^+ increase is in the lower millimolar range and causes plasma-membrane depolarization (Fig. 6b) which is possibly due to influx of positive charge through the NH_4^+ uniporter and to a minor extent through K^+ channels. The formed NH_4^+ (d[NH_4^+]) brought into relation with the altered pH (dpH] yields the apoplastic buffer capacity (ß) given above (see 3.2), according to

$$ß = d[NH_4^+]/dpH \qquad (1)$$

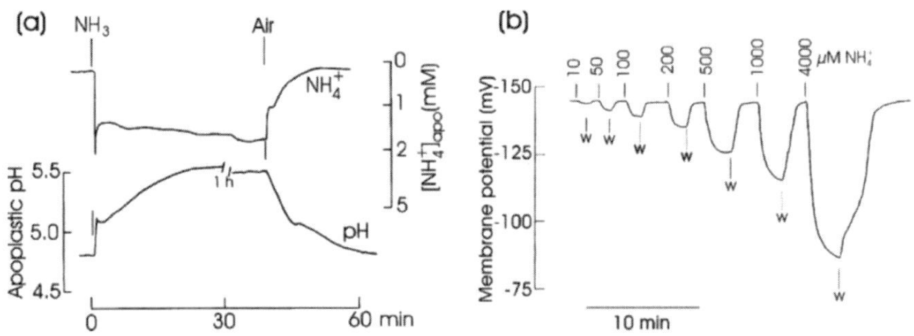

Fig. 6. Effect of NH_3 fumigation on apoplastic pH of Bromus erectus leaves. (a) Direct effect of NH_3 on pH and NH_4^+ formed (see text), and removal of NH_3 by air. (b) Effect of different NH_4^+ concentrations, as indicated, on the membrane potential of Bromus erectus leaf cells. W = removal of NH_4^+. From Hanstein and Felle (1999, Figs. 1 and 3), reprinted with permission of Blackwell Publ. Ltd.

4.2 CO_2

Apart from being a basic compound in photosynthesis, CO_2 is an important factor in stomatal regulation, whereby low CO_2 opens stomata. Even with homogeneous conditions there is a high variability in stomatal apertures. This stimulated us to build a CO_2 sensor (Fig. 1) that enables one to measure CO_2 in individual stomata. Fig. 7 gives the response of the sensor to CO_2

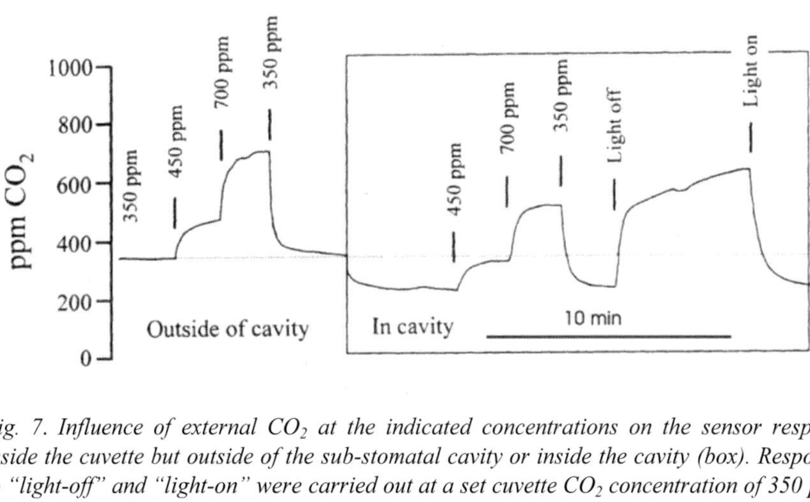

Fig. 7. Influence of external CO_2 at the indicated concentrations on the sensor response inside the cuvette but outside of the sub-stomatal cavity or inside the cavity (box). Responses to "light-off" and "light-on" were carried out at a set cuvette CO_2 concentration of 350 ppm. Reprinted from, Hanstein et al., 2001. With permission of Elsevier.

placed outside of the leaf as well as inside the leaf. Since the stoma investigated was open, the changes within the cavity are almost identical to the controlled CO_2 in the cuvette. Changes in light conditions, alter the CO_2 concentration within the leaf rapidly and reversibly, despite the fact that the stoma remains open during the relatively short time intervals in which "light-off" or "light-on" is superimposed.

Clearly, CO_2 dissolving in H_2O may change pH; it can be shown, however, that CO_2 shifts in the physiological relevant range per se have no effect on the apoplastic pH.

4.3 Combination of ion-selective and CO_2 microprobing

An early and reliable indicator of stomatal closure is the release of Cl^- from the guard cells which precedes stomatal closure and initiates it due to its depolarizing effect. As shown in Fig. 8, not every CO_2 pulse triggers Cl^- efflux: stomata remain open. Only prolonged presence of CO_2 initiates a

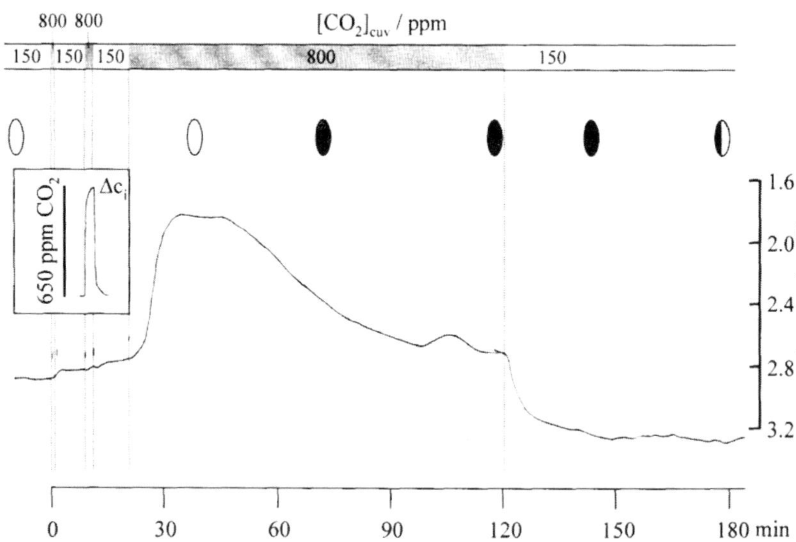

Fig. 8. Response of guard cell apoplastic Cl^- activity (pCl) to periods of elevated CO_2 of different duration. The CO_2 concentration within the cuvette ($[CO_2]_{cuv}$) is given in the bar on the top with shaded areas for periods of high $[CO_2]_{cuv}$. Inset shows the substomatal CO_2 change (Δci) during the 2-min CO_2 pulse, which was directly recorded by the CO_2 microsensor within the substomatal cavity. Stomatal openings microscopically observed (open ovals: stomata open; closed ovals: stomata closed). From Hanstein and Felle (2002, Fig. 3), reprinted with permission of "American Scociety of Biologists".

substantial increase in apoplastic Cl^- activity, in the shown case from a basic 1.5 mM to a 15 mM peak. This is fully in line with the transient Cl^- increase from 1 to 8 mM following ABA treatment (Fig. 4). As with ABA, the Cl^- activity recovers while the stomata close. Interestingly, the Cl^- activity is further reduced to 0.6 mM, as soon as the cuvette CO_2 is decreased. Further studies, using the same combination of methods, show that CO_2 concentration and light conditions interact and compensate each other in a complicated manner (Hanstein and Felle, 2002).

5. CONCLUSIONS

The direct investigation of the (ionic) apoplastic milieu by the use of microprobes is a novel and promising technique. It has three major advantages: (i) it can be used for the detection of a large number of ions, gases, and also redox compounds provided the sensors are available, (ii) with some experimental skill the sensors can be combined in one tip to measure several signals or components simultaneously at the same spot in the apoplast, (iii) the signal detection is entirely non-invasive and neither changes the compound to be tested nor does it interfere with the response(s) of the biological object. It is to be expected that the presented technique will give new insights in a variety of physiological problems.

REFERENCES

Ammann, D. (1986) Ion-selective micro-electrodes. principles, design and application. *Springer-Verlag*, Berlin, Heidelberg, New York, Tokyo.

Felle, H. and Bertl, A. (1986) The fabrication of H^+-selective liquid-membrane microelectrodes for use in plant cells. *J. Exp. Bot.*, 37, 1416–1428.

Felle, H.H. and Hanstein, S. (2002) The apoplastic pH of the substomatal cavity of *Vicia faba* leaves and its regulation. *J. Exp. Bot.*, 53, 73–82.

Felle, H.H., Hanstein, S., Steinmeyer, R. and Hedrich, R. (2000) Dynamics of ionic-activities in the apoplast of the substomatal cavity of intact *Vicia faba* leaves during stomatal closure evoked by ABA and darkness. *Plant J.*, 24, 297–304.

Felle, H.H., Herrmann, A., Hückelhoven, R. and K.-H. Kogel (2005) Root-to-shoot signalling: apoplastic alkalinization, a general stress response and defense factor in barley (*Hordeum vulgare*). *Protoplasma*, 227, 17–24.

Felle, H.H., Herrmann, A., Hanstein S., Hueckelhoven, R. and Kogel K.-H. (2004). Apoplastic pH signaling in barley leaves attacked by the powdery mildew fungus *Blumeria graminis* f.sp. hordei. Molecular Plant-Microbe interactions 17, 118–123.

Felle, H.H., Kondorosi, E., Kondorosi, A. and Schultze, E. (1998) The role of ion fluxes in Nod factor signaling in *Medicago sativa*. *Plant J.*, 13, 455–463.

Guern, J., Felle, H., Mathieu, Y. and Kurkdjian A. (1991) Regulation of intracellular pH in plant cells. Int. *Rev. Plant Cyt.*, 127, 111–173.

Hanstein, S. and Felle, H.H. (1999) The influence of atmospheric NH_3 on the apoplastic pH of green leaves: a noninvasive approach with pH-sensitive microelectrodes. *New Phytol.*, 143, 333–338.

Hanstein, S., De Beer, D. and Felle, H.H. (2001) Miniaturized carbon dioxide sensor designed for measurements within plant leaves. *Sensors and Actuators*, B 81, 107–114.

Hanstein, S. and Felle, H.H. (2002) CO_2 triggered anion release from guard cells in intact fava bean leaves: Kinetics of the onset of stomatal closure. *Plant Physiol.*, 130, 940–950.

Mühling, K.H. and Läuchli, A. (2000) Light-induced pH and K^+ changes in the apoplast of intact leaves. *Planta*, 212, 9–15.

Oja, V., Savchenko, G., Jakob, B. and Heber, U. (1999) pH and buffer capacities of apoplastic and cytoplasmic cell compartments in leaves. *Planta*, 209, 239–249.

Shabala, S. and Newman, I. (1999) Light-induced changes in hydrogen, calcium, potassium and chloride ion fluxes and concentrations from the mesophyll and epidermal tissues ofm bean leaves. Understanding the ionic basis of light-induced bioelectrogenesis. *Plant Physiol.*, 119, 1115–1124.

Stahlberg, R. and van Volkenburgh, E. (1999) The effect of light on membrane potential, apoplastic pH and cell expansion in leaves of *Pisum sativum* L. var. Argentum. *Planta*, 208, 188–195.

Szyroki, A., Ivashikina, N., Dietrich, P., Roelfsma, M.R., Ache, P., Reintanz, B., Deeken, R., Godde, M., Felle, H.H., Steinmeyer, R., Palme, K. and Hedrich, R. (2001) KAT1 is not essential for stomatal opening. *Proc. Nat. Acad. Sci. USA*, 98, 2917–2921.

THE ROLE OF THE LEAF APOPLAST IN MANGANESE TOXICITY AND TOLERANCE IN COWPEA (*VIGNA UNGUICULATA* L. WALP)

M.M. FECHT-CHRISTOFFERS, P. MAIER, K. IWASAKI, H.P. BRAUN and W.J. HORST

Institut für Pflanzenernährung, Leibniz Universität Hannover, Germany, horst@pflern.uni-hannover.de

Abstract. First visible Mn toxicity symptoms are brown spots on older leaves, followed by chlorosis, necrosis and leaf shedding. The brown spots represent local accumulations of oxidized Mn (Mn^{IV}) and oxidized phenols in the cell wall, especially of the epidermis. Differences in Mn resistance between cv TVu 91 (Mn-sensitive) and cv TVu 1987 (Mn-tolerant) are due to higher Mn tissue tolerance. The physiological mechanism of Mn toxicity and Mn tolerance are still poorly understood. The apoplast was proposed to be the most important compartment for development of Mn toxicity and Mn tolerance.

The detailed analysis and characterization of the proteome of the leaf apoplast confirm the particular role of PODs in the expression of Mn toxicity mediating H_2O_2 production/consumption and the oxidation of phenols in the leaf apoplast. The observed Mn-induced release of pathogenesis-related like proteins (PR-like) is attributed to a general stress response. Since PR-like proteins are induced by various other abiotic and biotic stresses, a specific physiological role of these proteins in response to excess Mn supply remains to be established. From the apoplastic metabolites, particular the composition of phenolic compounds seemed to be crucial for the development and avoidance of Mn toxicity. Phenolic compounds affect POD activities causing a stimulation or inhibition of PODs in the apoplast. Furthermore, sequestration of Mn by phenolic compounds and thus rendering Mn physiologically inactive might enhance Mn tolerance. The analysis of the release of organic acids into the apoplast and translocation of Mn into the vacuoles did not support the hypothesis, that sequestration of Mn by organic acids in the apoplast and the vacuoles is crucial for Mn tolerance. Silicon alleviated Mn toxicity symptoms not only by a decrease of the apoplastic Mn concentration and an increased adsorption of Mn to the cell walls but also by the soluble Si in the apoplast. Although the antioxidant ascorbic acid proved to be beneficial for protecting the leaf tissue from Mn toxicity, it is not considered as the most important factor in Mn tolerance.

The presented data confirm the importance of the apoplast for development and avoidance of Mn toxicity in the leaf tissue of cowpea. Conclusions about the chronology of Mn-induced physiological changes are difficult to draw. A more detailed study with emphasis

B. Sattelmacher† *and W.J. Horst (eds.), The Apoplast of Higher Plants: Compartment of Storage, Transport and Reactions*, 307–321.

on very early stages of Mn toxicity and a comparison of Mn-sensitive and Mn-tolerant leaves (genotype, Si nutrition, leaf age) is required.

Key words: cowpea, NADH peroxidase, proteome, manganese, tolerance, toxicity

1. INTRODUCTION

Manganese (Mn) toxicity occurs on acid and waterlogged soils, but was also observed under conditions such as drought (Khana and Michra, 1978), heat (Bartlett and James, 1980), and after steam sterilization of soils (Sonneveld and Voogt, 1975). Several developmental, environmental and nutritional factors influence Mn resistance e.g. leaf age (Horst, 1982), temperature (Marsh et al., 1989), light intensity (Wissemeier and Horst, 1992), silicon (Si) supply (Horst and Marschner, 1978) and nitrogen form (Langheinrich et al., 1992). Furthermore, large differences in Mn resistance exist between plant species and cultivars within species. Differences between cultivars of cowpea (*Vigna unguiculata* L. Walp.) were due to a higher Mn tolerance of the leaf tissue, represented by significantly different development of Mn toxicity symptoms at elevated Mn tissue contents (Horst, 1988). First visible Mn toxicity symptoms are brown spots in older leaves, followed by chlorosis, necrosis and leaf shedding. The brown spots represent local accumulations of oxidized Mn (Mn^{IV}) and oxidized phenols in the cell wall, especially of the epidermis (Wissemeier, 1988; Wissemeier and Horst, 1992).

The mechanism of Mn toxicity and tolerance are still unknown. To get a better understanding of Mn toxicity in plants, particularly changes in the leaf apoplast in response to Mn stress were investigated and their specific role on Mn toxicity and relevance for Mn tolerance are discussed.

2. MANGANESE-INDUCED CHANGES IN THE APOPLAST PROTEOME

2.1 Peroxidases

The oxidation of Mn^{II} in the apoplast has been proposed as a key reaction leading to Mn toxicity (Horst, 1988), because Mn^{III} may react as a powerful oxidant of proteins and lipids (Archibal and Fridovich, 1982). Kenten and Mann (1950) found a close relationship between the oxidation of Mn in the presence of peroxidase (POD) and phenols, and the activation of PODs by excess Mn was documented by Horst (1988), Horiguchi and Fukomoto

(1987), and Morgan et al. (1966). POD is often used as a physiological marker for plant-stress responses with an apparent lack in specificity. But particularly the stimulating effect of Mn on H_2O_2-producing PODs (Halliwell, 1978) indicates a specific effect of Mn on the function of PODs in the apoplast. A more detailed analysis of PODs in the leaf tissue was used to verify the specificity of the response of PODs to Mn excess.

PODs extracted from several fractions of the leaf tissue were significantly activated by Mn treatment in the Mn-sensitive cowpea cultivar TVu 91, whereas the Mn-tolerant cv. TVu 1987 showed no change in POD activities due to excess Mn supply (Fecht-Christoffers et al., 2003a). In TVu 91 (Mn-sensitive), particularly the soluble peroxidases of the leaf apoplast, extracted by collecting apoplastic washing fluid (AWF) using the vacuum-infiltration technique (Fecht-Christoffers et al., 2003a), were most affected by Mn treatment compared to PODs from the cytoplasm and bound to the cell wall (Fecht-Christoffers et al., 2003a). This is in agreement with results showing that extracellular PODs respond more sensitive to oxidative stress induced by e.g. ozone exposure, than cytosolic PODs (Castillo et al., 1984). Furthermore, activity of PODs in the AWF increased significantly and simultaneously with the formation of characteristic brown spots in leaves (Fecht-Christoffers et al., 2003a). These findings suggest a close relationship between free movable apoplastic PODs and the oxidation of Mn and phenols in the apoplast. Electrophoretic separation of proteins in the AWF by blue-native polyacryl-electrophoresis (BN-PAGE) demonstrated a strong release of PODs into the apoplast with increasing Mn-treatment duration. Released PODs were identified by nano LC-MS/MS as an acidic, anionic POD with a molecular mass of around 32kDa (Fecht-Christoffers et al., 2003b). In general, the extracellular space of plants contains acidic (anionic) and basic (cationic) PODs with differential affinities to substrates, where the acidic PODs are considered to be involved in the formation of the secondary cell wall and lignification (Campa, 1991; Ros Barcelo, 1997).

In tobacco, acidic PODs were strongly expressed in trichomes and the epidermis (Klotz et al., 1998) and not expressed in tissues or regions undergoing growth, probably due to the inhibitory effect of PODs on growth and elongation (MacAdam et al., 1992; de Souza and MacAdam, 1998). These observations are in agreement with own results, showing a strong development of brown depositions in the epidermal cell layer (cowpea, soybean) and at the base of trichomes (rape, tobacco and *Arabidopsis*). This was also observed in sunflower by Blamey et al. (1986). Peroxidases were also proposed to produce H_2O_2 in the apoplast necessary for lignification. Particularly basic PODs using NADH as reductant were proposed to catalyse H_2O_2 formation.

A two step control of PODs in the apoplast was described by Gaspar et al. (1985). Apoplastic PODs in the AWF of cowpea were able to oxidize NADH (Fig. 1) accompanied by the formation of H_2O_2 (Fig. 2). Both reactions increased with Mn treatment. The oxidation of NADH was enhanced in the presence of Mn and p-coumaric acid (Fig. 3). The stimulating effect on PODs of these co-factors has been documented previously (Halliwell, 1978), and they might be crucial for the development of Mn toxicity. The Mn concentrations in the apoplast range from 10 (control plants) to 160 μM (Mn-treated plants) (Fecht-Christoffers et al., 2003b). The drastic increase of Mn concentration in the apoplast might cause a direct activation of NADH-PODs with subsequent formation of H_2O_2. Since a Mn-induced H_2O_2-production was observed in washed intact leaf segments (Horst et al., 1999) formation of H_2O_2 by PODs stimulated by phenols and Mn might be the initiation of Mn toxicity with subsequent reduction of H_2O_2 by PODs, accompanied by oxidation of phenols and probably Mn^{II} .

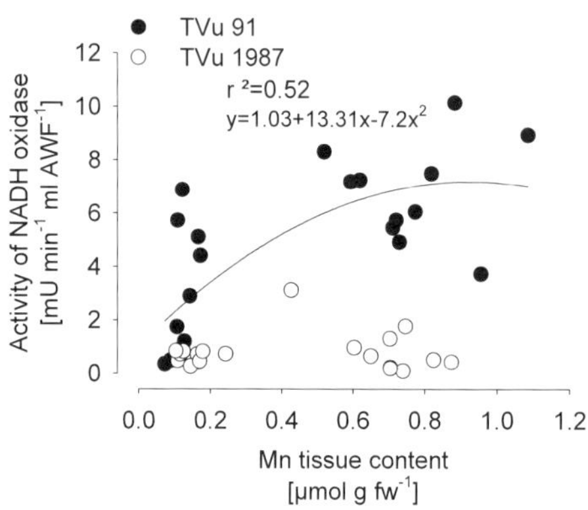

Fig. 1. Relationship between the NADH-oxidase activity in the apoplastic washing fluid (AWF) and Mn tissue content. NADH-oxidase activities and Mn tissue content were determined according to Fecht-Christoffers et al. (2006).

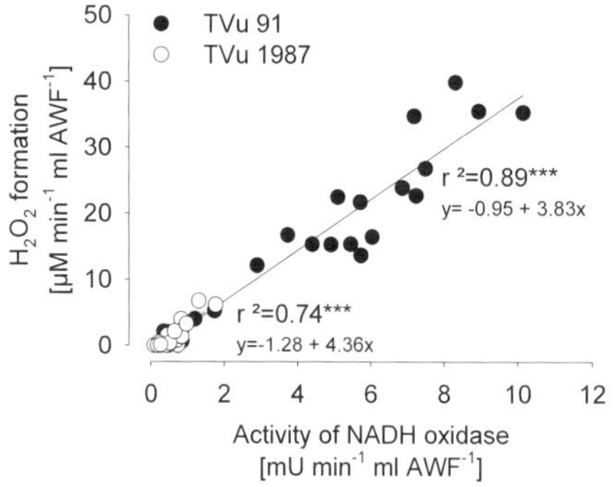

Fig. 2. Relationship between the potential H_2O_2 formation and NADH-oxidase activity in the leaf apoplastic washing fluid (AWF) of two cowpea cultivars. NADH-oxidase activity was detected first, followed by the combination of sample with guaiacol-POD-test mixture.

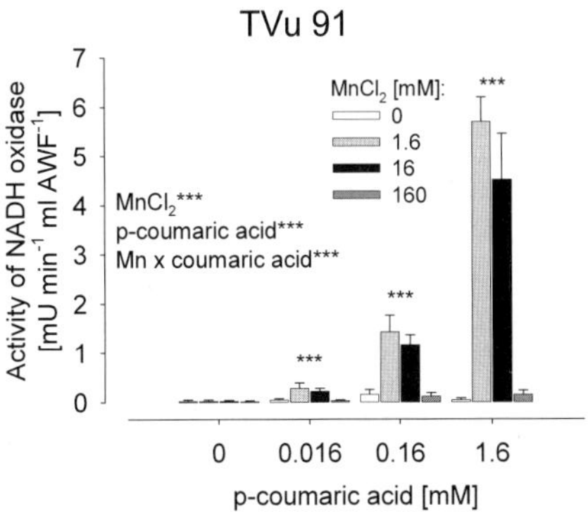

Fig. 3. Effect of the concentration of Mn and p-coumaric acid in vitro *on the NADH-oxidase activity in the leaf AWF of Mn-sensitive cowpea genotzpe TVu 91. For methods see Fecht-Christoffers et al., 2006.*

2.2 Pathogenesis-related-like proteins

The release of PODs in the apoplast was accompanied by the secretion of a range of further proteins into the apoplast (Fig. 4; Fecht-Christoffers et al., 2003b). Mn-induced proteins showed high sequence homologies to wound-induced proteins and pathogenesis-related proteins (PR), e.g. pathogenesis-related proteins class I (PR-1), glucanase (PR-2), chitinase class IV (PR-3), chitinase class III (PR-8) and thaumatin-like proteins (PR-5). Mn-induced expression of extracellular PR-like proteins, e.g. PR-1, PR-2 (glucanase), PR-3 (chitinase) and PR-5 (thaumatin-like) has previously been observed in

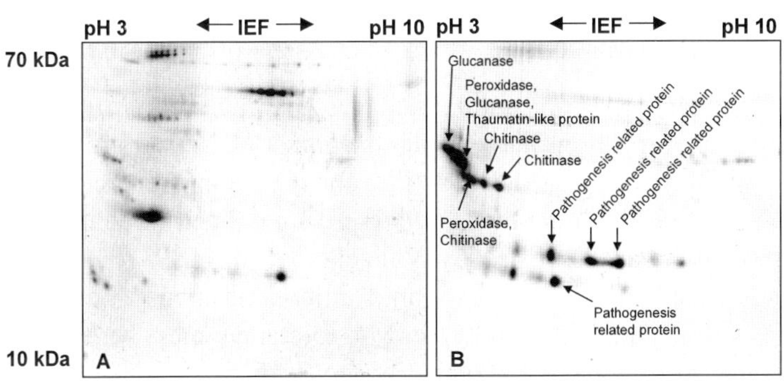

Fig. 4. 2D-resolution of water-soluble proteins from the leaf apoplast of cowpea by IEF/SDS-PAGE. Plants were treated with 50 µM Mn for 5 days (B), while control plants received 0.2 µM Mn continuously (A). Numbers on the top of the gels indicate the pH gradient, and the numbers on the left indicate molecular masses of proteins. Marked spots were identified by nano LC-MS/MS .

leaves of sunflower (Jung et al., 1995). PR-like proteins are not only induced by biotic stresses but also by a wide range of environmental factors and stresses (Didierjean et al., 1996). The expression of these proteins could often be related to the presence of hormones and plant signalling-molecules (Van Loon and Van Strien, 1999).

Different forms of stress applied to plants did not always result in similar transcriptional changes, indicating the presence of multiple pathways of gene regulation in response to abiotic stresses. Due to the complicated crosstalk in the signalling pathways within plants, the identification of primary effects of stresses is difficult. Since enhanced senescence, ethylene production and IAA-oxidation are typical features of advanced stages of Mn toxicity (reviewed by Horst, 1988; El-Jaoual and Cox, 1998) we conclude that induction of PR-like proteins and particular the Mn-induced release of PR-like proteins into the leaf apoplast are general stress responses of plants. A more detailed study of the kinetics of the protein exudation into the apoplast is required in order to assess specific roles of these PR-like proteins in Mn toxicity and Mn tolerance.

3. EFFECT OF MN ON APOPLASTIC METABOLITES

3.1 Ascorbic acid

Ascorbic acid (AA) is an important antioxidant in plants, as are tocopherol, carotinoides and phenols (Polle and Renneberg, 1993; Schmitz and Noga, 2000). Its role as an effective scavenger for oxidative compounds is well known and the regulatory effect of AA on the peroxidase-catalysed oxidation of phenols in the apoplast has been reported (Takahama and Oniki, 1992; Takahama, 1993).

The involvement of an antioxidant system, including an ascorbic acid regeneration-system, in protecting plants against oxidative stress induced by ozone (Mehlhorn et al., 1987, Castillo and Greppin, 1988), heavy metals (Chaoui et al., 1997, Gupta et al., 1999), and pathogen infection, especially in the apoplast, was described by Vanacker et al. (1998). González et al. (1998) suggested that Mn toxicity in common bean may be mediated by oxidative stress and that genotypic Mn tolerance may be related to the maintenance of higher AA levels in the leaf tissue under Mn excess. Takahama (1993) suggested that AA acts as a secondary electron donor by reducing phenoxyradicals, resulting in a complete inhibition of the radical chain-reaction. Since the expression of Mn toxicity has been proposed to be associated with the formation of phenoxy radicals and Mn^{III} in the extracellular space (Horst et al., 1999), the reduction of these highly reactive intermediates by AA in the apoplast could alleviate Mn toxicity.

In cowpea, Mn treatment caused a significant decrease of AA in the apoplast and cytoplasm, whereas at moderate expression of Mn toxicity, the AA level in the tissue was only slightly affected. This was only observed in Mn-sensitive leaf tissues, whereas the levels of AA were not affected in Mn-tolerant leaf tissues (Horst et al., 1999; Fecht-Christoffers et al., 2003a). In TVu 91 (Mn-sensitive), the leaf apoplastic but not the tissue AA concentration decreased as early as after one day of Mn treatment, whereas POD activity and formation of brown spots increased significantly only after at least 2 days of Mn treatment (Fig. 5).

A contributing role of apoplastic AA to Mn tolerance is being further indicated by alleviation of Mn toxicity through the application of AA to leaves of the Mn-sensitive cowpea cv. TVu 91 as expressed by a significant decrease of Mn-induced POD activity and density of brown spots (Fecht-Christoffers and Horst, 2005, Fig. 6). The beneficial effect of the application of antioxidants to plants on plant injury through abiotic stresses has been documented earlier (Noga and Schmitz, 1998).

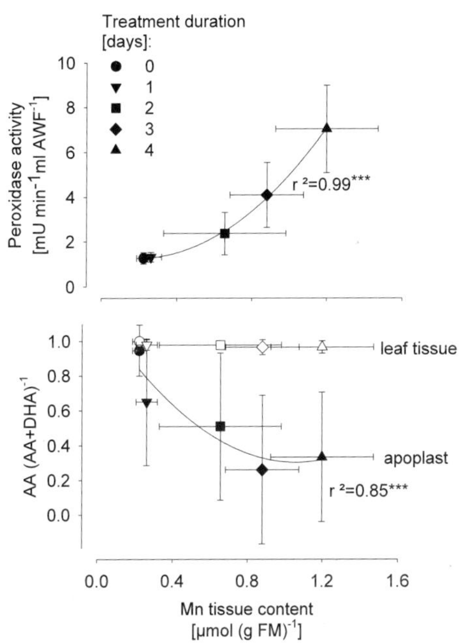

Fig. 5. Relationship between Mn tissue contents, the POD activity in the AWF and the ratio of reduced ascorbic acid (AA) to total ascorbate (AA+DHA) in the apoplast and leaf tissue. Plants of the cowpea cultivar TVu 91 (Mn-sensitive) were precultured hydroponically in a growth chamber. Mn supply was increased to 50 µM for 1, 2, 3 and 4 days , whereas control plants received 0.2 µM Mn, continuously.

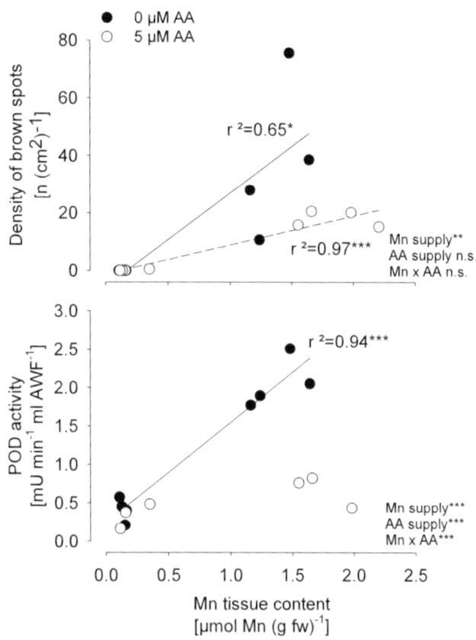

Fig. 6. Effect of ascorbic acid (AA) application on the development of Mn toxicity symptoms (brown spots, activity of apoplastic PODs). Plants of TVu 91 were grown hydroponically. Mn concentrations in nutrient solutions were increased to 50 µM Mn. Application of 5 µM AA or water via the petioles was started simultaneously with the Mn treatment.

3.2 Organic acids

The compartmentation of metals in the leaf tissue has been proposed as a key factor for leaf-tissue Mn tolerance (Wissemeier and Horst, 1990; Wang and Evangelou, 1995) Therefore, sequestration of Mn in the leaf apoplast and accumulation in the vacuole associated with organic anions was investigated (Maier, 1997; Horst et al., 1999). In cowpea, a close relationship between increasing Mn tissue contents and vacuolar Mn concentrations existed which was accompanied by increasing organic anion concentrations. This increase was especially steep for malate and oxalate in the Mn-sensitive cv. TVu 91 which does not support the hypothesis that accumulation in the vacuoles of organic anions confers genotypic Mn tolerance in cowpea. Also, the considerably enhanced Mn tolerance of cowpea plants supplied with Si could not be explained by sequestration of Mn by organic acids in the vacuoles (Maier, 1997; Horst and Maier, 1999; Horst et al., 1999). However, recent molecular biological studies show that in tobacco (Hirschi et al., 2000) and yeast (Schaaf et al., 2002) the expression of the vacuolar transporter AtCaX2 conferes enhanced Mn tolerance indicating that in these organisms the accumulation and sequestration of Mn in the vacuoles is an important component of Mn tolerance. Since vacuolar concentrations of Mn and organic acids did not satisfactorily explain the differences in Mn tissue tolerance in cowpea, the AWF was analysed for Mn and organic anions (Maier, 1997; Horst et al., 1999). Mn treatment induced the release of organic acid anions into the apoplast. But, due to the lack of differences between the cowpea cultivars differing in Mn tolerance and plants supplied with or without silicon, sequestration of Mn by organic acids in the leaf apoplast was not considered an important mechanism conferring Mn tolerance in this plant species (Maier, 1997; Horst et al., 1999; Horst and Maier, 1999).

3.3 Silicon

It is well established, that silicon (Si) greatly improves the Mn tolerance of many plant species including rice (Okuda and Takahashi, 1962; Horiguchi, 1988), barley (Williams and Vlamis, 1957; Horiguchi and Morita, 1987), bean (Horst and Marschner, 1978) cowpea (Horst et al., 1999), and pumpkin (Iwasaki and Matsumura, 1999). The physiological/molecular background of this effect is not yet well understood.

The relationship between the Mn and Si concentrations in the AWF and the severity of Mn toxicity symptoms were investigated in the leaves of the Mn-sensitive cowpea cultivar TVu 91 in solution-culture experiments (Iwasaki et al., 2001a,b). The expression of Mn toxicity symptoms was prevented when 1.44 mM Si was supplied together with 50 μM Mn. However, distinct Mn toxicity symptoms were observed in plants pre-treated with 1.44 mM Si and then exposed to 50 μM Mn without concurrent Si supply. In both Si treatments, plants had lower Mn concentrations in the AWF and higher amounts of adsorbed Mn in the cell walls than the plants treated at 50 μM Mn without Si supply. Inactivation of Mn in the cell walls by Si has been regarded as the main mechanism of Si-induced alleviation of Mn toxicity in cucumber (Rogalla and Römheld, 2002, Wiese et al., this volume, pp. 33–48). However, in cowpea the severity of Mn toxicity symptoms and the Mn-enhanced guaiacol POD activity in the AWF of these plants were not significantly correlated with the Mn concentrations in the AWF but were highly significantly correlated with the Si concentrations in the AWF. These results suggest that in cowpea, Si supply alleviate Mn toxicity symptoms not only by the decrease of apoplastic Mn concentration by an increased adsorption of Mn on the cell walls.

3.4 Phenolic compounds

The biosynthesis of phenolics is activated by a wide range of environmental, hormonal and nutritional factors, e.g. light, stress, growth regulators and the levels of nitrogen, phosphorous and boron and several functions were attributed to phenolic compounds in plant tissues (Rhodes, 1985). Among these, phenolic compounds were proposed to enhance metal tolerance by chelating metal ions (Heim et al., 2001). Furthermore, as already shown (Fig. 3), phenolic compounds have a stimulating effect on PODs (Halliwell, 1978). However, phenols were also proposed to suppress POD-catalysed reactions (Kenten and Mann, 1950).

A stimulation of the phenol metabolism by excess Mn was reported by Engelsma (1972) and Brown et al. (1984). The Mn-induced oxidation of phenols was proposed to be associated with an enhanced release of phenols into the apoplast (Wissemeier, 1988; Wissemeier and Horst, 1992). In cowpea, concentrations of apoplastic phenols increased due to Mn treatment and were positively correlated with Mn tissue contents (Fecht-Christoffers et al., 2006). A release of phenolic compounds into the extracellular space was also associated with the release of acidic PODs, assuming a POD-catalysed phenol oxidation and lignin formation as a response to stress (Castillo, 1986).

Horiguchi (1987) could not detect the oxidation of lignin precursors by PODs, but substrates like caffeic acid and chlorogenic acids were proposed to be oxidized by PODs leading to characteristic brown depositions. A significantly enhanced release of phenolics was only observed at high Mn supplies in cv. TVu 91 (Mn-sensitive), whereas cv. TVu 1987 (Mn-tolerant) was almost unaffected by Mn treatment (Fig. 7). However, at lower Mn

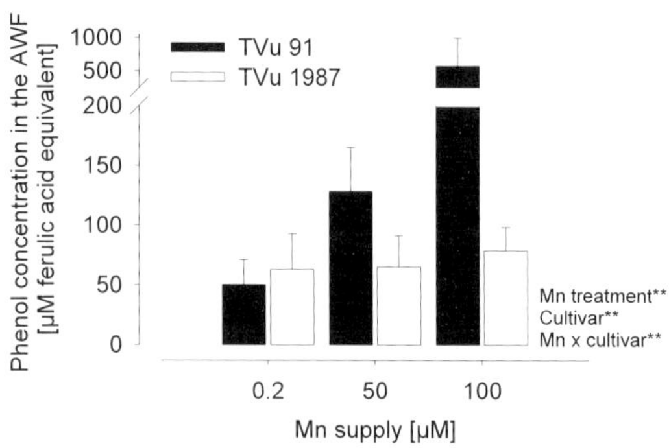

Fig. 7. Effect of Mn supply on phenol concentration in the AWF of two cowpea cultivars differing in Mn tolerance (TVu 91, Mn-sensitive; TVu 1987, Mn- tolerant).

supplies leading to less severe Mn toxicity symptoms in TVu 91 no differences between the genotypes existed (Fecht-Christoffers et al., 2006). Differences in phenol composition of the AWF between the genotypes could be identified suggesting that the phenol composition rather than the phenol concentration of the AWF is important for Mn toxicity and tolerance with regard to chelation and thus detoxification of Mn and/or as co-factor of NAHD-POD-catalysed H_2O_2 formation.

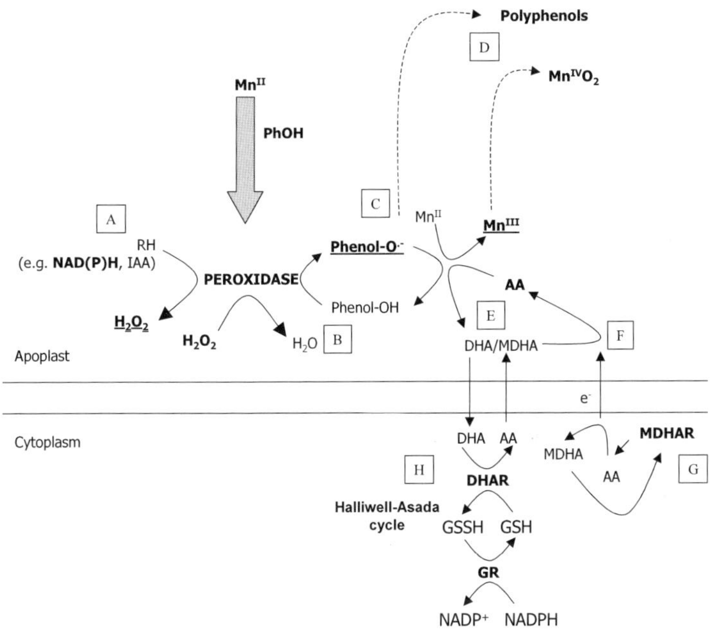

Fig. 8. Proposed reactions in the leaf apoplast of the Mn-sensitive cowpea cultivar TVu 91 caused by excess Mn. Peroxidases are directly stimulated by Mn^{II} and available apoplastic phenols with subsequent formation of H_2O_2 (A). H_2O_2 in the apoplast serves as signal inducing a cascade of mechanism in the apoplast and cytoplasm, leading to callose formation and the release of proteins, organic acids and phenols in the apoplast. Aliquots of H_2O_2 are reduced by peroxidase with subsequent oxidation of phenolic compounds (B). Intermediates of phenol oxidation (phenoxy radicals) oxidize Mn^{II}, causing the formation of Mn^{III} (C). Mn^{III} disproportionates to Mn^{II} and Mn^{IV}. Accumulation of Mn^{IV} and oxidized phenols in the cell wall causes the formation of brown spots (D). Ascorbate in the apoplast is involved in POD-catalysed redox reactions and is oxidized to monodehydroascorbate (MDHA) and dehydroascorbate (DHA) (E). For regeneration of MDHA in the apoplast (F), AA is oxidized to MDHA in the cytoplasm and reduction equivalents are transported into the apoplast. Regeneration of cytoplasmic MDHA occurred by MDHA reductase (MDHAR) (G). DHA is regenerated via the Halliwell-Asada cycle by the enzymes DHA reductase (DHAR) and glutathion reductase (GR) in the cytoplasm (H).

4. CONCLUSIONS

We propose a reaction chain in which apoplastic peroxidases producing and consuming H_2O_2 play a central role in the expression of Mn toxicity (Fig. 8). For the modulation of Mn tolerance three intervention paths appear to be important: (i) the characteristics of the peroxidases, (ii) the presence of cofactors or inhibitors of peroxidases, and (iii) the detoxification of oxygen/phenoxy radicals).

More detailed studies with emphasis on early stages of Mn toxicity and a comparison of Mn-sensitive and Mn-tolerant leaves (genotype, Si nutrition, leaf age) are required.

REFERENCES

Archibal, F.S. and Fridovich, I. (1982). The scavenging of superoxide radical by manganous complexes: In vitro. *Arch. Biochem. Biophys.*, 214, 452–463.

Bartlett, R.J. and James, B. (1980). Studying dried, stored soil samples-some pitfalls. *Soil Sci. Soc. Amer. J.*, 44, 721–724.

Blamey, F.P.C, Joyce, D.C., Edwards, D.G. and Asher, C.J. (1986). Role of trichomes in sunflower tolerance to manganese toxicity. *Plant Soil*, 91, 171–180.

Brown, P.H., Graham, R.D. and Nicholas, J.D. (1984). The effects of manganese and nitrate supply on the levels of phenolics and lignin in young wheat plants. *Plant Soil*, 81, 437–440.

Campa, A. (1991). Biological roles of plant peroxidases: Known and potential functions. In J. Evers, K. Evers and M.B. Grisham (eds.), Peroxidase in chemistry and biology Vol II (pp. 26–49). Boca Raton Fl: CRS Press.

Castillo, F.J., Penel, C. and Greppin, H. (1984). Peroxidase release induced by ozone in *Sedum album* leaves Involvement of Ca^{2+}. *Plant Physiol.*, 74, 846–851.

Castillo, F.J. (1986). Extracellular peroxidases as markers of stress? In H. Greppin, C. Penel, Th. Gaspar (eds.), Molecular and physiological aspects of plant peroxidases. (pp. 419–426). Switzerland: University of Geneva.

Castillo, F.J. and Greppin, H. (1988). Extracellular ascorbic acid and enzyme activities related to ascorbic acid metabolism in *Sedum album* leaves after ozone exposure. *Environ. Exp. Bot.*, 28, 231–238.

Chaoui, A., Mazhoudi, S., Ghorbal, M.H. and El Ferjani, E. (1997). Cadmium and zinc induction of lipid peroxidation and effects on antioxidant enzyme activities in bean (*Phaseolus vulgaris* L.). *Plant Sci.*, 127, 139–147.

de Souza, I.R.P. and MacAdam, J.W. (1998). A transient increase in apoplastic peroxidase activity precedes decrease in elongation rate of B73 maize (*Zea mays*) leaf blades. *Physiol. Plant.*, 104, 556–562.

Didierjean, L., Frendo, P., Nasser, W., Genot, G., Marivet, J. and Burkard, G. (1996). Heavy-metal-responsive genes in maize: Identification and comparison of their expression upon various forms of abiotic stress. *Planta*, 199, 1–8.

Engelsma, G. (1972). A possible role of divalent manganese ions in the photoinduction of phenylalanine ammonialyase. *Plant Physiol.*, 50, 599–602.

Fecht-Christoffers, M.M., Horst, W.J. (2005). Does apoplastic ascorbic acid enhance manganese tolerance of *Vigna unguiculata* and *Phaseolus vulgaris*? *J. Plant Nutr. Soil Sci.*, 168, 590–599.

Fecht-Christoffers, M.M., Maier, P. and Horst, W.J. (2003a). Apoplastic peroxidase and ascorbate are involved in manganese toxicity and tolerance of *Vigna unguiculata*. *Physiol. Plant.*, 117, 237–244.

Fecht-Christoffers, M.M., Braun, H.P., Lemaitre-Guillier, C., VanDorsselaer, A. and Horst, W.J. (2003b). Effect of manganese toxicity on the proteome of the leaf apoplast in cowpea (*Vigna unguiculata*). *Plant Physiol.*, 133, 1935–1946.

Fecht-Christoffers, M.M., Führs, H., Braun, H.P. and Horst, W.J. (2006). The role of H_2O_2-producing and H_2O_2-consuming peroxidases in the leaf apoplast of *Vigna unguiculata* L. in manganese tolerance. *Plant Physiol.*, 140, 1451–1463.

Gaspar, T., Penel, C., Castillo, FJ. and Greppin, H., (1985). A two-step control of basic and acidic peroxidases and its significance for growth and development. *Physiol. Plant.*, 54, 418–423.

González, A., Steffens, K.L. and Lynch, J.P. (1998). Light and excess manganese. Implications for oxidative stress in common bean. *Plant Physiol.*, 118, 493–504.

Gupta, M., Cuypers, A., Vangronsveld, J. and Clijsters, H. (1999). Copper affects the enzyme of the ascorbate-glutathioneglutathionee cycle and its related metabolites in the roots of *Phaseolus vulgaris*. *Plant Physiol.*, 106, 262–267.

Halliwell, B. (1978). Lignin synthesis. The generation of hydrogen peroxide and superoxide by horseradish peroxidase and its stimulation by manganese(II) and phenols. *Planta*, 140, 81–88.

Heim, A., Brunner, I., Frey, B., Fossard, E. and Luster, J. (2001). Root exudation, organic acids and element distribution in roots of Norway spruce seedlings treated with aluminum in hydrophonics. J. *Plant Nutr. Soil Sci.*, 164, 519–526.

Hirschi, K.D., Korenkov, V.D., Wilganowski, N.L. and Wagner, G.J. (2000). Expression of arabidopsis CAX2 in tobacco. Altered metal accumulation and increased manganese tolerance. *Plant Physiol.*, 124, 125–133.

Horemanns, N., Foyer, C.H., Potters, G. and Asard, H. (2000). Ascorbate function and associated transport system in plants. *Plant Physiol. Biochem.*, 38, 531–540.

Horiguchi, T. (1987). Mechanism of manganese toxicity and tolerance of plants II. Deposition f oxidized manganese and plant tissue. *Soil Sci. Plant Nutr.*, 33 (4), 595–606.

Horiguchi, T. (1988). Mechanism of manganese toxicity and tolerance of plants IV. Effect of silicon on alleviation of manganese toxicity in rice plants. *Soil Sci. Plant Nutr.*, 34, 65–73.

Horiguchi, T. and Fukomoto, T. (1987). Mechanism of manganese toxicity and tolerance of plants. III. Effect of excess manganese on respiration rate and peroxidase activity of various plant species. *J. Soil Sci. Plant Nutr.*, 58, 713–716.

Horiguchi, T. and Morita, S. (1987). Mechanism of manganese toxicity and tolerance of plants. VI. Effect of silicon on alleviation of manganese toxicity of barley. *J. Plant Nutr.*, 10, 229–2310.

Horst, W.J. (1982). Quick screening of cowpea genotypes for manganese tolerance during vegetative and reproductive growth. *Z. Pflanzenernähr. Bodenkd.*, 145, 423–435.

Horst, W.J. (1988). The physiology of manganese toxicity, In M.J. Webb, R.O. Nable, R.D. Graham, and R.J. Hannam (eds.), Mangenese in Soil and Plants. (pp. 175–188). Dodrecht/Boston/London: Kluwer Academic Publishers.

Horst, W.J. and Maier, P. (1999). Compartmentation of manganese in the vacuoles and in the apoplast of leaves in relation to genotypic manganese leaf-tissue tolerance in *Vigna unguiculata* (L.) Walp. In G. Gissel-Nielsen and A., Jensen (eds.), Plant Nutrition - Molecular Biology and Genetics. (pp. 223–234).Dordrecht: Kluwer Academic Publishers. ISBN 0-7923-5716-7

Horst, W.J. and Marschner, H. (1978). Symptome von Mangan-Überschuß bei Bohnen. *Z. Pflanzenernähr. Bodenkd.*, 141, 129–142.

Horst, W.J., Fecht, M., Naumann, A., Wissemeier, A.H. and Maier, P. (1999). Physiology of manganese toxicity and tolerance in *Vigna unguiculata* (L.) Walp. *J. Plant Nutr. Soil Sci.*, 162, 263–274.

Iwasaki, K. and Matsumura, A. (1999). Effect of silicon on alleviation of manganese toxicity in pumpkin (*Curcubita moschata Duch* cv. *Shintosa*). *Soil Sci. Plant Nutr.*, 45, 909–920.

Iwasaki, K., Maier, P., Fecht, M. and Horst, W.J. (2001a). Effects of silicon supply on apoplastic manganese concentrations in leaves and their relation to manganese tolerance in cowpea (*Vigna unguiculata* (L.) Walp.). *Plant Soil*, 238, 281–288.

Iwasaki, K., Maier, P., Fecht, M. and Horst, W.J. (2001b). Leaf apoplastic silicon enhances manganese tolerance of cowpea (*Vigna unguiculata*). *J. Plant Physiol.*, 159, 167–173.

Jung, JL., Maurel, S., Fritig, B. and Guenther, H. (1995). Different pathogenesis-related proteins are expressed in sunflower (*Helianthus annuus* L.) in response to physical, chemical and stress factors. *J. Plant Physiol.*, 145 (1-2), 153–160.

Kenten, R.H. and Mann, P.J.G. (1950). The oxidation of manganese by peroxidase systems. *Biochem J.*, 46, 67–73.

Khana, P.L. and Michra, B. (1978). Behavior of manganese in some acid soils in Western Germany in relation to pH and air-drying. *Geoderma.*, 20, 289–297.

Klotz, K.L., Liu, T.T.Y., Liu, L. and Lagrimini, L.M. (1998). Expression of the tobacco anionic peroxidase gene is tissue-specific and developmentally regulated. *Plant Mol. Biol.*, 36, 509–520.

Langheinrich, U., Tischner, R. and Goldbold, D.L. (1992). Influence of high Mn supply on Norway spruce (*Picea abies* (L.) *Karst.*) seedlings in relation to the nitrogen source. *Tree Physiol.*, 10, 259–271.

MacAdam, J.W., Sharp, R.E. and Nelson, C.J. (1992). Peroxidase activity in the leaf elongation zone of tall fescue II. Spatial distribution of apoplastic peroxidase activity in genotypes differing in length of the elongation zone. *Plant Physiol.*, 99, 879–885.

Maier, P. (1997). Bedeutung der Kompartimentierung von Mangan und organischen Säuren für die Mangantoleranz von Cowpea (*Vigna unguiculata* (L.) Walp.). Stuttgart: Verlag Ulrich E. Grauer. ISBN 3-86186-170-4.

Marsh, K.B., Peterson, L.S. and McCown, B.H. (1989). Gradient in Mn accumulation and changes in plant form for potato plants affected by Mn toxicity. *Plant Soil*, 121, 157–163.

Mehlhorn, H., Cottam, D.A., Lucas, P.W. and Wellburn, A.R. (1987). Induction of ascorbate peroxidase and glutathione reductase activities by interactions of mixtures of air pollutions. *Free Radical Res. Commun.*, 3, 1–5.

Morgan, P.W., Joham, H.E. and Amin, J.V. (1966). Effect of manganese toxicity on indoleacetic acid oxidase system of cotton. *Plant Physiol.*, 41, 718–724.

Noctor, G. and Foyer, C.H. (1998). Ascorbate and Glutathione: Keeping active oxygen under control. *Annu. Rev. Plant Physiol. Plant Mol. Biol.*, 49, 249–279.

Noga, G. and Schmitz, M. (1998). Antioxidants in higher plants: Biosynthesis, characteristics, actions and specific functions in stress defence. *Shaker Verlag*, Aachen. ISBN 3-8265-4418-8.

Okuda, A. and Takahashi, E. (1962). Studies on the physiological role of silicon in crop plants (Part 5). Effect of silicon supply on the injuries due to excessive amounts of FeII, MnII, CuII, AsII, CoII of barley and rice plant. *Jpn. J. Sci Soil Manure*, 33, 1–8.

Polle, A. and Renneberg, H. (1993). Significance of antioxidants in plant adaptation to environmental stress. In L. Fowden et al. (ed.) Plant adaption to Environmental Stress, (pp. 264–273). London: Capman and Hall.

Rhodes, M.J.C. (1985). The physiological significance of plant phenolic compounds. *Ann. Proc. Phyt. Soc.*, 25, 99–117.

Rogalla, H. and Römheld, V. (2002). Role of leaf apoplast in silicon-mediated manganese tolerance of *Cucumis sativus* L. *Plant, Cell Environ.*, 25 (4), 549–555.

Ros Barcelo, A. (1997). Lignification in plant cell walls. *Int. Rev. Cyt.*, 176, 87–132.

Schaaf, G., Catoni, E., Fitz, M., Schwacke, R., Schneider, A., von Wirén, N. and Frommer, W.B. (2002). A putative role for vacuolar calcium/manganese proton antiporter AtCAX2 in heavy metal detoxification. *Plant Biol.*, 2, 612–618.

Schmitz, M. and Noga, G. (2000). Ausgewählte pflanzliche Inhaltstoffe und ihr antioxidatives Potential in hydrophilen und lipophilen Extrakten von *Phaseolus vulgaris*-, *Malus domestica*- und *Vitis vinifera*-Blättern. *Gartenbauwissenschaften*, 65, 56–73.

Smirnoff, N. (2000). Ascorbic acid. Metabolism and functions of a multi-facetted molecule. Curr. Opin. *Plant Biol.*, 3, 229–235

Sonneveld, C. and Voogt, S.J. (1975). Studies on the manganese uptake of lettuce on steam-sterilised glasshouse soils. *Plant Soil*, 42, 49–64.

Takahama, U. (1993). Redox state of ascorbic acid in the apoplast of stems of *Kalanchoë daigremontiana Physiol. Plant.*, 89, 791–798.

Takahama, U. and Oniki, T. (1992). Regulation of peroxidase-dependent oxidation of phenolics in the apoplast of spinach leaves by ascorbate. *Plant Cell Physiol.*, 33, 379–387.

Vanacker, H., Harbinson, J., Ruisch, J., Carver, T.L.W. and Foyer, C.H. (1998). Antioxidant defence of the apoplast. *Protoplasma*, 205, 129–140.

Van Loon, L.C. and Van Strien, E.A. (1999). The families of pathogenesis-related proteins, their activities, and comparative analysis of PR-1 type proteins. *Physiol. Mol. Plant. Pathol.*, 55, 85–97.

Wang, J. and Evangelou, V.P. (1995). Metal tolerance aspects of plant cell wall and vacuoles. In M. Pessarakli. (ed.), Handvolume of plant and crop physiology. (pp. 695–717). New York: Marcel Dekker.

Williams, D.E. and Vlamis, J. (1957). The effect of silicon on yield and manganese-54 uptake and distribution in leaves of barley plants grown in culture solutions. *Plant Physiol.*, 32, 404–409.

Wissemeier, A.H. (1988). Beziehung zwischen Mangantoleranz und Oxidation von Mangan in den Blättern von Cowpea-Genotypen (*Vigna unguiculata* (L.) Walp.). Stuttgart: Verlag Ulrich E. Grauer. ISBN 3-9803063-6-4.

Wissemeier, A.H. and Horst, W.J. (1990). Manganese oxidation capacity of cowpea [*Vigna unguiculata* (L.) Walp.] leaves differing in manganese tolerance. *J. Plant Physiol.*, 136, 103–109.

Wissemeier, A.H. and Horst, W.J. (1992). Effect of light intensity on manganese toxicity symptoms and callose formation in cowpea (*Vigna unguiculata* (L.) Walp.). *Plant Soil*, 143, 299–309.

INTERACTION BETWEEN PHLOEM TRANSPORT AND APOPLASTIC SOLUTE CONCENTRATIONS

G. LOHAUS
Universität Göttingen, Biochemie der Pflanze, Germany, glohaus@gwdg.de

Abstract. In addition to several other physiological functions, the apoplast is involved in phloem loading of many plant species. Therefore, apoplastic solute concentrations influence phloem transport of solutes and *vice versa*. For studying this relationship it is necessary to know apoplastic solute concentrations as well as that in the phloem sap. Phloem sap was collected with the laser-aphid-stylet technique. Until now this is the best possibility to collect pure phloem sap from intact plants. The analysis of apoplastic fluids is more difficult because one of the major problems in any approach to study apoplastic ion relations is the method by with apoplastic solution is obtained. Several methods to analyse apoplastic fluids have been developed but all these methods have special advantages and disadvantages. We used the infiltration-centrifugation technique and made a critical evaluation of different parameters which influence the solute concentrations in the apoplast.

Plant growth and development are dependent on translocation of photoassimilates from the sites of synthesis to the sites of consumption or storage. In addition, substantial amounts of solutes transported to the leaves in the xylem are re-translocated in the phloem. This is also true under different stress conditions like salt stress. Approximately 13–36% of the Na^+ and Cl^- imported into the leaves through the xylem were exported by the phloem. It is concluded that phloem transport plays an important role in controlling the solute content of a leaf.

Key words: apoplastic solute concentration, infiltration-centrifugation technique, laser-aphid-stylet technique, phloem, salt stress, translocation, xylem

1. INTRODUCTION

The botanist, E. Münch (1930) coined the term apoplast. In leaves the apoplast constitutes all compartments beyond the plasmalemma, this include

B. Sattelmacher† and W.J. Horst (eds.), The Apoplast of Higher Plants: Compartment of Storage, Transport and Reactions, 323–336.

the interfibrillar space of the cell walls, the xylem and other gas- and water-filled spaces. The aqueous space of the leaf apoplast has an ion, metabolite and protein composition distinct from other cellular compartments. The most obvious functions associated with the apoplast are a) growth regulation, b) response to stress, c) exchange between plant and environment, d) interactions between pathogens and their hosts or non-hosts as well as uptake and transport of water, ions and metabolites (for reviews see Dietz, 1997; Sattelmacher, 2001).

However, in addition to these important functions, the apoplast is involved in apoplastic phloem loading in many species. Therefore, the metabolite and ion concentrations in the apoplastic fluid of leaves are of special interest. It is well known that ions and water are transported through xylem vessels, and sugars and amino acids through sieve tubes. In most plant species these metabolites do not entirely move through the cytoplasm via plasmodesmata from the mesophyll cells to the phloem, but via the apoplast (Giaquinta, 1983). The apoplast is not only involved in primary phloem loading of metabolites but also in the re-translocation of ions from the xylem via the leaf mesophyll cells into the phloem.

2. APOPLAST

2.1 Apoplastic water space and apoplastic gas space

Volumes of apoplastic water and gas space have been determined in leaves of several species (Cosgrove and Cleland, 1983; Speer and Kaiser, 1991; Winter et al., 1993; Lohaus et al., 2000; Lohaus et al., 2001). The volume of apoplastic air space differs widely among plant species. For example, the apoplastic air space of *Vicia faba* (885 µl g^{-1} FW) is 6-fold larger than that of *Zea mays* (151 µl g^{-1} FW; Lohaus et al., 2001) and that of *Hordeum vulgare* (343 µl g^{-1} FW; Winter et al., 1993) is two-fold larger than that of *Z. mays*. These data are in agreement with the results of Dengler et al. (1994) who showed that leaves of C3 plants reveal approximately 30% more apoplastic air space than C4 plants. In the former type of plants apoplastic air space represents up to 50% of the tissue volume. However, woody species such as beech have a smaller apoplastic gas space than herbaceous plants (Luwe and Heber, 1995). By contrast, the relative volume of the apoplastic water space differed only by a factor of three in different plant species (between 40 and 110 µl g^{-1} FW; Lohaus et al., 2001). Apoplastic air and water space of leaves vary depending on leaf age (Husted and Schjoerring, 1995). In *V. faba* apoplastic water and air space increase with leaf age (Lohaus et al., 2001).

2.2 Methods of studying apoplastic solute concentrations

One of the major problems in any approach to study apoplastic ion relations is the method by which apoplastic solution is obtained. Apart from the use of ion-selective microelectrodes, X-ray microanalysis, and ion-sensitive fluorochromes other methods have been developed to isolate apoplastic fluids, i.e. the application of pressure using a Scholander bomb and the infiltration-centrifugation method (Söding, 1941; Klement, 1965). Each of the above methods has specific advantages and disadvantages, e.g. fluorescence dyes have to be introduced via the petiole requiring severing the leaves from the plant which may result in an altered metabolism. However, Mühling and Läuchli (2002a) found similar concentrations of apoplastic Na^+ concentrations using the fluorescence-dye ratio-technique and analysis of apoplastic washing fluids.

2.2.1 Collection of apoplastic fluid

The standard infiltration-centrifugation technique for the collection of apoplastic fluids can be carried out following the instruction by Lohaus et al. (2001). Since infiltration and centrifugation procedure of the leaf material could damage the plasma membrane of the cells, cytosolic contamination of apoplastic fluids was quantified by comparing the activities of marker enzymes (e.g. malate dehydrogenase, hexosephosphate isomerase) in the apoplastic washing fluids with that of leaf extracts (Tetlow and Farrar, 1993; Lohaus et al., 1995; Lohaus et al., 2001). An underestimation of the contamination may occur, as these large enzymes do not diffuse out of the cells as easy as low molecular weight compounds. However, no low molecular weight metabolite is known which is exclusively present in the symplast.

2.2.2 Methodical performance influence the solute concentration in the apoplast

Due to the fact that the infiltration-centrifugation method allows a quick, inexpensive and easy collection of apoplastic fluid it is probably the most widely used in apoplast research. Despite this, it should be used with care because the method is influenced by several factors. One major constraint of the infiltration-centrifugation technique is the necessity to infiltrate a solution into the leaves. It cannot be excluded that the dilution of the apoplastic fluid caused by the infiltration procedure induces changes in the chemical equilibrium existing in the apoplast. For this reason, infiltration media with different osmolalities were tested, i.e. water, low osmolality (similar to the apoplastic solution), and high osmolality (similar to the cytosol of mesophyll cells). Within the osmolality range of 0 to 450 mOsm the measured apoplastic solute concentrations were similar (Lohaus et al., 2001).

Also the pH of the infiltrated medium could change the metabolite concentration in the apoplastic fluid. When leaves of *H. vulgare* and *Spinacia oleracea* were infiltrated with buffers of pH 4.5, 5.5 or 6.5 the concentration of hexoses decreased significantly (Lohaus et al., 2001). One reason for this could be the higher activity of extracellular acid invertase at decreased pH of the infiltration medium (Tetlow and Farrar, 1993; Lohaus et al., 1995). In most apoplastic phloem loaders, sucrose is translocated from the apoplast into the phloem by sucrose/H^+ symporters, and the transport is pH-dependent (Riesmeier et al., 1992). Higher apoplastic sucrose concentrations can also be caused by changes in the activities of such transporters due to the pH of the infiltration medium.

It cannot be excluded that damage induced by the centrifugation process leads to contamination of the apoplastic washing fluid with cytoplasmic or vacuolar components. Lohaus et al. (2001) showed increasing activities of the marker enzyme hexosephosphate isomerase in the apoplastic washing fluid of *V. faba* if centrifugal forces exceed 100 g and increasing cation concentrations in the apoplastic washing fluid if centrifugal forces exceeded 500 g. This finding, however, is contradictory to data from Husted and Schjoerring (1995) who showed that neither malate dehydrogenase activities nor NH_4^+ concentrations in the apoplastic fluids of young *Brassica napus* leaves were affected by the centrifugal force up to 12000 g.

2.3 Influence of the plant species and environmental conditions on solute concentration in the apoplast

2.3.1 Solute concentrations in the apoplast of different plant species

The knowledge of solute concentrations in the leaf apoplast is important for understanding several plant growth and transport processes such as phloem loading. Composition of the apoplastic fluid is highly variable and depends e.g. on plant species, age, time of day, nutritional status. The pH of the apoplast is usually lower than in the cytoplasm and ranges from 4.5 to 6.5 in various species (for review see Sakurai, 1998).

The total solute concentration in the apoplast was low and total contents of cations and anions, respectively, were between 10 and 20 mM (Lohaus et al., 2001). The lowest total metabolite and ion concentrations were found in *Z. mays*, and the highest in *H. vulgare* (Lohaus et al., 2001). The sucrose concentration was about 1–2 mM in several plant species, whereas the concentration of hexoses differed significantly between the species (Lohaus et al., 2001). The apoplastic sucrose represented 0.6% to 0.9% of the leaf sucrose content of the analyzed plants whereas hexoses in the apoplast represented 17%, 3%, 11%, and 9% of the total leaf hexose content in *V. faba*, *S. oleracea*, *H. vulgare*, and *Z. mays*, respectively (Lohaus et al., 2001).

The highest concentration of amino acids in the apoplast were found in *V. faba* (about 10 mM), the lowest in *Z. mays* (about 2 mM; Lohaus et al., 2001). Based on total amino acid concentration, the relative amounts of aspartate and glutamate differed between 9% in *V. faba* and 66% in *S. oleracea*. Amino acids in the apoplast made up about 1–2% of the total leaf content in *V. faba*, *H. vulgare*, and *Z. mays*, respectively and about 4% in *S. oleracea*.

The main anions in the apoplastic solution of the analyzed plant species were nitrate and chloride (Lohaus et al., 2001). In *S. oleracea*, oxalate was also an important anion (2.2 mM), while apoplastic fluid from *H. vulgare* contained high amounts of malate (3.5 mM). Lactate was also found in the apoplastic fluid of different plants (0.8–1.3 mM).

The main cation in the apoplastic solution was K^+ (7–16 mM) representing about 75% of the total apoplastic cation concentration (Lohaus et al., 2001). The apoplastic K^+ represented between 0.5 and 2.5% of the total K^+ content in leaves.

2.3.2 Influence of the sampling time

Diurnal variations in apoplastic solute concentrations are the results of changes of metabolic activity, caused, for example by processes involved in light-dark transition. The solute concentrations in the apoplastic fluid were not constant during the light and dark period (Lohaus et al., 2001). The concentrations of most apoplastic metabolites (sucrose, hexoses, malate, amino acids) were higher at the end of the light period than at the end of the dark period and correspond to the metabolite concentrations in the cytosol of mesophyll cells.

2.3.3 Influence of stress conditions, e.g. salt stress

It was suggested by Oertli (1968) that accumulation of salt in the leaf apoplast may be one factor for the salt toxicity syndrome. However, the increase of apoplastic solute concentration may be due to damage of membrane integrity rather than a primary response to salinity (Niu et al., 1995).

Solute concentrations in the apoplast of fully expanded leaves are mainly governed by import via the xylem, uptake into the symplast, as well as export through re-translocation into the phloem and leaching of solutes through rain. Thus, an increased salinity stress in roots of plants led to higher Na^+ and Cl^- concentrations in the shoot and in the leaf apoplast. *Z. mays* and *Ricinus communis* are generally considered to be salt-sensitive, whereas *H. vulgare* is more salt-tolerant. The influence of the salt treatment on Na^+ and Cl^- concentrations in the leaf apoplast of maize and *R. communis* was less pronounced than expected (Table 1) based on previous data suggesting an inverse relationship between apoplastic salt concentration and salt tolerance (Speer and Kaiser, 1991). In agreement with Mühling and Läuchli (2002b), we found that the Na^+ concentration in the apolastic fluid of maize leaves increased just to 4 mM and that of Cl^- to 5 mM with 100 mM external NaCl (Lohaus et al., 2000). In *H. vulgare* and *R. communis* we found also low apoplastic Na^+ concentrations.

Data by Lohaus et al. (2000) for *Z. mays*, *H. vulgare*, *R. communis*, and cotton (Mühling and Läuchli, 2002b) do not support the hypothesis by Oertli (1968).

2.3.4 Spatial concentration gradients in the leaf apoplast

If the solute concentrations in several parts of the apoplast are different is still a matter of debate. At least for C4 plant species bundle-sheath cells are connected to mesophyll cells via numerous plasmodesmata but their apoplastic compartments are separated by suberin lamellae (Evert et al., 1977). Thus, ionic conditions in the two apoplastic compartments may differ significantly. Since a similar situation has been described for wheat (Dietz, 1997) it may be anticipated that a separation of the apoplast of leaves into

Table 1. Apoplastic metabolite and ion concentrations (mM) of Ricinus communis, Hordeum vulgare, and Zea mays leaves. The infiltration medium was deionised water. Plants were treated with 100 mM external NaCl for 8 days. NaCl was increased stepwise, every 2 day 25 mM up to 100 mM. n = 8.

	Control			100 mM NaCl		
	Z. mays	*R. communis*	*H. vulgare*	*Z. mays*	*R. communis*	*H. vulgare*
K^+	7.7	13.7	16.0	6.7	16.1	17.0
Na^+	0.8	0.6	2.6	3.8	8.9	16.2
Cl^-	2.3	1.1	3.8	4.6	8.1	10.1
NO_3^-	1.6	0.3	6.5	1.1	0.3	7.1
Σ Amino acids	1.5	1.9	2.6	3.8	10.8	4.1
Sucrose	1.6	1.3	2.7	3.9	2.6	3.0

smaller compartments by diffusion barriers is a more common phenomenon in the plant kingdom. The movement of solutes in the apoplast of leaves has been intensely studied using tracer dyes and was reviewed by Canny (1990), but the movement of such dyes in leaves does not necessarily relate to the movement of solutes and water. There exist only few methods to analyze such differences in apoplast composition on a microscale. Mühling and Läuchli (2002a) used fluorescence dyes and have shown a higher apoplastic Na^+ concentration around epidermal and stomatal cells (20 mM) in comparison to leaf veins (10 mM) in cotton leaves treated with NaCl.

A disadvantage of the infiltration-centrifugation method is that it leads to an average apoplastic concentration of the leaf used. But it is possible to treat different parts of an intact leaf with different conditions and to collect apoplastic fluid from those treated leaf parts. Table 2 shows the apoplastic solute concentrations in spinach leaves in which two quarters of the leaf were shaded for 6 hours and the other two quarters were illuminated. Whereas the overall sucrose concentration in the leaf extracts from the shaded parts decreased by a factor of three it decreased only slightly in the apoplastic fluid. The hexose concentration generally low in spinach decreased slightly in shaded leaf extracts as well as in apoplastic fluids. This experiment shows that the spatial distribution gradient in the leaf apoplast is not well pronounced.

2.3.5 Ions homeostasis in the apoplast

Mimura et al. (1992) have shown that the phosphate concentration in the leaf apoplast is regulated by homeostatic mechanisms. It transiently decreased when transpiring excised leaves were placed into pure water, but returned to the initial concentration after some time. Nielson and Schjoerring (1998), demonstrated that NH_4^+ in the leaf apoplast is also highly regulated. Only these few examples exist to demonstrate an ion homeostasis in the apoplast although it has been suggested several times (Dietz, 1997).

Table 2. Metabolite concentrations in leaf extracts and in apoplastic fluids of Spinacea oleracea. Parts of the leaves were shaded for 6 hours whereas the other parts were illuminated.

	Shaded parts		Illuminated parts	
	Leaf extracts (μmol g^{-1} FW)	Apoplastic fluid (mM)	Leaf extracts (μmol g^{-1} FW)	Apoplastic fluid (mM)
Sucrose	2.6	0.9	9.5	1.2
Hexoses	0.3	0.6	0.4	0.9
Σ Amino acids	8.5	3.1	9.3	3.5

To analyse the ion and metabolite homeostasis in the apoplast we carried out the following experiment: The same leaves of *V. faba* were repeatedly (four times) infiltrated and centrifugated. After each centrifugation (4 min at 4°C, 80 g) about 750 µl apoplastic washing fluid (g^{-1} FW^{-1}) were obtained, which corresponds to 85% of the maximal value (whole apoplastic air volume). No decrease in apoplastic solute concentrations was observed during the first three steps (Table 3). The ion and metabolite concentrations in the apoplastic washing fluid for the first to the third infiltration-centrifugation step were nearly the same. This might be caused by a new solute equilibrium between apoplast and symplast immediately after or during centrifugation.

The adjustment of the equilibrium occurred within two minutes because an infiltration time of 30 min in comparison to two minutes did not affect the adjustment of the apoplastic solute concentrations (Lohaus et al., 2001). The concentrations of inorganic ions, malate and hexoses decrease after the fourth infiltration centrifugation step whereas the sucrose and amino acid concentrations were also similar or higher in comparison to the concentration after the first infiltration-centrifugation step. In comparison with the solute content per gram leaf extract each infiltration-centrifugation step withdraw from the *V. faba* leaves about 2.5% K^+, 0.3% Ca^{2+}, 4.8% NO_3^-, 7.3% Cl^-, 1.6% amino acids, 2.3% malate, 20% hexoses and 1.8% sucrose. With the exception of hexoses and to a smaller extent NO_3^- the solute loss from the leaves was low. The high solute loss through the infiltration-centrifugation steps led to lower solute concentrations in the whole leaf. On the other hand the lower solute concentrations in the whole leaf (mainly symplast) influenced the apoplastic concentrations.

Table 3. The same leaves of Vicia faba were fourfold infiltrated and centrifugated. Each step needs about 8 min. The solute concentration in the apoplast after the first infiltration-centrifugation step is defined as 100%. The values are given in percent of the concentration after the first infiltration and centrifugation step.

	Infiltration-centrifugation step				
	1.		2.	3.	4.
	mM	%	%	%	%
K^+	12.4 ± 0.5	(100)	113	105	107
Ca^{2+}	0.4 ± 0.1	(100)	92	80	79
NO_3^-	2.4 ± 1.0	(100)	92	92	71
Cl^-	10.7 ± 1.6	(100)	91	96	84
Amino acids	7.1 ± 1.6	(100)	109	127	123
Malate	1.1 ± 0.3	(100)	92	103	61
Hexoses	1.4 ± 0.6	(100)	90	59	38
Sucrose	1.5 ± 0.5	(100)	98	106	128

3. PHLOEM

Most crop plant species are apoplastic phloem loaders. In apoplastic assimilate export at least two membranes have to be passed by the ions to reach the sieve-element companion-cell complex: from the cytosol of bundle-sheath cells or minor vein parenchyma cells to the apoplastic space and subsequently from the apoplast to the sieve-element companion-cell complex. Several metabolite translocators have been localized in phloem cells that support such transport processes. Therefore, it is obvious that the solute concentration in the apoplast influences the solute concentration in the phloem. High apoplastic solute concentrations under stress conditions, e.g. salt stress, can influence the solute transport of these solutes into the phloem.

Soil salinity poses serious limitations to agriculture in many areas around the world and there are large differences in responses to salinity between plant species. Phloem retranslocation (Lessani and Marschner, 1978; Munns et al., 1986; Gouia et al., 1994) may contribute to avoid toxic ion concentrations in the leaf symplast and apoplast. It is still controversial whether high rates of ion retranslocation and low salt levels in the leaves indicate salt-tolerant or salt-sensitive plants (Lessani and Marschner, 1978; Läuchli and Wieneke, 1979; Cramer et al., 1994). On the one hand high ion retranslocation-activity prevents salt accumulation in fully expanded leaves, but on the other hand it represents a threat to younger leaves which are phloem sinks.

3.1 Apoplastic solute concentrations in comparison to those in the xylem sap and in the phloem sap

The xylem content is part of the apoplast but the solute concentrations in the xylem sap differ from those in the apoplastic washing fluid (Lohaus et al., 2000). The nitrate concentration was much higher in the apoplastic fluid than in the xylem sap (Table 4) and the sucrose concentration was quite low in the xylem sap (Table 4). Under salt-stress conditions the apoplastic K^+, NO_3^- and amino acid concentration increased much more than the corresponding concentrations in the xylem sap. The sucrose concentration shows again the opposite result.

One way of studying phloem transport is the direct measurement of metabolite and ion concentrations in sieve-tube sap obtained via the aphid-stylet technique (Barlow and Mc Cully, 1972; Lohaus et al., 1995). In this case the stylet of feeding aphids is severed with a laser beam. Using this method pure phloem sap can be obtained from intact plants. Based on data

Table 4. Metabolite and ion concentrations (mM) in the apoplastic fluid, in the xylem sap and in the phloem sap of leaves from Zea mays. Apoplastic fluids were collected with deionised water, xylem sap with a Passioura-typ pressure bomb (Passioura, 1980) and phloem sap with the laser aphid-stylet technique (Lohaus et al., 2000). Samples were taken after 9 hours of illumination. Plants were treated with 100 mM external NaCl for 8 days. NaCl was increased stepwise, every second day 25 mM up to 100 mM. Mean values of n = 8–10 independent measurements are given. n.d.=not detectable (below 0.1 mM).

	Control			100 mM NaCl		
	Apoplast	Xylem	Phloem	Apoplast	Xylem	Phloem
K^+	7.7	6.4	57	6.7	12.2	69
Na^+	1.0	0.2	0.7	3.8	2.2	11.9
Cl^-	2.3	1.0	8.7	4.6	6.3	31.6
NO_3^-	1.3	4.4	n.d.	1.1	7.7	n.d.
Σ Amino acids	1.3	2.0	59	3.8	5.8	74
Sucrose	2.6	0.2	822	4.8	0.1	879

from the literature, concentrations of Na^+ and Cl^- in the phloem sap of salt-treated plants can differ between 10 and 100 mM dependent on plant species (Downing, 1980 (*Aster tripolium*); Munns et al., 1986 (barley); Jeschke et al., 1987 (lupin)). Higher concentrations of Na^+ were observed in the phloem sap of salt-sensitive plants and lower concentrations in salt-tolerant plants. Jeschke et al. (1987) propose that a high Na^+ concentration in the phloem sap contributes substantially to the salt sensitivity of the plant. Unfortunately, in the available investigations phloem sap was obtained using different methods (manual incision or aphid stylet technique), and it cannot be ruled out that the different collection methods affect the concentrations measured. In experiments with salt-sensitive maize plants, Na^+ and Cl^- concentrations in the phloem never exceeded 14 and 33 mM, respectively, when 100 mM external NaCl was added (Tab. 4, Lohaus et al., 2000).

The concentrations of most other solutes increased slightly. Nitrogen was transported in the phloem sap exclusively in the form of amino-N and the nitrogen concentration increased 1.6-fold with salt treatment. The concentrations of NO_3^-, Ca^{2+} and hexoses in the phloem sap were below the detection limit. The ratio between apoplastic concentration and phloem concentration increased in the same order as the ratio between leaf and phloem concentrations, K^+ 0.10, Cl^- 0.14, and Na^+ 0.32 with 100 mM external NaCl (Lohaus et al., 2000). The lower ratio for K^+ than for Na^+ may be taken as an indication for a selectivity of K^+ over Na^+ by the systems transporting ions into the phloem. For salt-treated plants the ratio between leaf-tissue concentration and phloem concentration was 1.6 for K^+, 1.7 for Cl^- and 3.8 for Na^+ (Lohaus et al., 2000), which reflects the cellular compartmentation with most Na^+ immobilised in the vacuole.

3.2 Nutrient cycling: Relationship between xylem and phloem transport

Solute cycling between root and shoot has been established for several ions (Jeschke et al., 1987). However, the cycling must be under careful control of regulatory mechanisms since ions and growth substances function as essential nutrients and signals for development (Dietz, 1997). Therefore, the solutes must first spread in the leaves and may only secondarily be recycled by export in the phloem depending on the selective demand.

The rate of phloem retranslocation from maize leaves decreased in the order K^+, Cl^-, and Na^+ (Fig. 1), under 100 mM external NaCl. In barley Greenway et al. (1965) showed phloem export rates of K^+, Cl^-, and Na^+ to be in the range of 0.58, 0.24 and 0.08 μmol g^{-1} FW h^{-1}. Our data of maize for the same ions are 1.3, 0.6, and 0.22 μmol g^{-1} FW h^{-1}, respectively (Fig. 1). The higher translocation rates of ions found in maize may be due to the higher carbon translocation rate observed with C4 plants (about 200 μmol C g^{-1} FW h^{-1} in maize, (Lohaus et al., 2000) as opposed to C3 plants (about 60–70 μmol C g^{-1} FW h^{-1} in barley and spinach, Riens et al., 1994). In *Arabidopsis thaliana* the K^+ channel *AKT2/3* is expressed in both phloem and xylem tissues and is capable of mediating both influx and efflux of K^+ (Lacombe et al., 2000). A similar K^+ channel was also found in maize (Bauer

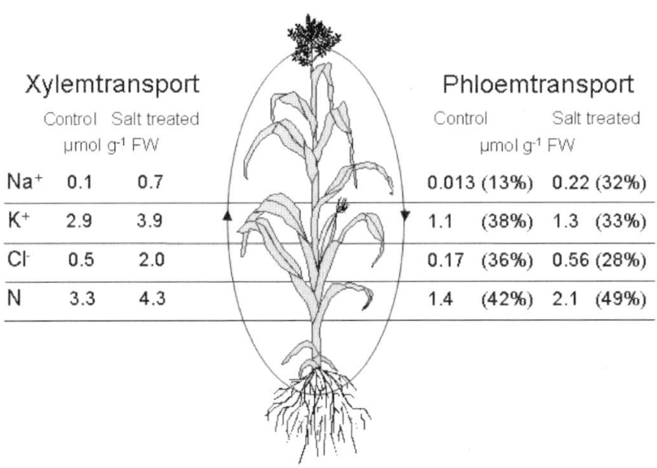

Fig. 1. Phloem export in relation to xylem import of solutes in maize leaves under control conditions and salt treatment (100 mM NaCl for 8 days).

et al., 2000). This could be one of the transport systems involved in exchange between phloem and xylem. It is concluded that phloem re-translocation of ions contribute significantly to the leaf solute balance, i.e. it prevents ion accumulation in apoplast and symplast. In *Z. mays* this is rather due to the high phloem transport rates than to the high ion concentrations in the phloem sap.

The ratio of phloem export of nitrogen (about 40%) and potassium (about 35%) to xylem import remained unaffected by salt treatment since the concentrations of amino-N and K^+ increased in phloem sap and also in the xylem sap (Fig. 1). Under controlled conditions, between 13% (Na^+) and 36% (Cl^-) of ions imported via the xylem are exported by the phloem again (Fig. 1). Similar results have been obtained in other studies (Munns et al., 1986; Gouia et al., 1994; Marschner et al., 1997). Upon exposure to 100 mM external NaCl the xylem import of Na^+ and Cl^- increased strongly (Fig. 1). Under the same conditions the phloem export of Na^+ was increased more strongly than xylem import, 13% of the imported Na^+ was re-translocated in control plants and 32% in salt treated plants. For Cl^- the xylem import to phloem export ratio was similar under both conditions. Na^+ phloem export is plant species-dependent. Approximately 10% of xylem-imported Na^+ is exported through the phloem in barley (Munns et al., 1986), 37% in cotton (Gouia et al., 1994), 40–70% in white lupin (Jeschke et al., 1987), 77% in bean (Gouia et al., 1994). It seems that high Na^+ retranslocation from the leaves is linked to high salt sensitivity of the plants.

4. CONCLUSION

The leaf apoplast has an ion and metabolite composition distinct from mesophyll cells and phloem sap. The determination of ion and metabolite concentrations in different plant-cell fluids like apoplastic fluid or phloem sap is of interest in several fields of plant biology, i.e. transport studies or stress response. The infiltration-centrifugation technique and the laser aphid-stylet technique allow collection of apoplastic fluid and phloem sap, respectively, but both methods have several advantages and disadvantages. Therefore, methods for the collection of pure fluids from different plant compartments should be improved.

Despite a high number of studies on salinity tolerance of plants, neither the physiologic processes at which salt stress damages plants nor the adaptive strategies of salt tolerance are fully understood. This is partly because physiological responses to salinity are very complex. Apoplastic ion and metabolite concentrations change upon application of salinity and this further influence ion accumulation in and transport from leaves. In the plants species analyzed the primary low apoplastic ion accumulation of NaCl was not different between salt-sensitive and salt-tolerant species, and the secondary strong accumulation in salt-sensitive plant species was not the cause of salt damage, but its consequence. Substantial amounts of solutes transported to the leaves in the xylem are retranslocated in the phloem. This is also true under different stress conditions like salt stress. Approximately 13–36% of the Na^+ and Cl^- imported into the leaves through the xylem were exported by the phloem. Therefore, retranslocation via the phloem is one strategy to prevent ion accumulation in mature leaves of plants grown at high concentration of salt. Nevertheless the problem of the plant continues to exist because Na^+ and Cl^- are further inside the plant only in other organs.

NOTES

The author is grateful to Prof. H.W. Heldt for stimulating discussions.

REFERENCES

Barlow, C.A. and Mc Cully, M.E. (1972) The ruby laser as an instrument for cutting the stylets of feeding aphids. *Can. J. Zool.*, 50, 1497–1499.

Bauer, C.S., Hoth, S., Haga, K., Philippar, K., Aoki, N. and Hedrich, R. (2000) Differential expression and regulation of K^+ channels in the maize coleoptile: molecular and biophysical analysis of the cells isolated from cortex and vasculature. *Plant J.*, 24, 139–145.

Canny, M.J. (1990) What becomes of the transpiration stream? *New Phytol.*, 114, 341–368.

Cosgrove, D.J. and Cleland, R.E. (1983) Solutes in the free space of growing stem tissues. *Plant Physiol.*, 72, 326–331.

Cramer, G.R., Alberico, G.J. and Schmidt, C. (1994) Salt tolerance is not associated with sodium accumulation of two maize hybrids. *Aust. J. Plant Physiol.*, 21, 675–692.

Dengler, N.G., Dengler, R.E., Donnelly, P.M. and Hattersley, P.W. (1994) Quantitative leaf anatomy of C3 and C4 grasses (*Poaceae*): Bundle sheath and mesophyll surface area relationships. *Ann. Bot.*, 73, 241–255.

Dietz, K.J. (1997) Functions and responses of the leaf apoplast under stress. *Prog. Bot.*, 58, 221–254.

Downing, N. (1980) Measurements of the osmotic concentrations of stylet sap, haemolymph and honeydew from aphid under osmotic stress. *J. Exp. Bot.*, 33, 557–573.

Evert, R.F., Eschrich, W. and Heyser, W. (1977) Distribution and structure of the plasmodesmata in mesophyll and bundle-sheath cells of *Zea mays* L. *Planta*, 136, 77–89.

Giaquinta, R.T. (1983) Phloem loading of sucrose. *Ann. Rev. Plant Physiol.*, 34, 347–387.

Gouia, H., Ghorbal, M.H. and Touraine, B. (1994) Effects of NaCl on flows of N and mineral ions and on NO_3^- reduction rate within whole plants of salt-sensitive bean and salt sensitive cotton. *Plant Physiol.*, 105, 1409–1418.

Greenway, H., Gunn, A., Pitman, M. and Thomas, D.A. (1965) Plant response to saline substrates. VI. Chloride, sodium, and potassium uptake and distribution within the plant during ontogenesis of Hordeum vulgare. *Aust. J. Biol. Sci.*, 18, 525–540.

Husted, S. and Schjoerring, J.K. (1995) Apoplastic pH and ammonium concentration in leaves of *Brassica napus* L. *Plant Physiol.*, 109, 1453–1460.

Jeschke, W.D., Pate, J.S. and Atkins, C.A. (1987) Partitioning of K^+, Na^+, Mg^{2+}, and Ca^{2+} through xylem and phloem component organs and nodulated white lupin under mild salinity. *J. Plant Physiol.*, 128, 77–93.

Klement, Z. (1965) Method of obtaining fluid from the intercellular spaces of foliage and the fluid`s merit as substrate for phytobacterial pathogens. *Phytopathology*, 55, 1033–1034.

Lacombe, B., Pilot, G., Michard, E., Gaymard, F., Sentenac, H. and Thibaud, J.B. (2000) A *skaker*-like K^+ channel with waek rectification is expressed in both source and sink phloem tissues of *Arabidopsis*. *Plant Cell*, 12, 837–851.

Läuchli, A. and Wieneke, J. (1979) Studies on growth and distribution of Na^+, K^+, and Cl^- in soybean varieties differing in salt tolerance. *Z. Pflanzenernähr. Bodenk.*, 142, 3–13.

Lessani, H. and Marschner, H. (1978) Relation between salt tolerance and long distance transport of sodium and chloride in various crop species. *Aust. J. Plant Physiol.*, 5, 27–37.

Lohaus, G., Hussmann, M., Pennewiss, K., Schneider, H., Zhu, J.J. and Sattelmacher, B. (2000) Solute balance of a maize (*Zea mays* L.) source leaf as affected by salt treatment with special emphasis on phloem retranslocation and ion leaching. *J. Exp. Bot.*, 51, 1721–1732.

Lohaus, G., Pennewiss, K., Sattelmacher, B., Hussmann M. and Muehling, K.H. (2001) Is the infiltration-centrifugation technique appropriate for the isolation of apoplastic fluid? A critical evaluation with different plant species. *Physiologia Plant.*, 111, 457–465.

Lohaus, G., Winter, H., Riens, B. and Heldt, H.W. (1995) Further studies of the phloem loading process in leaves of barley and spinach. The comparison of metabolite concentrations in the apoplastic compartment with those in the cytosolic compartment and in the sieve tubes. *Botanica Acta*, 108, 270–275.

Luwe, M. and Heber, U. (1995) Ozone detoxification in the apoplasm and symplasm of spinach, broad bean and beech leaves at ambient and elevated concentrations of ozone in air. *Planta*, 197, 448–455.

Marschner, H., Kirkby, E.A. and Engels, C. (1997) Importance of cycling and recycling of mineral nutrients within plants for growth and development. *Botanica Acta*, 110, 265–273.

Mimura, T., Dietz, K.-J., Kaiser, W., Schramm, M.J., Kaiser, G and Heber, U. (1992) Phosphate transport across biomembranes and cytosolic phosphate homeostasis in barley leaves. *Planta*, 180, 139–146.

Mühling, K.H. and Läuchli, A. (2002a) Determination of apoplastic Na^+ in intact leaves of cotton by *in vivo* fluorescence ratio-imaging. *Funct. Plant Biol.*, 29, 1491–1499.

Mühling, K.H. and Läuchli, A. (2002b) Effect of salt stress on growth and cation compartmentation in leaves of two plant species differing in salt tolerance. *J. Plant Physiol.*, 159, 137–146.

Münch, E. (1930) Die Stoffbewegungen in der Pflanze. *Fischer* Verlag, Jena, pp. 234.

Munns, R., Fisher, D.B. and Tonnet, M.L. (1986) Na^+ and Cl^- transport in the phloem from leaves of NaCl-treated barley. *Aust. J. Plant Physiol.*, 13, 757–66.

Nielson, K.H. and Schjoerring, J.K. (1998) regulation of apoplastic NH_4^+ concentration in leaves of oilseed rape. *Plant Physiol.*, 118, 1361–1368.

Niu, X., Bressan, R.A., Hasegawa, P.M., Pardo, J.M. and Niu, X.M. (1995) Ion homeostasis in NaCl stress environments. *Plant Physiol.*, 109, 735–742.

Oertli, J.J. (1968) Extracellular salt accumulation, a possible mechanism of salt injury in plants. *Agrochimica*, 12, 461–469.

Passioura, J.B. (1980) The transport of water from soil to shoot in wheat seedlings. *J. Exp. Bot.*, 31, 333–345.

Riens, B., Lohaus, G., Winter, H. and Heldt, H.W. (1994) Production and diurnal utilization of assimilates in leaves of spinach (*Spinacia oleracea* L.) and barley (*Hordeum vulgare* L.). *Planta*, 192, 497–501.

Riesmeier, J.W., Willmitzer, L. and Frommer, W.B. (1992) Isolation and characterization of a sucrose carrier cDNA from spinach by functional expression in yeast. *EMBO J.*, 11, 4705–4713.

Sakurai, N. (1998) Dynamic function and regulation of apoplast in the plant body. *J. Plant Res.*, 111, 133–148.

Sattelmacher, B. (2001) The apoplast and its significance for plant mineral nutrient. *New Phytogol.*, 149, 167–192.

Söding, H. (1941) Über den Nachweis einer aus dem Interzellularraum von Echeveria-Blättern auswaschbaren bakteriziden Substanz. *Bericht Deutsche Botanische Gesellschaft*, 59, 458–466.

Speer, M. and Kaiser, W.M. (1991) Ion relations of symplastic and apoplastic space in leaves from *Spinacia oleracea* L. and *Pisum sativum* L. under salinity. *Plant Physiol.*, 97, 990–997.

Tetlow, I.J. and Farrar, J.F. (1993) Apoplastic sugar concentration and pH in barley leaves infected with brown rust. *J. Exp. Bot.*, 44, 929–936.

Winter, H., Robinson, D.G. and Heldt, H.W. (1993) Subcellular volumes and metabolite concentrations in barley leaves. *Planta*, 191, 180–190.

INVESTIGATIONS OF THE MECHANISMS OF LONG-DISTANCE TRANSPORT AND ION DISTRIBUTION IN THE LEAF APOPLAST OF *VICIA FABA* L.

W. MERBACH[1], D. LÜTTSCHWAGER[2] and K. HÜVE[3]
[1]Institute of Soil Science and Plant Nutrition, Martin-Luther-University Halle-Wittenberg, Germany, merbach@landw.uni-halle.de; [2]Institute for Landscape Matter Dynamics, Leibniz Centre for Agricultural Landscape Research, Germany; [3]Department of Plant Physiology, Institute for Molecular and Cell Biology, University of Tartu, Estonia.

Abstract. The three important, but in their redistribution differently behaving plant nutrients phosphorus, iron and calcium were compared in their mobility within the xylem and the leaf apoplast. It was observed that xylem transport facilities were found to be large for all three nutrients and able to cope with bottleneck situations. The apoplast may serve various functions as site for reloading storage or even as deposit for excesses. While iron and in some cases even calcium may be precipitated, phosphorus is quickly taken up into the symplast and probably loaded to the phloem for redistribution.

Key words: calcium, distribution, iron, phloem, phosphorus, transfer, xylem

1. INTRODUCTION

The apoplastic transport and distribution of water and nutrients, especially of those, which due to their limited availability may become a limiting factor for plant growth, are of increasing interest in science as well as in practical aspects of plant nutrition (Mengel and Bübl, 1983, Horst and Göppel, 1986, Römheld and Marschner, 1986, Kochian, 1991, Steudle, 1992, Canny, 1995, Sattelmacher et al., 1998, Raghothama, 1999, Stephan,

B. Sattelmacher† and W.J. Horst (eds.), The Apoplast of Higher Plants: Compartment of Storage, Transport and Reactions, 337–352.

2002). While older publications concerning the application of dyes to leaves and their distribution are available (Rouschal, 1941, Schlafke, 1958, Altus and Canny, 1985), information about nutrient distribution within leaves is scarce (Marschner, 1995, Mimura, 1995, Grusak et al., 1999). Especially under conditions of limited water transport, only few results are published, although influences of water deficit can be found on plant as well as cell level, like changes in the redistribution of assimilates (Grzesiak et al., 1989) or modulation of ion flux across the plasmalemma by osmotic changes due to salt stress (Shabala, 2000). Water deficit was reported to cause an inhibition of calcium transport (Morard et al., 1996) and a decrease of the phosphate concentration in the xylem (Bahrun et al., 2002).

Three nutrients, which differ in their mobility within plants, are phosphorus (P), iron (Fe) and calcium (Ca). P can be limiting to plant growth due to restricted availability (e.g. Marschner, 1995, Smith, 2002) because it is needed in many substantial processes. It is part of nucleic acids and phospholipids, it transfers energy, it is substrate and regulator in photosynthetic and most other metabolic processes, and it regulates enzymes. P is highly mobile in plants, being transported in the xylem as well as in the phloem in appreciable amounts and may also be exchanged between the two pathways (Biddulph and Markle, 1944, van Bel, 1990).

Fe is well known to be not readily available (Mengel and Bübl, 1983, Römheld and Marschner, 1986, Römheld, 1987, Kochian, 1991, Guerinot and Yi, 1994). Fe is essential in redox processes as well as in the light harvesting process and chlorophyll synthesis, it occurs in heme-proteins, in iron-sulfur-proteins, and it has catalytic properties. Fe deficiency results in yellowing of leaves (chlorosis) and decrease in crop yield (Römheld and Marschner, 1991, Marschner, 1995). Symplastic Fe is usually complexed, as it on the one hand precipitates easily with e.g. phosphate, and on the other hand is toxic in higher amounts.

Ca is of immense importance in cell metabolism. Besides its function as a structural component stabilizing cell walls and membranes, it is essential in regulation and signal transduction (Marschner, 1995, Trewavas and Malhó, 1997, Zielinski, 1998). For example, Ca signals are involved in changing the activity of regulator enzymes (Poovaiah and Reddy, 1987) and influence gravitropism and phototropism (Gallagher et al., 1988). Due to its very low mobility in the phloem, Ca transport to growing shoot parts with low transpiration often becomes insufficient (Biddulph et al., 1959). Barta and Tibbitts (2000) found Ca deficiency symptoms when xylem transport was restricted, even though a small amount of Ca was transported in the phloem.

2. XYLEM TRANSPORT

2.1 Xylem transport and the leaf vein system

Xylem transport of water and solvents is taking place in dead wooden tubes, xylem vessels, of variable width. It is widely believed to be driven by leaf transpiration and root pressure, although the mechanism is still under discussion (Zimmermann et al., 1993, Steudle, 1995, Tyree, 1997, Wei et al., 2000, Wistuba et al., 2000, Zimmermann et al., 2002). Coming from the stem, bundles branch out, often in a complicated manner (Fritz, 1973), into the petiole and, in dicotyledon leaves, ramify in the leaf blade to form a two-dimensional network of anastomosing veins with large number of interconnected pathways (Jeje, 1985, Altus and Canny, 1985). The xylem part of these leaf veins forms part of the leaf apoplast, providing channels of low flow resistance within it. The main properties of this network are i) provision of water and nutrients to the leaf mesophyll, ii) redistribution of products of photosynthesis from the leaf mesophyll into other plant parts, and iii) mechanical support of the leaf blade. The size of leaf veins, i.e. the number and width of xylem elements is usually proportional to the leaf area supplied (Huber, 1928, Ewers et al., 1989). Veins smaller than a certain "critical" size allow only little axial flow compared to lateral losses (Rouschal, 1941, Canny, 1991) and serve mainly for local distribution of water and nutrients. Nevertheless, the possibility of conductive overload is under discussion, meaning a potential ability of veins to transport much larger volumes than they usually do. Early experiments of Wylie and Plymale (Wylie, 1938, Plymale and Wylie, 1944, Wylie, 1947) showed that leaf veins may have a high capacity to cope with leaf damage and still maintain xylem flux. Later, a high ability of plants to maintain xylem transport in spite of local wounding was demonstrated (e.g. Raschke, 1970, Altus and Canny, 1985, Omasa et al., 1991). The question gains more importance, when cavitation of xylem vessels resulting in local barriers to xylem flux is considered. During winter or dry periods in summer, the cross sectional area of conducting xylem is decreased by cavitation (Sperry et al., 1993, Hacke and Sauter, 1995, Linton et al., 1998, Cochard et al., 2001). Cavitation was also reported from the midrib of leaves (Kikuta et al., 1997, Cochard et al., 2002). Nardini et al. (2001) found only slightly decreasing leaf hydraulic conductivity with increasing cavitation and very little response of the stomatal conductivity. Cavitated vessels or cut veins could be bypassed making leaves relatively insensitive for local damage. Along the bypass, rather steep pressure gradients were calculated. Even after severing all visible veins of *Vicia faba* leaves, leaving intact only small veins and fine veins embedded in mesophyll, a dye fed to the cut leaf petiole reached the leaf tip within minutes (Hüve et al., 2002). Veins used to bypass the cuts had two to five xylem vessels of 3–9 μm in diameter. Flow velocity in the cut region was changed markedly especially in the small, still intact veins. It will be shown later (2.3) that such cuts were no barrier for the transport of nutrients as well.

Apart from this, veins within the leaf may serve the exchange of solutes between xylem and phloem due to the vicinity of transporting elements. Nutrients taken up by the roots normally enter the long-distance transport in the xylem sap. Sinks within the plant usually receive their nutrients via the phloem rather than from the xylem. This makes necessary either a transfer of nutrients from the xylem to the phloem (van Bel, 1984) or a redistribution from these leaves into other plant parts when nutrients have already been transported into transpiring leaves (Parsons et al., 1995, Zhang et al., 1995a).

2.2 Nutrients in the xylem

Xylem flow and transport of nutrients do not necessarily depend on transpiration (Tanner and Beevers, 1990, Köckenberger et al., 1997) but is strongly enhanced by it (Neumann and Stein, 1984, Hodson and Sangster, 1988) so that transport within xylem vessels will mainly be directed into transpiring leaves. Zhang et al. (1995a) found a transpiration-dependent primary distribution of iron, which later may change due to retranslocation. Covering of leaves or of parts of the leaf with plastic film inhibited xylem-transport of nutrients into this region (Grusak et al., 1999). In own experiments, phosphorus transport into such a region was only 0.43% of transport into an uncovered area after 2 h. Especially Ca is known for its strong dependence on transpiration when distributed within the plant, causing easily Ca deficiency in young leaves (Grusak and Pomper, 1999, Barta and Tibbitts, 2000).

In our experiments with bean plants, the largest part of root-fed phosphorus was translocated to leaves (Table 1). P was first transported to

Table 1. Radiolabeled phosphorus (P) was applied to the roots of bean plants. P detected in different parts of the shoot in percent of P translocated into the shoot.

	% of P translocated into the shoot	
	after 1 h	after 4 days
shoot axis	27.8	24.0
transpiring adult leaves	61.3	70.8
terminal bud	10.8	5.2

adult, transpiring leaves, and from there, supposedly, redistributed via the phloem, because it was found in appreciable amounts in the shoot apex already after 1 h. The influence of air humidity on P distribution within the plant proved to be low. Often, a different transport velocity in the xylem was observed for anions and cations (Wolterbeek et al., 1984). As the xylem walls are negatively charged - due to carboxyl groups – galacturonic or glucuronic acids of pectin and lignin, respectively, may act as a cation exchange matrix (Wolterbeek, 1987). When applied to the cut petiole, Ca was transported significantly at a slower rate than P and chelated Fe (Fig. 1). P in the xylem is transported mainly as phosphate (Bieleski, 1973). The slower movement of cations, especially of Ca, is therefore likely to be caused by adsorption at the xylem wall (Biddulph et al., 1961, Türk et al., 1994). In the leaf, higher calcium concentrations were found in veins than in the mesophyll (Barta and Tibbitts, 2000), as if a part of the Ca remained adsorbed in xylem vessels. Fe is oxidized (Fe^{II} to Fe^{III}) before entering the xylem vessels and is transported in the xylem as a complex presumably as Fe(III) citrate (Tiffin, 1966).

2.3 Nutrient transport in leaves with a partly impaired vein net

Severing part of the leaf vein net, affecting about 90% of the xylem cross sectional area on the level of the cuts, caused even stronger impacts on volume transport, as the remaining 10% of the xylem cross sectional area are vessels of rather small diameter. P, Fe, and Ca, all radiolabeled, were observed to bypass such cuts successfully and to enter the distal parts of the leaf blade. The successful bypass of Ca, moving at slower rate than P or Fe and being probably unable to use alternative symplastic pathways, was taken as evidence that small xylem vessels were the most likely bypass ways. Nutrients were transported in large veins up to a few mm from the cuts, then entered intact small and fine veins, bypassing the wounded area. Distal from the cuts, tracers were again loaded into large and primary veins (Fig. 2). Bypassing veins are thought to be only used for the local distribution of water and nutrients under normal conditions (Altus and Canny, 1985). Their axial flow resistance is much higher than in larger veins, but still lower than flow resistance in cell walls (Boyer, 1974, Steudle and Jeschke, 1983) or cytoplasm (Michael et al., 1997). Flow of P tracer into the leaf area distal from the cuts was 50–70% of the amount flowing into a similar area of an intact leaflet fed simultaneously.

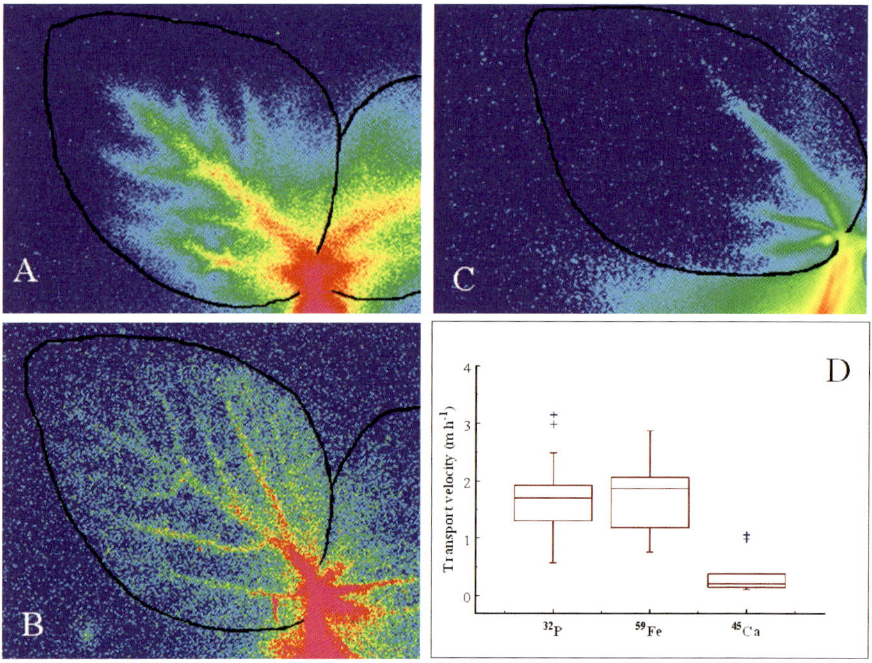

Fig. 1. Distribution of tracers after feeding to the cut petiole. A: ^{32}P applied as potassium phosphate; B: ^{59}Fe, chelated as Fe-EDTA or Fe-citrate; C: $^{45}Ca^{2+}$, applied as $CaCl_2$. A and B were fed for 2, C for 4 min. D: Boxplot of transport velocities of tracers within the midrib (P: n=25, Fe: n=7, Ca: n=11).

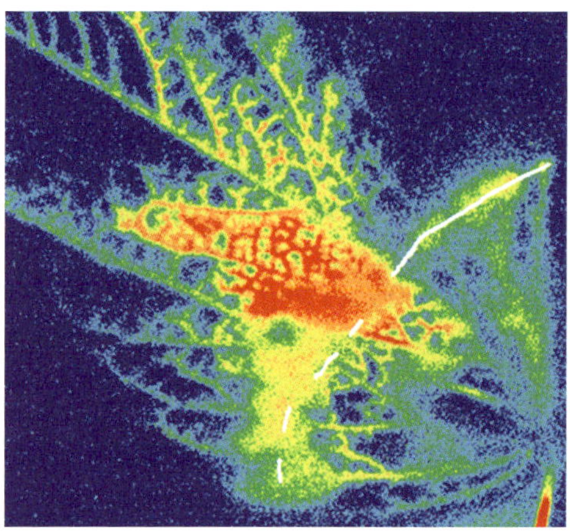

Fig. 2. Autoradiograph of a leaf fed with phosphate radiotracer (^{33}P) via the cut petiole for 5 min. Part of the leaf venation was severed before feeding (white lines).

3. NUTRIENTS IN THE APOPLAST AND UPTAKE INTO THE SYMPLAST

Information about the role of apoplastic transport of water and solutes in plant tissue is scarce, mainly due to technical limitations and to difficulties to separate the apoplastic from overall water flow in an intact plant tissue (Clarkson et al., 2000, Steudle, 2000). Nutrients reaching the apoplast of leaves may (i) be taken up into cells, leaving the apoplast, and be metabolised or transported to their destination within the symplast, (ii) continue to move on an apoplastic transport route, or (iii) be bound or precipitated. Already Levi (1968) observed that Ca entering the leaf along veins spreads slowly into the mesophyll, while P entered the mesophyll rapidly. While cations tend to accumulate in the cell walls, anions are more or less excluded due to negative charges (Grignon and Sentenac, 1991). A large part of the cations in the apoplast is bound (Mühling and Sattelmacher, 1995). Wolterbeek (1987) estimated the amount of negative charges as 0.055 mol kg^{-1} plant dry weight (DW), binding at a pH of 4.2 up to 0.06 mol cations per kg^{-1} DW. On the other hand, Michael and Ehwald (1996) supposed that the diffusion of ions is independent of ionic strength, and the diffusion coefficient in cell walls should be 50% of that of water. This value was found in potato storage parenchyma. Nevertheless, the resistance in plant tissues is still much higher, since the cross sectional area of the apoplast is small compared to the whole tissue (Michael and Ehwald, 1996).

So far, it is not sure whether a large part of ions is transported apoplastically from the xylem to the parenchyma. Keunecke et al. (1997) found a large number of different channels when patch-clamping xylem contact-cells. Among these, two were specific for K^+. It seems likely that ions are taken up from the xylem by bundle cells and transported symplastically. It is known for some ions like K^+ that the apoplast may serve as a dynamic reservoir. K is present mostly in soluble form, between 3 and 12 mM, which equals 1–4% of the whole leaf K content. In deficiency situations, K is kept at a constant minimal level of 2 mM (Mühling and Läuchli, 1999).

The P concentration in the apoplast is kept relatively constant between 0.5–2 mM depending on the concentration in the xylem sap (Mimura et al., 1990, Mimura, 1995, Raghothama, 1999). Xylem parenchyma tissue was shown to accumulate P from the xylem sap (Bieleski, 1966). P is readily taken up into the cytoplasm in an active transport process together with protons (Rebeille et al., 1983, Sakano et al., 1992, Okihara et al., 1995, Leggewie et al., 1997, Daram et al., 1998) or with sodium (Reid et al., 2000), and a large variety of different P transporters is currently characterized. Transporters usually have a high affinity for phosphate (e.g. Bieleski and Läuchli, 1992,

Bucher et al., 2001, Rausch and Bucher, 2002, Smith, 2002) and thus a high uptake capacity.

Also, concentrations within the cytosol are maintained at a constant value (Mimura, 1995), a surplus of P is stored in vacuoles, and can reach 85–95 % of the whole plant P content in well-supplied plants (Bieleski and Ferguson, 1983). Only a small portion of the whole plant P content will, therefore, be expected in the apoplast. The uptake into vacuoles is probably active, but the mechanism is not yet clearly identified (Mimura et al., 1990, Raghothama, 1999). Massonneau et al. (2000) found a low-affinity system with a K_m value of about 5 mM. Also, whether efflux of P from vacuoles is slow (Martinoia et al., 1986) or fast (Bligny et al., 1997) is not yet clear.

Fe is present in the symplast in complexed form, probably with nicotianamine as complexing agent (Becker et al., 1995, von Wirén et al., 1999, Stephan, 2002). Apoplastic Fe can be precipitated, and in this way stored, and may be remobilised (Stephan and Grün, 1989, Welch, 1995) in addition to the intracellular Fe storage protein ferritin. Recently, unidirectional Fe transport across the inner chloroplast envelope stimulated by the membrane potential was shown by a Fe^{2+}-sensitive dye (Shingles et al., 2002). Fe concentrations in the apoplast were found to be higher than in the symplast and variable (Becker et al., 1992), so that the apoplastic Fe pool was proposed as a mobile store. On the other hand, mobile Fe that was retranslocated in plants came from the symplast as well as the apoplast (Zhang et al., 1995b). The critical points in the re-use of precipitated apoplastic Fe is its pH-dependent solubility and the reduction process during uptake into the cytoplasm (Chaney et al., 1972, Moog and Brüggemann, 1994, Fox et al., 1996). Therefore, some authors conclude that precipitated apoplastic Fe is caught in the apoplast rather than stored, at least at high pH values, and can be re-used only with difficulties (Mengel, 1994, Kosegarten et al., 1999), while others reported no dependency of Fe uptake into cells at physiological pH values (Nikolič and Römheld, 1999, this volume, pp 369–388).

Ca is transported mostly along the apoplastic pathway when taken up into roots (Clarkson, 1993, Engels, 1999). It moves at a lower velocity than might be expected for free diffusion and forms a transport gradient after 2 min in the cortex of spruce roots (Kuhn et al., 2000). It probably binds to anionic groups, as was described also in xylem transport. Therefore, chelated Ca is more mobile than cationic Ca (Ferguson and Bollard, 1976, White et al., 1981). Furthermore, Ca can bind to the middle lamella and to the outside of the plasmalemma. The strength of these bindings is unclear; they may be completely reversible (Kuhn, 1993) or they may allow a partly remobilization under conditions of Ca deficiency (Morard et al., 2000). An excess of Ca is precipitated and detoxified in crystalline form, either in the vacuole of specialized cells (idioblasts) or, especially in gymnosperms, in the apoplast (Kinzel, 1989, Fink, 1991). Due to its regulator function and also to avoid precipitation of phosphate salts, the concentration of free Ca in the cytosol is kept as low as 0.1–0.2 μM (Felle, 1988, Evans et al., 1991, Clarkson, 1993). Concentrations of chelated or bound Ca are higher (1–10 mM, White et al., 1992). It is regulated by binding of Ca and by Ca^{2+}-ATPases and Ca^{2+}/H^{+}-antiporters which are located at the tonoplast, at the endoplasmatic reticulum, or at the plasma membrane (Piñeros and Tester, 1997, Geisler et al., 2000).

3.1 Apoplastic portion of phosphorus tracer

Apoplastic washing fluid (AWF), which is obtained by infiltration of a leaf under slight vacuum and retrieval of the liquid by mild centrifugation, is often used to obtain apoplastic concentrations of nutrients or tracers (Luwe et al., 1993; Mühling and Sattelmacher, 1995; Lohaus, this volume, pp 327–342). We applied this method to estimate the remainder of ^{32}P in the apoplast after feeding to the cut petiole. After 10 min, the leaf was infiltrated with distilled water and then centrifuged at 150 g.

During centrifugation, the petiole was placed at the bottom of the centrifugation tube, in contrast to the usual method, as we intended to obtain xylem sap besides the apoplastic fluid. After that, samples of the AWF as well as tissue samples of the whole, centrifuged leaf were analysed for radioactivity. The relatively small total amount of ^{32}P in the apoplast (Table 2) indicates a fast uptake of phosphorus into living cells, probably of the bundle sheath. A fast uptake into the symplast is likely, as high uptake rates into cells were found in suspensions cells (Mimura et al., 1990), in azolla plants (19.2 nmol P min^{-1} mg^{-1} FW, Bieleski and Läuchli, 1992) as well as for phosphate transporters expressed in an uptake-deficient yeast mutant (Bucher et al., 2001).

Table 2. Distribution of *^{32}P activity between AWF, leaf tissue, and petiole tissue after a feeding time of 10 min via the cut petiole. Damaged leaves according to Fig. 2.*

	percentage of ^{32}P activity in the apoplast	percentage of ^{32}P activity in the remaining leaf	percentage of ^{32}P activity in the petiole
intact leaf	3.87 ± 1.95%	77.8 ± 8.7%	18
"damaged" leaf	2.74 ± 1.22%	76.0 ± 5.8%	21

After 25 min of feeding of ^{33}P via the petiole into an intact leaf, primary and secondary veins and also fine veins were clearly marked, while transport of tracer into the intercostal tissue was still lower (Fig. 3A). At that time, most of the tracer was found in the vicinity (<0.5 mm) of veins. An activity profile along veins of decreasing size showed highest activity in the third order leaf veins. Quantification of this activity profile gave 76 times higher P concentrations in comparison to the activity calculated to fill the xylem and the water free space with undiluted solution. Furthermore, calculated activity differed from measured activity and tended to be highest in large veins instead of small veins with rather small vessel diameter between 4 and 8 μm.

Considering the high uptake rates of P into cells, it is assumed that after 25 min a substantial part of the radiotracer was already located within living cells, probably of the bundle sheath, but only small amounts were translocated into intercostal areas. If the mean distance between fine veins in leaves is 177 μm (Wylie, 1938), the finest veins in these leaves could not yet been reached by tracer after 25 min. The bundle sheath of small veins is a site of active uptake from the xylem (Keunecke et al., 2000) indicated also by the lower pH in the apoplast near bundle-sheath cells (Wilson et al., 1991). For an *Arabidopsis* phosphate transporter, especially high transcript levels were reported around vascular tissues of leaves (Bucher et al., 2001), supporting the idea of fast P uptake into the bundle sheath.

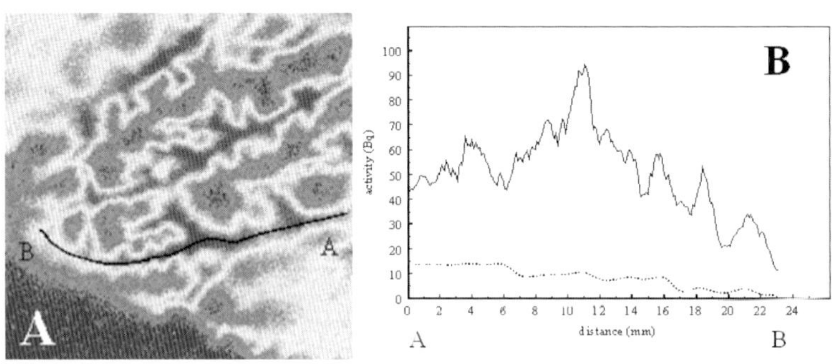

Fig. 3. A: Autoradiograph after 25 min of feeding of ^{33}P into an intact leaf. B: Activity profile along veins of decreasing size (solid line) and activity calculated to fill the xylem and water free space with undiluted solution (dashed line).

4. CONCLUSIONS

Different fates of nutrients in long distance transport and within the apoplast are compared and illustrated by experiments with P, Fe and Ca which are all three important nutrients, but differ largely in behaviour. Nutrients taken up at the roots enter transpiring leaves via the xylem flow. Transport of nutrients in the xylem takes place at different velocities for different nutrients, depending on charge and chemical form (free or chelated). Ca was transported significantly slower than P or chelated Fe due to negative charges in xylem walls, which act like a cation exchange matrix. For all three elements investigated, xylem flow (as driven by transpiration) was mainly directed into transpiring leaves. Xylem transport of nutrients was able to cope with bottleneck situations, induced in our case by large but local destruction of the leaf vein net. Such injuries could in nature occure by cavitation of large xylem vessels or by mechanical damage of the leaf. Such bottlenecks were bypassed by all three nutrients investigated, anions as well as cations, using mainly intact small and fine veins to bypass the wounded area. For P the flow into the leaf area distal from the wounded place was 50–70% of that into a comparable area of an undamaged leaf. Nutrients reaching the apoplast of leaves may (i) be taken up into cells, leaving the apoplast, (ii) continue on an apoplastic transport way, or (iii) be bound or precipitated. So besides transport, the apoplast may serve various functions as reloading site into cells or into the phloem, as store, or even as deposit for excess nutrients. The latter was observed in the case of Ca. While Fe and in some cases Ca may be precipitated, P is quickly taken up into the symplast, as was shown by the small percentage (3–4%) of P tracer, which was recovered in the apoplast after only 10 minutes of feeding. Especially small leaf veins are lined with cells highly efficient in nutrient uptake. Phloem-mobile nutrients such as P are likely to be loaded into the phloem for redistribution to sinks within the plant, which may not be sufficiently supplied by the xylem due to their low transpiration.

REFERENCES

Altus, D. P. and Canny, M. J. (1985). Water pathways in wheat leaves. I. The division of fluxes between different vein types. *Aust. J. Plant Physiol.*, 12, 173–181.

Bahrun, A., Jensen, C. R., Asch, F. and Mogensen, V. O. (2002). Drought-induced changes in xylem pH, ionic composition and ABA concentration act as early signals in field-grown Maize (*Zea mays* L.). *J. Exp. Bot.*, 53, 251–263.

Barta, D. J. and Tibbitts, T. W. (2000). Calcium localization and tipburn development in lettuce leaves during early enlargement. *J. Am. Soc. Hort. Sci.*, 125, 294–298.

Becker, R., Grün, M. and Scholz, G. (1992). Nicotianamine and the distribution of iron into the apoplasm and symplasm of tomato (*Lycopersicon esculentum* Mill.). *Planta*, 187, 48–52.

Becker, R., Fritz, E. and Manteuffel, R. (1995). Subcellular localization and characterization of excessive iron in the nicotianamine-less tomato mutant *Chloronerva*. *Plant Physiol.*, 108, 269–275.

Biddulph, O. and Markle, J. (1944). Translocation of radiophosphorus in the phloem of the cotton plant. *Am. J. Bot.*, 31, 65–70.

Biddulph, O., Cory, R. and Biddulph, S. (1959). Translocation of calcium in the bean plant. *Plant Physiol.*, 34, 512–549.

Biddulph, O., Nakayama, F. S. and Cory, R. (1961). Transpiration stream and the ascent of calcium. *Plant Physiol.*, 36, 429–436.

Bieleski, R. L. (1966). Sites of accumulation in excised phloem and vascular tissues. *Plant Physiol.*, 41, 447–454.

Bieleski, R. L. (1973). Phosphate pools, phosphate transport, and phosphate availability. *Annu. Rev. Plant Physiol.*, 24, 225–252.

Bieleski, R. L. and Ferguson, I. B. (1983). Physiology and metabolism of phosphate and its compounds. In A. Läuchli and R. L. Bieleski (eds.), Encyclopedia of Plant Physiology (422–449), Vol. 15A. Berlin, Heidelberg, New York: Springer.

Bieleski, R. L. and Läuchli, A. (1992). Phosphate uptake, efflux and deficiency in water fern, Azolla. *Plant, Cell Environ.*, 15, 665–673.

Bligny, R., Gout, E., Kaiser, W., Heber, U., Walker, D. and Douce, R. (1997). pH regulation in acid-stressed leaves of pea plants grown in the presence of nitrate or ammonium salts: Studies involving 31P-NMR spectroscopy and chlorophyll fluorescence. *Biochim. Biophys. Acta*, 1320, 142–152.

Boyer, J. S. (1974). Water transport in plants: Mechanism of apparent changes in resistance during absorption. *Planta*, 117, 187–207.

Bucher, M., Rausch, C. and Daram, P. (2001). Molecular and biochemical mechanisms of phosphorus uptake into plants. *J. Plant Nutr. Soil Sci.*, 164, 209–217.

Canny, M. J. (1991). The xylem wedge as a functional unit - speculations on the consequences of flow in leaky tubes. *New Phytol.*, 118, 367–374.

Canny, M. J. (1995). Apoplastic water and solute movement: New rules for an old space. *Annu. Rev. Plant Physiol. Plant Mol. Biol.*, 46, 215–236.

Chaney, R. L., Brown, J. C. and Tiffin, L. O. (1972). Obligatory reduction of ferric chelates in iron uptake by soybeans. *Plant Physiol.*, 50, 208–213.

Clarkson, D. T. (1993). Roots and the delivery of solutes to the xylem. *Phil. Trans. R. Soc. London B*, 341, 5–17.

Clarkson, D. T., Carvajal, M., Henzler, T., Waterhouse, R. N., Smyth, A., Cooke, D. T. and Steudle, E. (2000). Root hydraulic conductance: diurnal aquaporins expression and the effects of nutrient stress. *J. Exp. Bot.*, 51, 61–70.

Cochard, H. (2002). Xylem embolism and drought-induced stomatal closure in maize. *Planta*, 215, 466–471.

Cochard, H., Lemoine, D., Améglio, T. and Granier, A. (2001). Mechanisms of xylem recovery from winter embolism in Fagus sylvatica. *Tree Physiol.*, 21, 27–33.

Daram, P., Brunner, S., Persson, B. L., Amrhein, N. and Bucher, M. (1998). Functional analysis and cell-specific expression of a phosphate transporter from tomato. *Planta*, 206, 225–233.

Engels, C. (1999). Regulation of xylem transport of calcium from roots to shoot of maize by growth-related demand. *J. Plant Nutr. Soil Sci.*, 162, 287–294.

Evans, D. E., Briars, S.-A. and Williams, L. E. (1991). Active calcium transport by plant cell membranes. *J. Exp. Bot.*, 42, 285–303.

Ewers, F. W., Fisher, J. B. and Chiu, S. T. (1989). Water transport in the liana *Bauhinia fassoglensis* (*Fabaceae*). *Plant Physiol.*, 91, 1625–1631.

Felle, H. (1988). Cytoplasmic free calcium in *Riccia fluitans* L. and *Zea mays* L.: Interaction of Ca^{2+} and pH? *Planta*, 176, 248–255.

Ferguson, I. B. and Bollard, E. G. (1976). The movement of calcium in woody stems. *Ann. Bot.*, 40, 1057–1065.

Fink, S. (1991). The micromorphological distribution of bound calcium in needles of Norway spruce (*Picea abies*). *New Phytol.*, 119, 33–40.

Fox, T. C., Shaff, J. E., Grusak, M. A., Norvell, W. A., Chen, Y., Chaney, R. L. and Kochian, L. V. (1996). Direct measurement of ^{59}Fe labeled Fe^{2+} influx in roots of pea using a chelator buffer system to control free Fe in solution. *Plant Physiol.*, 112, 93–100.

Fritz, E. (1973). Microautoradiographic investigation on bidirectional translocation in the phloem of *Vicia faba*. *Planta*, 112, 169–179.

Gallagher, S., Short, T. S., Ray, P. M., Pratt. L. H. and Briggs, W. R. (1988). Light-mediated changes in two proteins found associated with plasma membrane fractions from pea stem sections. *Proc. Nat. Acad. Sci. USA*, 85, 8003–8007.

Geisler, M., Axelsen, K. B., Harper, J. F. and Palmgren, M. G. (2000). Molecular aspects of higher plant P-type Ca^{2+}-ATPases. *Biochim. Biophys. Acta*, 1465(1/2), 52–78.

Grignon, C. and Sentenac, H. (1991). pH and ionic conditions in the apoplast. *Annu. Rev. Plant Physiol. Plant Mol. Biol.*, 42, 103–128.

Grusak, M. A., DellaPenna, D. and Welch, R. M. (1999). Physiological processes affecting the content and distribution of phytonutrients in plants. *Nutrition Reviews*, 57(9), 27–33.

Grusak, M. A. and Pomper, K. W. (1999). Influence of pod stomatal density and pod transpiration on the calcium concentration of snap bean pods. *J. Am. Soc. Hort. Sci.*, 124, 194–198.

Grzesiak, S., Koscielniak, J., Filek, W. and Augustyniak, G. (1989). Effects of soil drought in the generative phase of growth of field bean (*Vicia faba* L.) on leaf water status, photosynthesis rate and growth of biomass. *J. Agron. Crop Sci.*, 162, 214–217.

Guerinot, M. L. and Yi, Y. (1994). Iron: nutritious, noxiuos, and not readily available. *Plant Physiol.*, 104, 815–820.

Hacke, U. and Sauter, J. J. (1995). Vulnerability of xylem to embolism in relation to leaf water potential and stomatal conductance in *Fagus sylvatica*, *F. purpurea* and *Populus balsamifera*. *J. Exp. Bot.*, 46, 1177–1183.

Hodson, M. J. and Sangster, A. G. (1988). Observations on the distribution of mineral elements in the leaf of wheat (*Triticum aestivum* L.) with particular reference to silicon. *Ann. Bot.*, 62, 463–471.

Horst, W. J. and Göppel, H. (1986). Aluminium-Toleranz von Ackerbohne (*Vicia faba*), Lupine (*Lupinus luteus*), Gerste (*Hordeum vulgare*) und Roggen (*Secale cereale*). *Z. Pflanzenern. u. Bodenk.*, 149, 83–109.

Huber, B. (1928). Weitere quantitative Untersuchungen über das Wasserleitungssystem der Pflanzen. *Jahrb. f. wiss. Botanik*, 67, 877–959.

Hüve, K., Remus, R., Lüttschwager, D. and Merbach, W. (2002). Water transport in impaired leaf vein systems. *Plant Biol.*, 4, 603–611.

Jeje, A. A. (1985). The flow and dispersion of water in the vascular network of dicotyledonous leaves. *Biorheology*, 22, 285–302.

Keunecke, M., Sutter, J. U., Sattelmacher, B. and Hansen, U. P. (1997). Isolation and patch clamp measurements of xylem contact cells for the study of their role in the exchange between apoplast and symplast of leaves. *Plant Soil*, 196, 239–244.

Keunecke, M., Lindner, B., Seydel, U., Schulz, A. and Hansen, U. P. (2000). Bundle-sheath cells of small veins in maize are the location of uptake from the xylem. *J. Exp. Bot.*, 52, 709–714.

Kikuta, S. B., Lo Gullo, M. A., Nardini, A., Richter, H. and Salleo, S. (1997). Ultrasound acoustic emissions from dehydrating leaves of deciduous and evergreen trees. *Plant, Cell Environ.*, 20, 1381–1390.

Kinzel, H. (1989). Calcium in the vacuoles and cell walls of plant tissue. Forms of deposition and their physiological and ecological significance. *Flora*, 182, 99–125.

Kochian, L. E. (1991). Mechanisms of micronutrient uptake and translocation in plants. In J. J. Mortvedt, F. R. Cox, L. M. Shuman and R. M. Welch (Eds.), Micronutrients in agriculture, 2nd Ed., Madison, Wisconsin, USA: Soil Science of America.

Köckenberger, W., Pope, J. M., Xia, Y., Jeffrey, K. R. Komor, E. and Callaghan, P. T. (1997). A non-invasive measurement of phloem and xylem water flow in castor bean seedlings by nuclear magnetic resonance microimaging. *Planta*, 201, 53–63.

Kosegarten, H., Hofmann, B. and Mengel, K. (1999). Apoplastic pH and Fe^{3+} reduction in intact sunflower leaves. *Plant Physiol.*, 121, 1069–1079.

Kuhn, A. J. (1993). Mikrosonden-Analysen zur Ionenaufnahme in Fichten (*Picea abies* (L.) Karst.). Berichte des Forschungszentrums Jülich 2744, (Diss. Univ. Hamburg).

Kuhn, A., Schröder, W. H. and Bauch, J. (2000). The kinetics of calcium and magnesium entry into mycorrhizal spruce roots. *Planta*, 210, 488–496.

Leggewie, G., Willmitzer, L. and Riesmeier, J. W. (1997). Two cDNAs from potato are able to complement a phosphate uptake deficient yeast mutant: Identification of phosphate transporters from higher plants. *Plant Cell*, 9, 381–392.

Levi, E. (1968). The distribution of mineral elements following leaf and root uptake. *Physiol. Plant.*, 21, 213–226.

Linton, M. J., Sperry, J. S. and Williams, D. G. (1998). Limits to water transport in *Juniperus osteosperma* and *Pinus edulis*: implications for drought tolerance and regulation of transpiration. *Funct. Ecol.*, 12(6), 906–911.

Luwe, M., Takahama, U. and Heber, U. (1993). Role of ascorbate in detoxifying ozone in the apoplast of spinach (*Spinacia oleracea* L.) leaves. *Plant Physiol.*, 101, 969–976.

Marschner, H. (1995). Mineral nutrition of higher plants. 2nd Ed., London, UK: Academic Press.

Martinoia, E., Schramm, M. J., Kaiser, G., Kaiser, W. M. and Heber, U. (1986). Transport of anions in isolated barley vacuoles. 1. Permeability to anions and evidence for a Cl-uptake system. *Plant Physiol.*, 80, 895–901.

Massonneau, A., Martinoia, E., Dietz, K. J. and Mimura, T. (2000). Phosphate uptake across the tonoplast of intact vacuoles isolated from suspension-cultured cells of *Catharanthus roseus* (L.) G. *Don. Planta*, 211, 390–395.

Mengel, K. (1994). Iron availability in plant tissues - iron chlorosis on calcareous soils. *Plant Soil*, 165, 275–283.

Mengel, K. and Bübl, W. (1983). Distribution of iron in vine leaves with HCO^{3-} [bicarbonate ion] induced chlorosis. Verteilung von Eisen in Blättern von Weinreben mit HCO^{3-} induzierter Fe-Chlorose. *Z. Pflanzenern. Bodenk.*, 146, 560–571.

Michael, W. and Ehwald, R. (1996). Exchange diffusion of alkali ions through the apoplast of the potato (*Solanum tuberosum* L.) storage parenchyma. *Plant, Cell Environ.*, 19, 243–246.

Michael, W., Schultz, A., Meshcheryakov, A. B. and Ehwald, R. (1997). Apoplasmic and protoplasmic water transport through the parenchyma of the potato storage organ. *Plant Physiol.*, 115, 1089–1099.

Mimura, T. (1995). Homeostasis and transport of inorganic phosphate in plants. *Plant Cell Physiol.*, 36, 1–7.

Mimura, T., Dietz, K.-J., Kaiser, W., Schramm, M. J., Kaiser, G. and Heber, U. (1990). Phosphate transport across biomembranes and cytosolic phosphate homeostasis in barley leaves. *Planta*, 180, 139–146.

Moog, P. R. and Brüggemann, W. (1994). Iron reductase systems on the plant plasma membrane. *Plant Soil*, 165, 241–260.

Morard, P., Pujos, A., Bernadac, A. and Bertoni, G. (1996). Effect of temporary calcium deficiency on tomato growth and mineral nutrition. *J. Plant Nutr.*, 19, 115–127.

Morard, P., Lacoste, L. and Silvestre, J. (2000). Effects of calcium deficiency on nutrient concentration of xylem sap of excised tomato plants. *J. Plant Nutr.*, 23, 1051–1062.

Mühling, K. H. and Sattelmacher, B. (1995). Apoplastic ion concentration of intact leaves of field bean (*Vicia faba*) as influenced by ammonium and nitrate nutrition. *J. Plant Physiol.*, 147, 81–86.

Mühling, K. H. and Läuchli, A. (1999). Effect of K^+ nutrition, leaf age and light intensity on apoplastic K^+ in leaves of *Vicia faba*. *J. Plant Nutr. Soil Sci.*, 162, 571–576.

Nardini, A., Tyree, M. T. and Salleo, S. (2001) Xylem cavitation in the leaf of *Prunus laurocerasus* and its impact on leaf hydraulics. *Plant Physiol.*, 125, 1700–1709.

Neumann, P. M. and Stein, Z. (1984). Relative rates of delivery of xylem solute to shoot tissues: possible relationship to sequential leaf senescence. *Physiol. Plant.*, 62, 390–397.

Nikolič, M. and Römheld, V. (1999). Mechanism of Fe uptake by the leaf symplast: Is Fe inactivation in leaf a cause of Fe deficiency chlorosis? *Plant Soil*, 215, 229–237.

Okihara, K., Kiyota, S., Mimura, T. and Sakano, K. (1995). Furosemide: a specific inhibitor of P_i transport across the plasma membrane of plant cells. *Plant Cell Physiol.*, 36, 53–58.

Omasa, K., Murayama, S., Matthews, M. A. and Boyer, J. S. (1991). Image diagnosis of photosynthesis in water-defecit plants. In Y. Hashimoto and W. Day, (Eds.), Mathematical and Control Applications in Agriculture and Horticulture (383–388) Oxford: Pergamon Press.

Parsons, R., Raven, J. A. and Sprent, J. I. (1995). Translocation of iron to the N_2-fixing stem nodules of *Sesbania rostrata* (Brem). *J. Exp. Bot.*, 46, 291–296.

Piñeros, M. and Tester, M. (1997). Calcium channels in higher plant cells: selectivity, regulation and pharmacology. *J. Exp. Bot.*, 48, 551–577.

Plymale, E. L. and Wylie, R. B. (1944). The major veins of mesomorphic leaves. *Am. J. Bot.*, 31, 99–106.

Poovaiah, B. W. and Reddy, A. S. N. (1987). Calcium messenger system in plants. *CRC Crit. Rev. Plant Sci.*, 6, 47–103.

Raghothama, K. G. (1999). Phosphate Acquisition. *Annu. Rev. Plant Physiol. Plant Mol.Biol.*, 50, 665–693.

Raschke, K. (1970). Stomatal response to pressure changes and interruptions in the water supply of detached leaves of *Zea mays*. *Plant Physiol.*, 45, 415–423.

Rausch, C. and Bucher, M. (2002). Molecular mechanisms of phosphate transport in plants. *Planta*, 216, 23–37.
Rebeille, F, Bligny, R., Martin, J.-B. and Douce, R. (1983). Relationship between the cytoplasm and vacuole phosphate pool in *Acer pseudoplatanus* cells. *Arch. Biochem. Biophys.*, 225, 143–148.
Reid, R. J., Mimura, T., Ohsumi, Y., Walker, N. A. and Smith, F. A. (2000). Phosphate uptake in Chara: membrane transport via Na/Pi cotransport. *Plant, Cell Environ.*, 23, 223–228.
Römheld, V. (1987). Different *strategies* for iron acquisition in higher plants. *Physiol. Plant.*, 70, 231–234.
Römheld, V. and Marschner, H. (1986). Mobilization of Fe in the rhizosphere of different plant species. *Adv. Plant Nutr.*, 2, 155–204.
Römheld, V. and Marschner, H. (1991). Function of micronutrients in plants. In J. J. Mortvedt, F. R. Cox, L. M. Shuman and R. M. Welch (Eds.), Micronutrients in agriculture (251–296), 2nd Ed., Madison, Wisconsin, USA: Soil Science of America.
Rouschal, E. (1941). Beiträge zum Wasserhaushalt von Gramineen und Cyperaceen. I. Die faszikuläre Wasserleitung in den Blättern und ihre Beziehung zur Transpiration. *Planta*, 32, 66–87.
Sakano, K., Yazaki, Y. and Mimura, T. (1992). Cytoplasmic acidification induced by inorganic phosphate uptake in suspension cultured *Catharanthus roseus* cells. *Plant Physiol.*, 99, 672–680.
Sattelmacher, B., Mühling, K. H. and Pennewiß, K. (1998). The apoplast – its significance for the nutrition of higher plants. *Z. Pflanzenern. Bodenk.*, 161, 485–498.
Schlafke, E. (1958). Kritische Untersuchungen zur Wanderung von Fluorochromen in Blättern. *Planta*, 50, 388–422.
Shabala, S. (2000). Ionic and osmotic components of salt stress specifically modulate net ion flux from bean leaf mesophyll. *Plant Cell Environ.*, 23(8), 825–837.
Shingles, R., North, M. and McCarty, R. E. (2002). Ferrous Iron transport across chloroplast inner envelope membranes. *Plant Physiol.*, 128, 1022–1030.
Smith, F. W. (2002). The phosphate uptake mechanism. *Plant Soil*, 245, 105–114.
Sperry J. S., Alder, N. N. and Eastlack, S. E. (1993). The effect of reduced hydraulic conductance on stomatal conductance and xylem cavitation. *J. Exp. Bot.*, 44, 1075–1082.
Stephan, U. W. (2002). Intra- and intercellular iron trafficking and subcellular compartmentation within roots. *Plant Soil*, 241, 19–25.
Stephan, U. W. and Grün, M. (1989). Physiological disorders of the nicotianamine-auxotroph tomato mutant chloronerva at different levels of iron nutrition. II. Iron deficiency response and heavy metal metabolism. *Biochem. Physiol. Pflanzen*, 185, 189–200.
Steudle, E. (1992). The biophysics of water: Compartmentation, coupling with metabolic processes, and flow of water in plant roots. In G. N. Somero, C. B. Osmond and C. L. Bolis (Eds.), Water and life: Comparative analysis of water relationships at the organismic, cellular, and molecular level, (173–204), Berlin: Springer-Verlag.
Steudle, E. (1995). Trees under tension. *Nature*, 378, 663–664.
Steudle E (2000). Water uptake by plant roots: an integration of views. *Plant Soil*, 226, 45–56.
Steudle, E. and Jeschke, W. D. (1983). Water transport in barley roots. *Planta*, 158, 237–248.
Tanner, W. and Beevers, H. (1990). Does transpiration have an essential function in long-distance ion transport in plants. *Plant, Cell Environ.*, 13, 745–750.
Tiffin, L. O. (1966). Iron translocation II. Citrate/iron ratios in plant stem exudates. *Plant Physiol.*, 41, 515–518.
Trewavas, A. J. and Malhó, R. (1997). Signal perception and transduction: the origin of the phenotype. *Plant Cell*, 9, 1181–1195.
Türk, S., Schimansky, C., Führ, F. and Zöttl, H. W. (1994). Transport und Einlagerung von Ca in sechsjährigen Fichten (*Picea abies* [L.] Karst.) nach Aufnahme von ^{45}Ca während der Maitrieb-bildung und zu Beginn der Depositionszeit. *AFJZ*, 165(10–11), 206–210.
Tyree, M. T. (1997). The cohesion-tension theory of sap ascent: current controversies. *J. Exp. Bot.*, 48, 1753–1765.
van Bel, A. J. E. (1984). Quantification of the xylem-to-phloem-transfer of amino acids by use of inulin(^{14}C)carboxylic acid as xylem transport marker. *Plant Sci. Lett.*, 35, 81–85.
van Bel, A. J. E. (1990). Xylem-phloem-exchange via the rays: the undervalued route of transport. *J. Exp. Bot.*, 41, 631–644.
von Wirén, N., Klair, S., Bansal, S., Briat, J. F., Khodr, H., Shioiri, T., Leigh, R. A. and Hider, R. C. (1999). Nicotianamine chelates both Fe^{III} and Fe^{II}. Implications for metal transport in plants. *Plant Physiol.*, 119, 1107–1114.

Wei, C. F., Steudle, E. and Tyree, M. T. (2000). Water ascent in plants: the ongoing debate. *Trends Plant Sci.*, 5(4), 145–146.
Welch, R. M. (1995). Micronutrient nutrition of plants. *Critical Rev. Plant Sci.*, 14, 49–82.
White, M. C., Chaney, R. L. and Decker, A. M. (1981). Metal complexation in xylem fluid. I. Chemical composition of tomato and soybean exudate. *Plant Physiol.*, 67, 292–300.
White, P. J., Banfield, J. and Diaz, M. (1992). Unidirectional Ca^{2+} fluxes in roots of rye (*Secale cereale* L.). A comparison of excised roots with roots of intact plants. *J. Exp. Bot.*, 43, 1061–1074.
Wilson, T. P., Canny, M. J. and McCully, M. E. (1991). Leaf teeth, transpiration and the retrieval of apoplastic solutes in balsam poplar. *Physiol. Plant.*, 83, 225–232.
Wistuba, N., Reich, R., Wagner, H.-J., Zhu, J. J., Schneider, H., Bentrup, F.-W., Haase, A. and Zimmermann, U. (2000). Xylem flow and its driving forces in a tropical liana: concomitant flow-sensitive NMR imaging and pressure probe measurements. *Plant Biol.*, 6, 579–582.
Wolterbeek, H. T. (1987). Cation exchange in isolated xylem cell walls of tomato. *Plant, Cell Environ.*, 10, 39–44.
Wolterbeek, H. T., Van Luipen, J. and De Bruin, M. (1984). Non-steady state xylem transport of fifteen elements into the tomato leaf: a model. *Physiol. Plant.*, 61, 599–607.
Wylie, R. B. (1938). Concerning the conductive capacity of the minor veins of foliage leaves. *Amer. J. Bot.*, 25, 567–572.
Wylie, R. B. (1947). Conduction in dicotyledon leaves. *Iowa Acad. Sci. Proc.*, 53, 195–202.
Zhang, C. D., Römheld, V. and Marschner, H. (1995a). Distribution pattern of root-supplied ^{59}iron in iron-sufficient and iron-deficient bean plants. *J. Plant Nutr.*, 18, 2049–2058.
Zhang, C. D., Römheld, V. and Marschner, H. (1995b). Retranslocation of iron from primary leaves of bean plants under iron deficiency. *J. Plant Physiol.*, 146, 268–272.
Zielinski, R. E. (1998). Calmodulin and calmodulin-binding proteins in plants. *Annu. Rev. Plant Physiol. Plant Mol. Biol.*, 49, 697–725.
Zimmermann, U., Haase, A., Langbein, D. and Meinzer, F. (1993) Mechanism of long-distance water transport in plants: reexamination of some paradigms in the light of new evidence. *Philos. Trans. R. Soc. Lond.*, 341, 19–31.
Zimmermann, U., Schneider, H., Wegner, L.H. and al. (2002). What are the driving forces for water lifting in the xylem conduit? *Physiol. Plant.*, 114, 327–335.

THE DYNAMICS OF IRON IN THE LEAF APOPLAST

Significance for the iron nutrition of plants

M. NIKOLIC[1,2] and V. RÖMHELD[1]

[1]Institut für Pflanzenernährung (330), Universität Hohenheim, Germany; [2]Centre for Multidisciplinary Studies of the Belgrade University, Serbia, mnikolic@cms.bg.ac.yu

Abstract. The leaf apoplast plays an important physiological role in nutrient transport and storage, however, its significance for the iron (Fe) nutrition is not sufficiently understood. There are only few studies in the literature primarily on the mechanism of Fe absorption by leaf cells; even less information is available on the mobility and binding forms of Fe in the leaf apoplast. This review summarizes current knowledge (sometimes very controversial) of the role of leaf apoplastic features (e.g. pH and organic acids) in modulating both the physiological availability of apoplastic Fe and reduction-mediated Fe uptake into the mesophyll cells. The conclusions drawn from our own studies contrast with the hypothesis of Fe inactivation in leaves induced by high bicarbonate or/and nitrate supply to roots.

Key words: apoplastic pH, bicarbonate, "chlorosis paradox", iron deficiency chlorosis, iron inactivation, nitrate

1. INTRODUCTION

Most plant species grown on calcareous soils suffer from iron (Fe) deficiency showing characteristic symptoms known as Fe chlorosis (yellowing) in young leaves, while older leaves remain green. The concentration of Fe frequently decreases in chlorotic leaves, although some woody plants grown under field conditions may sometimes show Fe chlorosis symptoms even with Fe concentration (i.e. expressed per leaf dry weight) higher than in green leaves. Over the past twenty years the explanation of this seemingly physiological paradox has been a matter of a very controversial debate.

B. Sattelmacher† and W.J. Horst (eds.), The Apoplast of Higher Plants: Compartment of Storage, Transport and Reactions, 353–371.

This paper summarizes the work within the DFG special research project "The Apoplast of Higher Plants" of a study of the mechanism of Fe uptake by the symplast of leaf cells, with special emphasis on the role of the leaf apoplast in modulating Fe deficiency chlorosis. In particular, our research has been focused on testing the hypothesis of Fe inactivation in the leaf apoplast as postulated by Mengel and co-workers (e.g. Mengel, 1994; Kosegarten et al., 2001).

2. IRON UPTAKE BY LEAF MESOPHYLL CELLS

2.1 Reductive Fe uptake

In contrast to roots, the mechanisms by which Fe enters leaf cells (Fig. 1) have been far less studied, focusing mainly on Fe^{III} chelate reduction by intact leaf tissue, protoplasts, or plasma membranes (Brüggemann et al., 1993; de la Guardia and Alcántara, 1996; Kosegarten et al., 1999; Nikolic and Römheld, 1999; González-Vallejo et al., 1999, 2000; Rombolà et al., 2000). It has also been demonstrated, using radioactive ^{59}Fe, that a reduction of Fe^{III} chelates in the apoplast precedes Fe uptake by the symplast of leaf mesophyll cells, since Fe^{2+} chelators [e.g. bathophenanthroline disulfonate (BPDS)] inhibit Fe uptake (Brüggemann et al., 1993; Nikolic and Römheld, 1999). Although reduction of Fe^{III} citrate in leaves is mediated by a plasma membrane-bound reductase, and probably supported by the formation of superoxide radicals, an exclusive role for a plasma membrane-associated Fe^{III} chelate redox-system in Fe uptake is still a matter of speculation (Schmidt, 1999, and refs. therein). Therefore, other, non-enzymatic mechanisms such as a photochemical reduction of Fe^{III} citrate by a shorter wavelength light or chemical reduction by reducing compounds present in the apoplast (e.g. phenolics and ascorbate) need further careful consideration.

As yet, there is no clear evidence that an increase in the expression of *FRO* genes, encoding a low-Fe-inducible Fe^{III} chelate reductase, and found in both Fe-deficient roots and shoots of *Arabidopsis thaliana* L. (*FRO*3, Robinson et al., 1997; *FRO*2, Connolly et al., 2003) and pea (*Pisum sativum* L.) (*FRO*1, Waters et al., 2002) is related to enhanced Fe uptake by mesophyll cells, and hence to an effective upregulation of this process in the chlorotic leaves. Also, in many of our studies we observed no significant differences either in reduction of Fe^{III} chelates or in uptake of ^{59}Fe by leaf discs between plants deficient or adequately supplied with Fe (e.g. Nikolic and Römheld, 1999; Nikolic et al., 2003). These and results of other authors (Brüggemann et al., 1993; de al Guardia and Alcántara, 1996; González-

Vallejo et al., 2000; Rombolà et al., 2000; Larbi et al., 2001) suggest that the Fe^{III} chelate redox-system in the leaf plasma-membrane may be present as isoforms different from those in roots, and which unlike those in roots, may not be inducible when Fe is limiting. Under light conditions de la Guardia and Alcántara (1996) and Rombolà et al. (2000) found that Fe^{III} reduction was even lower in chlorotic than in green leaves, suggesting that a plasma membrane-mediated Fe^{III} reduction directly depends on the level of cytosolic electron donors [NAD(P)H] generated through photosynthesis.

Besides the above mentioned stimulatory effect of light on plasma membrane-bound Fe^{III} reduction (Brüggemann et al., 1993; de la Guardia and Alcántara, 1996), light (e.g. UV and blue radiation) may directly affect Fe uptake into the symplast of leaf cells by non-enzymatic photoreduction of Fe^{III} citrate in the leaf apoplast (Fig. 1; Brown et al., 1979a, b; Krizek et al., 1982; Bienfait and Scheffers, 1992). For instance, the inhibitor of photosystem II dichlorophenyl-1, 1-dimethyl urea (DCMU) reduced the rates of both Fe^{III} reduction and Fe uptake under a high irradiance with red light (>630 nm) compared to the dark (Brüggemann et al., 1993). On the other hand, application of DCMU in both grapevine (*Vitis* sp.) and sunflower (*Helianthus annuus* L.) leaves exposed to a high photosynthetic photon flux density (PPFD; 500 μmol m^{-2} s^{-1}; cool white light, 350–700 nm), inhibited

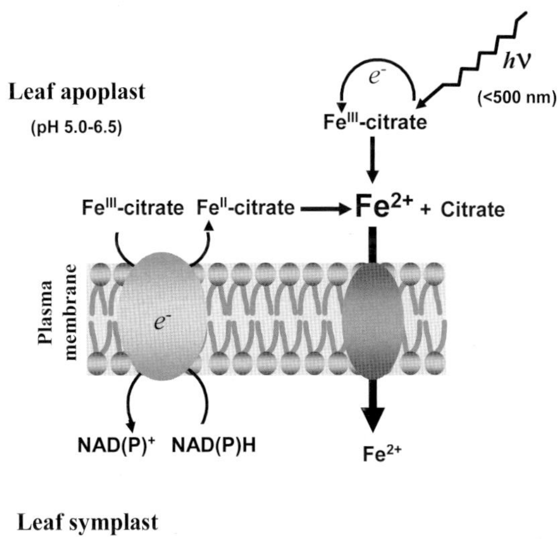

Fig. 1. Proposed model of Fe uptake by the symplast of leaf mesophyll cells.

uptake of ^{59}Fe by about 30%, which was still significantly higher compared to the uptake rate measured in darkness (V. Römheld and M. Nikolic, unpubl. data). Both these studies suggest that photochemical reduction of Fe^{III} citrate in the apoplast might be of great importance for Fe uptake in tissues that are more exposed to high irradiance, such as lamina of younger leaves. This would be of particular significance in leaves with low chlorophyll concentration (e.g. chlorotic leaves; Larbi et al., 2001), where, as a consequence of low photosynthetic rates, the amount of cytosolic reducing equivalents for a plasma membrane-associated Fe^{III} reduction may be decreased, and therefore less Fe^{III} citrate is reduced enzymatically.

Following a Fe^{III} reduction in the leaf apoplast, Fe^{2+} ions, released from ligands, cross the plasma membrane, most likely by means of a divalent cation transporter (Fig. 1). Indeed, ^{59}Fe uptake by the symplast of grapevine leaves was strongly inhibited by Cd^{2+} (V. Römheld and M. Nikolic, unpubl. data). Although Fe^{III} reduction seems to be a necessary step prior to Fe uptake, this reaction does not limit the uptake, suggesting that the uptake of Fe by mesophyll cells is regulated independently of the reduction activity (Schmidt, 1999). As far as we are aware, no plasma membrane Fe transporters have been clearly identified in leaves. Furthermore, the well-known Fe transporter IRT1 has been found to be exclusively expressed in roots of Arabidopsis (Connolly et al., 2002; Vert et al., 2002).

Leaf cells are capable of reducing and thereby taking up Fe from various chelated Fe^{III} forms, including plant-derived (e.g. citrate and malate), synthetic ligands such as ethylenediaminetetraacetic acid (EDTA) or ferric ethylenediaminedi(*o*-hydroxyphenyl) acetic acid (EDDHA) (Brüggemann et al., 1993; de la Guardia and Alcántara, 1996; Nikolic and Römheld, 1999; Rombolà et al., 2000; Larbi et al., 2001), as well as microbial siderophores (Fernandez et al., 2005). Furthermore, the ability of sunflower leaves to reduce Fe^{III} from a complex with humates has recently been demonstrated (Nikolic et al., 2003). However, the efficacy of Fe^{III} reduction and thus Fe uptake by leaf cells varies greatly between different Fe^{III} chelates, and is, for instance, higher (e.g. lower K_m) for Fe complexed with organic acids as compared with artificial chelates (e.g. Fe^{III}EDTA and Fe^{III}EDDHA; Nikolic and Römheld, 1999; Rombolà et al., 2000).

2.2 Apoplastic features that modulate Fe uptake

2.2.1 pH

In both roots and leaves Fe^{III} chelate reductase is pH-dependent (for review, see Moog and Brüggemann, 1994; Schmidt, 1999). To our knowledge, Brüggemann et al. (1993) were the first to report that a reduction of Fe^{III} citrate by plasma membrane vesicles, isolated from leaves of cowpea [*Vigna unguiculata* (L.) Walp.], is distinctly pH-dependent with a pH optimum in the range 6.5 to 6.8., a finding confirmed by Rombolà et al (2000) in the plasma membrane isolated from kiwifruit [*Actinidia deliciosa* (A. Chev.) C.F. Liang et A.R. Ferguson] leaves (pH optimum at 6.5).

On the other hand, our results did not reveal distinct pH dependence of *in vivo* reduction of Fe^{III} citrate by leaf discs of sunflower in the pH range from 5.0 to 6.5, typical of the leaf apoplastic fluid measured by various methods (Fig. 2A; Dannel et al., 1995; Mühling and Läuchli, 2001). Fe deficiency, induced by several factors in many plant species, does not seem to be accompanied by any significant change in the pH of the bulk apoplastic fluid, collected by centrifugation which remains in the range from 5.0 to 6.7 (Table 1). It must be taken into account, however, that the buffer capacity of

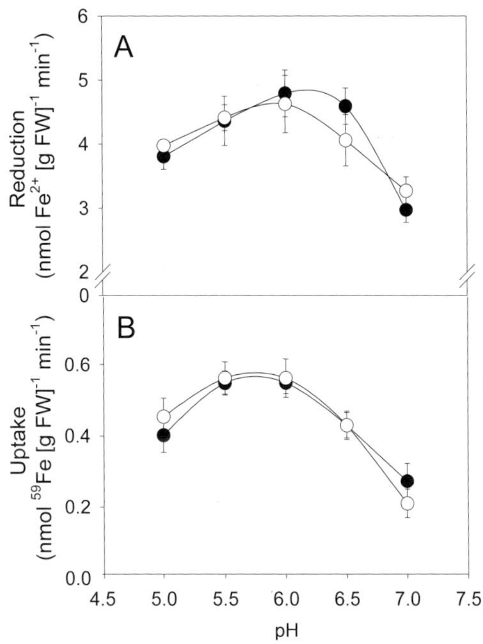

Fig. 2. pH dependence of Fe^{III} citrate reduction (A) and Fe uptake from $^{59}Fe^{III}$ citrate (B) in leaf discs of sunflower (redrawn from Nikolic and Römheld, 1999). After vacuum infiltration, leaf discs were incubated in the solutions [A, 20 μM Fe^{III} citrate, 1:1.1 molar ratio, 300 μM BPDS; and B, ^{59}Fe-labelled, 20 μM Fe^{III} citrate, 1:100 molar ratio, specific activity 4 μCi μmol^{-1} Fe] in darkness. Both solutions were buffered with 10 mM MES (pH 5.0, 5.5, 6.0, and 6.5, respectively) or 10 mM HEPES (pH 7.0). Plants were grown for 2 weeks in the nutrient solutions at high (black circle, 100 μM) and low (white circle, 0.2 μM) Fe, supplied as Fe^{III}EDDHA.

Table 1. Effect of different Fe nutrition and HCO_3^- or NO_3^- supply to roots on pH and concentrations of Fe and organic anions in the leaf apoplastic fluid of various plant species[a].

Plant species	Fe nutritional status	pH	Fe (μM)	Citrate (mM)	Malate (mM)
Helianthus annuus[b]	1 μM Fe; NO_3^-, pH 5.0 (G)	6.7	2.8	0.6	1.1
	1 μM Fe; NO_3^-, unbuffered (C)	6.7	1.3	1.6	2.4
	1 μM Fe; NH_4NO_3, pH 7.5 (C)	6.4	0.7	4.3	3.3
	10 μM Fe + HCO_3^- (C)	6.5	1.3	2.4	9.6
Vicia faba[c]	100 μM Fe (G)	5.5	31.7	3.4	2.7
	1 μM Fe + HCO_3^- (C)	5.3	11.1	9.2	3.9
Vitis sp.[d]	10 μM Fe (G)	5.3	4.1	0.2	2.5
	10 μM Fe + HCO_3^- (C)	5.2	1.8	0.5	4.5
Beta vulgaris[e]	45 μM Fe (G)	6.5	6.0	0.7	0.7
	0 μM Fe (C)	6.3	2.5	4.4	2.2

[a] Apoplastic fluid was collected from intact leaves by the centrifugation method; the pH measurements were carried out by using a glass microelectrode; Fe was supplied as Fe^{III}EDDHA, with exception of the experiment with sugar beet where it was supplied as Fe^{III}EDTA.
[b] Compiled data from Nikolic and Römheld (2001, 2003), and V. Römheld and M. Nikolic (unpubl. data). Plants were grown in nutrient solutions supplied with 4 mM N (NO_3^-or NH_4NO_3); the pH of nutrient solutions was buffered either with 2-(N-morpholino)ethanesulfonic acid (MES; pH 5.0) or with HEPES (pH 7.5), both at final concentration of 5 mM; HCO_3^- was supplied at a concentration of 10 mM.
[c] Compiled data from Nikolic and Römheld (1999), and V. Römheld and M. Nikolic (unpubl. data).
[d] From V. Römheld and M. Nikolic (unpubl. data).
[e] Based on data from López-Millàn et al. (2000).
G, green; C, chlorotic.

cell walls and the presence of CO_2 in the apoplastic space can contribute to modulation of bulk pH (an average of parenchyma, adaxial, and abaxial epidermal cells) during centrifugation. Moreover, the major limitation of this method of obtaining apoplastic pH is that it does not reveal possible spatial pH changes within the whole apoplast.

It has been shown that the presence of the H^+-ATPase inhibitor vanadate depresses Fe^{III} reduction and thus Fe uptake into the symplast of broad bean (*Vicia faba* L.) leaf discs (V. Römheld and M. Nikolic, unpubl. data). Furthermore, a marked increase in an ATPase-mediated proton pumping, found in the leaf cells of Arabidopsis as an immediate response to Fe deficiency (Schmidt, 2003), may play a pivotal role in maintaining optimal pH for Fe^{III} reductase activity in the vicinity of the apoplastic side of the plasma membrane. That a decrease in reduction of Fe^{III} citrate occurs only at high apoplastic pH (e.g. 7.0), caused by the infiltration of 4-(2-hydroxyethyl)-1-piperazineethanesulfonic acid (HEPES) buffer into the apoplast of sunflower leaves, has been demonstrated by two different experimental approaches: 1) in leaf discs by measuring the absorbance of Fe^{II}BPDS complex in the reaction medium (Nikolic and Römheld, 1999), and 2) in excised leaves determined by microscope image analysis in the formation of the Fe^{II} ferrozine complex (Kosegarten et al., 1999).

The pH dependence of Fe uptake by leaf discs of sunflower infiltrated into pH-controlled uptake medium showed a similar tendency to that of Fe^{III} reduction; the pH optimum was also in the physiologically relevant pH range (Fig. 2B). This finding was recently confirmed using a different experimental approach by infiltration of pH buffers together with labelled $^{59}Fe^{III}$ citrate into the leaf apoplast of intact leaves (e.g. sunflower) by the transpiration stream (Fig. 3). Infiltration of various buffer solutions into the excised leaves has been shown to be very efficient in controlling the apoplastic pH of xylem vessels allowing only minor pH changes (Kosegarten et al., 1999). This approach is also suitable in studying the effect of light, without any risk of photochemical reduction and thus breaking Fe citrate complexes in the uptake solution during the uptake period. From these observations it would appear that only $^{59}Fe^{III}$ citrate transported to the apoplast of mesophyll cells via the xylem apoplast can be reduced photochemically by exposing the leaf lamina to light. Furthermore, in vitro photoreduction of Fe^{III} citrate by a high PPFD (500 μmol m^{-2} s^{-1}; cool white light, 350–700 nm) has been found to be much less pH-regulated in the range from 5.0 to 7.5 as compared with the reduction by leaf

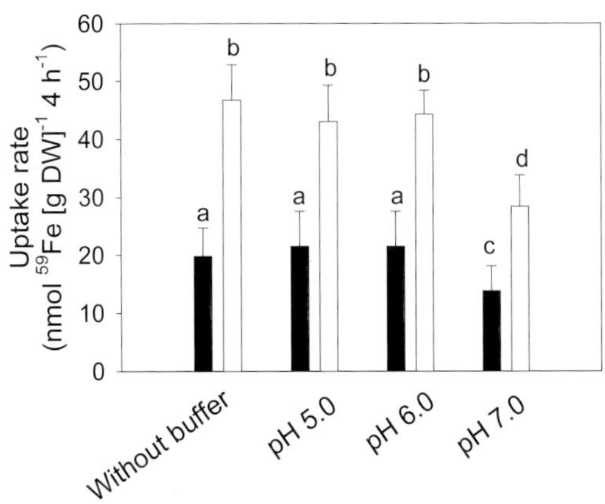

Fig. 3. Effect of apoplastic pH on ^{59}Fe uptake by the symplast of intact sunflower leaves under light and dark conditions (from Nikolic and Römheld, 2003). The ^{59}Fe-labelled solution (10 μM Fe^{III} citrate, 1:100 molar ratio, specific activity 10 μCi μmol^{-1} Fe) without or with various pH-buffers (50 mM MES, pH 5.0 and 6.0, respectively; 50 mM HEPES, pH 7.0) was infiltrated into the apoplast of excised leaves via the petiole by the transpiration stream. Leaves were further incubated in darkness or under light (500 μmol m^{-2} s^{-1}). Black bars, dark; white bars, light. Different letters denote significant differences at P<5%.

tissue in darkness (V. Römheld and M. Nikolic, unpubl. data; Nikolic and Römheld, 1999).

As has been shown in the recent study of Nikolic and Römheld (2003), apart from the effect of high apoplastic pH, high light intensity strongly increased Fe uptake rates into the symplast of leaf cells of sunflower. Hence, the Fe uptake rate at high apoplastic pH (7.0) under high irradiance was even higher than that found at low apoplastic pH (5.0 and 6.0, respectively) in the dark (Fig. 3). Increase in apoplastic pH under dark conditions may probably be enhanced by inhibiting light-induced H^+ extrusion into the leaf apoplast (Mühling and Läuchli, 2001). Then inhibition of Fe^{III} reduction can be more prominent, taking into account the prevailing enzymatic Fe^{III} reduction by plasma membrane-bound reductase.

2.2.2 Organic acids

Iron deficiency has been generally shown to cause increases in the concentration of organic anions (mainly citrate and malate) in roots, stem exudates, and leaves where predominantly citrate appears to accumulate (Abadía et al., 2002). However, the relative proportion of soluble Fe present in the apoplastic fluid as various complexes with organic anions is very low, containing only up to 1% of total leaf Fe (Nikolic and Römheld, 1999, 2001, 2003). As a consequence of increases in organic anions concentrations but decreases in the total soluble apoplastic Fe fraction, the molar ratio of e.g. citrate:Fe dramatically increases in the leaf apoplastic fluid of various plant species under Fe deficiency (Table 1; see Abadía et al., 2002, and refs. therein). Since $[FeCitOH]^-$, and sometimes $[FeCit_2OH]^{3-}$, have been assumed to be the major Fe species in the apoplastic fluid, the citrate:Fe molar ratio could be one of the most important factors controlling this speciation (López-Millàn et al., 2000, 2001).

Increasing in citrate : Fe molar ratio from the normal range (up to 100) found in the apoplastic fluid of Fe-adequate plants (for review see Abadía et al., 2002) to 1000 significantly decreased Fe uptake by mesophyll cells of sunflower leaves (Nikolic and Römheld, 1999). Since the citrate : Fe molar ratio found in the leaf apoplastic fluid of Fe-deficient plants is above 1000 (Nikolic and Römheld, 1999, 2002; López-Millàn et al., 2000, 2001), Fe uptake in chlorotic leaves seems to be impaired. The activity of the plasma membrane-bound Fe^{III} chelate reductase has also been shown to decrease significantly when the citrate:Fe ratio increases from 100 to 500 (González-Vallejo et al., 1999). The inhibition of in vitro light-dependent reduction of Fe^{III} citrate at citrate : Fe ratio of 1000 was caused by a fast re-oxidation of Fe^{2+} by high excess citrate (Bienfait and Scheffers, 1992).

3. MOBILITY AND BINDING FORMS OF IRON IN THE LEAF APOPLAST

In contrast to roots, there is little information in the literature on Fe distribution between apoplast and symplast in leaves, mainly because of a lack of reliable methods for assessing this distribution. Pich and Scholz (1991) were the first, to our knowledge, who published an alternative method for estimation of apoplastic leaf Fe by calculating the difference between total Fe content in intact leaves and that present in isolated protoplasts. Another method was described by Becker et al. (1992) in which apoplastic leaf Fe was determined from leaf discs after continuous shaking in a solution containing 10 mM Na_2EDTA. Further, Nikolic and Römheld (2002) modified the method of Bienfait et al. (1985) and adapted it for determination of apoplastic Fe in leaf discs: Fe was removed as Fe^{II}[bipyridyl]$_3$ complex during vacuum infiltration and incubation of leaf discs with bipyridyl (Fe^{2+} scavenger) and dithionite (reductant) under N_2 atmosphere. No significant contamination of the assay solution neither by Fe leakage from the symplast nor by Fe released from membranes and organelles of broken cells on disc cut edges was indicated during 30 min reductive remobilization of Fe^{II} (Nikolic and Römheld, 2002). This simple method is suitable for rapid partitioning between apoplastic and symplastic Fe in the leaves. A method to determine different Fe-binding forms in the leaf apoplast, combining a collection of apoplastic washing fluid (AWF) by an infiltration/centrifugation technique with reductive extraction of isolated cell walls in the presence of bipyridyl and dithionite has recently been reported (Nikolic and Römheld, 2003).

Once taken up by the roots, Fe is transported to the shoots via the xylem, mainly in the form of negatively charged Fe^{III} citrate complexes (Tiffin, 1966; White et al., 1981; Cataldo et al., 1988). The xylem stream enters the leaf via the veins from the petiole and moves towards the sites of high evaporation such as leaf margins (Sattelmacher, 2001). In agreement with this most accepted concept, Fe^{III} can be bound to polymeric structural components of cell walls, presumably to free carboxyl groups of acidic polysaccharides (e.g. pectins rich in galacturonate residues; Fry et al., 2002, and refs. therein), before being taken up by the symplast of mesophyll cells through a plasma membrane-associated transporter. Thus, the relative proportion of soluble Fe present in the leaf apoplastic fluid as various complexes with organic anions (e.g. citrate and malate) seems to be very low; most of the apoplastic Fe is bound in the cell walls (over 95%) with different binding strengths (Nikolic and Römheld, 2001, 2003). However, even this relatively small soluble Fe fraction might be of particular significance for Fe uptake into the symplast.

According to the method of Nikolic and Römheld (2003), Fe removed during reductive extraction of the isolated cell wall-material represents a weakly wall-bound fraction, mainly attached to the pectin, while Fe remaining after extraction can be considered a strongly wall-bound fraction. Because of spatial limitation for cations *in vivo*, the cation exchange capacity of isolated cell-wall material is much higher than that of the cell walls of intact plant tissue (Sattelmacher, 2001). Thus, during the isolation procedure, the cell walls are able to fix a certain portion of symplastic Fe (e.g. soluble Fe-containing proteins and Fe in vacuoles), which may lead to an overestimation of the proportion of Fe located in the cell walls (Nikolic and Römheld, 2003). Nevertheless, the relative proportions of insoluble, wall-bound Fe is not dependent on Fe status. On the other hand, the relative amount of soluble Fe in the apoplastic fluid, and sometimes mobile Fe in the cell walls (weakly wall-bound), decrease several fold in chlorotic leaves (Nikolic and Römheld, 2001, 2003).

It can be concluded that availability of Fe in the leaf apoplast might be caused by a dynamic equilibrium between Fe^{III} from cell-wall pools (which can easily be remobilised by organic anions) and soluble Fe^{III} complexes in the apoplastic solution. In principal, irrespective of what could be the inducing factor of Fe deficiency, concentration of apoplastic Fe (e.g. soluble Fe in apoplastic fluid and insoluble wall-bound Fe) is always lower in chlorotic than in green leaves, with a tendency to follow changes in concentration of symplastic Fe.

4. IRON INACTIVATION IN LEAVES: PRO AND CONTRA

4.1 The "chlorosis paradox"

The main support for the postulated Fe inactivation in leaves (e.g. Bergmann, 1992; Mengel, 1994) comes from the findings that the concentration of total leaf Fe expressed per leaf dry weight basis in chlorotic leaves can sometimes be the same or even higher than in green one of field-grown woody plants e.g. pear (*Pyrus communis* L.) and grapevine ("the chlorosis paradox"; e.g. Morales et al., 1998; Römheld, 2000). Earlier explanation of this phenomenon was a long-distance effect of HCO_3^- on the inhibition of Fe transfer from vascular bundles into intercostal leaf cells (Bergmann, 1992). More detailed explanation, put forward in the past decade, is that the higher Fe concentrations in the leaves of chlorotic plants could be caused by an increase in the pH of the leaf apoplast induced by high bicarbonate (HCO_3^-) or/and nitrate (NO_3^-) supply in the soil or nutrient solutions (e.g. Mengel, 1994). It has been hypothesized that since a high leaf apoplastic pH depresses the activity of plasma membrane-located Fe^{III} reductase, less Fe^{2+} can be transported across the plasma membrane into the leaf symplast, resulting in Fe deficiency chlorosis (Kosegarten et al., 2001). Presumably, this non-utilized Fe^{III} should be accumulated outside the plasma membrane and most probably trapped in the cell walls, thus implying a relative increase in Fe concentration. Despite this hypothesis, however, chlorotic plants with higher concentrations of Fe in younger leaves have never, so far, been found in plants grown in nutrient solutions under controlled environmental conditions (Römheld, 2000; Nikolic and Römheld, 2002).

Another explanation is that various soil factors (e.g. CO_2, ethylene, low temperature, high water content, and drought), which result in severe inhibition of root growth, might be responsible for triggering a restriction in leaf expansion that, in turn, elevates the Fe concentration in these chlorotic leaves as a consequence of the diminished dilution (Römheld, 2000). This explanation would appear to account for the phenomenon of the "chlorosis paradox", which occurs occasionally under field conditions and is always associated with inhibited leaf growth (Häussling et al., 1985; Römheld, 2000; Nikolic and Römheld, 2002).

4.2 Can high HCO_3^- root supply cause Fe inactivation in leaves?

It is well known that at high concentration in the soil solution, HCO_3^- penetrates the root apoplast, and thereby neutralizes protons and lowers Fe uptake by root cells as a consequence of inhibition of the plasma membrane-bound Fe^{III} chelate reductase at high apoplastic pH (Römheld and Marschner, 1986; Toulon et al., 1992). However, according to the hypothesis of Mengel (1994) Fe deficiency chlorosis on calcareous soils is not caused by low Fe acquisition by roots, but by restricted Fe uptake from the leaf apoplast into the leaf symplast. There is no evidence, however, that free dissolved HCO_3^- can be transported from root to shoot via the xylem to a an extent high enough to cause a substantial increase in the pH of the xylem sap (Bialczyk and Lechowski, 1995; Peiter, et al., 2001). López-Millán et al. (2000) have recently confirmed that most of the dissolved HCO_3^- taken up is utilized in the processes of organic carbon fixation [e.g. phospohoenol-pyruvate (PEP)-carboxylase] by roots of sugar beet (*Beta vulgaris* L.). Therefore, an increase in organic anions (mainly citrate and malate) in the apoplastic fluid of chlorotic leaves of various plant species grown at elevated HCO_3^- concentration (Table 1) is mainly due to enhanced PEP-carboxylase activity in roots and the export of organic acids to the leaves via the xylem, rather than to an increase in leaf carbon-fixation. Furthermore, the presence of high HCO_3^- in the nutrient or soil solutions was shown to affect neither the bulk pH of apoplastic fluid obtained by centrifugation of intact leaves of various plant species (Table 1; López-Millán et al., 2001; Nikolic and Römheld, 2002) nor the apoplastic pH estimated by using a pH-sensitive fluorescent dye, i.e. fluorescein isothiocyanate (FITC)-dextran (Kosegarten et al., 1999). Therefore, the considerable increase in the apoplastic pH of sunflower leaves, reported in the earlier study of Mengel et al. (1994), seems to be an artefact of the experiment with excised leaves where diffusion of HCO_3^- from a solution infiltrated into the leaf apoplast was not restricted. It has recently been demonstrated that at extremely high HCO_3^- supply of 20 mM (above the environmental concentration in calcareous soils), the xylem pH of maize (*Zea mays* L.) seedlings, as measured in situ with a novel xylem pH-probe, raised by about 1 pH unit, however without any Fe deficiency symptoms (Wegner and Zimmermann, 2004).

The concentration of symplastic and all forms of apoplastic Fe [i.e. soluble (apoplastic fluid) and insoluble (weakly and strongly wall-bound)] were significantly decreased in chlorotic leaves of HCO_3^--supplied sunflower plants (Nikolic and Römheld, 2001). Although the relative amount of Fe in the apoplastic fluid decreased about three times, relative proportions of Fe in other apoplastic fractions were not affected by HCO_3^-.

In the field environment, higher concentrations of Fe were found only in the leaves of the grapevine genotypes in which there was severe inhibition of leaf expansion (Häussling et al., 1985; Nikolic and Römheld, 2002). However, relative apoplastic Fe did not increase in chlorotic leaves, even in the plants with severe growth inhibition (Table 2). Also, the concentration of Fe in the leaf apoplastic fluid was about 2-fold lower in chlorotic grapevine

Table 2. Chlorophyll content, growth characteristics, and concentrations of total and relative apoplastic Fe (share of total Fe) in green and chlorotic leaves of grapevine[a] (data adapted from Nikolic and Römheld, 2002).

Leaves	Chlorophyll (SPAD index)	Leaf dry weight (g leaf^{-1})	Leaf area (cm^2 leaf^{-1})	Total Fe (μg^{-1} DW)	Relative apoplastic Fe (%)
		Site I (cv. Pinot Noir, rootstock 5BB)			
Green	24a	0.6a	136a	75a	30a
Chlorotic	5b	0.2b	53b	82a	28a
		Site II (cv. Würzer, rootstock SO4)			
Green	29a	0.6a	130a	114a	33a
Chlorotic	9b	0.3b	70b	78b	28a

[a] Plants were grown under field conditions on calcareous soil in the wine growing area of the river Nahe (Laubenheim, Germany); different letters denote significant differences at P<5%.

leaves, irrespectively of whether or not the expansion of young leaves was restricted (V. Römheld and M. Nikolic, unpubl. data). Furthermore, it has clearly been shown in short-term ^{59}Fe-uptake experiments with grapevine plants, that the presence of HCO_3^- in the uptake root medium strongly inhibits both root uptake and root-to-shoot translocation of Fe (Nikolic et al., 2000; Römheld, 2000).

It can be summarized that HCO_3^- does not appear to physiologically inactivate Fe in the leaf apoplast, nor does it cause an inhibition of Fe uptake from the apoplast into the symplast of leaf mesophyll. This is in agreement with the conclusions of Kosegarten et al. (1999, 2001).

4.3 Can nitrate nutrition result in Fe inactivation in leaves?

It has been hypothesized that inactivation of Fe in the leaves is caused primarily by NO_3^-, which, supplied from the roots via xylem transport, can increase the pH in the leaf apoplast by proton co-transport of NO_3^- into the leaf cells, and thus inhibit plasma membrane-associated Fe^{III} reduction (Mengel, 1994; Kosegarten et al., 1999). In our experiments, the bulk pH of the apoplastic fluid from sunflower leaves of NO_3^--fed plants did not significantly differ from those grown in nutrient solution supplied with sole ammonium (NH_4^+) or NH_4NO_3 (Nikolic and Römheld, 2003), which is in agreement with the findings of Kosegarten et al. (1999) showing mean data of the apoplastic pH in the excised leaves of sunflower (Table 3). The pH values obtained by using a pH-sensitive fluorescence dye with lower molecule size (i.e. 5-carboxyfluorescein) were very similar to those found in direct pH measurements of leaf apoplastic fluid (López-Millán et al., 2000). Hence, the lower pH values (about 1 pH unit) reported by Kosegarten et al. (1999) could have resulted from the restricted access of the relatively large molecules of FITC-dextrane to the whole apoplastic space. In Kosegarten's experiments the increase in apoplastic pH above 7.0 in the small interveinal areas of young sunflower leaves were interpreted as a consequence of high uptake rates of NO_3^- by rapidly expanding mesophyll cells (Kosegarten et al., 1999). Considering, however, that such a high pH accounts for less than 10% of the total interveinal leaf apoplast of NO_3^--fed plants (see Kosegarten et al., 1999), it seems difficult to explain how a so small proportion of the apoplastic space could depress Fe uptake to an extent inducing leaf chlorosis. In particular, this intercostal area of increased apoplastic pH could not be identified as a main area of leaf growth.

Increasing the N supply to roots solely as NO_3^- (up to 40 mM) did not change the relative distribution of Fe between leaf apoplast and symplast, if the external pH of the root medium is kept constantly low (Nikolic and Römheld, 2003). On the other hand, NO_3^- supply to the unbuffered nutrient

Table 3. Effect of N nutrition on the apoplastic pH of sunflower leaves obtained with different methods[a].

Nutritional N source	Chlorophyll (SPAD index)	Apoplastic pH	Method	Reference
NH_4NO_3 (3 mM)	n.d (G)	5.4	Fluorescent dye (intact tissue)	Kosegarten et al. (1999)[c]
NO_3^- (6 mM)	n.d. (C)	5.5		
NH_4NO_3 (2 mM)	36 (G)	6.5	Centrifugation, pH-microelectrode (apoplastic fluid)	Nikolic and Römheld (2003)[d]
NO_3^- (4 mM)	17 (C)	6.7		
NO_3^- (40 mM, pH 5.0[b])	32 (G)	6.6		
NH_4^+ (4 mM, pH 7.5[b])	4 (C)	6.5		

[a] In both studies, the Fe concentration in the various nutrient solutions was 1 μM.
[b] The pH of nutrient solutions was buffered either with MES (pH 5.0) or with HEPES (pH 7.5).
[c] Data are mean pH values from the interveinal area of young leaves after infiltration with xylem sap.
[d] Data represent the bulk pH of apopalstic fluid corresponding to the whole leaf apoplast.
n.d., not determined; G, green; C, chlorotic.

solution resulted in significant decreases in concentration of leaf Fe in both compartments, symplast and apoplast, however, with a tendency to maintain the relative proportion of apoplastic Fe unchanged (Nikolic and Römheld, 2003). The relative proportions of symplastic and apoplastic (i.e. insoluble wall-bound Fe) leaf Fe did not differ between green (low external pH) and chlorotic (high external pH) leaves in both NH_4NO_3- and NO_3^--fed plants (Fig. 4). Exceptionally, the relative proportion of soluble Fe in the apoplastic fluid was lower in chlorotic leaves irrespective of N form. These results, however, showed the contrary to what would have been expected according to the hypothesis of Mengel and co-workers: Both binding forms of leaf cell-wall Fe were significantly lower in all chlorotic plants grown in the high pH nutrient solutions regardless of N form (see Nikolic and Römheld, 2003).

Although it has been reported that the perfusion of excised sunflower leaves with NO_3^- resulted in microsites with apoplastic pH of around 7.0 (Hoffmann and Kosegarten, 1995), loading of 6 mM NO_3^- into the apoplast of both sunflower and grapevine leaves did not depress ^{59}Fe uptake into the leaf symplast nor did it induce an increase in ^{59}Fe trapped into the isolated cell walls (Nikolic and Römheld, 2003; V. Römheld and M. Nikolic, unpubl. data). The lack of correlation between leaf apoplastic NO_3^- concentration and Fe concentration in the cell walls (Fig. 5) also provides further evidence against the postulated hypothesis of Fe inactivation in leaves.

In the nutrient solutions without pH control, high NO_3^- supply decreased Fe uptake due to an increase in root surface pH from 5.6 (initial pH in the renewed solution) to about 7.0 as measured after two days (Nikolic and Römheld, 2003), most likely because of protons consumption by root cells via H^+/NO_3^- symport (Meharg and Blatt, 1995). In the same study, the depression of Fe uptake even by solely NH_4^+ -fed sunflower plants has also

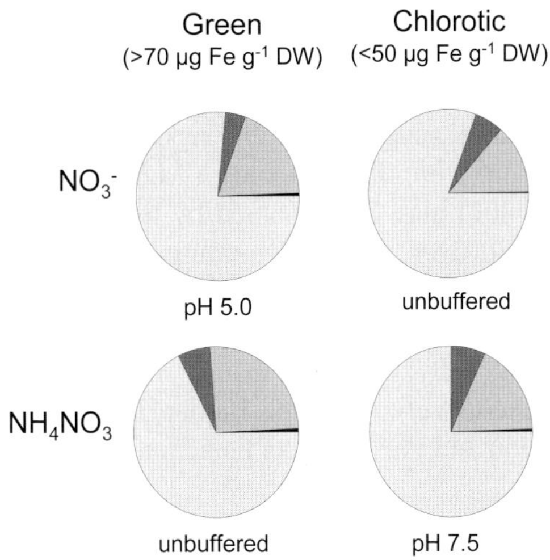

Fig. 4. Effects of N forms and external pH on relative distribution of Fe between symplastic and apoplastic fractions and apopalstic Fe binding forms of young sunflower leaves (from Nikolic and Römheld, 2003). Plants were grown for 2 weeks in the nutrient solutions at 4 mM N supplied with either $Ca(NO_3)_2$ or NH_4NO_3. In all treatments Fe was applied as Fe^{III}EDDHA at 1 µM. Symplastic Fe, light gray. Apoplastic Fe: weakly bound (cell walls), gray; strongly bound (cell walls), dark gray; soluble (AWF), black.

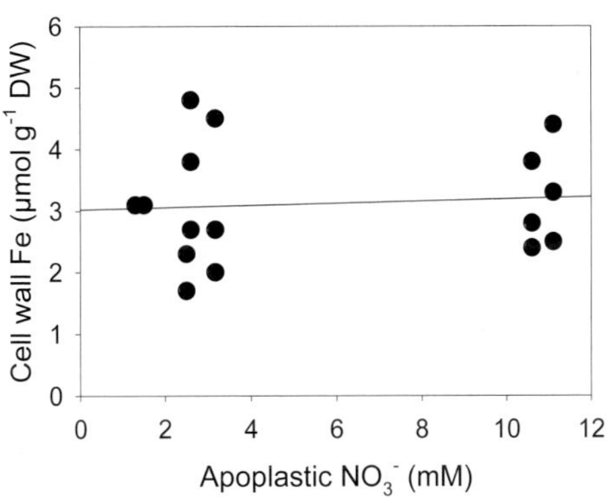

Fig. 5. Cell wall Fe concentration in relation to NO_3^- concentration in the leaf apoplastic fluid from young leaves of sunflower (from Nikolic and Römheld, 2003).

been found when the pH of nutrient solution was buffered at 7.5. This was confirmed using a similar experimental approach with grapevine plants (V. Römheld and M. Nikolic, unpubl. data). Thus, regardless of the presence of high concentration of NO_3^- in the leaf apoplast, Fe chlorosis did not appear if sufficient Fe was taken up by the roots, which was possible in the low pH-buffered nutrient solution (Nikolic and Römheld, 2003).

According to Kosegarten et al. (2001), Fe trapped in the leaf apoplast at high pH can be mobilized by artificial lowering of the apoplastic pH (e.g. foliar acid spraying) to favour Fe^{III} reduction and thus Fe uptake into the symplast. We therefore suggest that the reported regreening of chlorotic leaves after spraying with diluted acids (Tagliavini et al., 1995, 2000; Kosegarten et al., 2001) may be explained by the effect of acids (e.g. citric acid) in extracting Fe from cell wall binding sites and thereby increasing soluble Fe in the leaf apoplastic fluid, rather than by the postulated enhanced reduction of Fe^{III} due to lowering of the apoplastic pH.

We conclude that NO_3^- nutrition results in Fe deficiency chlorosis in Strategy I plants exclusively from the high pH at the root surface which inhibits Fe^{III} reduction and thus restricts Fe acquisition.

5. CONCLUDING REMARKS

The results obtained in our contribution to this research project have allowed more light to be shed on the dynamics of Fe in the leaf apoplast, necessary for a better understanding of Fe chlorosis. Although the pH of the leaf apoplast seems to be important in regulating the external reduction of Fe^{III} compounds (e.g. by plasma membrane-bound reductase and/or less pH-dependent photochemical reduction) and thereby the uptake of Fe from the apoplast into the symplast of leaf cells, our findings show that within the physiologically relevant pH range (as measured in various Fe-deficient plant species) this effect is only of marginal significance. In particular our work on the so far controversially discussed Fe inactivation in the leaf (i.e. induced by NO_3^- nutrition) does not support this well-known hypothesis. On the contrary, we found that if the pH increase of the nutrient solution supplied with solely NO_3^- is prevented by pH buffers, the form of N nutrition neither induced Fe deficiency nor did it influence the partitioning of Fe between apoplast and symplast in the leaves. The results obtained in our studies clearly show that occurrence of Fe deficiency chlorosis induced by either high HCO_3^- or NO_3^- supply to roots is exclusively caused by inhibited uptake and translocation of Fe from roots to shoots as a consequence of high pH at the root surface. When supply of Fe to the shoots is low, some features in the leaf apoplast (e.g. organic acids) may influence the distribution of Fe between the soluble (apoplastic fluid) and insoluble, cell wall-bound fractions and thereby the uptake of Fe into the symplast of mesophyll cells. Iron-deficiency chlorosis might be thus modulated to a certain extent.

ACKNOWLEDGEMENTS

We thank Ernest A. Kirkby (University of Leeds, UK) for improving the English of the manuscript. The first author is grateful to the Serbian Ministry of Science and Environmental Protection (grant no. 143020B).

REFERENCES

Abadía, J., López-Millàn, A.F., Rombolà, A. and Abadía, A. (2002). Organic acids and Fe deficiency: a review. *Plant Soil*, 241, 75–86.

Becker, R., Grün, M. and Scholz, G. (1992). Nicotianamine and the distribution of iron into the apoplasm and sympalsm of tomato (*Lycopersicon esculentum* Mill.). I. Determination of the apoplasmic and symplasmic iron pools in roots and leaves of the cultivar Bonner Beste and its nicotianamine-less mutant chloronerva. *Planta,* 187, 48–52.

Bergmann, W. (1992). Nutritional disorders of plants - Development, visual and analytical diagnosis (p. 233). Jena: Gustav Fisher Verlag.

Bialczyk, J. and Lechowski, Z. (1995). Chemical composition of xylem sap of tomato grown on bicarbonate containing medium. *J. Plant Nutr.*, 18, 2005–2021.

Bienfait, H.F. and Scheffers, M.R. (1992). Some properties of ferric citrate relevant to the iron nutrition of plants. *Plant Soil*, 143, 141–144.

Bienfait, H.F., van den Briel, W. and Mesland-Mul, N.T. (1985). Free space iron pools in roots. Generation and mobilization. *Plant Physiol.*, 78, 596–600.

Brown, J.C., Cathey, H.M., Bennett, J.H. and Thimijan, R.W. (1979a). Effect of light quality and temperature on Fe^{3+} reduction, and chlorophyll concentration in plants. *Agron. J.*, 71, 1015–1021.

Brown, J.C., Foy, C.D., Bennett, J.H. and Christiansen, M.N. (1979b). Two light sources differentially affected ferric iron reduction and growth of cotton. *Plant Physiol.*, 63, 692–695.

Brüggemann, W., Maas-Kantel, K. and Moog, P.R. (1993). Iron uptake by leaf mesophyll cells: The role of the plasma membrane-bound ferric-chelate reductase. *Planta*, 190, 151–155.

Cataldo, D.A., McFadden, K.M., Garland, T.R. and Wildung, R.E. (1988). Organic constituents and complaexation of nickel (II), iron (III), cadmium (II), and plutonium (IV) in soybean xylem exudates. *Plant Physiol.*, 50, 208–213.

Connolly, E.L., Campbell, N.H., Grotz, N., Prichard, C.L. and Guerinot, M.L. (2003). Overexpression of the FRO2 ferric chelate reductase confers tolerance to growth on low iron and uncovers posttranscriptional control. *Plant Physiol.*, 133, 1102–1110.

Connolly, E.L., Fett, J.P. and Guerinot, M.L. (2002). Expression of the IRT1 metal transporter is controlled by metals at the levels of transcript and protein accumulation. *Plant Cell*, 14, 1347–1357.

Dannel, F., Pfeffer, H. and Marschner, H. (1995). Isolation of apoplasmic fluid from sunflower leaves and its use for studies on influence of nitrogen supply on apoplasmic pH. *J. Plant Physiol.*, 146, 273–278.

de la Guardia, M.D. and Alcántara, E. (1996). Ferric chelate reduction by sunflower (*Helianthus annuus* L.) leaves: influence of light, oxygen, iron-deficiency and leaf age. *J. Exp. Bot.*, 47, 669–675.

Fernandez ,V., Ebert, G. and Winkelmann. G. (2005). The use of microbial siderophores for foliar iron application studies. *Plant Soil*, 272, 245–252.

Fry, S.C., Miller, J.G. and Dumville, J.C. (2002). A proposed role for copper ions in cell wall loosening. *Plant Soil*, 247, 57–67.

González-Vallejo, E.B., González-Reyes, J.A., Abadía, A., López-Millán, A.F., Yunta, F., Lucena, J.J. and Abadía, J. (1999). Reduction of ferric chelates by leaf plasma membrane preparations from Fe-deficient and Fe-sufficient sugar beet. *Austr. J. Plant Physiol.*, 26, 601–611.

González-Vallejo, E.B., Morales, F., Cistué, L., Abadía, A. and Abadía, J. (2000). Iron deficiency decreases the Fe(III)-chelate reducing activity of leaf protoplasts. *Plant Physiol.*, 122, 1–8.

Häussling, M., Römheld, V. and Marschner, H. (1985). Beziehungen zwischen Chlorosegrad, Eisengehalten und Blattwachstum von Weinreben auf verschiedenen Standorten. *Vitis*, 24, 158–168.

Hoffmann, B. and Kosegarten, H. (1995). FITC-dextran for measuring apoplast pH and apoplastic pH gradients between various cell types in sunflower leaves. *Physiol. Plan.*, 95, 327–335.

Kosegarten, H.U., Hoffmann, B. and Mengel, K. (1999). Apoplastic pH and Fe^{3+} reduction in intact sunflower leaves. *Plant Physiol.*, 121, 1069–1079.
Kosegarten, H., Hoffmann, B. and Mengel, K. (2001). The paramount influence of nitrate in increasing apoplastic pH of young sunflower leaf to induce Fe deficiency chlorosis, and the re-greening effect brought about acid foliar sprays. *J. Plant Nutr. Soil Sci.*, 164, 155–163.
Krizek, D.T., Bennett, J.H., Brown, J.C., Zaharieva, T. and Norris, K.H. (1982). Photochemical reduction of iron. I. Light reactions. *J. Plant Nutr.*, 5, 323–333.
Larbi, A., Morales, F., López-Millán, A.F., Gogorcena, Y., Abadía, A., Moog, P.R. and Abadía, J. (2001). Technical advance: reduction of Fe(III)-chelates by mesophyll leaf discs of sugar beet. Multi-component origin and effects of Fe deficiency. *Plant Cell Physiol.*, 42, 94–105.
López-Millán, A.F., Morales, F., Abadía, A. and Abadía, J. (2000). Effects of iron nutrition on the composition of the leaf apoplastic fluid and xylem sap in sugar beet: implication for iron and carbon transport. *Plant Physiol.*, 124, 873–884.
López-Millàn, A.F., Morales, F., Abadía, A. and Abadía, J. (2001). Iron deficiency-associated changes in the composition of the leaf apoplastic fluid from field-grown pear (*Pyrus communis* L.) trees. *J. Exp. Bot.*, 52, 1489–1498.
Meharg, A.A. and Blatt, M.R. (1995). NO_3^- transport across the plasma membrane of *Arabidopsis thaliana* root hairs: kinetic control by pH and membrane voltage. *J. Membr. Biol.*, 145, 49–66.
Mengel, K. (1994). Iron availability in plant tissues - iron chlorosis on calcareous soils. *Plant Soil*, 165, 275–283.
Mengel, K., Plänker, R. and Hoffmann, B. (1994). Relationship between leaf apoplast pH and iron chlorosis of sunflower (*Helianthus annuus* L.). *J. Plant Nutr.*, 17, 1053–1065.
Moog, P.R. and Brüggemann, W. (1994). Iron reductase systems on the plant plasma membrane - a review. *Plant Soil*, 165, 241–260.
Morales, F., Grasa, R., Abadía, A. and Abadía, J. (1998). Iron chlorosis paradox in fruit trees. *J. Plant Nutr.*, 21, 815–825.
Mühling, K. H. and Läuchli, A. (2001). Influence of chemical form and concentration of nitrogen on apoplastic pH of leaves. *J. Plant Nutr.*, 24, 399–411.
Nikolic, M. and Römheld, V. (1999). Mechanism of Fe uptake by the leaf symplast: Is Fe inactivation in leaf a cause of Fe deficiency chlorosis? *Plant Soil*, 215, 229–237.
Nikolic, M. and Römheld, V. (2001). The role of the leaf apoplast in iron nutrition of plants. In W. J. Horst et al. (Eds.), Plant nutrition - Food security and sustainability of agro-ecosystems (pp. 274–275). Dordrecht: Kluwer Academic Publishers.
Nikolic, M. and Römheld, V. (2002). Does high bicarbonate supply to roots change availability of iron in the leaf apoplast? *Plant Soil*, 241, 67–74.
Nikolic, M. and Römheld, V. (2003). Nitrate does not result in iron inactivation in the apoplast of sunflower leaves. *Plant Physiol.*, 132, 1303–1314.
Nikolic, M., Cesco, S., Römheld, V., Pinton, R. and Varanini, Z. (2003). Uptake of iron (^{59}Fe) complexed to water-extractable humic substances by sunflower leaves. *J. Plant Nutr.*, 26, 2243–2252.
Nikolic, M., Römheld, V. and Merkt, N. (2000). Effect of bicarbonate on uptake and translocation of ^{59}Fe in grapevine rootstocks differing in their resistance to Fe deficiency chlorosis. *Vitis*, 39, 145–149.
Peiter, E., Yan, F. and Schubert, S. (2001). Lime-induced growth depression in Lupinus species: Are soil pH and bicarbonate involved? *J. Plant Nutr. Soil Sci.*, 164, 165–172.
Pich, A. and Scholz, G. (1991). Nicotinamine and the distribution of iron into apoplast and symplast of tomato (*Lycopersicon esculentum* Mill.) II. Uptake of iron by protoplasts from the variety Bonner Beste and its nicotianamine-less mutant chloronerva and the compratmentation of iron in leaves. *J. Exp. Bot.*, 42, 1517–1523.
Robinson, N.J., Sadjuga, J. and Groom, Q.J. (1997). The froh gene family from *Arabidopsis thaliana*: Putative iron-chelate reductases. *Plant Soil*, 196, 245–248.
Rombolà, A.D., Brüggemann, W., Tagliavini, M., Marangoni, B. and Moog, P. (2000). Iron source affects iron reduction and re-greening of kiwifruit (*Actinidia deliciosa*) leaves. *J. Plant Nutr.*, 23, 1751–1765.
Römheld, V. and Marschner, H. (1986). Mobilization of iron in the rhizosphere of different plant species. In B. Tinker and A. Läuchli (Eds.), Advances in plant nutrition, Vol. 2 (pp. 155–192). New York: Praeger Publishers.
Römheld, V. (2000). The chlorosis paradox: Fe inactivation as a secondary event in chlorotic leaves of grapevine. *J. Plant Nutr.*, 23, 1629–1643.
Sattelmacher, B. (2001). The apoplast and its significance for plant mineral nutrition. *New Phytol.*, 149, 167–192.

Schmidt, W. (1999). Mechanism and regulation of reduction-based iron uptake in plants. *New Phytol.*, 141, 1–26.

Schmidt, W. (2003). Iron homeostasis in plants: sensing and signaling pathways. *J. Plant Nutr.*, 26, 2211–2230

Tagliavini, M., Abadía, J., Rombolà, A.D., Abadía, A., Tsipouridis, C. and Marangoni, B. (2000). Agronomic means for the control of iron deficiency chlorosis in deciduous fruit trees. *J. Plant Nutr.*, 23, 2007–2022.

Tagliavini, M., Scudellari, D., Marangoni, B. and Toselli, M. (1995). Acid-spray regreening of kiwifruit leaves affected by lime-induced chlorosis. In J. Abadía (Ed.), Iron nutrition in soils and plants (pp. 389–396). Dordrecht: Kluwer Academic Publishers.

Tiffin, L.O. (1966). Iron translocation. II. Citrate/iron ratios in stem exudates. *Plant Physiol.*, 41, 515–518.

Toulon, V., Sentenac, H., Thibaud, J.-B., Davidian, J.-C., Moulineau, C. and Gringon, C. (1992). Role of apoplast acidification by the H^+ pump: effect on the sensitivity to pH and CO_2 of iron reduction by roots of *Brassica napus* L. *Planta*, 186, 212–218.

Vert, G., Grotz, N., Dédaldéchamp, F., Gaymard, F., Guerinot, M.L., Briat, J.-F. and Curie, C. (2002). IRT1, an *Arabidopsis* transporter essential for iron uptake from the soil and for plant growth. *Plant Cell*, 14, 1223–1233.

Waters, B.M., Blevins, D.G. and Eide, D.J. (2002). Characterization of FRO1, a pea ferric-chelate reductase involved in root iron acquisition. *Plant Physiol.*, 129, 85–94.

Wegner, L.H. and Zimmermann, U. (2004). Bicarbonate-induced alkalinization of the xylem sap in intact maize seedlings as measured in situ with a novel xylem pH probe. *Plant Physiol.*, 136, 3469–3477.

White, M.C., Decker, A.M. and Chaney, R.L. (1981). Metal complexation in xylem fluid: I. Chemical composition of tomato and soybean stem exudate. *Plant Physiol.*, 67, 292–300.

SELF-REPORTING *ARABIDOPSIS THALIANA* EXPRESSING pH- AND [CA^{2+}] - INDICATORS UNVEIL APOPLASTIC ION DYNAMICS

C. PLIETH[1], D. GAO[1], M.R. KNIGHT[2], A.J. TREWAVAS[3] and B. SATTELMACHER†

[1]Zentrum für Biochemie und Molekularbiologie, Christian-Albrechts-Universität, Kiel, Germany, cplieth@zbm.uni-kiel.de; [2]Department of Plant Sciences, Durham University, School of Biological and Biomedical Sciences, UK; [3]Institute for Cell and molecular Biology, University of Edinburgh, Scotland.

Abstract. The complex network of metabolic and signal transduction pathways in a living cell is ruled by a plethora of different compounds. In particular inorganic ions are of importance for cellular homeostasis and signal transduction and, therefore, tightly controlled. How ion flux and regulation is perturbed by abiotic stimuli has not yet been worked out in detail in whole intact plant systems.

For non-invasive *in vivo* measurements of intra- and extracellular ion concentrations we tried a novel approach. We produced transgenic *Arabidopsis thaliana* expressing fused pH- and calcium indicators in the cytoplasm and in the apoplast. The transgenic *Arabidopsis* lines provide new information how calcium and pH are regulated extracellularly and intracellularly in *Arabidopsis* under abiotic stress. By direct comparison of all four measurands (cytoplasmic free calcium concentration, $[Ca^{2+}]_{cyt}$; apoplastic free calcium concentration, $[Ca^{2+}]_{apo}$; cytoplasmic pH; pH_{cyt}; and apoplastic pH, pH_{apo}) taken under identical experimental conditions new general conclusions can be drawn on the impact of unfavourable situations on activity, flux, and crosstalk of second messenger ions like Ca^{2+} and H^+.

Key words: aequorin, apoplast, calcium, cytoplasm, drought, pH, pH-sensitive GFP, salinity

B. Sattelmacher† and W.J. Horst (eds.), The Apoplast of Higher Plants: Compartment of Storage, Transport and Reactions, 373–392.

1. INTRODUCTION

1.1 Plants respond sensitively to changes in their environment

Plants cannot escape their environment and must withstand multiple abiotic and biotic stresses. Hence, they have evolved sophisticated strategies which ensure survival and reproduction under threat of all relevant environmental factors (relevant environmental factors are for example wind, light, cold, CO_2, O_2, water supply, and abiotic stresses like mineral nutrient deficiency, drought, UV, ozone, as well as biotic factors like parasitic and symbiotic organisms). These strategies are founded on a network of molecular signal transduction pathways enabling efficient reponsiveness to environmental stimuli. The questions about the causal chains and their cross-linking, i.e. how plants perceive, compile, and respond to environmental stimuli are belaboured in many labs worldwide. Many cellular compounds such as Ca^{2+}, lipids, H^+, cyclic nucleotides, and inositolphosphates are listed as messengers used by plants to forward and compile cellular signals (Sanders et al., 1999, 2002). Therefore, novel experimental strategies which unveil interference of cellular parameters and communication of transduction pathways are required to understand this complex network, and techniques are needed to directly visualize cellular transducing components *in vivo*.

1.2 Ions regulate cellular events

The cellular ion milieu has significant impact on signal transduction. Many metabolic processes in plants are dependent on ions like H^+, Ca^{2+}, as well as K^+, Na^+, Mg^{2+}, Cl^-, and NO_3^-. Physiological responses can only be produced with "well-balanced" ion concentrations. The cytoplasmic free Ca^{2+} concentration ($[Ca^{2+}]_{cyt}$), in particular, has been found to play an outstanding role in a wide range of cellular events (Trewavas and Malhó, 1998; Knight H., 2000; Sanders et al. 2002) although its' alleged universality has recently been critically questioned (Plieth, 2001, 2005). Ca, is believed to be a second messenger able to activate proteins (enzymes, transcription factors) and thus influence enzyme activity and gene expression directly or indirectly via calcium-binding proteins.

The proton activity ($[H^+]$), is an important factor as well. $[H^+]$ is involved in cell signalling either directly or in cross-talk with plant hormones or calcium (Gilroy and Trewavas, 1994; Ward et al., 1995; Blatt and Grabov, 1997; Roos, 2000; Felle, 2001, 2005). Therefore, cells exert tight control over $[H^+]$ (Felle, 1987) as is the case for $[Ca^{2+}]$.

1.3 The apoplastic ion milieu is involved in signal transduction as well

The apoplast consisting of the cell wall and the intercellular space is the first plant compartment encountering environmental signals. It has been suggested that the apoplast – in cooperation with the plasma membrane - is involved not only in the response but also in the perception and transduction of various environmental signals (Hoson, 1998). The apoplastic ion mileu has direct contact to the environment – as far as there is no cuticle – and it is not bordered by a membrane like the cytoplasm. Nevertheless, its homeostasis is well regulated by buffer and transport mechanisms (Peters and Felle, 1991; Grignon and Sentenac, 1991; reviews: Sakurai, 1998a, Sattelmacher, 2001) and sensitively responds to environmental stimuli (Figs. 10–13) (Lee and Satter, 1989; Taylor et al., 1996; Felle et al., 2000). In particular calcium seems to perform different tasks in the apoplast. DeMarty et al. (1984) demonstrated that calcium activates cell-wall phosphatases. Moreover, it has been reported several times that apoplastic calmodulin plays an important role in signal transduction (Sun et al., 1994, 1995; Tang et al., 1996; Ma and Sun, 1997; Ma et al., 1999). Thus, Ca seems to have a signalling function in the apoplast as well.

2. METHODS AND TECHNIQUES

2.1 Measuring ion concentrations *in vivo*

One indispensable pre-requisite to study ion signalling is the capability to measure free ion concentrations *in vivo*. This was quite difficult so far. On the one hand ion-selective microelectrodes have been used for a wide range of different ion species. This technique provides excellent linearity and easy calibration over a broad range of ion concentrations (Felle, 1991; Felle and Hanstein, this volume, pp. 295–306). However, this technique requires a lot of practise in micromanipulation. On the other hand optical methods using fluorescent indicators have also been widely used. They provide both, spatial information about ion distribution and kinetic information with good temporal resolution, however, at the price of a narrow dynamic concentration range in which the indicators respond approximately linearly. Additionally, indicator dyes have been found difficult to introduce into plant tissue (Fricker et al., 1999, 2001).

Now, molecular biology techniques permit to circumvent at least the loading problems previously encountered with the indicator technique.

Reporter plants expressing their own indicator proteins proved to be an elegant way. Genetically encoded indicator proteins have the advantage that they can be targeted to many different organelles, compartments, and tissues by fusion with specific promoters, signal sequences, or by trapped enhancers (Kiegle et al., 2000; list in Plieth, 2001). About 15 years ago this method was first demonstrated in tobacco (Knight et al., 1991). These so-called "selfreporting plants" are able to monitor their cytoplasmic Ca^{2+} ($[Ca^{2+}]_{cyt}$) concentration. Since that time efforts are being made to develop new indicator proteins for a broad range of cellular components that can then be visualized *in vivo* (Zhang et al., 2002; Miyawaki, 2003).

There are actually two types of recombinant indicators available: Luminescent and fluorescent proteins. Luminescent proteins, so-called luciferases, catalyse a redox reaction, by which light is produced (Fig. 1). Fluorescent proteins, however, glow only when excited by light with a suitable wavelength.

2.2 Luciferases

Many animals use bioluminescence for different purposes. Luciferases catalysing light generating reactions have been evolved separately more than thirty times. The different origins make luciferases a very heterogenic group of enzymes with light generation being the only property they have in common. Coelenterate luciferases, in particular, are oxidoreductases attacking the CH-OH-group of an organic electron donor (luciferin) while consuming oxygen Fig. 1. Aequorin (Fig. 1), is a unique light emitting protein, originally characterized in the luminescent jellyfish *Aequorea victoria*. It has two–in other luminescent organisms usually separated - properties: First, it is a luciferase, able to catalyse oxidization of special low molecular weight luciferins called, coelenterazines. Second, it is a luciferin binding protein and hence called "photoprotein". The third and important property is the Ca^{2+} dependency of its luciferase activity.

Aequorin bioluminescence is directly correlated with the actual free Ca^{2+} ion concentration ($[Ca^{2+}]$). This makes aequorin an excellent and widely used $[Ca^{2+}]$ reporter protein (Campbell et al., 1989). It can be fused with other reporter proteins such as the green fluorescent protein (GFP) (Baubet et al., 2000; Kiegle et al., 2000; Moore, 2000). This eases the location of expressed recombinant aequorin by fluorescence microscopy Fig. 2. Aequorin has been used in a broad range of plant species (listed in Plieth, 2001) and thus many new aspects of Ca^{2+} mediated signal transduction have been discovered.

2.3 GFP

Fluorescent proteins are mainly used as markers to visualize cellular structures or particular tissues (Dixit et al., 2006). They emit green light when excited with blue light. The best characterized marker is the green fluorescent protein also from *Aequorea victoria*. It consists of a fluorophor (= three aminoacids) surrounded by eleven so-called "β-coils" forming a "lantern" and shielding the fluorophor from the environment (http://www.tsienlab.ucsd.edu/). The *Aequorea*-GFP has been modified (mutated) many times by exchange of different aminoacid residues near the fluorophor to give fluorescent proteins of different colour and to develop other properties (e.g. Griesbeck et al., 2001; Tsien, 2005). The slight pH-dependence of GFP fluorescence has been developed further by directed *in-vitro* evolution and DNA shuffling. This yielded the so-called "pHluorins" with excellent indicator properties (Miesenböck et al., 1998). The fluorescence intensity is mainly correlated to the indicator concentration. An organism expressing wild-type GFP can therefore only report relative pH changes but not the absolute pH value. The pHluorins, however, are ratio indicators with a spectral band where the fluorescence is increasing with increasing pH and another band with antiparallel behaviour (Figs. 5A, 6A). The isosbestic point designates the wavelength where the fluorescence is independent from pH. The ratio of fluorescence data taken left and right from the isosbestic point is a measure for the absolute pH. It is dependent on the logarithm of the ion concentration and calibration curves of such ratio indicators are typically of sigmoidal shape (Figs. 5B, 6B).

The use of GFP in higher plants was initially limited. After alteration of the codon usage and removal of a cryptic intron GFP expression in *Arabidopsis thaliana* was improved (Haseloff et al., 1997). Other modifications increased GFP solubility for better expression in the cytoplasm. These variants were called soluble-modified GFP (smGFP; Davis and Vierstra, 1998). These amendments necessary for sufficient expression of GFP in plants (i.e. removal of the cryptic intron, changes to *A. thaliana* codon usage and improvement of solubility) have been combined with the properties of pHluorins. The resulting pH indicators (At-pHluorins) have been successfully expressed in plants (Moseyko and Feldman, 2001; Plieth et al., 2001; Gao et al., 2004).

2.4 Measuring ion concentrations in the apoplast

pH and Ca concentrations in the apoplast have been measured in many plants either by selective electrodes (Felle, 1998; Felle et al., 2000, Felle and

Hanstein, this volume, pp. 295–306), by collecting apoplastic fluid (Lohaus et al., 2001; Lohaus, this volume, pp. 323–336, Dannel et al., 1995, Mühling and Sattelmacher, 1995; Fecht-Christoffers et al., this volume, pp. 307–322) or by impermeable fluorescent dyes (Sakurai, 1998a, b; Mühling et al., 1995; Hoffmann and Kosegarten, 1995). However, more *in vivo* information on the change of apoplastic pH and [Ca^{2+}] is desired from whole and undisturbed (i.e. not impaled or infiltrated) plants under abiotic stress. Such knowledge can help to better understand the role of the apoplast.

We started a novel approach and produced transgenic plants which simultaneously express their own [Ca^{2+}] and pH indicators at the desired location (i.e. the cytoplasm and the apoplast). For this study, we expressed soluble modified pHluorins with aequorin in fusion in the cytoplasm of *Arabidopsis*. We also targeted pHluorins and an aequorin variant with low Ca affinity (Kendall et al., 1992) as fusion proteins to the apoplast by using an *Arabidopsis* chitinase signal sequence. Series of experiments were performed to demonstrate that these self-reporting plants are able to monitor changes of pH and [Ca^{2+}] in the cytoplasm and in the apoplast. Salt and drought stress were studied in particular with this novel technique to discover new aspects of ion flux, regulation, and signalling.

For [Ca^{2+}] and pH measurements different experimental set-ups have to be used: [Ca^{2+}] changes are detected with a luminometer (Fig. 3A) whereas pH data are obtained by fluorescence ratio imaging (Fig. 3B). Fig. 3 illustrates the main difference between both techniques: Due to the extremely low light output the luminescence signal (Fig. 3A) is taken from the whole root system whereas the fluorescence signal (Fig. 3B) is taken from a representative root segment. This must be taken into account when interpreting obtained data.

3. RESULTS AND DISCUSSION

Before expression in *Arabidopsis* pHluorins (Miesenböck et al., 1998) were modified as described in Plieth et al. (2001) and in Gao et al. (2004) and fused with aequorin variants. To verify that modifications did not alter the indicator properties, all indicator and fusion cDNAs were also cloned into bacterial expression vectors. Proteins were expressed in bacteria and assessed luminometrically (Fig. 4) and fluorometrically (Figs. 5, 6):

3.1 Characterization of aequorins

Wild-type aequorin (AQ) is a useful Ca^{2+} indicator for measuring cytoplasmic Ca^{2+} concentration in the range between 10 and 10,000 nM. In

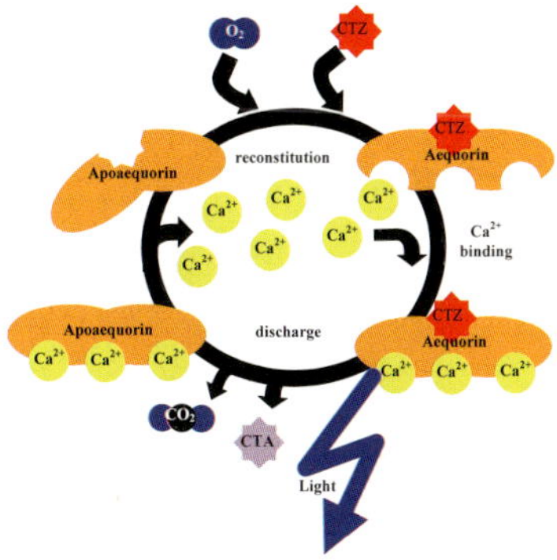

Fig. 1. Active aequorin is reconstituted in vivo *from apoaequorin and a low molecular weight luminophore called coelenterazine (CTZ) in the presence of oxygen (O_2). When calcium is bound, coelenterazine is oxidised to coelenteramide (CTA) and the protein undergoes a conformational change accompanied by the release of carbon dioxide (CO_2) and emission of blue (469 nm) light (after Prasher et al., 1985).*

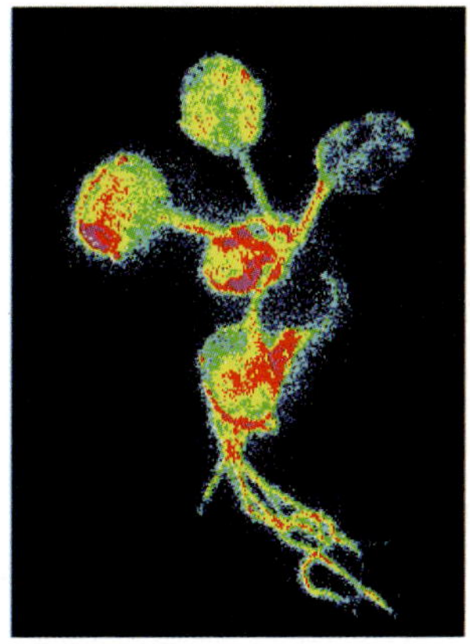

Fig. 2. False-colour image of a transgenic Arabidopsis plant expressing the calcium indicator aequorin in the cytoplasm. Image was taken with a photon-counting CCD camera.

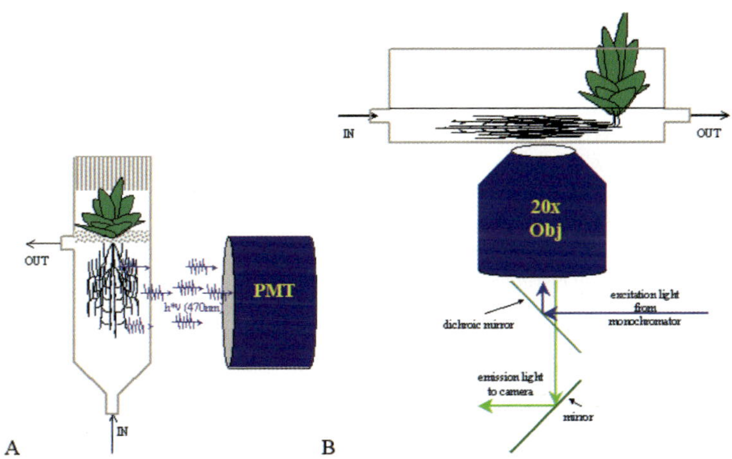

Fig. 3. Scheme of experimental set-ups to study [Ca^{2+}] by means of aequorin luminescence (A) and pH by means of fluorescence imaging (B) in transgenic reporter plants under influence of different changing environmental factors; i.e. perifusion with different buffer solutions. PMT = Photomultiplier tube; Obj = Objective.

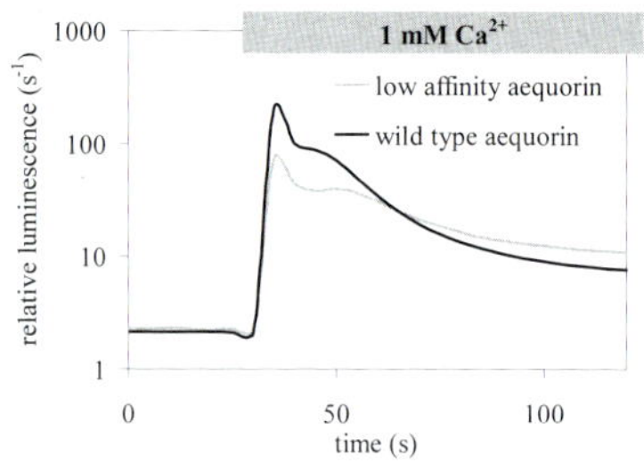

Fig. 4. Response of bacteria expressing aequorin variants to external $[Ca^{2+}]$. Curves are representative of three separate experiments. Relative luminescence (x 10000) is luminescence of each integration interval divided by luminescence still remaining in the specimen.

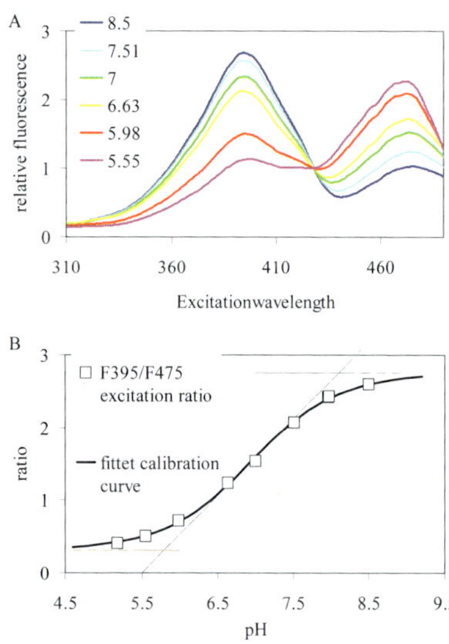

Fig. 5. *A: Excitationspectra of ratiometric GFP at different pHs (emission wavelength = 508 nm). Curves were normalized by F(428ex; 508em) B: The corresponding calibration curve for fluorescence excitation-ratio R (395ex/475ex; 508em); dissociation constant, i.e. midpoint of the fittet curve = kd = 6.9 +/– 0.03; optimal dynamic range for pH- measurements is between 5.6 and 8.0.*

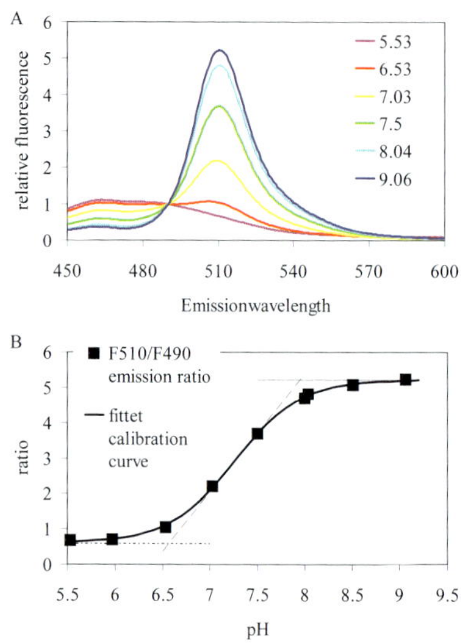

Fig. 6. *A: Emissionspectra of ecliptic GFP at different pHs (excitation wavelength = 400 nm). Curves were normalized by F(400ex; 490em); B: The corresponding calibration curve for fluorescence emission ratio R(400ex; 490em/510em); dissociation constant, i.e. midpoint of the fittet curve = kd = 7.25 +/– 0.02; optimal dynamic range for pH- measurements is between 6.5 and 8.0.*

the apoplast however, higher concentrations are expected which would saturate and discharge the indicator luciferase quickly. Therefore, we introduced an aequorin variant with lower Ca^{2+} binding affinity (D119A = LAAQ; Kendall et al., 1992) and thus with the capability to report higher

Ca^{2+} concentrations. We expressed both (AQ and LAAQ) in bacteria and studied their light response when treated with 1 mM $CaCl_2$. The kinetics (Fig. 5) were similar to what has previously been shown (Jones et al., 2002). Relative luminescence of initial response was lower in LAAQ expressing bacteria (Fig. 5, grey curve) in comparison to native AQ (Fig. 11, black curve). The measured relative luminescence indicates that less LAAQ than AQ is discharged during Ca^{2+} entry into the cells. This verifies that LAAQ is, due to its lower Ca^{2+} binding affinity, the better choice for a higher $[Ca^{2+}]$ range as is expected in the apoplast.

3.2 Characterization of pHluorins

Ratiometric At-pHluorin (ratioGFP) has pH-dependent spectra with isoexcitation point at 428 nm (Fig. 6A). The excitation maxima of ratioGFP are at 395 nm and 475 nm and the emission maximum is at 508 nm. Its' kd is 6.73 +/– 0.03 and its' optimal dynamic range is in the interval $5.6 < pH < 7.8$ which makes it suitable for both cytoplasmic and apoplastic pH measurements (Fig. 6B). Ecliptic At-pHluorin (eclipticGFP) displayed a ratiometric behaviour in the emission mode with an isoemission point at 490 nm (Fig. 7A). The excitation maxima of eclipticGFP are at 398 nm and 477 nm and the emission maximum is at 510 nm. Its' kd is 7.25 +/– 0.02 and its' optimal dynamic range is in the interval $6.5 < pH < 8.0$ which makes it suitable especially for cytoplasmic pH measurements (Fig. 7B). The spectra (Figs. 6 and 7) show that all modifications made to produce At-pHluorins hardly altered spectral properties and/or pH-dependency when compared with original pHluorins from Miesenböck et al. (1998). Additionally, when fused to aequorin the emission spectra were negligibly broadened by 3 nm in HBW and peaks were shifted by about 1 nm towards the blue. Also the pH-dependence of At-pHluorins did not significantly alter when fused with

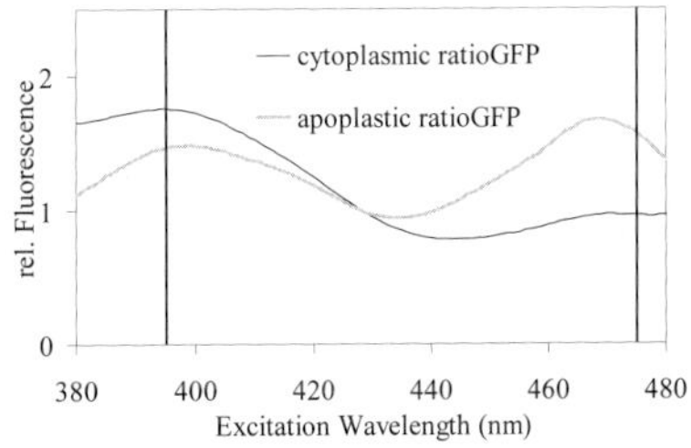

Fig. 7. In vivo *spectra of ratioGFP expressed in roots. Curves were normalized by F(428ex; 530em).*

aequorin and chitinase propeptide. Moreover, salt concentrations (0–200 mM NaCl) at constant pH had negligible influence on the fluorescence ratios (data not shown).

3.3 Plant material

We used the CaMV 35S-promoter to express indicator fusions in the cytoplasm. An Arabidopsis chitinase signal (22 aa) was added to deliver the indicator complex to the apoplast. In summary, we produced five different transgenic lines expressing

1. GFP5 and aequorin in the apoplast
2. ratiometric pHluorin and low-affinity aequorin in the apoplast
3. ecliptic pHluorin and low-affinity aequorin in the apoplast
4. soluble modified ratiometric pHluorin and aequorin in the cytoplasm
5. soluble modified ecliptic pHluorin and aequorin in the cytoplasm

The fact that none of our transgenic lines differ in phenotype from the wild-type verifies that expression of the indicators do not interfere with signal transduction, growth, and development. It suggests that neither cellular ion regulation in general nor ion buffering in particular is affected by the alien Ca^{2+} and H^+ binding indicator proteins.

3.4 Fluorescence: Expression and localization of the indicator complex

The expression level of indicator protein – as scrutinized by fluorescence – seemed to be higher in the apoplast-expressing lines than in the cytoplasmic-expressing lines. This can be explained with the permanent export of indicator in apoplast-targeted lines and extracellular accumulation.

The *in vivo* spectra (Fig. 7) taken from dissected roots expressing ratiometric pHluorin in the cytoplasm (black curve) and in the apoplast (grey curve) are markedly different and their F395/F475 ratios indicate $pH_{cyt} \approx 7.2$ and $pH_{apo} \approx 6.3$ which is close to what has been expected. The spectra indicate compartments of different acidity: pH ≈ 6.3 in apoplast and pH ≈ 7.2 in cytoplasm. The apoplastic pH has been reported many times from different species and the majority of values varied between 5.3 (Kosegarten and Englisch, 1994) and 6.7 (Dannel et al., 1995).

A change of external pH induced no change in fluorescence ratio of cytoplasmic expressed At-pHluorin (Fig. 8, black curve). The pH indicator expressed in the apoplast, in contrast, reported a strong dependence of apoplastic pH on external pH (Fig. 8, grey curve). This demonstrates that the

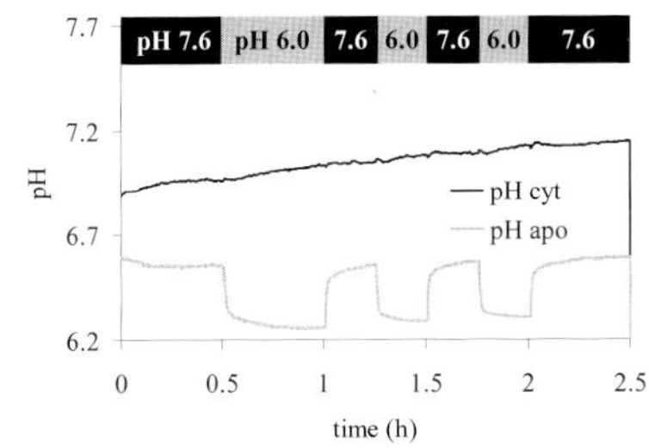

Fig. 8. pH changes in the apoplast (grey curve) and the cytoplasm (black curve) of roots in response to changes in external pH. External pH was adjusted with 10 mM HEPES (pH 7.6) and 10 mM MES (pH 6.0) in standard medium (KCl, $CaCl_2$, $MgCl_2$ – 1 mM each).

indicator complex is located in a compartment which has direct access to the outer medium and confirms successful targeting of the protein to the extra-cellular space.

3.5 Luminescence: *In vivo* reconstitution of aequorins

Plants expressing the GFP5:AQ fusion in the cytoplasm and in the apoplast were *in vivo* reconstituted with two different coelenterazine (CTZ) derivatives. Fig 9 shows absolute luminescence during the first hours of

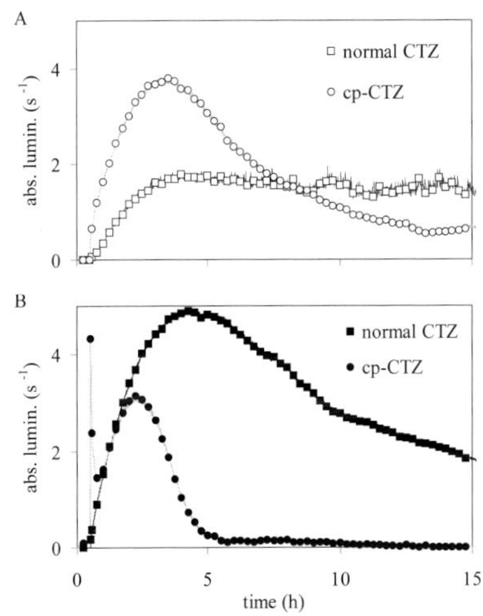

Fig. 9. Development of basal luminescence during in vivo *reconstitution with different coelenterazine (CTZ) derivatives. Squares indicate reconstitution with native CTZ and circles indicate reconstitution with cp-CTZ. Addition of CTZ at t = 0.6 h to 10 μM final concentration. Note: Absolute luminescence (x 10000) is given here, which is luminescence of each integration interval divided by total luminescence produced by the specimen. Relative luminescence (x 10000) is given in all other figures, which is luminescence of each integration interval divided by luminescence still remaining in the specimen. A: Reconstitution of aequorin expressed in the cytoplasm. B: Reconstitution of aequorin targeted to the apoplast.*

reconstitution. The time courses demonstrate that maximal basal luminescence in the cytoplasm (Fig. 9A) is reached after 4 h. The apparent binding constant of cp-aequorin is $pk_{Ca} \approx 6.4$ compared to $pk_{Ca} \approx 5.9$ of native aequorin (Plieth and Trewavas, 2002). Consequently basal luminescence produced by cp-aequorin decays even with the low resting level of free $[Ca^{2+}]$ in the cytoplasm (Fig. 9A; circles) whereas basal luminescence from native aequorin is stable here (Fig. 9A; squares). A much higher $[Ca^{2+}]$ is present in the apoplast.

Hence even luminescence from native aequorin decays and cp-aequorin luminescence dissipates within 5 h in this compartment (Fig. 9B). We found that basal luminescence even from low affinity aequorin (LAAQ) is not stable in the apoplast when reconstituted with native CTZ and also decays below minimum detection level within two days (data not shown). Nevertheless, there is a time window of more than 12 h for $[Ca^{2+}]$ measurements after *in vivo* reconstitution of targeted aequorin with native CTZ.

3.6 Ion responses during mimicked stress situations

Changes of $[Ca^{2+}]$ and $[H^+]$ in the cytoplasma and the apoplast were determined in response to cold and repeated periods of salt stress (i.e. 100 mM NaCl) and "drought". The latter was mimicked by isosmotic mannitol.

3.6.1 $[Ca^{2+}]$ response to cold in the cytoplasm and in the apoplast

The typical $[Ca^{2+}]$ responses to periods of cold obtained from apoplastic expressing lines (Fig. 10B) were completely different from cytoplasmic $[Ca^{2+}]$ responses (Fig. 10A; Plieth et al., 1999a). These responses, too, showed that aequorin has successfully been targeted to a compartment different from the cytoplasm. The data suggest, that $[Ca^{2+}]$ is slightly lowered in this compartment during cold period.

3.6.2 Osmotic ("drought") stress

The apoplastic $[Ca^{2+}]$ remained almost unaffected by mannitol treatment (Fig. 11A, grey curve). Apart from tiny responses during mannitol wash-out there are no significant changes in the $[Ca^{2+}]_{apo}$ signal. This is surprising since we expected that a significant amount Ca^{2+} ions is shifted from the extracellular space during hypo-osmotic treatment into the cytoplasm. This should give a significant transient fall in the apoplastic luminescence signal. As the latter failed to appear we conclude that the cellular $[Ca^{2+}]_{cyt}$ transients during mannitol-washouts (Fig. 11A black curve) were mainly produced by

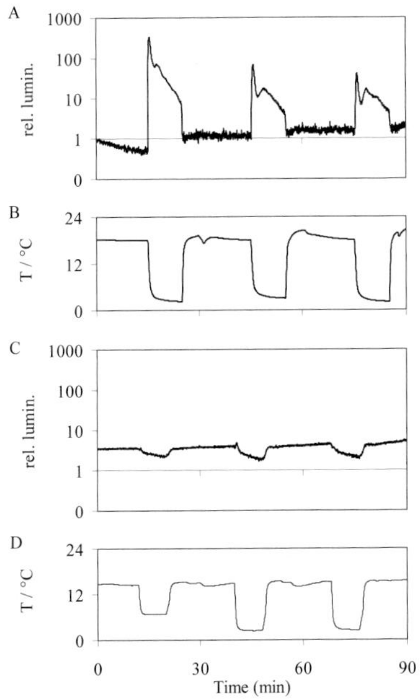

Fig. 10. Luminescence response to cold from plants expressing aequorin in the cytoplasm (A, B) and in the apoplast (C, D). A and C represent relative luminescence and B and D the corresponding temperature measured in parallel.

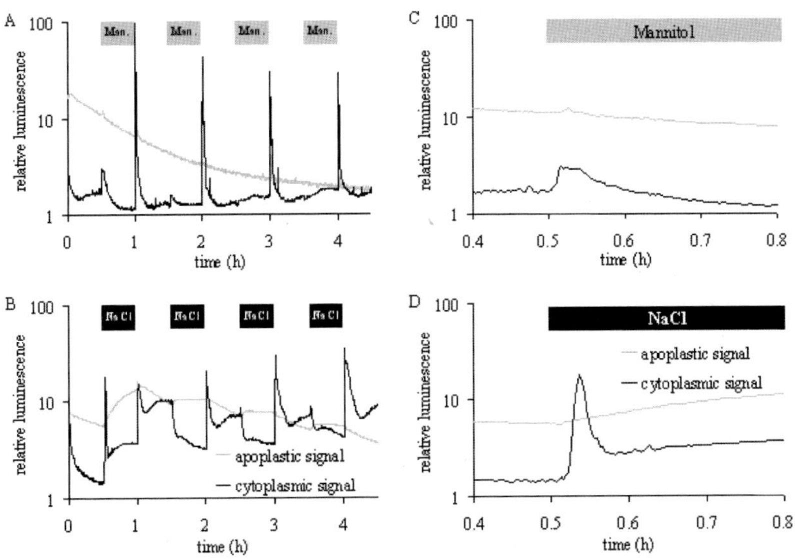

Fig. 11. A Cytoplasmic and apoplastic [Ca^{2+}] in response to "drought" stress (200 mM mannitol vs H_2O) B Cytoplasmic and apoplastic [Ca^{2+}] in response to salt stress (100 mM NaCl vs H_2O). C and D are close-ups of A, and B respectively.

release of Ca^{2+} ions from internal stores (e.g. vacuole) rather than influx from the outside. The overall continuing decay of the apoplastic signal (Fig. 11A grey curve) indicates permanent washout of Ca^{2+} ions by the perfusion medium which was not supplemented with extra Ca^{2+}.

The cytoplasmic free calcium concentration ($[Ca^{2+}]_{cyt}$), in contrast, was drastically affected and pronounced $[Ca^{2+}]_{cyt}$ transients were observed (Fig. 11A black curve) during hyper-osmotic and hypo-osmotic stimuli. The hypo-osmotic triggered $[Ca^{2+}]_{cyt}$ response was much more pronounced. This has been observed before (Pauly et al., 2001) indicating that this stimulus is more critical than the hyper-osmotic one. In both cases (hyper- and hypo-osmotic shock) the $[Ca^{2+}]_{cyt}$ is brought back to base level after the transient. $[Ca^{2+}]_{cyt}$ did not stay on elevated levels as is the case for salt treatment (described below).

The F395/F475-fluorescence-excitation ratio of the pH-sensitive GFP revealed that "drought" (mannitol) treatment (Fig. 13A) does not influence both, cytoplasmic pH (pHcyt) and apoplastic pH (pHapo). This is in contrast to NaCl stress (see below).

3.6.3 NaCl ("salt") stress

Salt, in contrast, produces $[Ca^{2+}]$ and pH changes in both compartments (Figs. 11 to 13): The finding that $[Ca^{2+}]_{cyt}$ is increased during salt stress is in line with observations from Lynch et al. (1989) who observed a $[Ca^{2+}]_{cyt}$ increase in maize root protoplasts during salt stress. The finding that $[Ca^{2+}]_{cyt}$ responses to salt is markedly different from "drought" has been anticipated by Shi et al. (2002) who showed that root growth inhibition is different under salt and drought stress. In order to discriminate salt stress and osmotic response, we switched between mannitol and NaCl solutions with the same osmotic pressure (Fig. 12). This brings about a uniform osmotic stress and

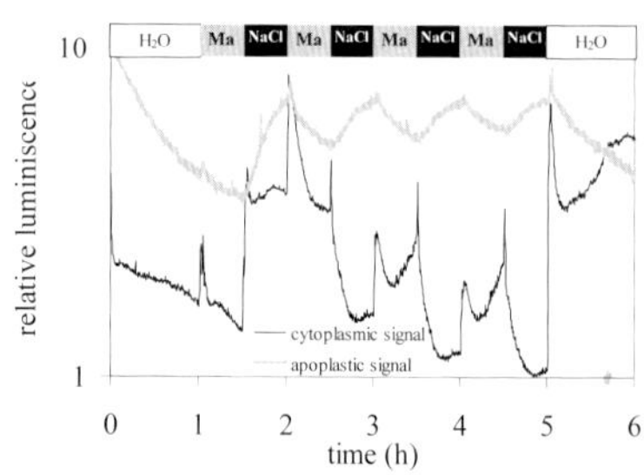

Fig. 12. Calcium responses upon both, "drought" and salt stress (200 mM Mannitol vs 100 mM NaCl vs H O) in Arabidopsis roots.

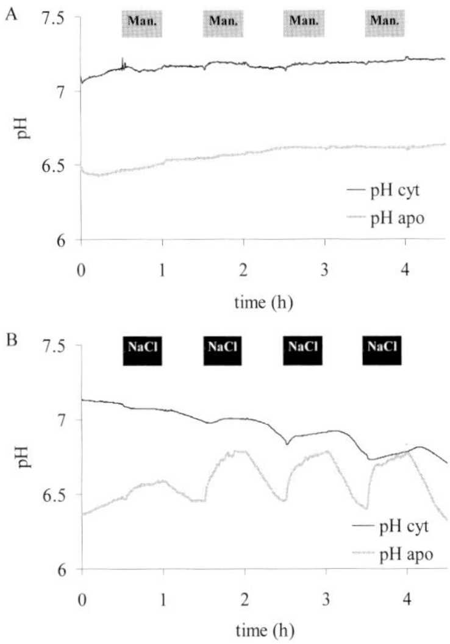

Fig. 13. A Cytoplasmic and apoplastic pH responses upon "drought" stress (H_2O vs 200 mM mannitol) in Arabidopsis roots. B Cytoplasmic and apoplastic pH responses upon salt stress (H_2O vs 100 mM NaCl) in Arabidopsis roots.

confirmed that external Na caused a drastic increase in $[Ca^{2+}]_{apo}$ and gave a similar kinetic pattern (Fig. 12 grey curve) as with NaCl – H_2O treatment (Fig. 11 grey curve).

There were two differences between salt and "drought" $[Ca^{2+}]_{cyt}$ responses: First, the whole $[Ca^{2+}]_{cyt}$ was permanently shifted by salt towards a higher level (Figs. 11B and 12 black curves). Second, the short-term response to "drought" was prolonged when compared to the $[Ca^{2+}]_{cyt}$ kinetic under salt treatment (Fig. 11C, D). In particular these short-term differences are well in line with Knight et al. (1997).

Prolonged $[Ca^{2+}]_{cyt}$ elevations have been seen with other studies where the cytoplasm has been challenged with an excess of other monovalent ions, namely H^+ (Plieth et al., 1997, 1999b). The main conclusion drawn in these studies is that Ca acts as protecting and ameliorative agent. Our assumption is that this can hold under Na^+ stress too.

Salt stress (NaCl), also resulted in significant alterations of pH in the apoplast and to a minor extent in the cytoplasm (Fig. 13B). The pH responses became more pronounced with number of salt treatments in both compartments. This sort of "sensitisation mechanism" (i.e. increasing response amplitudes with increasing number of treatments) is opposite to "adaptation" mechanisms (i.e. decreasing response amplitudes with increasing number of treatments) usually observed with other abiotic stimuli (e.g. cold – Fig. 10A).

4. CONCLUSIONS

For a better understanding of cellular ion regulation, ion interaction, transport, and signalling it is important to be able to measure free ion concentrations in different compartments (e.g. apoplast, ER, vacuole), organelles (e.g. mitochondria, chloroplasts, nucleus), or tissues (e.g. stomata, epidermis, vascular tissue) of intact plants and in response to a variety of environmental stimuli. With the approach presented here we went a big stepforward towards this aim.

We joined cDNA of pH-sensitive GFPs and aequorin variants and expressed the pH-pCa-indicators under control of the 35S promotor with and without a chitinase signal propeptide. Control experiments demonstrate that the propeptide delivers the gene product to an acidic compartment with high $[Ca^{2+}]$ which has direct access to the surrounding medium. Therefore, we conclude that the expression of the indicator complex in the apoplast is sufficient for reliable pH and pCa studies. Lines expressing the indicator without propeptide allow cytoplasmic pH and pCa measurements.

The main advantages are:

1. The indicators (GFPs, AQs) are produced by the transgenic plant itself. Thus, loading problems, as is the case for low molecular weight indicator-dyes (Plieth and Hansen, 1996) do not exist and artefacts resulting from loading procedures are avoided. Plants remain unperturbed before the experiment.
2. The indicator complex can be targeted specifically to almost any compartment, tissue, organ, or organelle of interest (Kiegle et al., 2000; Plieth, 2001).
3. Measurements can be performed in whole intact plant systems and under "physiological" conditions; i.e. conditions similar to what plants experience in the wild. Thus, any study which makes use of "self-reporting plants" relates more closely to conditions in the field and deserves to be called "biologically relevant".

5. PERSPECTIVES

Very first experiments provided here give new information how Ca behaves extracellularly and intracellularly in Arabidopsis. However, much more detailed experimentation is needed to obtain significant data which underline or invalidate current hypotheses of stress-induced ion signaling.

Furthermore ion mapping (Fricker et al., 1999; Roos, 2000), i.e visualisation of ion distribution, gradients, and long-distance ion transport can now be made possible by imaging aequorin luminescence and/or GFP fluorescence in whole intact plants and/or organs. Another aim is to modify the spectral characteristics of the indicators such that cytoplasmic and apoplastic signals can be distinguished and hence simultaneously be measured in the same plant. This, supplemented with an electrophysiological approach, could permit direct visualisation of ion fluxes from one to the other compartment, direct quantification of buffer capacities, and evaluation of ion regulation and crosstalk between compartments and interaction of different ion species.

ACKNOWLEDGEMENTS

We thank Gero Miesenböck and James Rothman (Memorial Sloan-Kettering Cancer Centre, New York) for the generous gift of pHluorin cDNA. We thank Cathy Moore (Oxford University) for the generous gift of plasmid pCM2, and Jim Haselhoff (Cambridge University) for the binary vector pBINm-gfp5-ER. This study was supported by grants to BS and DG (Deutsche Forschungsgemeinschaft SA 359/12-3) and to CP (Deutsche Forschungsgemeinschaft PL253/1-1 and European Commission BIO4-CT97-5080)

REFERENCES

Baubet, V., Mouellic, H.L., Campbell, A.K., Lucas-Meunier, E., Fossier, P. and Brulet, P. (2000) Chimeric green fluorescent protein-aequorin as bioluminescent Ca^{2+} reporters at the single cell level. *Proc. Nat. Ac. Sci. USA*, 97, 7260–7265.

Blatt, M.R. and Grabov, A. (1997) Signal redundancy gates and integrationn in the control of ion channels for stomatal movement. *J. Exp. Bot.*, 48, 529–537.

Campbell, A.K., Patel, A., Houston, W.A.J., Scolding, N.J., Frith, S., Morgan, B.P, and Compston, D.A.S. (1989) Photoproteins as indicators of intracellular free Ca^{2+}. *J. Bioluminescence and Chemiluminescence*, 4, 463–474.

Dannel, F., Pfeffer, H. and Marschner, H. (1995) Isolation of apoplasmic fluid from sunflower leaves and its use for studies on influence of nitrogen supply on apoplasmic pH. *J. Plant Physiol.*, 146, 273–278.

Davis, S.J. and Vierstra, R.D. (1998) Soluble, high fluorescent variants of green fluorescent protein (GFP) for use in higher plants. *Plant Mol. Biol.*, 36, 521–528.

DeMarty, M., Morvan, C. and Thellier, M. (1984) Calcium and the Cell Wall. *Plant Cell Environ.*, 7, 441–448.

Dixit, R., Cyr, R. and Gilroy, S. (2006) Using intrinsically fluorescent proteins for plant cell imaging. *Plant J.*, 45, 500–615.

Felle, H.H. (1987) Proton transport and pH control in *Sinapis alba* root hairs: a study carried out with double barreld pH microelectrodes. *J. Exp. Bot.*, 340, 354.

Felle, H.H. (1991) Ca^{2+} selective microelectrodes and their application to plant cells and tissues. *Plant Physiol.*, 91, 1239–1242.

Felle, H.H. (1998) The apoplast pH of the *Zea mays* root cortex as measured with pH-sensitive microelectrodes: aspects of regulation. *J. Exp. Bot.*, 49, 987–995.

Felle, H.H. (2001) pH: Signal and messenger in plant cells. *Plant Biol.*, 3, 577–591.

Felle, H.H. (2005) pH regulation in anoxic plants. *Ann. Bot.*, 96, 519–532.

Felle, H.H., Hanstein, S., Steinmeyer, R. and Hedrich, R. (2000) Dynamics of ionic activities in the apoplast of the sub-stomatal cavity of intact *Vicia faba* leaves during stomatal closure evoked by ABA and darkness. *Plant J.*, 24, 297–304.

Fricker, M.D., Parsons, A., Tlalka, M., Blancaflor, E., Gilroy, S., Meyer, A. and Plieth, C. (2001) Fluorescent probes for living plant cells. In: C. Hawes and B. Satiai-Jeunemaitre (eds.) Plant Cell Biology: A Practical Approach - 2nd Ed, *Oxford University Press*, Oxford, pp. 35–84.

Fricker, M.D., Plieth, C., Knight, H., Blancaflor, E., Knight, M.R., White, N. and Gilroy, S. (1999) Fluorescence and luminescence techniques to probe ion activities in living plant cells. In: B. Mason (ed.) Fluorescent and luminescent probes for biological activity, *Academic Press*, San Diego, pp. 569–596.

Gao, D., Trewavas, A.J.T., Knight, M.R., Sattelmacher, B. and Plieth, C. (2004) Self-reporting *Arabidopsis thaliana* expressing pH- and [Ca^{2+}]- indicators unveil ion dynamics in the cytoplasm and in the apoplast under abiotic stress. *Plant Physiol.*, 134, 898–908.

Gilroy, S. and Trewavas, A. (1994) A decade of plant signals. *BioEssays*, 16, 677–682.

Griesbeck, O., Baird, G.S., Campbell, R.E., Zacharias, D.A. and Tsien, R.Y. (2001) Reducing the environmental sensitivity of yellow fluorescent protein. J. Biol. Chem., 276, 29188–29194.

Grignon, C. and Sentenac, H. (1991) pH and ionic conditions in the apoplast. *Ann. Rev. Plant Physiol. Plant Mol. Biol.*, 42, 103–128.

Haseloff, J., Siemering, K.R., Prasher, D.C. and Hodge, S. (1997) Removal of a cryptic intron and subcellular localization of green fluorescent protein are required to mark transgenic *Arabidopsis* plants brightly. *Proc. Nat. Ac. Sci. USA*, 94, 2122–2127.

Hoffmann, B. and Kosegarten, H. (1995) FITC-dextran for measuring apoplast pH and apoplast pH gradients between various cell types in sunflower leaves. *Physiologia Plantarum*, 95, 327–335.

Hoson, T. (1998) Apoplast as site of response to environmental signals. *J. Plant Res.*, 111, 167–177.

Jones, H.E., Holland, I.B. and Campbell, A.K. (2002) Direct measurement of free Ca^{2+} shows different regulation of Ca^{2+} between the periplasm and the cytosol of *Escherichia coli*. *Cell Calcium*, 32, 183–192.

Kendall, J.M., Sala-Neby, G., Ghalaut, V., Dormer, R,L. and Campbell, A.K. (1992) Engineering the Ca^{2+}-activated photoprotein aequorin with reduced affinity for calcium. *Biochem. and Biophys. Res. Comm.*, 187, 1091–1097.

Kiegle, E., Moore, C.A., Haseloff, J., Tester, J. and Knight, M.R. (2000) Cell-type-specific calcium responses to drought salt and cold in Arabidopsis roots. *Plant J.*, 23, 267–278.

Knight, H. (2000) Calcium signaling during abiotic stress in plants. *Int. Rev. of Cytol.*, 195, 269–324.

Knight, H., Trewavas, A.J. and Knight, M.R. (1997) Recombinant aequorin methods for measurement of intracellular calcium in plants. *Plant Mol. Biol. Manual*, C4, 1–22.

Knight, M.R., Campbell, A.K., Smith, S.M. and Trewavas, A.J. (1991) Transgenic plant aequorin reports the effects of touch and cold-shock and elicitors on cytoplasmic calcium. *Nature*, 352, 524–526.

Kosegarten, H. and Englisch, G. (1994) Effect of various nitrogen forms on the pH of leaf apoplast and on iron chlorosis of *Glycine max* L. *Zeitschrift für Pflanzenernährung und Bodenkunde*, 157, 401–405.

Lee, Y. and Satter, R.L. (1989) Effects of white blue red light and darkness on pH of the apoplast in *Samanea pulvinus*. *Planta*, 178, 31–40.

Lohaus, G., Pennewis, K., Sattelmacher, B., Hussmann, M. and Mühling, K.H. (2001) Is the infiltration-centrifugation technique appropriate for the isolation od apoplastic fluid? A critical evaluation with different plant species. *Physiologia Plantarum*, 111, 457–465.

Lynch, J., Polito, V.S. and Läuchli, A. (1989) Salinity stress increases cytoplasmic Ca^{2+} activity in maize root protoplasts. *Plant Physiol.*, 90, 1271–1274.

Ma, L. and Sun, D.-Y. (1997) The effects of extracellular calmodulin on initiation of *Hippeastrum rutilum* pollen germination and tube growth. *Planta*, 202, 336–340.

Ma, L., Xu, X., Cui, S. and Sun, D. (1999) The presence of a heterotrimeric G protein and its role in signal transduction of extracellular calmodulin in pollen germination and tube growth. *Plant Cell*, 11, 1351–1363.

Miesenböck, G., Angelis, D.A. de and Rothman, J.E. (1998) Visualizing secretion and synaptic transmission with pH-sensitive green fluorescent protein. *Nature*, 394, 192–195.

Miyawaki, A. (2003) Visualization of the spatial and temporal dynamics of intracellular signaling. *Developmental Cell*, 4, 295–305.

Moore, C.A. (2000) An investigation into cellular and molecular aspects of calcium-based specificity of signalling in *Arabidopsis thaliana*. D. Phil. thesis, University of Oxford.

Moseyko, N. and Feldman, L.J. (2001) Expression of pH-sensitive green fluorescent protein in *Arabidopsis thaliana*. *Plant, Cell Environ.*, 24, 557–563.

Mühling, K.-H., Plieth, C. and Hansen, U.-P. and Sattelmacher, B. (1995) Apoplastic pH of intact leaves of *Vicia faba* as influenced by light. *J. Exp. Bot.*, 46, 377–382.

Mühling, K.H. and Sattelmacher, B. (1995) Apoplastic ion concentration of intact leaves of field bean (*Vicia faba*) as influenced by ammonium and nitrate nutrition. *J. Plant Physiol.*, 147, 81–86.

Pauly, N., Knight, M.R., Thuleau, P., Graziana, A., Muto, S., Ranjeva, R. and Mazars, C. (2001) The nucleus together with the cytosol generates patterns of specific cellular calcium signatures in tobacco suspension culture cells. *Cell Calcium*, 30, 413–421.

Peters, W. and Felle, H.H. (1991) Control of apoplast pH in corn coleoptile segments. I: The endogeneous regulation of cell wall pH. *J. Plant Physiol.*, 1367, 655–661.

Plieth, C. (2001) Plant calcium signaling and monitoring - pros and cons and recent experimental approaches. *Protoplasma*, 218, 1–23.

Plieth, C. (2005) Calcium: Just another regulator in the machinery of life? *Annals of Botany*, 96, 1–8.

Plieth, C. and Hansen U.-P. (1996) Methodological aspects of pressure loading of Fura-2 into Characean cells. *J. Exp. Bot.*, 47, 1601–1612.

Plieth, C., Hansen, U.-P., Knight, H. and Knight, M.R. (1999a) Temperature sensing by plants: The primary mechanisms of signal perception and calcium response. *Plant J.*, 18, 491–497.

Plieth, C., Sattelmacher, B. and Hansen, U.-P. (1997) Cytoplasmic Ca^{2+}-H^{+}-exchange buffers in green algae. *Protoplasma*, 198, 107–124 and Vol 199, 223 (correction).

Plieth, C., Sattelmacher, B., Hansen, U.-P. and Knight, M.R. (1999b) Low pH-mediated elevations in cytosolic calcium are inhibited by aluminium: A potential mechanism for aluminium toxicity. *Plant J.*, 18, 643–650.

Plieth, C., Sattelmacher, B., Trewavas, A.J., Hansen, U.P. and Knight, M.R. (2001) Engineering plants expressing calcium and pH indicators in the cytoplasm and the apoplast. In: Horst, W.J. et al. (eds.) "Plant Nutrition - Food security and sustainability of agro-ecosystems through basic and applied research (Proceedings of XIV International Plant Nutrition Colloquium, Hannover, Germany, 2001) - Developments in Plant and Soil Sciences - Series", Vol. 92, *Kluwer Academic Publishers*, Dordrecht Netherlands, pp. 252–253.

Plieth, C. and Trewavas, A.J. (2002) Reorientation of seedlings in the earth's gravitational field induces cytosolic calcium transients. *Plant Physiol.*, 129, 786–796.

Prasher, D., McCann, R.O. and Cormier, M.J. (1985) Cloning and expression of the cDNA coding for aequorin a bioluminescent calcium-binding protein. *Biochemical and Biophysical Research Communications*, 126, 1259–1268.

Roos, W. (2000) Ion mapping in plant cells - methods and applications in signal transduction research. *Planta*, 210, 347–370.

Sakurai, N. (1998a) Dynamic function and regulation of apoplast in the plant body. *J. Plant Res.*, 111, 133–148.

Sakurai, N. (1998b) Dynamism of apoplast. *J. Plant Res.*, 111, 131.

Sanders, D., Brownlee, C. and Harper, J.F. (1999) Communicating with calcium. *Plant Cell*, 11, 691–706.

Sanders, D., Pelloux, J., Brownlee, C. and Harper, J.F. (2002) Calcium at the crossroads of signaling. *Plant Cell*, 14, S401–S417.

Sattelmacher, B. (2001) The apoplast and its significance for plant mineral nutrition. *New Phytol.*, 149, 167–192.

Shi, H., Quintero, F.J., Pardo, J.M. and Zhu, J.-K. (2002) The putative plasma membrane Na^{+}/H^{+} antiporter SOS1 controls long distance Na^{+} transport in plants. *Plant Cell*, 14, 465–477.

Sun, D.Y., Bian, Y.Q., Zhao, B.H., Zhao, L.Y., Yu, X.M. and Shengjun, D. (1995) The effects of extracellular calmodulin on cell wall regeneration of protoplasts and cell division. *Plant Cell Physiol.*, 36, 133–138.

Sun, D.Y., Li, H.B. and Cheng, G. (1994) Extracellular calmodulin accelerates the proliferation of suspension-cultured cells of *Angelica dahurica*. *Plant Science* (Limerick), 99, 1–8.

Tang, J., Wu, S., Bai, J. and Sun, D.-Y. (1996) Extracellular calmodulin-binding proteins in plants: purification of a 21-kDa calmodulin-binding protein. *Planta*, 198, 510–516.

Taylor, D.P., Slattery, J. and Leopold, A.C. (1996) Apoplastic pH in corn root gravitropism: a laser scanning confocal microscopy measurement. *Physiologia Plantarum*, 97, 35–38.

Trewavas, A.J. and Malhó, R. (1998) Ca^{2+} signalling in plant cells: The big network! *Curr. Opin. Plant Biol.*, 1, 428–433.

Tsien, R.Y. (2005) Building and breeding molecules to spy on cells and tumors. *FEBS Letters*, 579, 927–932.

Ward, J.M., Pei, Z.-M. and Schroeder, J.I. (1995) Roles of ion channels in initiation of signal transduction in higher plants. *Plant Cell*, 7, 833–844.

Zhang, J., Campbell, R.E., Ting, A.Y. and Tsien, R.Y. (2002) Creating new fluorescent probes for cell biology. *Nat. Rev. Mol. Cell Biol.*, 3, 906–918.

Section 6

The Apoplast Compartment for Plant–Microbe Interactions

CONSTRAINTS FOR ENDOPHYTIC BACTERIA

T. HUREK
Allgemeine Mikrobiologie, Fakultät für Biologie und Chemie, Universtät Bremen, Germany, thurek@uni-bremen.de.

Key words: anoxia, cortical root death, endophyte, endosymbiont, ethanol, nutrient deficiency, malate, stress

1. BACTERIA INHABITING THE APOPLAST

1.1 Root exudation into the apoplast

Plants relocate up to 40% of net carbon assimilation from plant shoots as soluble and insoluble materials to roots and into adjacent soil (Whipps, 1984; Whipps and Lynch, 1986; Van Veen et al., 1991). This rhizodeposition is an important source of energy for microorganisms (Van Veen et al., 1989) and may consist up to 50% of root exudates (Marschner, 1995). Root exudates represent low-M_r compounds like organic acids, amino acids, sugars, phenolics, etc. and high-M_r compounds such as proteins or mucilage, of which the latter are believed to represent the minority of materials exuded by roots. Plants have the potential to synthesize up to 100,000 compounds (Verpoorte, 2000) of which many may be secreted, and of which an unknown number is assumed to play important roles in root-root, root-microbe, and root-insect communications (Walker et al., 2003).

It has been known for long that root exudation depends on physiological and abiotic conditions such as age and soil, respectively, varies with plant species, and increases in the presence of microorganisms in the rhizosphere (Curl and Truelove, 1986; Lynch, 1990). But it has been firmly established only recently that nutrient deficiencies (Lopez-Bucio et al., 2000a),

B. Sattelmacher† and W.J. Horst (eds.), The Apoplast of Higher Plants: Compartment of Storage, Transport and Reactions, 395–403.

inorganic ion stresses (De La Fuente et al., 1997), and anoxia (Rivoal and Hanson, 1993) represent major environmental stimuli for the release of organic acids (Lopez-Bucio et al., 2000b; Ryan et al., 2001). Under these conditions the exudation of organic acid anions like citrate and malate may show effluxes in the micromolar range per g fresh weight h^{-1} and may account for more than 20% of the total plant dry weight in plants (Ryan et al., 2001).

Due to their high affinity to di- and tri-valent cations, organic acids benefit plants by either increasing or decreasing the availability of nutrients (e.g. P and Fe^{3+}) or toxic cations (particularly Al^{3+}), respectively. Edaphic stress of Al poisoning occurs typically in acid soils whereas that of P- and Fe-deficiencies is characteristic for alkaline calcareous soils. In alkaline soils, representing more than 25% of the earth's surface, P forms sparsely soluble compounds with calcium and magnesium, whereas in acidic soils, which are typical for 40% of the world's arable land, P forms low solubility complexes with Fe and Al. Organic acids replace P from these compounds by forming stronger complexes with Fe or Al than P does. Thereby, P solubility and availability for the plant is increased. Since at the close to neutral pH in the cytosol organic acids are almost fully negatively charged and cell membranes are largely impermeable to ions, organic acids have to be transported mainly as anions from the cytosol into the medium. For their release into the apoplast, membrane-bound transporters such as anion channels are considered to be most important (Ryan et al., 2001). However, transport of organic compounds into the apoplast by exocytosis, gains increasing attention, too (Walker et al., 2003). Nitrogen deficiency is also known to increase root exudation (see e.g. Darwent et al., 2003), but here, the composition of the exudates is largely unknown.

Keeping in mind the enhanced exudation of organic material during nutrient stress, it is probably not coincidential that very high numbers of heterotrophic, apoplast-residing bacteria have been obtained from internal root tissues of plants challenged by severe nutrient limitations (Hurek and Reinhold-Hurek, 2006). The best example is probably *Azoarcus* sp. BH72, a natural diazotrophic grass endophyte of Kallar grass growing in a monoculture stand in a low fertility, alkaline wetland soil where P and N availability was low (Hurek et al., 2002). There, this bacterium was cultured out in huge numbers (approx 10^9 cells g^{-1} root fresh weight) from surface-sterilized roots of symptom-less Kallar grass plants (Reinhold et al., 1986). This bacterium could never be isolated from that location again which might indicate that *Azoarcus* sp. BH72 is a random isolate. But in situ hybridisations with a homologous antisense *nifH* probe, sequence analysis of *nifH* mRNA, and estimation of *nifH* transcription levels showed over a time period of 10 years that this bacterium is indigenous to this habitat and metabolically highly active there (Hurek et al., 2002; Hurek and Reinhold-Hurek, 2006). Numbers obtained e.g. for *Azoarcus* sp. BH72 in 1984 are in the range of densities known for pathogenic bacteria (10^8–10^{10} bacteria per g root fresh weight) and are much higher than those usually reported for crops grown under standard fertilization regimes (10^5 bacteria per g root fresh weight) (Hallmann, 2001).

The speculation that under these conditions organic acids play an important role not only in plant but also in endophyte nutrition gets more momentum if one considers the observations that typical root colonizers show half-saturation constants for growth on organic acids in the micromolar to submicromolar range (Hurek et al., 1987), are strongly attracted to organic acids by chemotaxis (Reinhold et al., 1985), and colonize roots preferentially at sites (Reinhold-Hurek and Hurek, 1998) where a high release of organic material has been frequently observed (Curl and Truelove, 1986; Hoffland et al., 1989; Darwent et al., 2003). Furthermore, root endophytic bacteria typically grow on organic acids. Some apoplast-colonizing microbes such as *Azoarcus* sp. BH72 show even a considerable degree of nutritional dependence on these carbon sources, since they can use for growth only few organic acids, some alcohols such as ethanol, and no carbohydrates (Reinhold-Hurek et al., 1993). It may be interesting to note here, that organic acids like malate do not only play an important role in tolerance to nutrient deficiencies or inorganic iron stresses, but also in bacterial plant endosymbiosis. In nodules, dicarboxylic acids represent the preferred substrate to sugars for nitrogen fixation by bacteroids (Werner, 1992).

1.2 Anoxia

Ethanol is another potentially important nutrient for endophytic bacteria. This alcohol is produced in plants growing at low-oxygen conditions e.g. in wetland soils or sediments, which are permanently or periodically flooded. Other organic compounds such as lactate, malate, amino acids or glycerol are accumulated during anoxia as well (Crawford, 1978; Henzi and Brändle, 1993). But the major pathway in plants for the generation of energy (ATP) and oxidation power (e.g. NAD^+) under oxygen starvation appears to be glycolysis linked to ethanol fermentation (Summers et al., 2000). Since ethanol can pass directly through cell membranes by simple diffusion, this compound would almost instantly appear in the apoplast after its synthesis through alcohol dehydrogenase. Similar to P-stressed plants exuding organic acid anions, O_2-deprived plants produce ethanol in the micromolar range per g fresh weight h^{-1} as well (Studer and Brändle, 1984; Summers et al., 2000). Clearly, the release of such high amounts of potential substrates for microorganisms into the apoplast should be sufficient to support high numbers of bacteria there.

1.3 Cortical root death

Plants exposed naturally to oxygen-deprived conditions form longitudinally interconnected intercellular gas spaces in the root cortex that facilitate gas diffusion between roots and the aerial environment (Armstrong, 1979). These so called aerenchymas have a high potential for root aeration by longitudinal O_2 transport and for oxidising the rhizosphere by radial oxygen loss in situ (Bodelier, 2003). In many but not all plant species, aerenchyma is formed by a spatially distinct process during which cells die selectively in the cortex. Cortical cell death appears to be the consequence of a pathway which is either constitutive or can be triggered by stress, such as oxygen, N or P deficiency. It is generally initiated by ethylene and starts behind the root apex from where it progresses radially and tangentially into adjacent cells (for a recent comprehensive review see Evans, 2003). It is assumed that the cellular contents of low molecular weight compounds from the lysing cells are usually taken up by their living neighbours. However, cell debris would constitute an abundant, readily available source of nutrients for microbes as well, residing in the apoplast of the root cortex.

1.4 K^+ efflux / H^+ influx like exchange response

Beyond these largely stress-induced plant responses leading to high efflux of nutrients into the apoplast, information on nutrient availability in this compartment in the absence of these processes is scarce. Nutrient concentrations in the intercellular spaces are usually considered to be too low and variable to support large microbial populations (Hallmann, 2001). However, evidence has accumulated that growth and activity of some endophytes in internal plant tissues seem to be much less limited by the availability of nutrients than growth and activity of others (Hurek and Reinhold-Hurek, 2006). Perhaps, some bacteria elicit a K^+ efflux / H^+ influx like exchange response by partial degeneration of the host-cell membrane, as has been suggested by Hallmann (2001). Such a process would increase the pH of the apoplast and would stimulate the release of organic compounds together with bacterial growth (Atkinson and Baker, 1987a, b). The abilitity to elicit a K^+ efflux / H^+ influx like exchange response in undiseased plants may thus be a mechanism through which endophytes gain access to nutrients, when substrate concentrations in the apoplast are too low.

1.5 Solute exchange with the atmosphere

But the apoplast is colonized not only by heterotrophic bacteria which depend solely on nutrients provided by the plant. Rennenberg et al. (1998) and Papen et al. (2002) first reported on the presence of chemolithoautotrophic nitrifiers residing in the intercellular space of needles from *Picea abies* in a spruce ecosystem exposed to high nitrogen deposition (Teuber et al., this volume, pp. 405–426). The unequivocal detection of physiologically active, chemolithoautotrophic nitrifiers in plant tissue is a surprise, since their occurrence is believed to be restricted to soils and waters only. There, these bacteria are known to play a crucial role in the N cycle by oxidizing ammonium to nitrate. Nitrifiers were not detected in needles from plants growing in an environment with low N precipitation, which suggests that a high concentration of fixed nitrogen (NO_y and NH_x) in the ambient air is a prerequisite for substantial colonization of the needle apoplast. This would make sense, since plant-associated nitrification would critically depend on nitrite and ammonia, which can be provided by the atmosphere at a site with high nitrogen deposition. N often limits plant growth in terrestrial ecosystems but may become a pollutant if applied in high amounts. High N deposition has been implicated with many effects related to plants (Nihlĝard, 1985; Aber et al., 1989), including alterations in foliar nutritional status (Näsholm and Ericsson, 1990), increased microbial attack resulting from changes of plant chemistry, or even forest decline. Possibly, plants become more susceptible for colonization of chemolithoautotrophic nitrifiers when they are subjected to high amounts of atmospheric N.

1.6 Genetic traits of endophytic bacteria colonizing the apoplast

Obviously, there are many situations prevailing in natural ecosystems, where in the apoplast nutrients are plentifully available for instant microbial consumption. Since bacteria are common at this site, and considerable microbial activities have been identified here from heterotrophic (Hurek and Reinhold-Hurek, 2006) as well as autotrophic bacteria (Teuber et al., this volume, pp. 405–426), it is reasonable to assume that the apoplast is sometimes inhabited by a plethora of microorganisms. While this assumption may be true with respect to numbers (Hurek and Reinhold-Hurek, 2006), it is certainly not true with respect to diversity. From the thousands of different microorganisms occuring in soil (Torsvik et al., 1990), only very few interact with plants (Kowalchuk et al., 2002). In order to achieve considerable population densities in tissues of symptomless plants in natural environments, bacteria probably have to acquire genetic traits which are useful for a life in plants but not necessarily in soil. The detection of traits suitable for an endophytic lifestyle has been considerably hampered by the fact that bacteria differ in their capabilities to interact with plants at the species or even strain level. Accordingly, a certain capability of a particular bacterium can not be generalized for the genus, and each bacterium has to be studied separately (Reinhold-Hurek and Hurek, 2006). Furthermore, many bacteria colonizing plants seem to defy cultivation efforts and could not be examined yet (Hurek and Reinhold-Hurek, 2006). In spite of these drawbacks, some factors could already be identified to be important for bacterial endophytes thriving in the apoplast of plants. While type IV pili (Dörr et al., 1998) and cellulase (Reinhold-Hurek et al., this volume, pp. 427–444) correlate positively with plant colonization, formation of flagella, and type III secretion appear to be detrimental to it (Iniguez et al., 2005).

2. ENDOSYMBIOTIC ENDOPHYTES

We do not know much about the apoplast as a stable environment for bacterial endophytes and the precise distribution of nutrients for endophytic microbial activity there. But we know from high nitrogenase transcription in Kallar grass or several other grasses by the apoplast colonizing endophyte *Azoarcus* sp. BH72 or by unknown bacteria, respectively, that nutrient pools required for these activities must be considerable (Hurek and Reinhold-Hurek, 2006). These microbial processes, like probably all endophytic bacterial processes in the apoplast, typically remain symptomless. But this is not the case with all endophytes. Plant endosymbiosis with nonphotosynthetic bacteria is generally accompanied by morphological changes culminating in the formation of a nodule. The nodule represents the specifically generated and optimized organ for the heterotrophic, N_2-fixing cell organelle, the symbiosome (Werner, 1992) and is particularly well characterized in legumes. Symbiosomes reside in infected plant cells and contain the bacterial microsymbiont. Schubert (this volume, pp. 445–454) provided now evidence that there is an important apoplastic step in the nutrient transport to infected cells in indeterminate *Vicia faba* root nodules. While the route of assimilate transport into the nodule is probably symplastic as it is in general to growing root cells (Pritchard et al., 2000), the solute uptake by the infected cells seems to be mainly apoplastic. This combines an unselective bulk flow of assimilates to the location where the highly energy demanding nitrogen fixation takes place with a selective apoplastic uptake by the infected plant cells containing the symbiosomes. The data indicate that a substantial bulk flow of solutes between infected and neighboring uninfected cells may occur. But a symplastic transport is not important for the nutrient uptake of the majority of infected cells, since the proportion of infected cells which are connected by plasmodesmata to neighboring uninfected cells is rather low, and infected cells are not connected with each other (Abd-Alla et al., 2000). Rather, cells, which are connected to the bulk flow of solutes pouring into the nodule, may accumulate organic compounds under the oxygen starving conditions prevailing in the nodule (see 1.2) and release them into the apoplast. From there, metabolites are likely to be taken up by infected cells, emphasising the role of the apoplast in solute supply to heterotrophic bacterial endophytes whether they are symbiotic or not.

REFERENCES

Abd-Alla, M. H., Koyro, H-W., Yan, F., Schubert, S., Peiter, S. (2000). Functional structure of the indeterminate *Vicia faba* L. root nodule: implications for metabolic transport. *Plant Physiol.*, 157, 335–343.

Aber, J. D., Nadelhoffer, K. J., Steudler, P. and Melillo, J. M. (1989). Nitrogen saturation in northern forest ecosystems. *Bioscience*, 39, 378–386.

Armstrong, W. (1979). Aeration in higher plants. *Adv. Bot. Res.*, 7, 225–332.

Atkinson, M. M. and Baker, C. J. (1987a). Alteration of plasmalemma sucrose transport in *Phaseolus vulgaris* by *Pseudomonas syringae* pv. *syringae* and its association with K^+ / H^+ exchange. *Phytopathology*, 77, 1573–1578.

Atkinson, M. M. and Baker, C. J. (1987b). Association of host plasma membrane K^+ / H^+ exchange with multiplication of *Pseudomonas syringae* pv. *syringae* in *Phaseolus vulgaris*. *Phytopathology*, 77, 1273–1279.

Bodelier, P. L. E. (2003) Interactions between oxygen-releasing roots and microbial processes in flooded soils and sediments. In: H. de Kroon and E. J. W. Visser (Eds.) Ecological Studies. (Vol. 168., pp. 331–362). Berlin: Springer .

Crawford, R. M. M. (1978). Metabolic adaptations to anoxia. In: R. M. M. Crawford and D. D. Hook (Eds.) Plant Life in Anaerobic Environments. (pp. 119–136). Michigan, USA: Ann Arbor Science.

Curl, E. A. and Truelove, B. (1986). The rhizosphere. Berlin: Springer.

Darwent, M. J., Paterson, E., McDonald, A. J. S., Tomos, A. D. (2003). Biosensor reporting of root exudation from *Hordeum vulgare* in relation to shoot nitrate concentration. *J. Exp. Bot.*, 54, 325–334.

De La Fuente, J. M., Ramírez-Rodríguez, V., Cabrera-Ponce, J. L., Herrera-Estrella, L. (1997). Aluminium tolerance in transgenic plants by alteration of citrate synthesis. *Science*, 276, 1566–1568.

Dörr, J., Hurek, T. and Reinhold-Hurek, B. (1998). Type IV pili are involved in plant-microbe and fungus-microbe interactions. *Mol. Microbiol.*, 30, 7–17.

Evans, D. E. (2003). Aerenchyma formation. *New Phytol.*, 161, 35–49.

Hallmann, J. (2001). Plant interactions with endophytic bacteria. In: M. J. Jeger and N. J. Spence (Eds.), Biotic interactions in plant-pathogen associations. (pp. 87–119). CABI Publishing.

Henzi T, Brändle R. (1993). Long-term survival of rhizomatous species under oxygen deprivation. In: M. B. Jackson, C. R. Black, (Eds.) Interacting Stresses on Plants in a Changing Climate. (pp. 305–314). Berlin: Springer-Verlag.

Hoffland, E., Findenegg, G. R., Nelemans, J. A.(1989). Solubilization of rock phosphate by rape. II. Local root exudation of organic acids as a response to P-starvation. *Plant Soil*, 113, 161–65.

Hurek, T., Handley, L., Reinhold-Hurek , B., Piché, Y. (2002). Azoarcus grass endophytes contribute fixed nitrogen to the plant in an unculturable state. *Mol. Plant-Microbe Interact.*, 15, 233–242.

Hurek, T., Reinhold, B., Fendrik, I. and Niemann, E. G. (1987). Root-zone-specific oxygen tolerance of *Azospirillum* spp. and diazotrophic rods closely associated with Kallar grass. *Appl. Environ. Microbiol.*, 53, 163–169.

Hurek, T. and Reinhold-Hurek, B. (2006). Molecular ecology of N_2-fixing microbes associated with gramineous plants: hidden activities of unknown bacteria. In: D. Werner and W. E. Newton (Eds.).

Agriculture, Forestry, Ecology and the Environment. Dordrecht, The Netherlands: Kluwer Academic Publishers, (in press).

Iniguez, A. L., Dong, Y., Carter, H., Ahmer, B., Stone, J. and Triplett, E. W. (2005). Regulation of enteric endophytic bacterial coloniazation by plant defenses. *Mol. Plant-Microbe Interact.*, 18, 169–178.

Kowalchuk, G. A., Buma, D. S., de Boer, W., Klinkhamer, P. G., van Veen, J. A. (2002). Effects of above-ground plant species composition and diversity on the diversity of soil-borne microorganisms. *Antonie Van Leeuwenhoek*, 81, 509–520.

López-Bucio, J., De La Vega, O. M., Guevara-García, A., Herrera-Estrella, L. (2000a). Enhanced phosphorus uptake in transgenic tobacco plants that overproduce citrate. *Nature Biotechnology*, 18, 450–453.

López-Bucio, J., Nieto-Jacobo, M. F., Ramírez-Rodríguez, V., Herrera-Estrella, L. (2000b). Organic acid metabolism in plants: from adaptive physiology to transgenic varieties for cultivation in extreme soils. *Plant Sci.*, 160, 1–13.

Lynch, J. M. (1990). The rhizosphere. Chichester, England : John Wiley and Sons.

Marschner, H. (1995). Mineral nutrition of higher plants, Ed. 2. London, England: Academic Press.

Näsholm, T. and Ericsson, A. (1990). Seasonal changes in amino acids, protein and total nitrogen in needles of fertilized Scots pine trees. *Tree Physiology*, 6, 267–281.

Nihlgård, B., (1985). The ammonium hypothesis - an additional explanation to the forest dieback in Europe. *Ambio*, 14, 2–8.

Papen, H., Geßler, A., Zumbusch, E. and Rennenberg, H. (2002). Chemolithoautotrophic nitrifiers in the phyllosphere of a spruce ecosystem receiving high atmospheric nitrogen input. *Current Microbiol.*, 44, 56–60.

Pritchard, J., Winch, S. and Gould, N. (2000). Phloem water relations and root growth. *Austr. J. Plant Physiol.*, 27, 539–548.

Reinhold, B., Hurek, T. and Fendrik, I. (1985). Strain-specific chemotaxis of *Azospirillum* spp. *J. Bacteriology*, 162, 190–195.

Reinhold, B., Hurek, T., Niemann, E.-G. and Fendrik., I. (1986). Close association of *Azospirillum* and diazotrophic rods with different root zones of Kallar grass. *Appl. Environ. Microbiol.*, 52, 520–526.

Reinhold-Hurek, B. and Hurek, T. (1998). Life in grasses: diazotrophic endophytes. *Trends Microbiol.*, 6, 139–144.

Reinhold-Hurek, B. and Hurek, T. (2006). Endophytic associations of *Azoarcus* spp. In: D. Werner and W. E. Newton (Eds.). Agriculture, Forestry, Ecology and the Environment. Dordrecht, The Netherlands: Kluwer Academic Publishers, in press.

Reinhold-Hurek, B., Hurek, T., Gillis, M., Hoste, B., Vancanneyt, M., Kersters, K. and De Ley, J. (1993). *Azoarcus* gen. nov., nitrogen-fixing Proteobacteria associated with roots of Kallar grass (*Leptochloa fusca* (L.) *Kunth*), and description of two species, *Azoarcus indigens* sp. nov. and *Azoarcus communis* sp. nov. *Int. J. Syst. Microbiol.*, 170, 1445–1451.

Rennenberg, H. Kreutzer, K. Papen, H. and Weber, P. (1998) Consequences of high loads of nitrogen for spruce (*Picea abies*) and beech (*Fagus sylvatica*) forests. *New Phytol.*, 139, 71–86.

Rivoal, J., Hanson, A. D. (1993). Evidence for a large and sustained glycolytic flux to lactate in anoxic roots of some members of the halophytic genus Limonium. *Plant Physiol.*, 101, 553–560.

Ryan, P. R., Delhaize, E. and Jones, D. L. (2001). Function and mechanism of organic anion exudation from plant roots. *Annu. Rev. Plant Physiol. Plant Mol. Biol.*, 52, 527–560.

Summers, J. E., Ratcliffe, R. G., Jackson, M. B. (2000). Anoxia tolerance in the aquatic monocot *Potamogeton pectinatus*: absence of oxygen stimulates elongation in association with an unusually large Pasteur effect. *J. Exp. Bot.*, 51, 1413–1422.

Studer, C., Brändle, R. (1984). Oxygen consumption and availability in the rhizomes of *Acorus calamus* L., *Glyceria maxima Holmb.*, *Menyanthes trifoliata* L., *Phalaris arundinacea* L., *Phragmites communis Trin.* and *Typha latifolia* L. *Botanica Helvetica*, 94, 23–31.

Torsvik, V., Goksoyr, J. and Daae, F. L. (1990). High diversity in DNA of soil bacteria. *Appl. Environ. Microbiol.*, 56, 782–787.

Van Veen, J. A., Merckx, R. and Van de Geijn, S. C. (1989). Plant and soil related controls of the flow of carbon from roots through the soil microbial biomass. *Plant Soil*, 115, 179–188.

Van Veen, J. A., Liljeroth, E., Lekkerkerk, L. J. A. and Van de Geijn, S. C. (1991). Carbon fluxes in plant-soil systems at elevated atmospheric CO_2. *Ecol. Appl.*, 1, 175–181.

Verpoorte, R. (2000). Plant secondary metabolism. In: R. Verpoorte, A. W. Alfermann (Eds.). Metabolic Engineering of Plant Secondary Metabolism. (pp. 1–29). Dordrecht, The Netherlands: Kluwer Academic Publishers.

Walker, T. S., Bais, H. P. Grotewold, E. and Vivanco, J. M. (2003). Root exudation and rhizosphere biology. *Plant Physiol.*, 132, 44–51.

Werner, D. (1992). Physiology of nitrogen-fixing legume nodules: compartments and functions. In: G. Stacey, R. H. Burris, H. J. Evans (Eds.). Biological Nitrogen Fixation. (pp. 399–431). New York, USA: Chapman and Hall.

Whipps, J. M. (1984) Environmental factors affecting the loss of carbon from the roots of wheat and barley seedlings. *J. Exp. Bot.*, 35, 767–773.

Whipps, J. M. and Lynch, J. M. (1986). The influence of the rhizosphere in crop productivity. *Adv. Microb. Ecol.*, 9, 187–244.

THE APOPLAST OF NORWAY SPRUCE (*PICEA ABIES*) NEEDLES AS HABITAT AND REACTION COMPARTMENT FOR AUTOTROPHIC NITRIFIERS

M. TEUBER[1], H. PAPEN[1], R. GASCHE[1], T.H. EßMÜLLER[2] and A. GEßLER[2]

[1]Forschungszentrum Karlsruhe GmbH, Institute for Meteorology and Climate Research, Atmospheric Environmental Research (IMK-IFU), Department Terrestrial Biosphere under Global Change, Germany, hans.papen@imk.fzk.de; [2]Albert-Ludwigs-University of Freiburg, Institute for Forest Botany and Tree Physiology, Germany.

Abstract. By the combination of both, the molecular biological fluorescence in situ hybridization-technique with the technique of confocal laser scanning-microscopy (cLSM) it could be unequivocally demonstrated for the first time at the genetic level that autotrophic nitrifiers are present inside spruce needles of a spruce forest ecosystem (The "Höglwald") exposed to high levels of atmospheric nitrogen deposition and that they are located within the apoplastic space of the needle leaves (sub-stomatal cavity). In contrast, autotrophic nitrifers could not be detected in needles of adult trees at a spruce forest site ("Villingen") exposed to low levels of atmospheric N input. When needles from adult spruce trees at the Högwald site were exposed to 10 Pa acetylene – an inhibitor of ammonia monooxigenase of chemolithoautotrophic ammonia oxidisers (CAO) – the sink strength of the needles for NH_3 decreased significantly. Since the reduction of NH_3 deposition due to acetylene-induced inhibition of the ammonia monooxigenase was greatest when stomata were open and only minute when stomata were closed, it is concluded that physiologically active CAO are located inside the needles rather than on the needle surface. On the other hand, a reduction of NH_3 uptake when applying acetylene was not observed with adult spruce trees from the nitrogen limited stand at Villingen. From the results obtained it is concluded that the observed NH_3 flux from the atmosphere into the needle leaves in N-polluted forests is not exclusively a plant physiological process as has been assumed in the past, but is the result of both plant physiological plus microbial processes. A seasonal pattern of the colonization of the needles by nitrifiers at the Höglwald site could not be demonstrated. For gaining first insights into the pathway by which needles might be colonized by autotrophic nitrifiers, sterile spruce seedlings fumigated with both ammonia and air were inoculated in the laboratory with nitrifier cultures. A successful establishment of the autotrophic nitrifiers in the phyllosphere

B. Sattelmacher† and W.J. Horst (eds.), The Apoplast of Higher Plants: Compartment of Storage, Transport and Reactions, 405–425.

of the spruce seedlings could not be achieved by a single inoculation event, however, was successful after multiple inoculations. It is concluded that autotrophic nitrifiers obviously are only able to colonize spruce needles in a later stage of spruce development.

Key words: chemolithoautotrophic nitrifiers, ammonia oxidizers, nitrite oxidizers, ammonia deposition, ammonia monooxigenase, apoplast, stomatal cavity, confocal laser scanning microscopy, spruce, needle leaves

1. INTRODUCTION

Chemolithoautotrophic ammonia oxidizers (CAO) and nitrite oxidizers (CNO) play a crucial role in the nitrogen cycle (Bock et al., 1989). They are not only responsible for the biological oxidation of ammonium to nitrate in soils and waters, but are also significant sources and/or sinks of atmospheric N trace gases like e.g. NH_3, NO_2, N_2O and NO (e.g. Gasche and Papen, 1999; Papen and Butterbach-Bahl, 1999). Hitherto identified habitats of CAO and CNO are the pedo-, hydro- and lithosphere (Ward, 1986; Bock et al., 1989). Microbiological studies at the Höglwald Forest, Bavaria, Germany, a spruce ecosystem exposed to high loads of N and high atmospheric NH_3 and NO_2 concentrations (Rennenberg et al., 1998) had provided preliminary evidence that autotrophic nitrifiers are present in the phyllosphere, most probably within the apoplast of the needles, a hitherto not identified habitat for chemolithoautotrophic nitrifiers (Papen et al., 2002). In addition, in this previous study fumigation experiments of spruce twigs with NH_3 and with NH_3 plus 10 Pa C_2H_2 (an inhibitor of bacterial ammonia monooxigenase) gave additional preliminary evidence that the observed in-situ NH_3 uptake by spruce needles may not only be due to the physiological activity of the plant but that the activity of autotrophic nitrifiers significantly contributes to this uptake at the plant/atmosphere interface (Papen et al., 2002). However, up to now, there was a lack of a systematic assessment of the localization of nitrifiers within the needles or on the needle surface. In addition, there was no information whether colonization of the spruce phyllosphere occurs only under the particular conditions of a forest ecosystem exposed to high loads of atmospheric N or is a wide spread phenomenon. Geßler et al. (2002) have demonstrated that NH_3 deposition flux to needles of adult spruce trees at the N-affected field site Höglwald was significantly higher as compared to the fluxes observed for adult spruce trees grown under clean air conditions, giving a first preliminary indication that differences in abundance or activity of nitrifiers within the needles or on the needles surface may depend on the magnitude of atmospheric N input.

Another question not addressed so far was, how colonization of needles with nitrifiers takes place.

The key objectives of the project were (i) to determine the localization of autotrophic nitrifiers within the needles (apoplast) or on the needle surface by applying combined molecular biological and microscopical (cLSM) techniques, (ii) to quantify cell numbers of autotrophic nitrifiers in/on the needles of spruce trees exposed to different amounts of atmospheric N deposition (iii) to clarify the pathway of bacterial colonization of spruce needles and (iv) to quantify the bacterial contribution to the observed NH_3 uptake of spruce needles from the atmosphere at field sites exposed to different levels of atmospheric N deposition.

2. MATERIALS AND METHODS

2.1 Field sites

Quantification of cell numbers of CAO and CNO and of their contribution to atmospheric NH_3 uptake into needle leaves was performed at two forest ecosystems dominated by spruce, differing in the extent of atmospheric N input.

2.1.1 N-affected field site (Höglwald)

The field site Höglwald is located 50 km west-north-west of Munich (Germany, longitude 11°10' E; latitude 48°30' N) in the pre-alpine region at 540 m a.s.l. The forest stand is dominated by c. 90 year old spruce (*Picea abies*). Pedospheric and atmospheric N availability is high. During the present investigation wet deposition of inorganic N, measured in the throughfall, was 30 kg N ha^{-1} a^{-1} (10 kg NO_3^-–N ha^{-1} $year^{-1}$ and 20 kg NH_4^+-N ha^{-1} $year^{-1}$) (Rennenberg et al., 1998). Mean annual NH_3 and NO_2 concentrations in ambient air of the canopy region amounted to 4.1 (Huber, 1997) and 6.92 nmol mol^{-1}, respectively, in 1995 (Gasche and Papen, unpublished data). For a detailed site description, see Kreutzer and Weiss (1998).

2.1.2 N-limited field site (Villingen)

The spruce forest ecosystem near Villingen (Germany, longitude 8°22' E; latitude 48°3' N; 900 m a.s.l.) is a clean-air site, largely unaffected by agricultural activities with total N input from the atmosphere below 10 kg N ha^{-1} $year^{-1}$ (Feger et al., 1992). NH_3 concentrations in ambient air are supposed to be significantly below the average NH_3 concentration for Germany amounting to 3.5 nmol mol^{-1} (Asman and Van Jaarsfeld, 1990). The stand is fully stocked with 85–125 years-old spruce trees. As a consequence of the intensive long term forest management the site is characterized as N-limited. For a detailed site description see Rennenberg et al. (1998).

2.2 Sampling of spruce needles

As described before (Papen et al., 2002), spruce needles were collected from the canopy of spruces in spring and late summer/autumn of the years 2000 till 2002. In the years 2000 and 2001 the spruce needles were taken from the lower parts of the canopy from a height of 2–10 m, in the year 2002 the samples were taken from the sun crown of the spruces from a height of nearly 35 m.

The needles of spruce seedlings were taken directly after the inoculation with the nitrifiers and in the following every 2 months.

2.3 Microbiological studies

2.3.1 Determination of cell numbers of CAO and CNO within spruce needles by MPN technique

Spruce needles from adult spruce trees of the field sites as well as from spruce seedlings grown in the lab under controlled conditions (see below) were harvested and processed in the lab under sterile conditions to needle extracts as described (Papen et al., 2002) using a Waring blender and 1 M ice-cold potassiumphosphate as an extraction buffer (pH = 7.6). Cell numbers were determined by the MPN technique as described (Papen and von Berg, 1998; Papen et al., 2002) using published MPN tables (de Man, 1983).

2.3.2 Detection of CAO and CNO in MPN test tubes by PCR

DNA extraction: DNA from bacterial cells obtained from MPN test tubes which had been tested positive for the presence of nitrite/nitrate (CAO) or disappearance of nitrite/nitrate (CNO), was extracted and purified according to the method described by Pellicer et al. (1978).

Oligonucleotide primers: for the PCR priming of CAO and CNO oligonucleotides reported in the literature were used (Hermansson and Lindgren, 2001; Degrange and Bardin, 1995). Additionally, *Nitrobacter*-specific oligonucleotide primers were selected by comparing the 16S rDNA sequences of *Nitrobacter* species, *Rhodopseudomanas palustris* and *Bradyrhizobium japonicum*. The two *Nitrobacter*-specific primers finally selected were: forward primer Nw f (5'-CGG AGC ATG GAG CAC AGG-

3') and reverse primer Nw r (5'-CCC CTT TGC TTC CCA TTG-3') (Teuber, 2003). The oligonucleotides were synthesized and obtained from MWG Biotech AG, Ebersberg, Germany.

PCR amplification, cloning and sequencing of the 16S rDNA gene: PCR-amplification was performed using Taq-DNA-polymerase (Invitrogen, Karlsruhe, Germany) and a specifically adapted PCR protocol as described by Teuber (2003). Cloning and sequencing of the 16S rDNA gene was performed using the TOPO TA Cloning kit (Invitrogen GmbH, Karlsruhe, Germany) and the BigDye RR Terminator Cycle Sequencing kit (PE Applied Biosystems, Weiterstadt, Germany), respectively.

2.3.3 Bacterial strains/cultures used and inoculation of spruce seedlings with nitrifiers

To test and to ensure the specificity of oligonucleotide probes for in situ hybridization experiments (see below) the following bacterial strains/cultures were used: *Nitrosomas europaea* ATCC 25978, *Nitrobacter winogradskyi* ATCC 25391, *Escherichia coli* TG1, *Rhodococcus rodochrous* (IMK-IFU isolate) and mixed cultures of nitrifiers obtained from isolation experiments of the organic layer of the Höglwald site. The culture conditions for the specific strains were: for *N. europaea* and *N. winogradskyi* according to the "National Collections of Industrial and Marine Bacteria" (1994), *for E. coli* according to Sambrook et al. (1989), for *R. rodochrous* according to Atlas and Parks (1993), and for the enriched mixed culture of nitrifiers according to Schmidt and Belser (1982).

For inoculation of spruce seedlings either a mixed culture of CAO and CNO from the Höglwald site or a pure culture of *N. europaea* ATCC 25978 were used. The strain/cultures were grown to the early lag phase in Erlenmeyer flasks containing 200 ml of respective medium and thereafter centrifuged (10.000 rpm, 1 h, 4°C), washed in 1 mM cold potassium phosphate buffer (pH = 7.6), pelleted again by centrifugation, and finally resuspended at a concentration of 106 cells per ml. This cell suspension was stored at 4°C and used for inoculation of seedlings by spreading the bacteria cells homogeneously over the seedlings using a sterile spray flacon.

2.3.4 In situ hybridization experiments

2.3.4.1. Selection of oligonucleotide probes for specific detection of autotrophic nitrifiers

For fluorescence in situ hybridization (FISH) the following 16S rRNA targeted oligonucleotide probes were used: Nso190 (Mobarry et al., 1996) complementary to a region specific for the group of ammonia-oxidizing β-Proteobacteria, NIT3 (Wagner et al., 1996) complementary to a region specific for the Genus *Nitrobacter* and CNIT3 (Wagner et al., 1996) complementary to a region specific of *Bradyrhizobium japonicum*, *Rhodopseudomonas palustris*, *Afipia clevelandis* and *Afipia felis*. Oligonucleotide probes were labeled with the following fluorescent dyes: 5-carboxy-X-rhodamin (ROX for Nso190) and fluoresceine-5-isothiocyanate (FITC for NIT3). The synthetically labeled oligonucleotides were obtained from MWG Biotech AG, Ebersberg, Germany.

2.3.4.2. Fixation and preparation of bacteria and slices of spruce needles

Bacteria cells or intact spruce needles, respectively, were fixed in 4% formaldehyde for 16 h at 4°C. After removal of the fixative the samples were washed and stored in 1 x PBS buffer (pH 7.4) at 4°C for nearly 8 weeks (Amann et al., 1990). Bacterial cell suspensions were spread onto glass slides and dehydrated as reported (Schramm et al., 1996). Prior to cutting the needles were frozen in the embedding compound Tissue-Tek at –20°C for nearly 20 min and cut into longitudinal slices (thickness: 80 μm) using a microtome. Needle slices obtained were suspended into a drop of 1 x PBS buffer (pH = 7.4) on glass slides covered with aminoalcylsilane and a thin film of albumen from chicken eggs for a stronger mechanical fixation of the slices. After immobilization (by drying at 70°C for 15 min) the slices were dehydrated with 50, 80 and 100% ethanol at 50°C (each step: 30 min) and could be stored until analysis under dry and clean conditions at room temperature.

2.3.4.3. In situ hybridization and DAPI counterstain

In situ hybridization was performed according to Gieseke et al. (2001): The hybridization buffer contained 0.9 M NaCl, x% formamide (55% for Nso190, 40% for NIT3), 20 mM Tris-HCl (pH = 7.4), and 0.1% sodium dodecyl sulfate (SDS). Probe concentrations were 10 ng/μl. Probe NIT3 was used with the competitor oligonucleotide probe CNIT3 in the same concentration. The slides were covered with a thin plastic film, and after denaturation of the samples for 10 min at 95°C the slides were incubated for 1 h at exactly 46°C in a humidity chamber. For removal of unbound probes the slides were washed for 20 min at 48°C in a buffer containing 20 mM Tris, x mM NaCl (20 mM for Nso190, 56 mM for NIT3), 5 mM EDTA and 0.01% SDS (Manz et al., 1992). Slides were rinsed briefly with distilled water and air-dried. Multiple probe hybridization was performed in subsequent steps by first hybridizing with the probe of higher stringency.

After in-situ hybridization the samples were counterstained with the DNA-specific dye DAPI. For staining the DAPI stock solution (concentration: 100 $\mu g\ ml^{-1}$ water) was diluted to 2 $\mu g\ ml^{-1}$, applied onto glass slides, which were incubated for 10 min in the dark (Manz et al., 1992). After short washing with 4 x SSC-buffer (pH = 7.0) the slides were mounted with the anti-bleaching agent n-propyl gallate (70% (v/v) glycerol, 25% (w/v) 0.5 M Tris (pH = 9.0) and 5% (w/v) n-propyl gallate) (Giloh and Sedat, 1982).

2.3.5 Confocal Laser Scanning Microscopy (cLSM)

The samples were analyzed by a LSM 510 scanning confocal laser microscope (Zeiss, Jena, Germany) equipped with an UV, an Argon and a HeNe laser supplying excitation at wavelengths of 366, 488 and 543 nm, and a 63x water immersion lens was used. Confocal images were captured and merged using the Zeiss LSM software (Zeiss, Jena, version 2.3).

2.4 Plant physiological studies

Adult trees: At both experimental sites (Höglwald and Villingen) twigs from the sun-exposed crown of dominant trees were used for the experiments. Elemental analyses of the leaves of both stands showed no nutrient deficiencies or imbalances (Rennenberg et al., 1998), while at the Höglwald site leaves and phloem of twigs contained high amounts of soluble arginine (Geßler et al., 1998) indicating excessive N supply (Näsholm and Ericson, 1989).

Young trees: At both sites 4–5 years-old spruce trees from natural succession of both experimental sites were used for additional microbiological and plant physiological experiments.

Seedlings: In order to characterize colonization of needles with nitrifiers and to assess the effect of possible colonization on NH_3 gas exchange, spruce seedlings were grown from seeds under sterile conditions. Surface-sterilized seedlings were grown in sterilized incubators described in detail by Muller et al. (1996). 10 weeks after seed germination (when primary leaves had been developed) the seedlings in two of four incubators were fumigated with sterile air containing 50 nmol mol^{-1} NH_3, whereas the other two incubators were flushed with NH_3-free air. From both fumigation treatments plants from one incubator were inoculated with nitrifying bacteria (see above), whilst the plants in the other incubator remained untreated. In May 2000 plants were inoculated with mixed cultures of autotrophic nitrifiers (CAO and CNO) obtained from needles of spruce trees from the field site Höglwald, whereas in May 2001 a pure culture of *Nitrosomonas europaea* ATCC 25978 was used for inoculation of the seedlings. Immediately before and after inoculation as well as 8 and 16 weeks later, plant material was sampled for quantification of cell numbers of colonizing bacteria. NH_3 exchange experiments with seedlings were performed 8–16 weeks after inoculation. Light intensity and air temperature during the incubation was 175 μmol m^{-2} s^{-1} and 20°C.

2.5 Gas exchange measurements

Twigs from adult and young spruce trees from the two field sites and whole seedlings grown in the laboratory experiments were exposed to defined NH_3 concentrations applying the dynamic chamber technique (detailed description see Geßler et al., 2000; 2002). In the field experiments with adult trees environmental conditions within the chambers were adapted to ambient conditions. Young trees from the two field sites were excavated with intact root systems and the surrounding undisturbed soil and transferred into the laboratory. Gas-exchange measurements were performed with twigs with two to three needle age-classes still attached to the trees under controlled conditions in the laboratory (air temperature 20°C, light intensity 175 μmol m^{-2} s^{-1}). During at least 12 hours young and adult trees were exposed to NH_3 concentrations that are typically observed after slurry application to the surrounding fields. Thereafter, acetylene as an inhibitor of the ammonia monooxigenase of the CAO was added to the fumigation air in addition to NH_3 (Papen et al., 2002). During fumigation NH_3 flux, transpiration and stomatal conductance were determined and calculated as 60 min mean value (Geßler et al., 2000; 2002). Fluxes from the atmosphere to the leaves are given as negative values.

Seedlings were also fumigated under controlled conditions (air temperature 20°C, light intensity 175 μmol m^{-2} s^{-1}) with different NH_3 concentrations (Geßler et al., 2000).

3. RESULTS AND DISCUSSION

Table 1 summarizes the results obtained from the MPN studies for the quantification of CAO and CNO in extracts of needles taken from the adult trees at the two experimental sites. At the highly N-affected Höglwald site CAO and CNO were detected at all sampling dates. Cell numbers of CAO varied between 8.3 x 10^3 and 6.6 x 10^5 cells g^{-1} needle fresh weight (NFW), those of CNO between 8.3 x 10^3 and 4.2 x 10^4 cells g^{-1} NFW. A pronounced seasonality of cell numbers could not be demonstrated (Table 1). These results are in good agreement with earlier findings at the Höglwald site

Table 1. Most Probable Numbers (MPN) of chemolithoautotrophic ammonia and nitrite-oxidizers [g^{-1} needle fresh weight] in needle extracts of mature spruce trees of the two different experimental sites Höglwald (N-affected) and Villingen (N-limited) for the period September 2000 – October 2002. Numbers given represent average values from three independent samples after an incubation time of 9 weeks each.

	Höglwald		*Villingen*	
Sampling date	*Ammonia oxidizers*	*Nitrite oxidizers*	*Ammonia oxidizers*	*Nitrite oxidizers*
Sep 2000	1.19×10^4	1.19×10^4	–	1.10×10^4
May 2001	8.28×10^3	8.28×10^3	–	9.60×10^3
Oct 2001	9.47×10^3	9.47×10^3	–	6.98×10^3
May 2002	1.34×10^5	4.13×10^4	–	4.32×10^4
Oct 2002	6.59×10^5	4.16×10^4	–	4.24×10^4

demonstrating unequivocally that CAO and CNO are present in the phyllosphere of adult spruce trees at this site (Papen et al., 2002). In contrast, at the N-limited site (Villingen) CAO could not be detected in needle extracts at any sampling date (Table 1). However, CNO were detected using the MPN method (nitrate-positive MPN test tubes). In order to clarify, whether the nitrate detected in these MPN test tubes derived indeed from CNO or was a result of chemical oxidation of nitrite to nitrate or from other biological sources, e.g. heterotrophic nitrifiers, PCR technique was applied for specific detection of CNO. Applying this technique revealed that CNO were not present in the tested MPN test tubes inoculated with extracts from needles of the Villingen site. Thus, most likely abiotic or other biotic factors must have been responsible for the nitrate production in these MPN test tubes. In contrast, results obtained from applying the PCR technique for the specific detection of CAO to MPN test tubes incubated with needle extracts from the Höglwald site, which had been positively tested for the presence of CAO, were in good agreement with the results obtained by the MPN technique, provided bacterial DNA sufficient in quantity for PCR assays could be isolated from the MPN test tubes. We thus conclude – supported by the results obtained from the in situ hybridization experiments in combination with confocal laser scanning microscopy (see below) – that autotrophic nitrifiers are not present in needles of the spruce trees at the N-limited site Villingen.

In order to gain informations (i) about one possible pathway by which colonization of spruce needles might proceed (i.e. atmospheric transport), (ii) whether nitrifiers deposited to needles will exclusively colonize the needle surface or are even able to spread inside the needles (apoplast), and (iii) whether colonization of spruce needles by CAO and CNO is dependent on the exposure to NH_3 concentration in ambient air, spruce seedlings, which had been grown up under sterile conditions in the lab, were inoculated either with a mixture of CAO and CNO or with a pure culture of *Nitrosomonas europaea* ATCC 25978 and then fumigated with/without NH_3 in ambient air. The results obtained from these experiments are summarized in Table 2.

Determination of cell numbers of CAO and CNO of needles sampled immediately after inoculation with a mixture of CAO and CNO (April 2000) revealed cell numbers of 1.4 103 cells g^{-1} NFW (CAO) and 5.3 103 cells g^{-1} NFW (CNO), respectively (data are only available for the non fumigated

Table 2. Most Probable Numbers (MPN) of chemolithoautotrophic ammonia and nitrite-oxidizers [g^{-1} needle fresh weight] in needle extracts of spruce seedlings, which had been inoculated with autotrophic nitrifiers. In the year 2000 spruce seedlings were inoculated with mixed cultures of nitrifiers isolated from needles of spruce trees from the field site Höglwald whereas in May 2001 a pure culture of Nitrosomonas europaea ATCC 25978 was used. Numbers given represent average values from three independent samples after an incubation time of 9 weeks each.

	NH_3 -fumigated		*non-fumigated*	
Sampling date	*Ammonia oxidizers*	*Nitrite oxidizers*	*Ammonia oxidizers*	*Nitrite oxidizers*
		Year 2000		
Apr 2000	–	–	*1.4×10^3*	*$5.3 X 10^3$*
Jun 2000	–	*2.1×10^4*	–	*1.4×10^4*
Sep 2000	–	*9.1×10^4*	–	*8.7×10^3*
		Year 2001		
May 2001	–	–	–	–
Jun 2001	*1.32×10^4*	–	*9.95×10^4*	–
Sep 2001	–	–	*1.22×10^4*	–

seedlings, not for the NH_3 fumigated seedlings). Whereas at the following sampling dates (June and September 2000) CAO could not be detected anymore, CNO seemed to be present at low cell numbers (MPN method). However, PCR assays of these MPN test tubes were negative for CNO indicating that the nitrate detected in these MPN test tubes most likely cannot be attributed to CNO activity (see above). Thus, it has to be concluded that inoculation of spruce seedlings with a mixed culture of CAO and CNO did not result in the establishment of a stable nitrifier population on/within the needles of spruce seedlings. In a second series of experiments, in which spruce seedlings were inoculated twice (May and June 2001) with a pure culture of the ammonia oxidizer *N. europaea* ATCC 25978, it was found that at least part of the initially applied population could be retrieved c. 8 weeks after the second inoculation (September 2001) in the experimental series with seedlings that had not been fumigated with NH_3, though cell numbers had markedly declined from 9.95×10^4 to 1.22×10^4 cells g^{-1} NFW (Table 2). However, in extracts from needle seedlings which had been exposed to NH_3 fumigation, such a temporarily stabilization of the population was not observed.

In summary, the results indicate that colonization of spruce needles seems to be restricted to spruce forest sites which are exposed to high loads of atmospheric N deposition. However, experiments with needles from young spruce trees (4–5 years old) did not show any colonization of the needles with autotrophic nitrifiers (data not shown). Obviously, colonization of the needles must take place in a later stage of spruce development. On the other hand, artificial inoculation of spruce seedlings with autotrophic ammonia oxidizers resulted in an at least temporarily colonization of the needles. A possible explanation for these – at a first glance – conflicting results is that young spruce trees, in contrast to the adult trees, are too small to be infected with nitrifiers via the atmosphere and to be supplied with ammonia from the atmosphere (due to the filtering effect of the atmosphere by the canopy of the adult trees), provided the infection pathway is, indeed, via atmospheric transport of the microbes.

The results obtained from these microbiological studies are supported by the results obtained from experiments using a combination of in situ hybridization and confocal laser scanning microscopy. The objective of these experiments was to unequivocally demonstrate the exact localization of the autotrophic nitrifiers, i.e. whether they are localized inside the needles (in the apoplast) as was earlier concluded from the combination of microbiological and plant physiological studies (see: Papen et al., 2002), or whether they are located outside the needles, i.e. on the needle surface.

In a series of experiments it was tested whether the oligonucleotide probes choosen for specific detection of CAO and CNO fulfilled the criterion of specificity. Figure 1 gives a representative example for results obtained from

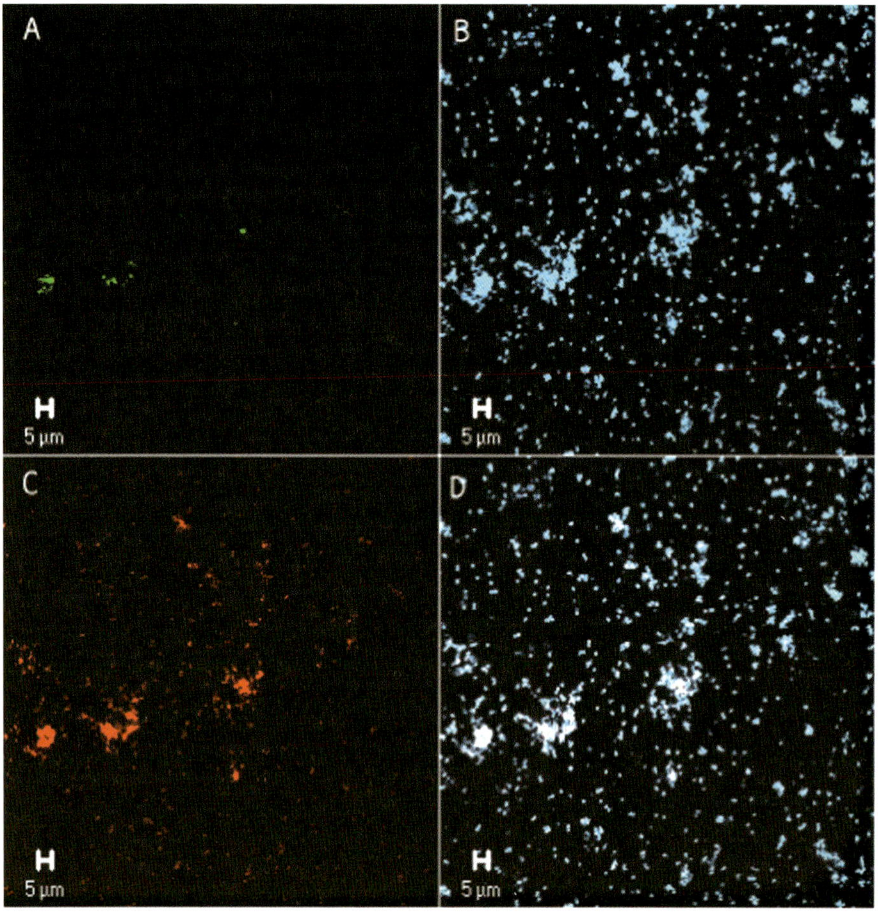

Fig. 1. Representative example from a series of experiments performed in order to validate the specificity of the choosen oligonucleotide probes for the specific detection of CAO and CNO by multiple hybridization of a cell mixture obtained from pure cultures of R. rhodochrous and the autotrophic ammonia oxidizer N. europaea and the autotrophic nitrite oxidizer N. winogradskyi by combination of fluorescence in situ *hybridization (FISH) with confocal Laser Scanning Microscopy (cLSM). All images were recorded simultaneously. (A): signal of the autotrophic nitrite oxidizer N. winogradskyi hybridized with FITC-labeled probe NIT3 (Ar-laser: 488 nm, colored green by image analysis); (B): staining with DAPI (UV-laser: 364 nm, colored blue by image analysis); (C): signal of the autotrophic ammonia oxidizer N. europaea hybridized with ROX-labeled probe Nso190 (HeNe-laser: 543 nm, colored red by image analysis) (D): combination of A+B+C.*

these series of experiments. Obviously, in amixture of cells containing the autotrophic ammonia oxidizer *N. europaea*, the autotrophic nitrite oxidizer *N. winogradskyi* and the heterotrophic actinomycete *R. rhodochrous*,the choosen 16S rRNA-targeted oligonucleotide probes did exclusively bind to their specific target organisms: *N. winogradskyi* (Fig. 1A: green signal of bound NIT3 oligonucleotide probe labeled with FITC) and *N. europaea* (Fig. 1C: red signal of bound Nso190 oligonucleotide probe labeled with ROX).

Figure 1B shows the unspecific binding of DAPI to DNA of all cells present within the mixture (blue signal). Fig. 1D shows the combined signal of A + B + C (white signal), thus allowing identification and localization of the autotrophic nitrifiers within the cell mixture.

Having validated the specificity of the choosen oligonucletide probes for binding in their specific target organisms the probes were applied for the detection and localization of CAO and CNO in needle tissue by combination of in situ hybridization with confocal laser scanning microscopy in longitudinal sections of intact needles taken from (i) adult trees from the experimental sites Höglwald and Villingen and (ii) seedlings that were grown under controlled conditions in the lab. Initial difficulties with the autofluorescence of the needle tissue (e.g. chloroplasts) that overlays the fluorescence signals of probes labeled with ROX and FITC, respectively, were solved by modifying dehydration steps in fixation and preparation procedure of the needles (see above). By this means the prerequisites were achieved for the detection and exact localization of autotrophic nitrifiers in sections of spruce needles by combining fluorescence in situ hybridization with confocal laser scanning microscopy.

Figure 2 gives a representative example for results of the experimental series performed by applying the combination of FISH with cLSM on longitudinal sections of needles taken from the field sites and from the lab.

In the given example needles of adult trees taken from the N-affected site Höglwald were *in situ* hybridized with both the oligonucleotide probe Nso190 (labeled with ROX, red signal in Fig. 2A) for the detection and localization of autotrophic ammonia oxidizers and probe NIT3 (labeled with FITC, green signal in Fig. 2C) for the detection and localization of autotrophic nitrite oxidizers. Fig. 2B shows the signal of DAPI, which stained all DNA present in the sample and Fig. 2D the combined signal of A + B + C, thus allowing specific identification and localization of the autotrophic nitrifiers within longitudinal sections of needle samples. Thus, it could unequivocally be demonstrated for the first time that CAO as well as CNO are, indeed, located inside the needle and colonize the needle apoplast (stomatal cavity). Furthermore, the figures illustrate that both CAO and CNO occurred closely aggregated in form of micro-colonies/cell-clusters

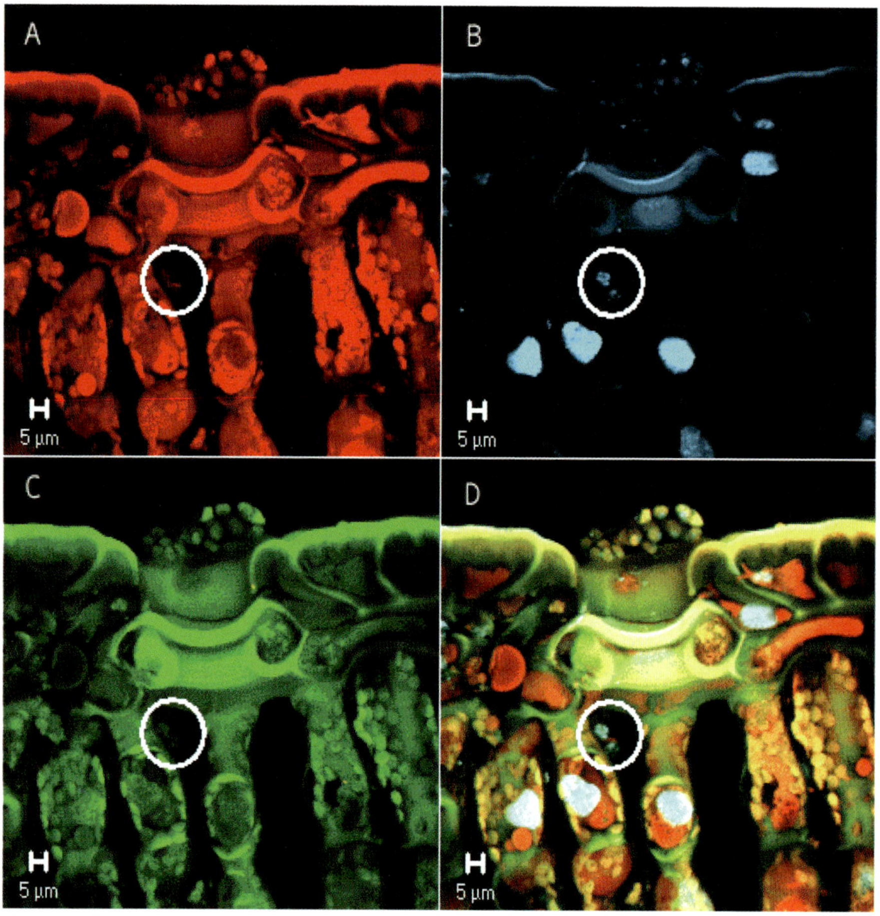

Fig. 2. Representative example from a series of analyses of in situ identification and localization of autotrophic nitrifiers within the apoplast (marked by white circle) from longitudinal sections of needles taken from twigs of mature spruce trees at the Höglwald experimental site by combination of fluorescence in situ *hybridization (FISH) with confocal Laser Scanning Microscopy (cLSM). All images were recorded simultaneously from one single longitudinal section of one spruce needle (33 stacked optical sections with a thickness of 0.58 µm each). (A): autotrophic ammonia oxidizers hybridized with ROX-labeled probe Nso190 (HeNe-laser: 543 nm, colored red by image analyis); (B): staining with DAPI (UV-laser: 364 nm, colored blue by image analyis); (C): autotrophic nitrite oxidizers hybridized with FITC-labeled probe NIT3(Ar-laser: 488 nm, colored green by image analysis); (D): combination of images A+B+C.*

within the apoplast of spruce needles (Fig. 1A, C, D). This co-occurrence is not unexpected, since in soils both autotrophically nitrifying bacterial groups are e.g. closely socialized and physiologically directly dependent on each other. The CAO produce the nitrite, which in nature is immediately further oxidized by CNO to nitrate, thereby preventing accumulation of nitrite to toxic concentrations.

For all sampling dates at the N-affected site Höglwald autotrophic nitrifiers were exclusively detected and localized inside the needles (the needle apoplast), but were never detected on the needle-surface. In contrast to these findings regarding needles from adult trees from the N-affected site (Höglwald), in needle samples taken from adult trees of the Villingen site (N-limited) CAO and CNO could never be detected neither inside the needles (apoplast) nor on the needle surface. This finding also applies for any needle samples taken at any sampling date from the spruce seedlings that had been grown under controlled conditions in the lab (fumigated as well as non-fumigated).

The results obtained from the simultaneously performed plant physiological experiments on NH_3 exchange between the phyllosphere of spruce trees and the atmosphere are presented in Figs. 3–6.

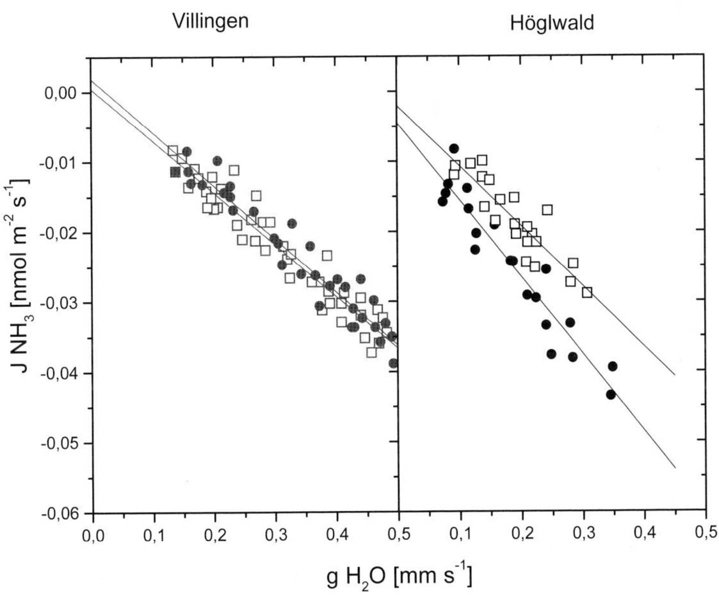

Fig. 3. Correlation between NH_3 flux (J_{NH3}) and stomatal conductance (g_{H2O}) at the N-limited site Villingen and the N-affected site Höglwald with (□) and without (•) inhibition of bacterial ammonia monooxigenase with 10 Pa acetylene. At both field sites twigs were exposed to c. 49 nmol NH_3 mol^{-1}. Negative values for J_{NH3} indicate fluxes from the atmosphere to the needles.

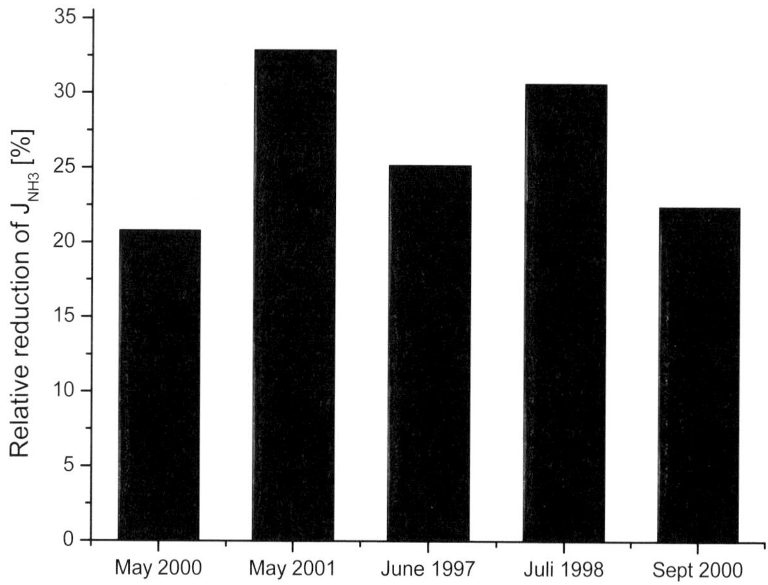

Fig. 4. Relative reduction of J_{NH3} for NH_3-fumigated twigs of adult trees at the field site Höglwald due to addition of acetylene at different times during the growing seasons 1997 to 2001. Reduction was calculated for $g_{H2O} = 0.3$ mm s^{-1} (cp. Fig. 3). NH_3 concentrations in the different fumigation experiments ranged between 49 and 60 nmol NH_3 mol^{-1}.

Figure 3 shows a typical result of a NH_3 -exchange experiment performed with adult trees at the N-affected field site Höglwald in comparison to the N-limited field site Villingen. In order to compare measurements performed under different environmental conditions, NH_3 flux (J_{NH3}) was plotted against stomatal conductance (g_{H2O}), since g_{H2O} is the major determinant for J_{NH3} in spruce (Geßler et al., 2002). It is clear from Fig. 3 that at the Villingen site the addition of 10 Pa acetylene – an inhibitor of the ammonia monooxigenase of CAO (Hynes und Knowles, 1978) – did not change the relation between J_{NH3} and g_{H2O} indicating that at this site nitrifier activity did not contribute to the observed uptake of NH_3 by plant leaves. This finding is in agreement with the results obtained from MPN assays and the in situ hybridization experiments (Table 1, Fig. 2) which showed that no autotrophic nitrifiers could be detected in needle samples taken from the Villingen site. NH_3 exposition experiments performed with twigs from adult trees at the Höglwald site revealed that (i) J_{NH3} was higher at a given g_{H2O} as compared to the Villingen site and (ii) that the addition of 10 Pa acetylene

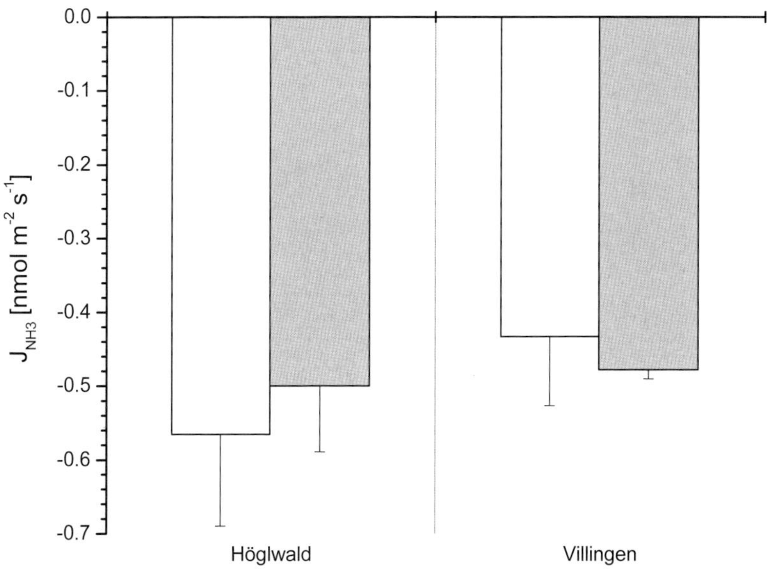

Fig. 5. Mean J_{NH3} for 4–5 years-old spruce trees from the field sites Höglwald and Villingen. White bars show J_{NH3} without, grey bars with the addition of acetylene. NH_3 concentration was c. 56 nmol NH_3 mol^{-1}. g_{H2O} was constant during the measurements and amount to c. 0.2 mm s^{-1}. Data shown are mean values from three trees.

decreased J_{NH3} significantly. This result demonstrates clearly that the CAO present in the needles (Table 1) are physiologically active and contribute significantly to the sink strength of needles for NH_3.

Since the reduction of NH_3 deposition due to acetylene-induced inhibition of the ammonia monooxigenase was greatest when stomata were open (high g_{H2O}) and only minute when stomata were closed ($g_{H2O} = 0$) it must be concluded that physiologically active CAO are located inside the needles rather than on the surface. Depending on g_{H2O} the acetylene-induced reduction of J_{NH3} amounted to between 15 and 25%. This conclusion is strongly supported by the results obtained from the *in situ* hybridization experiments reported above according to which CAO and CNO were exclusively found within the apoplast of the needles (stomatal cavity) and could never be detected on the needle surface.

In order to characterize seasonal and interannual patterns of NH_3 exchange, experiments were performed at different times during the growing season in different years. Figure 4 shows the relative reduction of J_{NH3} caused

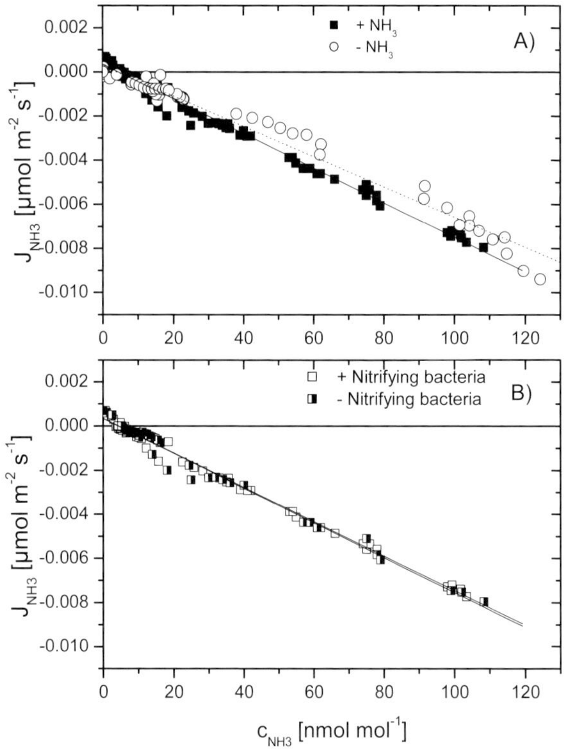

Fig. 6. Effect of long-term NH_3 fumigation and inoculation with nitrifying bacteria on NH_3 deposition flux to the needles of spruce seedlings. A) shows NH_3 deposition flux to eight seedlings fumigated for 10–16 weeks with 50 nmol NH_3 mol^{-1} and to six seedlings fumigated with filtered air. B) shows deposition flux to seedlings (n = 6) inoculated with N. europaea ATCC 25978 for 10–16 weeks compared to controls without inoculation. Both, inoculated and sterile seedlings were grown in an NH_3-enriched atmosphere (50 nmol NH_3 mol^{-1}).

by the addition of acetylene calculated for g_{H2O} = 0.3 mm s^{-1}. Neither pronounced seasonal nor interannual differences in inhibition and, thus, in physiological activity of CAO could be observed. This result is in agreement with the finding that cell numbers of nitrifying bacteria also did not reveal pronounced seasonal fluctuations (Table 1). In agreement with the MPN assays no acetylene-induced inhibition of NH_3 uptake and, thus, no physiological activity of CAO could be observed at the N-limited Villingen site.

Figure 5 shows the results of the gas-exchange measurements performed with 4–5 years-old spruce trees from the natural succession of both field sites. Trees from both origins showed no significant reduction in J_{NH3} when acetylene was added to the fumigation air, thus, confirming the results obtained from the MPN assays and the *in situ* hybridization experiments (see above).

Figure 6 shows that seedlings growing for several weeks in a NH_3-enriched atmosphere exhibit increased sink strength for NH_3 at higher NH_3 concentrations. This finding is in agreement with observations made by Sutton et al. (1995) who observed increased NH_3 uptake capacity in plants exposed to long-term elevated NH_3. It may be assumed that increased NH_4^+ concentrations in the apoplast as a consequence of NH_3 fumigation induces the expression of NH_4^+ transporters (Pearson et al., 2002). The inoculation with and the presence of CAO (Table 2) did, however, not contribute to changes in the sink strength of the leaves. Possible reasons may be the comparably low cell numbers and/or lacking or reduced physiological activity of the bacteria present.

4. CONCLUSIONS

In this paper for the first time evidence is presented from combined microbiological and plant physiological studies that CAO and CNO at a N-affected spruce forest (Höglwald) colonize the apoplast of spruce needles and are physiologically active within the needles as demonstrated by a 10–25% reduction in NH_3 uptake by the needles in the presence of a gaseous inhibitor of CAO. Thus, the observed NH_3 flux from the atmosphere into the needle leaves is not exclusively a plant physiological process, as has been assumed in the past, but is the result of both plant physiological plus microbial processes. However, the colonization of the needle apoplast seems to be restricted to adult spruce trees exposed to elevated levels of atmospheric N, whereas the needle apoplast of young and adult spruce trees from N-limited field sites are not colonized by CAO and CNO. The pathway by which the needles of N-affected spruce are colonized could not be clarified with certainty. However, the results obtained from inoculation experiments of spruce seedlings with nitrifiers indicate that nitrifiers spread onto the needle surface are able to survive at least temporarily on the needles, but were unable to enter the apoplast of the needles e.g. via the stomata, most likely due to intact epistomatal wax tubules on the needles of the young trees (seedlings) preventing the bacteria from entering the stomatal cavity. The demonstration of the presence and activity of CAO and CNO within the apoplast of N-affected adult spruce trees has potential ecological consequences: since nitrification is an acid producing process (nitrous and nitric acid), the question arises, whether acid-induced needle injuries are not only due to acid rain – as hitherto believed – but might be at least partially be due to colonization with/activity of nitrifiers within the needle apoplast. Furthermore, nitrifiers also produce and consume the direct and indirect climate-relevant trace gases NO and N_2O. Thus, it is most likely that these microorganisms do not only significantly contribute to the exchange of these N trace-gases between soils and the atmosphere as is very well documented by the literature, but also to the exchange of these gases between the phyllosphere and the atmosphere. This would imply consequences of our present understanding of the complex air-chemical processes and reactions within the forest canopy.

REFERENCES

Amann, R.I., Binder, B.J., Olson, R.J., Chisholm, S.W., Devereux, R. and Stahl, D.A. (1990) Combination of ^{16}S rRNA-targeted oligonucleotide probes with flow cytometry for analyzing mixed microbial populations. *Appl. Environ. Microbiol.*, 56, 1919–1925.

Asman, W.A.H. and Van Jaarsfeld, H.A. (1990) Regionale und europaweite Emission und Verfrachtung von NH_3-Verbindungen. In: Ammoniak in der Umwelt, Kreisläufe, Wirkungen, Minderung. *Landwirtschaftsverlag*, Münster-Hiltrup, Germany, pp. 2.1–2.35.

Atlas, R.M. and Parks, L.C. (eds) (1993) Handvolume of microbiological media. *CRC Press*, Boca Raton, Ann Arbor, London, Tokyo.

Bock, E., Koops, H.-P. and Harms, H. (1989) Nitrifying Bacteria. In: Schlegel, H.G. and Bowien, B. (eds) Autotrophic bacteria. Science Tech Publishers, Madison WI and *Springer-Verlag*, Berlin, pp. 81–96.

Degrange, V. and Bardin, R. (1995) Detection and counting of Nitrobacter populations in soil by PCR. *Appl. Environ. Microbiol.* 61: 2093–2098.

de Man, J.C. (1983) MPN tables, corrected. *Eur. J. Appl. Microbiol.*, 17, 301–305.

Feger, K.-H., Brahmer, G. and Zöttl, H.W. (1992) Projekt ARINUS VI. Stickstoffumsatz und Auswirkungen der experimentellen Ammonsulfatgabe. *KfK-PEF-Berichte*, 94, 199–211.

Gasche, R. and Papen, H. (1999) A 3-year continuous record of nitrogen trace gas fluxes from untreated and limed soil of a N-saturated spruce and beech forest ecosystem in Germany. 2. NO and NO_2 fluxes. *J. Geophys. Res.*, 104, 18505–18520.

Geßler, A., Rienks, M. and Rennenberg, H. (2000) NH_3 and NO_2 fluxes between beech trees and the atmosphere - correlation with climatic and physiological parameters. *New Phytol.*, 147, 539–560.

Geßler, A., Rienks, M. and Rennenberg, H. (2002) Stomatal uptake and cuticular adsorption contribute to dry deposition of NH_3 and NO_2 to needles of adult spruce (*Picea abies*) trees. *New Phytol.*, 156, 179–194.

Geßler, A., Schneider, S., Weber, P., Hanemann, U. and Rennenberg, H. (1998) Soluble N compounds in trees exposed to high loads of N: A comparison between the roots of Norway spruce (*Picea abies*) and beech (*Fagus sylvatica*) trees. *New Phytol.*, 138, 385–399.

Gieseke, A., Purkhold, U., Wagner, M., Amann, R. and Schramm, A. (2001) Community structure and activity dynamics of nitrifying bacteria in a phosphate-removing biofilm. *Appl. Environ. Microbiol.*, 67, 1351–1362.

Giloh, H. and Sedat, J.W. (1982) Fluorescence microscopy: Reduced photobleaching of rhodamine and fluorescein protein conjugates by n-propyl gallate. *Science*, 217, 1252–1255.

Hermansson, A. and Lindgren, P.E. (2001) Quantification of ammonia-oxidizing bacteria in arable soil by real-time PCR. *Appl. Environ. Microbiol.*, 61, 972–976.

Huber, C. (1997) Untersuchungen zur Ammoniakimmission und zum Stoffhaushalt auf ungekalkten und neugekalkten Flächen in einem stickstoffübersättigten Fichtenökosystem (Höglwald). PhD thesis, published in: Reihe Ökologie, *Hieronymus-Verlag*, München, 1–183.

Hynes, R.K. and Knowles, R. (1978) Inhibition by acetylene of ammonia oxidation in Nitrosomonas europaea. *FEMS Microbiol. Lett.*, 4, 319–321.

Kreutzer, K. and Weiss, T. (1998) The Höglwald Field Experiments - aims, concept and basic data. *Plant Soil*, 199, 1–10.

Manz, W., Amann, R., Ludwig, W., Wagner, M. and Schleifer, K.-H. (1992) Phylogenetic oligodeoxynucleotide probes for the major subclasses of proteobacteria: Problems and solutions. System. *Appl. Microbiol.*, 15, 593–600.

Mobarry, B.K., Wagner, M., Urbain, V., Rittmann, B.E. and Stahl, D.A. (1996) Phylogenetic probes for analyzing abundance and spatial organization of nitrifying bacteria. *Appl. Environ. Microbiol.*, 62, 2156–2162.

Muller, B., Touraine, B. and Rennenberg, H. (1996) Interaction between atmospheric and pedospheric nitrogen nutrition in spruce (*Picea abies* L. *Karst*) seedlings. *Plant Cell Environ.*, 19: 345–355.

Näsholm, T. and Ericson, A. (1989) Seasonal changes in amino acids, protein, and total nitrogen in needles of fertilized Scots pine trees. *Tree Physiol.*, 6, 267–281.

Papen, H. and Butterbach-Bahl, K. (1999) A 3-year continuous record of nitrogen trace gas fluxes from untreated and limed soil of a N-saturated spruce and beech forest ecosystem in Germany. 1. N_2O emissions. *J. Geophys. Res.*, 104, 18487–18503.

Papen, H., Geßler, A., Zumbusch, E. and Rennenberg, H. (2002) Chemolithoautotrophic nitrifiers in the phyllosphere of a spruce ecosystem receiving high atmospheric nitrogen input. *Curr. Microbiol.*, 44, 56–60.

Papen, H. and von Berg, R. (1998) A Most Probable Number method (MPN) for the estimation of cell numbers of heterotrophic nitrifying bacteria in soil. *Plant Soil*, 199, 123–130.

Pearson, J.N., Finnemann, J. and Schjoerring, J.K. (2002) Regulation of the high-affinity ammonium transporter (BnAMT1; 2) in the leaves of *Brassica napus* by nitrogen status. *Plant Mol. Biol.*, 49, 483–490.

Pellicer, A., Wiger, M., Axel, R. and Silvenstein, S. (1978) The transfer and stable integration of the HSV thymidine kinase gene into mouse cells. *Cell*, 14, 133–141.

Rennenberg, H., Kreutzer, K., Papen, H. and Weber, P. (1998) Consequences of high loads of nitrogen for spruce (Picea abies) and beech (Fagus sylvatica) forests. *New Phytol.*, 139, 71–86.

Sambrook, J., Fritsch, E.F. and Maniatis, T. (1989) Molecular cloning, a laboratory manual. *Cold Spring Harbor Laboratory Press*, Cold Spring Harbor, 2nd edition, USA.

Schmidt, E.L. and Belser, L.W. (1982) Nitrifying bacteria. In: Page, A.L., Miller, R.H. and Keeney, D.R. (eds.) Methods in Soil Analysis. Agronomy 9, part 2, 2nd edition, *ASA, SSSA*, Madison, WI, USA, pp. 1027–1042.

Schramm, A., Larsen, L.H., Revsbech, N.P., Ramsing, N.B., Amann, R. and Schleifer, K.-H. (1996) Structure and Function of a Nitrifying Biofilm as Determined by In Situ Hybridization and the Use of Microelectrodes. *Appl. Environ. Microbiol.*, 62, 4641–4647.

Sutton, M.A., Fowler, D., Burkhardt, J.K. and Milford, C. (1995) Vegetation atmosphere exchange of ammonia: Canopy cycling and the impacts of elevated nitrogen inputs. *Water Air Soil Pollut.*, 85, 2057–2063.

Teuber, M. (2003) Nachweis, Lokalisation und Quantifizierung von autotrophen Nitrifizierern im Kronenraum der Fichte (*Picea abies* (L.) *Karst.*). PhD thesis, University Freiburg (in press).

Wagner, M., Rath, G., Koops, H-P., Flood, J. and Amann, R. (1996) In situ analysis of nitrifying bacteria in sewage treatment plants. *Wat. Sci. Tech.*, 34, 237–244.

Ward, B.B. (1986) Nitrification in marine environments. In: Prosser, J.I. (ed) Nitrification. Special Publications of the Society of General Microbiology 20, *IRL Press*, Oxford, pp. 157–184.

THE RICE APOPLAST AS A HABITAT FOR ENDOPHYTIC N_2-FIXING BACTERIA

B. REINHOLD-HUREK, A. KRAUSE, B. LEYSER, L. MICHÉ and T. HUREK

Laboratory of General Microbiology, University of Bremen, Germany, breinhold@uni-bremen.de

Abstract. The only biological reaction counterbalancing the loss of N from soils or ecosystems is biological nitrogen fixation, the enzymatic reduction of N_2 to ammonia carried out by prokaryotes. Moreover, N is one of the most widely used fertilizer nutrient. Although there are no special symbiotic structures, it has been shown that some graminaceous crops such as certain Brazilian sugar cane cultivars can derive a substantial part of their N from biological nitrogen fixation. This raises the question for the microbial diazotrophic partner(s) and for the mechanisms of interactions with their hosts. The current research on diazotrophs associated with graminaceous plants is summarized here, with special focus on the interactions between the proteobacterium *Azoarcus* sp. strain BH72 and rice. The root cortex is the prominent colonization site for endophytic diazotrophs, however bacteria occur mainly in the apoplast and not inside living plant cells. Approaches to probe their activity and the plant environment and their results are outlined.

Key words: endophytes, nitrogen fixation, nitrogenase, gene expression

1. INTRODUCTION

The availability of nitrogen (N) often limits plant growth in terrestrial ecosystems. In agriculture, N is one of the most widely used fertilizer nutrients, with a still increasing global input. For example in developing Asia, the total input of N fertilizer was less than 1.5 million tons per year in 1961, whereas it has increased to almost 47 million tons in 1996 (Dawe, 2000). The only biological reaction counterbalancing the loss of N from soils

B. Sattelmacher† and W.J. Horst (eds.), The Apoplast of Higher Plants: Compartment of Storage, Transport and Reactions, 427–443.

or ecosystems is biological nitrogen fixation, the enzymatic reduction of N_2 to ammonia. This process is unique to *Bacteria* and *Archaea*, and is estimated to contribute globally between 100–200 million tons of fixed N per year: estimates for terrestrial systems are 90–130 Tg N per year (Galloway et al., 1995), and for marine systems 100–200 Tg N per year (Karl et al., 2002).

Animals and plants profit from biologically fixed nitrogen directly if they are in association or symbiosis with N_2-fixing prokaryotes, or indirectly after mineralization of these bacteria. One of the best-studied interactions is the root nodule symbiosis between rhizobia and legumes (Schubert, this volume, pp. 445–454), which may contribute 100–450 kg N per ha and year (Peoples et al., 1995). However, the most important crops worldwide, wheat, rice and maize, belong to the *Graminaceae* that do not naturally form these specialized symbiotic structures. Nevertheless it has been shown that some graminaceous crops can substantially profit from biological nitrogen fixation (BNF). Estimates for agriculturally used rice (*Oryza sativa*) cultivars range up to 21% of plant nitrogen contributed by BNF. Certain Brazilian sugar cane cultivars can derive a large part (up to 70%) of their N from BNF (Boddey et al., 2003; Lima et al., 1987). This raises the question for the microbial diazotrophic partner(s) and for the mechanisms of interactions with their hosts. The current research on diazotrophs associated with graminaceous plants focuses on endophytic bacteria (Boddey et al., 2003; Hurek and Reinhold-Hurek, 2003; James and Olivares, 1998; Reinhold-Hurek and Hurek, 1998b): Endophytic microorganisms multiply and spread inside plants without causing damage of the host plant or conferring an ecological threat to the plant; while facultative endophytes are also prevalent in the soil (McInroy and Kloepper, 1995), most diazotrophic grass endophytes do not survive well in soil and appear to be ecologically dependent on living plants (Reinhold-Hurek and Hurek, 1998b). Although they invade plants, they cannot be regarded as typical pathogens or endosymbionts (see below). Moreover, it has been demonstrated recently that some of them contribute fixed nitrogen to their grass host (Hurek et al., 2002; Sevilla et al., 2001). However, for an improvement of this process or its application to a wider range of agricultural crops, more knowledge on the host-bacteria interactions of endophytes is required.

2. COLONISATION OF ROOTS

Circumstantial evidence – the isolation from surface-sterilized roots – first suggested that some nitrogen-fixing bacteria may be colonizing the interior of grass roots (see e.g. McClung et al., 1983; Patriquin et al., 1983; Reinhold et al., 1986). Sound microscopic evidence for internal colonization came from resin-embedded material, to avoid displacing of bacteria from the root surface during the process of sectioning or washing. This was first shown in electron microscopic (Umali-Garcia et al., 1980), immuno gold electron microscopic (Bashan and Levanony, 1988), or light microscopic studies (Reinhold and Hurek, 1988), where colonies of *Azospirillum* sp., *Azoarcus* sp. or other diazotrophs were detected in the cortex of grass roots. More detailed studies revealed for the first time that diazotrophic endophytes were even capable of invading the stele of grasses (Hurek et al., 1991, 1994), a microhabitat which was previously thought to be invariably sterile in healthy plants. Colonization of root aerenchymatic tissues, cortex regions, more rarely stele and xylem or shoots of grasses and cereals has meanwhile been demonstrated unequivocally for several diazotrophs such as *Azoarcus* sp. strain BH72, *Herbaspirillum seropedicae*, *Gluconacetobacter diazotrophicus*, *Rhizobium leguminosarum* or *Klebsiella pneumoniae*, and has been extensively reviewed (James and Olivares, 1998; Reinhold-Hurek and Hurek, 1998a; Yanni et al., 1997). In summary, colonization occurs mainly apoplastically. Bacteria occur in the intercellular spaces, there is no evidence for intracellular colonization of living plant cells. Intracellular colonization can be observed; however, these cells are dead. In some cases it has been observed that diazotrophs are localized between the plant cell-wall and the plant cytoplasmic membrane, which has retracted from the cell wall (Hurek et al., 1994), suggesting that these endophytes invade dying cells, or that plant cell-death is induced after bacterial invasion. Tissues which have been found to be colonized are outer root cell-layers, mainly cortex regions – often microcolonies are close to the stele in aerenchyma -, parenchymatic tissues in the stele of roots and in the culm of grasses, and the interior of xylem cells. It has been speculated that the colonization of xylem vessels may assist in systemic spreading of bacteria into shoots, where they can be detected, too (Gyaneshwar et al., 2001; Hurek et al., 1994; James and Olivares, 1998; James et al., 1997).

These colonization sites which were mainly studied in gnotobiotic cultivation systems can also be found in soil-grown, uninoculated plants. For example in the wild rice species *Oryza longistaminata*, plant cells filled with bacteria were detected in the second outer cell layer of roots (Fig. 1A). Microcolonies of bacteria can also be found adjacent to the stele in the aerenchyma of roots (Fig. 1 B, C).

Internal colonization of healthy grass roots by bacteria raises the question where these bacteria enter the plant. Microscopic studies on roots grown in

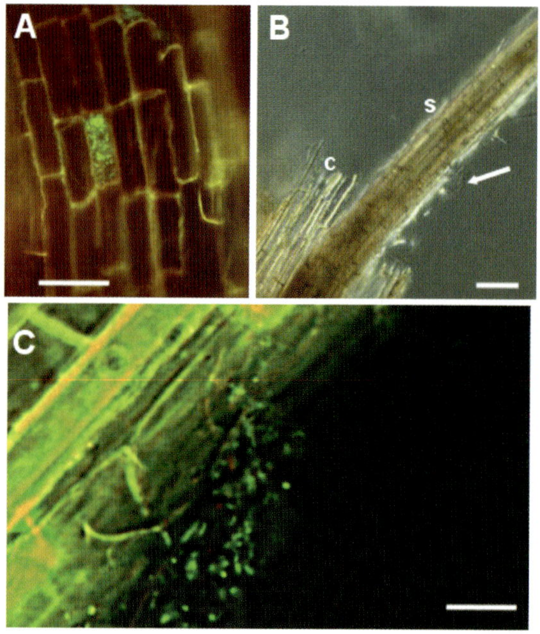

Fig. 1. Bacteria colonizing roots of uninoculated, soil-grown wild rice Oryza longistaminata. Bacteria appear green fluorescing (A, C) after staining with the Live-Dead Bacterial Viability kit (Dörr et al., 1998). (A) Bacteria inside a plant cell, image focus is in the second outer cell layer. (B) Overview of root, cortex (c) being partially removed from the stele (s). Arrow points to microcolony attached to the stele. Phase contrast image. (C) Close-up of B showing fluorescent single bacterial cells after deconvolution of the image. Bars 40 µm (A), 45 µm (B), or 10 µm (C), respectively.

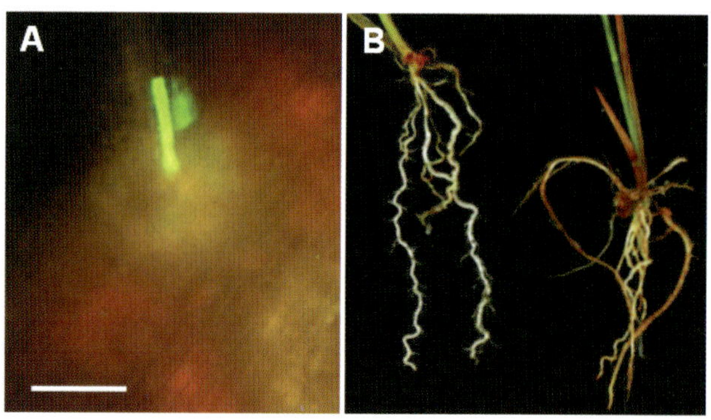

Fig. 2. Inoculation response of different cultivars of Oryza sativa. (A) Fluorescence microscopy image of roots of cultivar IR36 grown in gnotobiotic culture with a nifH::gfp transcriptional fusion reporter strain of Azoarcus sp. BH72 (strain BHGN35 ,Egener et al., 1998). Plant cell of emerging lateral root filled with fluorescing bacteria expressing nitrogenase genes. Bar 40 µm. (B) Response of cultivar IR42 to inoculation with Azoarcus sp. BH72. Left, uninoculated roots; right, browning of inoculated roots.

gnotobiotic systems, e.g. of rice and Kallar grass (Hurek et al., 1991, 1994), provided evidence that there are two main sites of ingress into roots. In the zone of elongation and differentiation above the root tip, root surfaces are densely colonized, as reporter gene activity-based studies suggest. In this zone where root cells and cell walls are not yet fully differentiated, bacteria were found even in the inner root, suggesting that they might invade central tissues prior to formation of the stele and the endodermis.

Another site of primary colonization are emergence points of lateral roots, where also ingress of bacteria into the roots can be detected. These observations suggested that diazotrophic endophytes actively invade healthy roots (see also Section 3).

3. BACTERIAL ACTIVITY IN THE RICE APOPLAST

The cortex region of roots or the aerenchymatic tissue of flood-tolerant plants such as rice, are the main colonization sites for endophytic diazotrophs. This raises the question whether these sites enable high microbial activity, especially nitrogen fixation, when compared to true endosymbioses inside living plant cells as in the rhizobium-legume root nodules. Since aerenchymatic tissue consists mainly of cell walls of lysed cortex cells, this may not be likely at first sight. To study endophytic activities, mRNA-based studies may give good evidence for localization of nitrogenase activity, since the transcription of the structural genes *nifHDK* is tightly regulated in accordance with conditions favorable for nitrogen fixation (Arcondéguy et al., 2001; Dixon, 1998; Egener et al., 2002). Moreover, bacterial mRNA is fairly unstable, thus preventing false positive results due to carry-over effects. Using in situ hybridization with fluorescent probes against nitrogenase genes of *Azoarcus* sp. strain BH72 on sections of resin-embedded roots, active nitrogenase gene expression was shown in the aerenchyma of soil-grown Kallar grass roots. The stringency of hybridization suggested that these nitrogen-fixing bacteria were highly related to *Azoarcus* sp. BH72 (Hurek et al., 1997).

Also reporter gene studies using *nifH::gus* or *nifH::gfp* fusions showed that apoplastic, active nitrogen fixation occurs in roots of rice seedlings in gnotobiotic culture with *Azoarcus* sp. (Egener et al., 1998, 1999; Reinhold-Hurek and Hurek, 1998b). These transcriptional fusions with reporter genes allow detection of the reporter protein activity when the promoter of the target gene is activated and thus the reporter gene cotranscribed with the target gene (*nifH* encoding the iron protein of nitrogenase). Bacteria located in the aerenchyma in large microcolonies attached to the outer cell layers or to the stele were stained in GUS assays for activity of the reporter enzyme ß-glucuronidase. Since the green fluorescent protein (GFP) variant used by us allows single-cell fluorescence only when the promoter is very strong in *Azoarcus* sp. BH72 or cells are very actively fixing nitrogen, respectively, (Egener et al., 1998), utilization of this reporter gene allowed a rough estimation of promoter activity in roots. Endophytic nitrogenase gene expression was highly active inside roots of rice seedlings, as judged by single cell fluorescence of *nifH::gfp* fusions (Egener et al., 1999; Reinhold-Hurek and Hurek, 1998b). Fluorescent cells or microcolonies were detected in intercellular spaces of exodermis layers and in the aerenchymatic air spaces. *NifH* was also strongly expressed intracellularly, e.g. inside plant cells close to emergence points of lateral roots (Fig. 2A). Therefore, the root cortex of rice appears to provide suitable conditions for nitrogen fixation, including sufficiently low concentrations of oxygen and combined N. Infection and expression rates were higher when external carbon sources for the bacteria were added, but they were also detected when concentrations were negligible (5 mg L^{-1} of malic acid) or absent (Egener et al., 1999). Addition of low amounts of carbon sources may be required as a "starter" for survival and fitness of the inoculum prior to accumulation of root exudates. The question remains open which carbon sources are available for the endophytes in the apoplast of rice roots.

4. MECHANISMS OF ROOT COLONIZATION

The above-mentioned observations suggest that diazotrophic endophytes actively colonize healthy roots, a process which may involve specific mechanisms of recognition, bacterial invasion and establishment in endophyte-grass interactions. Unfortunately, there are only very few data on mechanisms of invasion or establishment for diazotrophic endophytes, while more data have accumulated for *Azospirillum* spp. (Steenhoudt and Vanderleyden, 2000) which are typically rhizoplane colonizers. However, also this genus harbors strains which have been repeatedly detected inside roots, e.g. strain Sp245 (Baldani et al., 1987). Which genes are required for establishment of endophytes on or in roots? One of the few examples are type IV pilin genes of *Azoarcus* sp. strain BH72. The first crucial event in the infection process of symbiotic or pathogenic bacteria is the attachment to epithelial cells of the host. Type IV pili play an essential role in mediating bacterial adherence to and colonization of host cell-surfaces and are virulence factors in many human or animal pathogenic bacteria. They are filamentous cell appendages that are much thinner than flagella, of about 6 nm thickness. They are assembled primarily from a single protein subunit, pilin, which consists of approx. 150 aa residues. A short positively charged leader peptide that is cleaved off the prepilin, and a highly conserved N-terminal region are characteristic for type IV pilins (Strom and Lory, 1993). The hydrophobic N-terminal region is postulated to form the center of the pilus by hydrophobic interactions of 54-aa α-helices of the pilin subunits, whereas the C-terminal part is exposed to the pilus surface (Forest and Tainer, 1997; Parge et al., 1995). This hypervariable region is the most highly exposed portion (Forest and Tainer, 1997) and the source of strain-specific antigenic variation and interaction with host surfaces (Nassif et al., 1993). Despite their importance in animal systems, up to now type IV pili have not been found to be essential virulence factors in plant pathogenic bacteria.

In *Azoarcus* sp. BH72, genes encoding type IV pili were detected and analyzed (Dörr et al., 1998). Formation of pili on solid media was dependent on the *pilAB* locus. *PilA* encodes an unusually short (6.4 kDa) putative pilin precursor showing 100% homology to the conserved N terminus of the *Pseudomonas aeruginosa*-type IV pilin. *PilB* encodes for a 14.2 kDa polypeptide showing weak similarity to FimF, a component of type I fimbriae of *E. coli*, however up to now there was no homologue detected in any other bacterium including sequenced genomes. Both genes are transcribed in an operon, translated and essential for pilus formation. Moreover, *pilB* was found to be extruded beyond the cell surface by immuno fluorescence studies. Thus *pilB* is likely to be part of the pilus itself (Dörr et al., 1998), which is highly unusual since type IV pili consist usually only of one pilin. However, *pilB* might compensate for the small size of *pilA* in pilus function.

Emergence points of lateral roots are among the primary sites of colonization of grass roots by *Azoarcus* sp. BH72. When gnotobiotic cultures of rice seedlings were inoculated with wild-type *Azoarcus*, or a *pilAB*, a *pilA* or *pilB* mutant, respectively, the degree of colonization of lateral emergence points was strongly reduced in all mutants. Complementation with *pilAB* partially restored the wild-type phenotype (Dörr et al., 1998). Thus, type IV pili or *pilAB* genes, respectively, are essential for efficient adhesion, colonization of and ingress into rice roots. Interestingly, these pili are also involved in the attachment of *Azoarcus* sp. BH72 to fungal mycelium (Dörr et al., 1998) in co-culture with an ascomycete strain 2003 (related to *Acremonium alternatum*), which was isolated from surface-sterilized roots of Kallar grass (Hurek and Reinhold-Hurek, 1999). Thus, type IV pili in *Azoarcus* sp. BH72 appear to be important determinants for the adhesion to eukaryotic hosts. The fact that interactions with both, plants and fungi, are affected, underlines the hypothesis that in nature, both types of eukaryotes may be important hosts for *Azoarcus*.

Also other findings indicate that the endophytic colonization of grass roots is a process involving active bacterial strategies. For the ingress of bacteria into the root, plant boundaries have to be overcome, such as middle lamella (mainly pectins) for intercellular colonization or secondary cell walls (mainly cellulose and hemicellulose) for intracellular ingress. It is tempting to speculate that this might be mediated by enzymes degrading plant-cell wall-polymers, as has been shown for many phytopathogens (Herron et al., 2000; Toth et al., 2003). Plant cell-wall-degrading enzymes have also been detected in *Azoarcus* sp. strain BH72: two types of cellulolytic enzymes are present (Reinhold-Hurek, Hurek, Claeyssens et al., 1993). An exoglycanase was characterized which has cellobiohydrolase and ß-glucosidase activity on ß-1,4 cellooligosachharides, and also on a wider substrate spectrum including xylosides. The endoglucanase preferably attacks oligosaccharides larger than cellobiose and releases large oligomers from substrates such as carboxymethylcellulose, and is thus likely to loosen larger cellulose fibers. In contrast to most phytopathogens, *Azoarcus* sp. BH72 does not metabolize the substrates or breakdown products (cellulose, cellobiose, glucose); these bacteria do not grow on any carbohydrate tested (Reinhold-Hurek and Hurek, 2000; Reinhold-Hurek, Hurek, Claeyssens et al., 1993; Reinhold-Hurek, Hurek, Gillis et al., 1993). Unlike in pathogens, the enzymes are not efficiently excreted into the culture supernatant, but remain bound to the cell surface (Reinhold-Hurek, Hurek, Claeyssens et al., 1993). This might cause a less aggressive attack of plant cells by an endophyte in comparison to plant pathogens. An isogenic mutant was constructed in which the gene encoding the endoglucanase was inactivated. Indeed, the mutant showed a strong decrease in intracellular infection of root epidermis cells of rice seedlings. In addition, systemic spreading into the rice shoot was strongly decreased below the detection level in this mutant (Reinhold-Hurek et al., 2006). This suggests that endoglucanase may be important for endophytic infection of rice roots by strain BH72.

5. PLANT RESPONSE TO COLONISATION OF THE APOPLAST

The lack of a truly endosymbiotic association and the occurrence of deep penetration into roots, even into the stele, suggest that the host interactions of diazotrophic grass endophytes lies between symbiotic and phytopathogenic lifestyle (Reinhold-Hurek and Hurek, 1998a). Since the rice genome sequence is known (Goff et al., 2002; Yu et al., 2002), and the genome of strain BH72 is currently sequenced (Battistoni et al., 2005; Hurek and Reinhold-Hurek, 2003), mechanisms of initial interactions can be studied by functional genomic and proteomic approaches in this system. Information about the response of rice roots to microbial colonization is scarce, as the important pathogens in paddy rice culture mostly attack the aerial part of the plant.

In isotopic and N-balance studies it has been observed that *Oryza sativa* varieties differ in the promotion of nitrogen fixation in inoculated soil, indicating genotypic differences with respect to host-bacteria interactions (Shrestha and Ladha, 1996). Variety-specific differences of rice-associated nitrogenase expression have also been observed in gnotobiotic culture systems inoculated with *Herbaspirillum seropedicae* (Gyaneshwar et al., 2002). Evidence for varietal differences comes also from our survey of rice varieties for apoplastic colonization in gnotobiotic culture systems. In previous studies with *O. sativa* cv. IR36, endophytic colonization by strain BH72 had been shown, however significant nif expression had not been detected by immunogold electron microscopy (Hurek et al., 1994). Application of quartz sand instead of agar media and the omission of a nitrogen source for the plant (proline) in the culture medium later lead to the detection of root-associated nitrogenase activity in *Oryza sativa*. Transcriptional activation of bacterial nitrogenase genes in or on roots is visualized by reporter gene activity of a transcriptional *nifH::gusA* fusion in *Azoarcus* sp., and is used by us to indicate a physiologically successful root colonization. This has been shown for the japonica-type rice variety *O. sativa* cv. nipponnbare (Egener et al., 1999). Application of this culture system for screening of several indica-type cultivars of rice originating from the IRRI (International Rice Research Institute, Los Banos, Philippines) breeding program revealed that also cv. IR36 supports nitrogenase gene expression of *Azoarcus* sp. strain BH72 (Fig. 2A). Interestingly, cultivars differed in their interaction with *Azoarcus* sp., even when they were closely related: IR36 and IR42 are sister lines derived from the same crossing. However, cv. IR42 was not successfully colonized under these conditions. It developed a slightly brownish root colour upon inoculation with *Azoarcus* sp. (Fig. 2 B, cv. IR42), and a less intense GUS staining of the *nifH::gus* fusion (not shown), similar to another cultivar IR72. Thus the interaction between *Azoarcus* sp. and these rice cultivars appeared to be less compatible. The putative accumulation of phenolic compounds causing root browning resembled morphologically a defense response of roots to pathogens. This indicates that the rice apoplast does not simply provide a suitable microenvironment for endophytic growth, but that there is a subtle balance in compatibility and defense.

In general, jasmonic acid (JA) and salicylic acid (SA) are key signalling phytohormones in numerous plant responses to stresses such as pathogen attack and exposure to fungal elicitors (Berger, 2002; Kunkel and Brooks, 2002); recent studies suggest that in rice especially JA plays an important role in defense mechanisms (Kim et al., 2003; Rakwal and Komatsu, 2000). In order to study whether these known signaling molecules may be involved in these morphological changes of roots, jasmonic acid (JA) (100 μM) or salicylic acid (1 mM) were added to the plant medium of cv. IR42 without bacterial inoculation. Salicylic acid did not induce visible changes of root color, whereas addition of jasmonate resulted in brownish roots, with outer cell layers adjacent to sclerenchyma showing increased cell-wall browning (Miché et al., 2006). This suggested that plant defense might be involved in bacterial host specificity, and that JA might participate in the signaling pathway leading to rice-root changes in less compatible interactions with the endophyte.

To elucidate whether a JA -induced defense response is involved in the apoplastic rice-root infection by *Azoarcus* sp. and the differential colonization of cultivars, we compared root proteomes. Both *O. sativa* cultivars IR36 and IR42 were studied in response to application of JA and to inoculation with the *nifH::gus* reporter strain *Azoarcus* sp. BHNG3.1. Plants were grown in gnotobiotic culture for 2 weeks in plant medium free of combined nitrogen, containing quartz sand. Among about 1000 spots detected on the different two-dimensional polyacrylamide gels (2-D-PAGE), 56 showed reproducible induction patterns after bacterial inoculation. Surprisingly, relatively few spots were detected that were induced both, by endophyte colonization and JA; as could be expected for the more compatible interaction with cv. IR36, significantly less spots were induced in cv. IR36 (two spots) than in cv. IR 42 (seven spots). Some of these proteins were identified by mass spectrometry (Miché et al., 2006). The two spots induced by bacterial inoculation as well as by JA treatment in both rice varieties showed homology to SalT, a protein that was found to be induced by salt and drought stress in rice roots (Claes et al., 1990). SalT induction has also been detected in rice seedlings (Moons et al., 1997) demonstrating an overlap between salt- and JA induction of proteins. Recent studies determined that SalT is a mannose-binding rice lectin that responds to a wide range of stresses, suggesting its implication in a global mechanism of response to environmental stresses (de Souza Filho et al., 2003). Being cytoplasmic, those particular lectins might be involved in specific endogenous protein-carbohydrate interactions involved in cellular regulation and signaling (Van Damme et al., 2004). In conclusion, our studies revealed that jasmonic acid-induced stress responses or pathogenesis-related proteins are not relevant in a compatible interaction, i.e. when a rice variety is physiologically successfully colonized by the bacteria. However, in future it needs to be investigated what are the bacterial effectors causing the response in incompatible systems, and what is suppressing the defense response in a compatible variety to allow an efficient endophytic colonization. Moreover, additional proteins that were induced by bacterial inoculation of rice were identified revealing more insights in the nature and physiology of this endophytic interaction (Miché et al., manuscript in preparation).

6. APPROACHES TO PROBING THE ROOT APOPLAST

It is evident that endophytic diazotrophs, e.g. *Azoarcus* sp. strain BH72, are growing and are physiologically highly active inside rice roots although they show apoplastic colonization and not an endosymbiosis (see above). The question remains open which carbon sources are available to and utilized by *Azoarcus* sp. in the rice apoplast. A simple chemical analysis of apoplastic fluid of rice roots is rendered difficult by the formation of large aerenchymatic air spaces in mature roots of flooded plants and thus by occurrence of very fragile tissue and rapid disruption of cells upon mechanic manipulation. An experimental approach to probing this microenvironment is the application of reporter genes expressed in endophytic bacteria. When bacterial promoters inducible in the presence of specific carbon sources are transcriptionally coupled to a reporter gene such as *gfp*, the detection of fluorescence of the protein can be regarded as an indication of gene induction and thus of presence of a certain concentration of the respective carbon source. Bacteria carrying such a fusion can be used for probing the microenvironment in the rice apoplast, when bacterial GFP fluorescence is analyzed during endophytic colonization of roots. An example for a similar approach of probing the microenvironment was described above in section 3 for *nifH::gus* or *nifH::gfp* fusions.

The overview of experimental approach applied in our study is depicted in Fig. 3. For generation of promoter fusions in *Azoarcus* sp., a vector was used that is a suicide plasmid in strain BH72, which, however, contains mob genes to allow mobilization into *Azoarcus* sp. by conjugation (pK18mob2) (Tauch et al., 1998). Into the multiple cloning site a cartridge was inserted which contained two reporter genes, *gfp* and *gusA*, in tandem without promoter, but with translational stops in all three reading frames upstream of the genes, and Shine-Dalgarno sequences commonly used in *Azoarcus* sp. strain BH72 (pIMMG). To generate a promoter library transcriptionally fused with the reporter genes, random DNA fragments of strain BH72 were

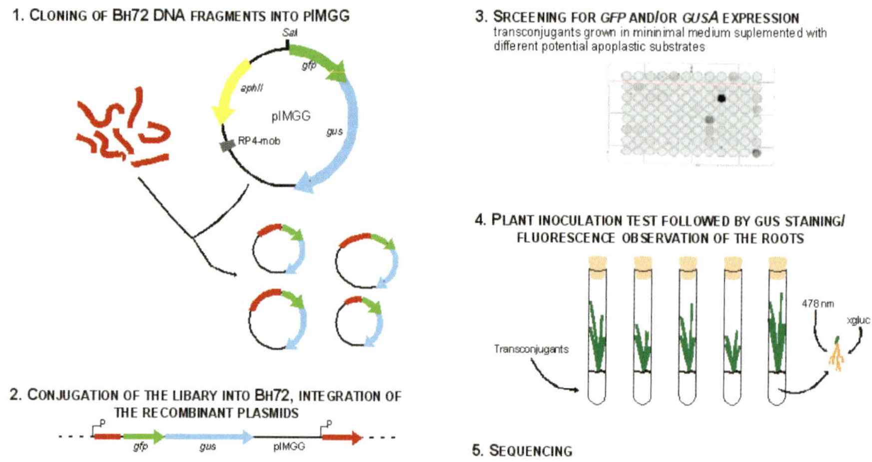

Fig. 3. Experimental approach for obtaining a promoter fusion library with reporter genes in Azoarcus sp. BH72 and for screening of expression.

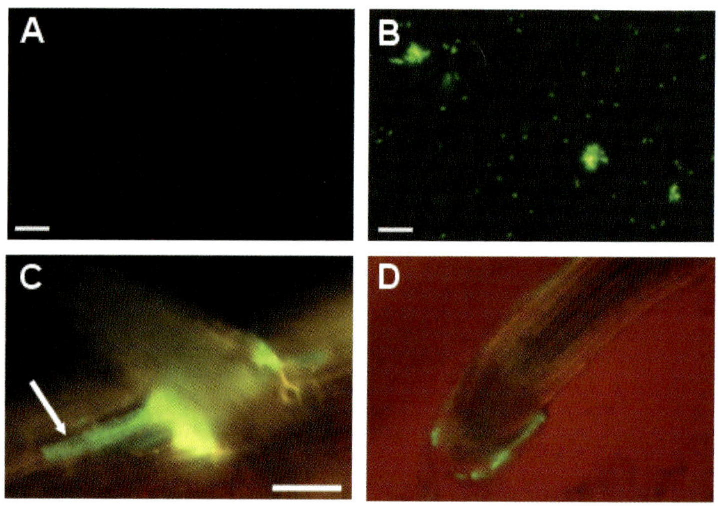

Fig. 4. Expression of alcohol dehydrogenase by Azoarcus sp. BH72. Reporter strain BH12c9 carrying a transcriptional fusion of the PQQ-dependent alcohol dehydrogenase gene with gfp (exaA3::gfp-gus). Fluorescence microscopy images showing cells grown on malic acid (A) or on ethanol (B) or in gnotobiotic culture with O. sativa cv. IR36 (C–D). Bacterial cells expressing gfp in a plant cell close to the emergence point of lateral roots (C, arrow) or under the calyptra at the root tip (D). Bars 10 µm (A, B) or 25 µm (C), respectively.

inserted into the SalI restriction site of this vector pIMMG. Fragments of 2–5 kb in size had been generated by partial digestion of chromosomal DNA of strain BH72 and size fractionation. This promoter fusion library was transferred to wild-type BH72 by conjugation in triparental mating experiments.

Transconjugants in which plasmids had integrated into the chromosome by homologous recombination were selected by antibiotic resistance and stored in microtiter plates. This mutant library was screened for promoter activity on different carbon sources by quantifying GFP fluorescence of cell suspension in a fluoroimager (Typhoon, Amersham Biosciences). Mutants with interesting expression profiles were characterized by sequence analysis of the DNA adjacent to the *gfp* reporter gene.

Several transconjugants were found to have increased *gfp* expression upon addition of alcohols. Sequence analysis revealed that they contained genes encoding putative PQQ-dependent alcohol dehydrogenases. We focused on mutant BH12c9 in which a transcriptional *eaxA3::gfp-gus* fusion was analyzed in more detail. In ß-glucuronidase (GUS) assays, cells grown on ethanol as a carbon source showed a 5-fold increased gene expression in comparison to cells grown on malic acid under aerobic conditions. Gene expression was further elevated (8-fold) under microaerobic conditions. Visual inspection for GFP fluorescence corroborated these findings: bacteria grown on malic acid did not show any fluorescence (Fig. 4A), while single-cell fluorescence and thus strong gene expression was detected in ethanol-grown bacteria (Fig. 4 B). This mutant was used for probing the apoplastic environment in roots of rice seedlings. *Oryza sativa* cv. IR36 was inoculated with mutant BH12c9, and roots grown under flooded conditions inspected for GFP fluorescence after 4 days, without external addition of any carbon source. Single-cell fluorescence was detected close to root tips, especially under the calyptra (Fig. 4C). It was also detected in bacteria colonizing roots endophytically, in intercellular spaces or intracellularly in outer cell layers (Fig. 4D).

Therefore, it is likely that ethanol occurs in sufficiently high concentration in rice roots to cause alcohol dehydrogenase gene induction. Thus it is also likely to act as a carbon source for endophytic growth of *Azoarcus* sp. in roots. Ethanol might be of importance in the root as habitat for endophytes. When aeration of roots is reduced by flooding, ethanol concentrations in roots can rise immediately (Crawford and Baines, 1977). In root tips, ethanol can always be detected, even under atmospheric oxygen pressure (Betz, 1957). Alcoholic fermentation is important for survival of plants under anaerobic conditions, and alcohol dehydrogenase activity as well as ethanol accumulation are increased in rice root seedlings upon oxygen-deficiency stress (Mustroph and Albrecht, 2003). Expression and mutational analyses of alcohol dehydrogenases in *Azoarcus* sp. BH72 will unravel the importance of this substrate for colonization of the rice apoplast.

7. CONCLUSION

Current results indicate that endophytic colonization of roots of graminaceous plants is an active process. As a prerequisite for ingress, the bacteria appear to require adhesion molecules for establishment on the root surface, such as pili. Degradative enzymes such as cellulases aid invasion and systemic spreading inside the plant, and nutrients available in the apoplast, e.g. ethanol, support bacterial activity. These mechanisms resemble virulence mechanisms of pathogenic bacteria. However, it is not clear yet why the endophytic interactions are mutualistic: how is compatibility achieved which results in high colonization densities without harming the plant? Initial studies showed that a more compatible rice cultivar up-regulates less stress-related proteins than a less compatible cultivar. This indicates that in certain endophyte-plant interactions, plant-defense reactions may be overcome by the bacteria. Several genome sequencing projects have been launched for diazotrophic grass endophytes: *Herbaspirillum seropedicae*, *Gluconacetobacter diazotrophicus*, *Azospirillum* sp. and *Azoarcus* sp. strain BH72. These data and follow-up studies such as functional genomic analyses will give important insights into the mechanisms of interaction and the molecular cross-talk between grasses and of these fascinating bacteria.

ACKNOWLEDGEMENTS

This work was supported in part by grant RE756/6-1 from the DFG, and by grants from the BMBF (0311946 and GenoMik 031U213D) to B. R.-H.

REFERENCES

Arcondéguy, T., Jack, R. and Merrick, M. (2001). PII signal transduction proteins, pivotal players in microbial nitrogen control. *Microbiol. Mol. Biol. Rev.*, 65, 80–105.

Baldani, V. L. D, Baldani, J. I., and Döbereiner, J. (1987). Inoculation of field grown wheat with *Azospirillum brasilense* spp. *Biol. Fertil. Soils*, 4, 37–40.

Bashan, Y. and Levanony, H. (1988). Interaction between *Azospirillum brasilense* Cd and wheat root cells during early stages of root colonization. In W. Klingmüller (Ed.), Azospirillum IV (pp. 166–173). Berlin: Springer Verlag.

Battistoni, F., Bartels, D., Kaiser, O., Reamon-Buettner, S. M., Hurek, T. and Reinhold-Hurek, B. (2005) Pysical map of the *Azoarcus* sp. strain BH72 genome based on a bacterial artificial chromosome (BAC) library as a platform for genome sequencing and functional analysis. *FEMS Microbiol. Lett.*, 249, 233–240.

Berger, S. (2002). Jasmonate-related mutants of *Arabidopsis* as tools for studying stress signaling. *Planta*, 214, 497–504.

Betz, A. (1957). Zur Atmung wachsender Wurzelspitzen. *Planta*, 50, 122–143.

Boddey, R. M., Urquiaga, S., Alves, B. J. R. and Reis, V. (2003). Endophytic nitrogen fixation in sugarcane: present knowledge and future applications. *Plant Soil*, 252, 139–149.

Claes, B., Dekeyser, R., Villarroel, R., Van den Bulcke, M., Bauw, G., Van Montagu, M. and Caplan, A. (1990). Characterization of a rice gene showing organ-specific expression in response to salt stress and drought. *Plant Cell*, 2, 19–27.

Crawford, R. M. M. and Baines, M. (1977). Tolerance of anoxia and ethanol metabolism in tree roots. *New Phytol.*, 79, 519–526.

Dawe, D. (2000). The potential role of biological nitrogen fixation in meeting future demnad for rice and fertilizer. In J. K. Ladha and P. M. Reddy (Eds.), The quest for nitrogen fixation in rice (pp. 1–9). Makati City: International Rice Research Institute.

de Souza Filho, G. A., Ferreira, B. S., Dias, J. M., Queiroz, K. S., Branco, A. T., Bressan-Smith, R. E., Oliveira, J. G. and Garcia, A. B. (2003). Accumulation of SALT protein in rice plants as a response to environmental stresses. *Plant Sci.*, 164, 623–628.

Dixon, R. (1998). The oxygen-responsive NIFL-NIFA complex: a novel two-component regulatory system controlling nitrogenase synthesis in gamma- Proteobacteria. *Archives of Microbiology*, 169, 371–380.

Dörr, J., Hurek, T. and Reinhold-Hurek, B. (1998). Type IV pili are involved in plant-microbe and fungus-microbe interactions. *Mol. Microbiol.*, 30, 7–17.

Egener, T., Hurek, T. and Reinhold-Hurek, B. (1998). Use of green fluorescent protein to detect expression of nif genes of *Azoarcus* sp. BH72, a grass-associated diazotroph, on rice roots. *Mol. Plant-Microbe. In.*, 11, 71–75.

Egener, T., Hurek, T. and Reinhold-Hurek, B. (1999). Endophytic expression of nif genes of *Azoarcus* sp. strain BH72 in rice roots. *Mol. Plant-Microbe. In.*, 12, 813–819.

Egener, T., Sarkar, A., Martin, D. E. and Reinhold-Hurek, B. (2002). Identification of a NifL-like protein in a diazotroph of the beta-subgroup of the Proteobacteria, *Azoarcus* sp. strain BH72. *Microbiol.*, 148, 3203–3212.

Forest, K. T. and Tainer, J. A. (1997). Type-4 pilus-structure: outside to inside and top to bottom - a minireview. *Gene*, 192, 165–169.

Galloway, J. N., Schlesinger, W. H., Levy, H. I., Michaels, A. F. and Schnoor, J. L. (1995). Nitrogen fixation: Anthropogenic enhancement-environmental response. *Global Biogeochem. Cycles*, 9, 235–252.

Goff, S. A., Ricke, D., Lan, T. H., Presting, G., Wang, R., Dunn, M., Glazebrook, J., Sessions, A., Oeller, P., Varma, H., Hadley, D., Hutchison, D., Martin, C., Katagiri, F., Lange, B. M., Moughamer, T., Xia, Y., Budworth, P., Zhong, J., Miguel, T., Paszkowski, U., Zhang, S., Colbert, M., Sun, W. L., Chen, L., Cooper, B., Park, S., Wood, T. C., Mao, L., Quail, P., Wing, R., Dean, R., Yu, Y., Zharkikh, A., Shen, R., Sahasrabudhe, S., Thomas, A., Cannings, R., Gutin, A., Pruss, D., Reid, J., Tavtigian, S., Mitchell, J., Eldredge, G., Scholl, T., Miller, R. M., Bhatnagar, S., Adey, N., Rubano, T., Tusneem, N., Robinson, R., Feldhaus, J., Macalma, T., Oliphant, A. and Briggs, S. (2002). A draft sequence of the rice genome (*Oryza sativa* L. ssp. *japonica*). *Science*, 296, 92–100.

Gyaneshwar, P., James, E. K., Mathan, N., Reddy, P. M., Reinhold-Hurek, B. and Ladha, J. K. (2001). Endophytic colonization of rice by a diazotrophic strain of *Serratia marcescens*. *J. Bacteriol.*, 183, 2634–2645.

Gyaneshwar, P., James, E. K., Reddy, P. M. and Ladha, J. K. (2002). Herbaspirillum colonization increases growth and nitrogen accumulation in aluminium-tolerant rice varieties. *New Phytol.*, 154, 131–145.

Herron, S. R., Benen, J. A., Scavetta, R. D., Visser, J. and Jurnak, F. (2000). Structure and function of pectic enzymes: virulence factors of plant pathogens. *Proc. Natl. Acad. Sci. USA*, 97, 8762–8769.

Hurek, T., Egener, T. and Reinhold-Hurek, B. (1997). Divergence in nitrogenases of *Azoarcus* spp., Proteobacteria of the ß-subclass. *J. Bacteriol.*, 179, 4172–4178.

Hurek, T., Handley, L., Reinhold-Hurek , B. and Piché, Y. (2002). *Azoarcus* grass endophytes contribute fixed nitrogen to the plant in an unculturable state. *Mol. Plant-Microbe In.*, 15, 233–242.

Hurek, T. and Reinhold-Hurek, B. (1999). Interactions of *Azoarcus* sp. with rhizosphere fungi. In A. Varma and B. Hock (Eds.), Mycorrhiza (2nd ed., pp. 595–614). Berlin: Springer Verlag.

Hurek, T. and Reinhold-Hurek, B. (2003). *Azoarcus* sp. strain BH72 as a model for nitrogen-fixing grass endophytes. *J. Biotechnol.*, 106, 169–178.

Hurek, T., Reinhold-Hurek, B., Van Montagu, M. and Kellenberger, E. (1991). Infection of intact roots of Kallar grass and rice seedlings by *Azoarcus*. In M. Polsinelli and R. Materassi and M. Vincenzini (Eds.), Nitrogen fixation (pp. 235–242). Dordrecht, Netherlands: Kluwer academic publishers.

Hurek, T., Reinhold-Hurek, B., Van Montagu, M. and Kellenberger, E. (1994). Root colonization and systemic spreading of *Azoarcus* sp. strain BH72 in grasses. *J. Bacteriol.*, 176, 1913–1923.

James, E. K. and Olivares, F. L. (1998). Infection and colonization of sugar cane and other graminaceous plants by endophytic diazotrophs. *Crit. Rev. Plant Sci.*, 17, 77–119.

James, E. K., Olivares, F. L., Baldani, J. I. and Döbereiner, J. (1997). Herbaspirillum, an endophytic diazotroph colonizing vascular tissue in leaves of *Sorghum bicolor* L. *Moench. J. Exp. Bot.*, 48, 785–797.

Karl, D., Bergman, B., Capone, D., Carpenter, E., Letelier, R., Lipschultz, F., Paerl, H., Sigman, D. and Stal, L. (2002). Dinitrogen fixation in the world's oceans. *Biogechem.*, 57/58.

Kim, S. T., Cho, K. S., Yu, S., Kim, S. G., Hong, J. C., Han, C. D., Bae, D. W., Nam, M. H. and Kang, K. Y. (2003). Proteomic analysis of differentially expressed proteins induced by rice blast fungus and elicitor in suspension-cultured rice cells. *Proteomics*, 3, 2368–2378.

Kunkel, B. N. and Brooks, D. M. (2002). Cross talk between signaling pathways in pathogen defense. *Curr. Opin. Plant Biol.*, 5, 325–331.

Lima, E., Boddey, R. M. and Döbereiner, J. (1987). Quantification of biological nitrogen fixation associated with sugar cane using a ^{15}N-aided nitrogen balance. *Soil Biol. Biochem.*, 19, 165–170.

McClung, C. R., Van Berkum, P., Davis, R. E. and Sloger, C. (1983). Enumeration and localization of N_2-fixing bacteria associated with roots of *Spartina alterniflora Loisel. Appl. Environ. Microbiol.*, 45, 1914–1920.

McInroy, J. A. and Kloepper, J. W. (1995). Survey of indigenous bacterial endophytes from cotton and sweet corn. *Plant Soil*, 173, 337–342.

Miché, L., Battistoni, F., Gemmer, S., Belghazi, M. and Reinhold-Hurek, B. (2006). Upregulation of jasmonate-inducible defense proteins and differential colonization of roots of *Oryza sativa* cultivars with the endophyte *Azoarcus* sp. *Mol. Plant-Microbe Interact.*, 19, 502–511.

Moons, A., Prinsen, E., Bauw, G. and Van Montagu, M. (1997). Antagonistic effects of abscisic acid and jasmonates on salt stress-inducible transcripts in rice roots. *Plant Cell*, 9, 2243–2259.

Mustroph, A. and Albrecht, G. (2003). Tolerance of crop plants to oxygen deficiency stress: fermentative activity and photosynthetic capacity of entire seedlings under hypoxia and anoxia. *Physiol. Plant.*, 117, 508–520.

Nassif, X., Lowy, J., Stenberg, P., O'Gaora, P., Ganji, A. and So, M. (1993). Antigenic variation of pilin regulates adhesion of *Neisseria meningitidis* to human epithelial cells. *Mol. Microbiol.*, 8, 719–725.

Parge, H. E., Forest, K. T., Hickey, M. J., Christensen, D. A., Getzoff, E. D. and Tainer, J. A. (1995). Structure of the fibre-forming protein pilin at a 2.6 A resolution. *Nature*, 378, 32–38.

Patriquin, D.G., Döbereiner, J. and Jain, D.K. (1983) Sites and processes of the association between diazotrophs and grasses. *Can. J. Microbiol.*, 29, 900–915.

Peoples, M. B., Herridge, D. F. and Ladha, J. K. (1995). Biological nitrogen fixation: an efficient source of nitrogen for sustainableagricultural production? *Plant Soil*, 174, 3–28.

Rakwal, R. and Komatsu, S. (2000). Role of jasmonate in the rice (*Oryza sativa* L.) self-defense mechanism using proteome analysis. *Electrophoresis*, 21, 2492–2500.

Reinhold, B. and Hurek, T. (1988). Light microscopical visualization of diazotrophs on and in roots by immunogold-silver staining. *Biol. Fertil. Soils*, 6, 298–301.

Reinhold, B., Hurek, T., Niemann, E.-G. and Fendrik, I. (1986). Close association of *Azospirillum* and diazotrophic rods with different root zones of Kallar grass. *Appl. Environ. Microbiol.*, 52, 520–526.

Reinhold-Hurek, B. and Hurek, T. (1998a). Interactions of gramineous plants with *Azoarcus* spp. and other diazotrophs: Identification, localization and perspectives to study their function. *Crit. Rev. Plant Sci.*, 17, 29–54.

Reinhold-Hurek, B. and Hurek, T. (1998b). Life in grasses: diazotrophic endophytes. *Trends Microbiol.*, 6, 139–144.

Reinhold-Hurek, B. and Hurek, T. (2000). Reassessment of the taxonomic structure of the diazotrophic genus *Azoarcus sensu lato* and description of three new genera and species, *Azovibrio restrictus* gen. nov., sp. nov., *Azospira oryzae* gen. nov., sp. nov and *Azonexus funguphilus* gen. nov., sp. nov. *Int. J. Syst. Evol. Microbiol.*, 50, 649–659.

Reinhold-Hurek, B., Hurek, T., Claeyssens, M. and Van Montagu, M. (1993). Cloning, expression in *Escherichia coli,* and characterization of cellulolytic enzymes of *Azoarcus* sp., a root-invading diazotroph. *J. Bacteriol.*, 175, 7056–7065.

Reinhold-Hurek, B., Hurek, T., Gillis, M., Hoste, B., Vancanneyt, M., Kersters, K. and De Ley, J. (1993). *Azoarcus* gen. nov., nitrogen-fixing proteobacteria associated with roots of Kallar grass (*Leptochloa fusca* (L.) *Kunth*) and description of two species *Azoarcus indigens* sp. nov. and *Azoarcus communis* sp. nov. *Int. J. Syst. Bacteriol.*, 43, 574–584.

Reinhold-Hurek, B., Maes, T., Van Montagu, M. and Hurek, T. (2006). An endoglucanase is involved in infection of rice roots by the not cellulose-metabolising endophyte *Azoarcus* sp. BH72. *Mol. Plant-Microbe In.*, 19, 181–188.

Sevilla, M., Burris, R. H., Gunapala, N. and Kennedy, C. (2001). Comparison of benefit to sugarcane plant growth and $^{15}N_2$ incorporation following inoculation of sterile plants with *Acetobacter diazotrophicus* wild-type and nif- mutant strains. *Mol. Plant-Microbe In.*, 14, 358–366.

Shrestha, R. K. and Ladha, J. K. (1996). Genotypic variation in promotion of rice dinitrogen fixation as determined by nitrogen-15 dilution. *Soil Sci. Soc. Am. J.*, 60, 1815–1821.

Steenhoudt, O. and Vanderleyden, J. (2000). *Azospirillum*, a free-living nitrogen-fixing bacterium closely associated with grasses: genetic, biochemical and ecological aspects. *FEMS Microbiol. Rev.*, 24, 487–506.

Strom, M. S. and Lory, S. (1993). Structure-function and biogenesis of the type IV pili. *Annu. Rev. Microbiol.*, 47, 565–596.

Tauch, A., Zheng, Z., Pühler, A. and Kalinowski, J. (1998). *Corynebacterium striatum* chloramphenicol resistance transposon Tn5564: genetic organization and transposition in *Corynebacterium glutamicum*. *Plasmid*, 40, 126–139.

Toth, I. K., Bell, K. S., Holeva, M. C. and Birch, P. B. (2003). Soft rot erwiniae: from genes to genomes. *Mol. Plant Pathol.*, 4, 17–30.

Umalia-Garcia, M., Hubell, D. H., Gaskins, M. and Dazzo, F. (1980). Association of *Azospirillum* with grass roots. *Appl. Environ. Microbiol.*, 39, 219–226.

Van Damme, E. J., Zhang, W. and Peumans, W. J. (2004). Induction of cytoplasmic mannose-binding jacalin-related lectins is a common phenomenon in cereals treated with jasmonate methyl ester. *Comm. Appl. Biol. Sci.*, 69, 23–31.

Yanni, Y. G., Rizk, R. Y., Corich, V., Squartini, A., Ninke, K., Philip-Hollingsworth, S., Orgambide, G., De Bruijn, F., Stoltzfus, J., Buckley, D., Schmidt, T. M., Mateos, P. F., Ladha, J. K. and Dazzo, F. B. (1997). Natural endophytic association between *Rhizobium leguminosarum* bv. *trifolii* and rice roots and assessment of its potential to promote rice growth. *Plant Soil*, 194, 99–114.

Yu, J., Hu, S., Wang, J., Wong, G. K., Li, S., Liu, B., Deng, Y., Dai, L., Zhou, Y., Zhang, X., Cao, M., Liu, J., Sun, J., Tang, J., Chen, Y., Huang, X., Lin, W., Ye, C., Tong, W., Cong, L., Geng, J., Han, Y., Li, L., Li, W., Hu, G., Li, J., Liu, Z., Qi, Q., Li, T., Wang, X., Lu, H., Wu, T., Zhu, M., Ni, P., Han, H., Dong, W., Ren, X., Feng, X., Cui, P., Li, X., Wang, H., Xu, X., Zhai, W., Xu, Z., Zhang, J., He, S., Xu, J., Zhang, K., Zheng, X., Dong, J., Zeng, W., Tao, L., Ye, J., Tan, J., Chen, X., He, J., Liu, D., Tian, W., Tian, C., Xia, H., Bao, Q., Li, G., Gao, H., Cao, T., Zhao, W., Li, P., Chen, W., Zhang, Y., Hu, J., Liu, S., Yang, J., Zhang, G., Xiong, Y., Li, Z., Mao, L., Zhou, C., Zhu, Z., Chen, R., Hao, B., Zheng, W., Chen, S., Guo, W., Tao, M., Zhu, L., Yuan, L. and Yang, H. (2002). A draft sequence of the rice genome (*Oryza sativa* L. ssp. *indica*). *Science*, 296, 79–92.

THE APOPLAST OF INDETERMINATE LEGUME NODULES: COMPARTMENT FOR TRANSPORT OF AMINO ACIDS, AMIDES AND SUGARS?

S. SCHUBERT

Institute of Plant Nutrition, Interdisciplinary Research Center (IFZ), Justus Liebig University Giessen, Germany, sven.schubert@ernaehrung.uni-giessen.de

Abstract. In efficiently dinitrogen fixing legumes total nitrogen (N) concentrations in the plant tissue are maintained at a comparatively constant level irrespective of various stresses and mineral N supply. This suggests tight regulation of nitrogenase activity in symbiosomes. Two competing hypotheses have been proposed to explain regulation by a variable diffusion barrier or a feedback mechanism, respectively. For the efficiently dinitrogen-fixing nodules of *Vicia faba* the microstructural and chemical findings suggest that import of assimilates occurs via symplastic pathways, preventing selective uptake of carbon compounds. The unselective inflow of reduced N compounds from phloem allows a feedback control to regulate nitrogenase activity. This concept is also supported by the fact that isolated infected protoplasts in contrast to uninfected protoplasts have no uptake capabilities for sucrose or glucose. It is concluded that feedback regulation is the primary mechanism, which regulates dinitrogen fixation according to host-plant demand. Variations in nitrogenase activity, however, result in strong fluctuations of respiration requiring tight control of oxygen supply. Negative oxygen effects on the nitrogenase enzyme are avoided by strict control of oxygen diffusion by means of the variable diffusion barrier and oxygen buffering by leghemoglobin.

Key words: apoplastic transport, legume nodule, metabolite transport, protoplast, *Rhizobium leguminosarum*, symplastic transport, ultrastructure, *Vicia faba*

B. Sattelmacher† *and W.J. Horst (eds.), The Apoplast of Higher Plants: Compartment of Storage, Transport and Reactions*, 445–454.

1. INTRODUCTION

Biological dinitrogen fixation is restricted to specialized procaryotic species, which possess the enzymatic machinery to produce NH_3 from atmospheric N_2. Free-living organisms attain only a poor productivity of dinitrogen fixation resulting in a few kilograms per hectare and year. Some species, however, form symbiotic associations with plants improving the productivity significantly. In some cases, several hundred kilograms of nitrogen (N) may then be fixed per hectare and year (LaRue and Patterson, 1981).

Three major reasons have been identified for the high productivity of symbioses relative to free-living organisms. First, the high sensitivity of the nitrogenase enzyme may restrict efficient dinitrogen fixation under conditions of high partial oxygen concentrations (Giller and Wilson, 1991). In the most efficient symbiotic systems, such as the symbioses between rhizobia and legumes, permanent and variable oxygen diffusion barriers control oxygen transport into the tissues with N_2-fixing capacity (Iannetta et al., 1993). Consequently, oxygen concentrations in these tissues are tightly regulated and maintained at a constant low level allowing both protection of the nitrogenase enzyme and sufficient oxygen supply to the respiration chain. Second, whereas free-living organisms only fix N_2 for their own metabolism, symbiosomes export the overwhelming portion of reduced N to the host plant. In the former case, nitrogenase activity is down-regulated (Kennedy et al., 1994), in the latter case, reduced N is exported allowing continuous nitrogenase activity. Third, the high energetic demand of dinitrogen fixation limits this process under most natural conditions. Symbiotic systems allow generous supply of assimilates from photosynthesis avoiding energetic limitation (Hunt and Layzell, 1993).

Large differences have been reported for the biological dinitrogen fixation of various legume species and cultivars. Whereas in some cases almost the total N demand of the host plant can be met by dinitrogen fixation (except for N required for the establishment of the symbiosis and N delivered by the seed), some legumes are known as rather inefficient. Although in some cases the physiological background for these differences seems clear (Merbach and Schilling, 1980), in other cases it remains obscure (Hansen et al., 1993, Becher et al., 1997). Efficient systems such as the *Vicia faba-Rhizobium leguminosarum* biovar. *viciae* symbiosis (Hardarson, 1993) show remarkably constant N concentrations in the plant tissue even when stressed with drought (Gallacher and Sprent, 1978, Plies-Balzer et al., 1995), acidity (Schubert et al., 1990), or when differentially supplied with mineral N (Schubert, 1995). Evidently, the process of dinitrogen fixation is tightly controlled.

Two competing hypotheses have been proposed to explain the control of dinitrogen fixation in legumes. According to Minchin et al. (1985), Hunt and Layzell (1993), and Witty and Minchin (1998) the variable oxygen barrier in the outer cortex of legume nodules regulates the diffusion of oxygen into the infected tissues and thus controls respiration and dinitrogen fixation. Alternatively, the regulation of dinitrogen fixation is explained in terms of a feedback control (Pate, 1976, Oti-Boateng and Silsbury, 1993; Parsons et al., 1993, Schubert, 1995, Schulze, 2003). In this model, reduced N compounds, which are not consumed for growth or storage purposes in the shoot, are retranslocated via phloem into the nodules. Thus, the ratio of sucrose and reduced N compounds is decisive for the activity of the nitrogenase enzyme, which may be inhibited by the accumulation of reduced N compounds.

Whereas export of amides and amino acids (ureides in determinate nodules) occurs via xylem, it is not known whether the import of sucrose into the infected tissues occurs via apolastic or symplastic pathways. The functioning of the feedback control of dinitrogen fixation requires symplastic pathways without the possibility to discriminate between sucrose and reduced N compounds. On the other hand, apoplastic pathways would require membrane transport processes thus allowing selective uptake of sucrose. This would not allow feedback control of nitrogenase activity by import of reduced N compounds. Therefore, to improve the understanding of the regulation of dinitrogen fixation, the fine-structure of the indeterminate nodule type of *Vicia faba* was studied in more detail. Additionally, to characterize the uptake capacity of infected and non-infected cells of the *Vicia faba* nodule, protoplasts were prepared to distinguish between the plasmamembrane properties of these two cell types.

2. MICROSTRUCTURAL IMPLICATIONS FOR METABOLITE TRANSPORT

Upon infection by effective strains of *Rhizobium leguminosarum* biovar. *viciae*, faba bean (*Vicia faba* L.) forms new organs, the root nodules. Light microscopy of a longitudinal cross-section through a nodule reveals a central zone (consisting of an infected and a senescent zone) with both infected and uninfected cells (Abd-Alla et al., 2000, Fig. 1). Infected cells responsible for dinitrogen fixation are filled with specialized organelles, the symbiosomes, which comprise one or more bacteroids surrounded by a peribacteroid membrane. The latter is derived from the host plant plasma membrane (Whitehead and Day, 1997). The infected zone is surrounded by cortical tissues. An outer cortex is separated from an inner cortex by endodermal cells.

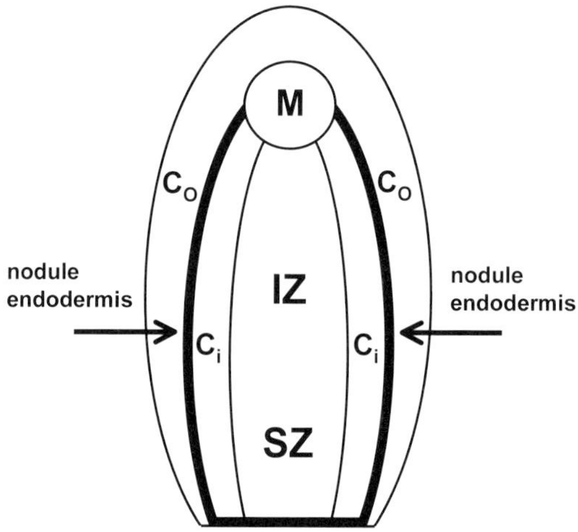

Fig. 1. Schematic representation of a longitudinal cross section through a Vica faba root nodule. The infected zone (IZ) and the senescent zone (SZ) are surrounded by an endodermis separating outer (C_o) and inner (C_i) cortex. Vascular bundles are located between inner and outer cortex. Meristematic tissue (M) allows continuous growth of the nodule.

Transversal cross-sections show vascular bundles with phloem and xylem elements between outer and inner cortex directly connected to the long-distance transport systems of the root (Figs. 2, 3). Morphologically, the indeterminate faba bean nodule can be distinguished from a determinate soybean nodule by an apical meristematic tissue of densely packed small cells (Brown and Walsh, 1996, Abd-Alla et al., 2000) which allows continual growth of the nodule. The meristem interrupts the nodule vascular bundles that (like in determinate nodules: Streeter, 1993) form a dead end.

Staining with berberine followed by counterstaining with aniline blue identified Casparian bands both in the nodule endodermis (Fig. 1) and in the vascular endodermis (Abd-Alla et al. 2000, Figs. 2, 3). Radial and transversal cell walls of the nodule endodermis are incrusted with lignin and suberin (Hartmann et al., 2002). At the proximal end, the nodule endodermis fuses with the root endodermis thus forming a nodule interior that is apoplastically isolated from the outer cortex. Ultrastructural and chemical analyses provide evidence for apoplastic barriers for metabolite movement between vascular bundles and the infected tissue (Abd-Alla et al., 2000, Hartmann et al., 2002). Transport of ions and metabolites into and out of the nodules thus appears to be restricted to symplastic movement.

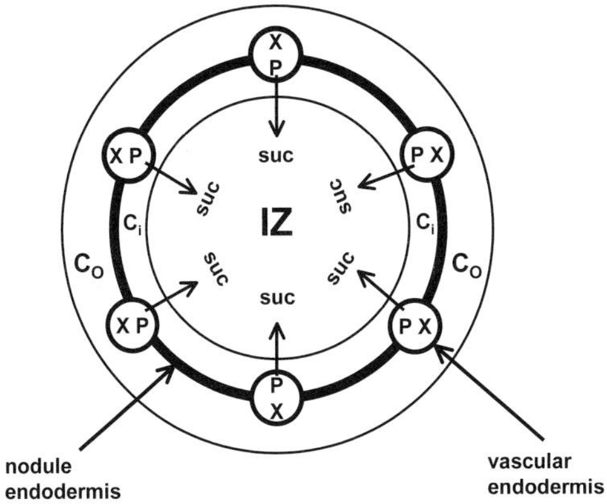

Fig. 2. Assimilate import into faba bean nodules. Schematic representation of a transversal cross-section through a Vicia faba root nodule in the actively dinitrogen-fixing region. The infected zone (IZ) is surrounded by a nodule endodermis separating outer (C_o) and inner cortex (C_i). Vascular bundles with phloem (P) and xylem (X) elements are surrounded by a vascular endodermis. Sucrose (suc) delivered by phloem is imported into the infected zone.

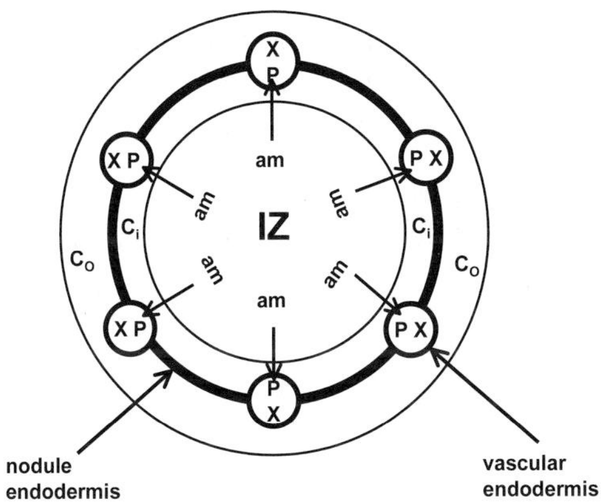

Fig. 3. Nitrogen export from faba bean nodules. Schematic representation of a transversal cross section through a Vicia faba root nodule in the actively dinitrogen-fixing region. The infected zone (IZ) is surrounded by a nodule endodermis separating outer (C_o) and inner cortex (C_i). Vascular bundles with phloem (P) and xylem (X) elements are surrounded by a vascular endodermis. Amino acids and amides (am) are exported via xylem.

Table 1. Plasmodesmatal frequencies at different cell interfaces of five week-old faba bean nodules. Values are means (± SE). From Abd-Alla et al., 2000.

Cell type interface	*Plasmodesmatal frequency (m^{-6})*
Vascular endodermis cell – inner cortex cell	0.78 (±0.02)
Inner cortex cell – uninfected cell (central tissue)	0.66 (±0.03)
Inner cortex cell – infected cell (central tissue)	0.21 (±0.01)
Uninfected cell – uninfected cell	0.71 (±0.03)
Uninfected cell – infected cell	0.53 (±0.02)
Infected cell – infected cell	0.02 (±0.01)

Plasmodesmatal frequencies between the various cell types support the concept of symplastic rather than apoplastic movement of metabolites (Table 1). Whereas between infected cells hardly any plasmodesmatal connections were observed, all other cell types showed abundant plasmodesmatal frequencies. These findings suggest that apoplastic transport may only occur between infected cells. Furthermore, Abd-Alla et al. (2000) proposed that uninfected cells (whose function in indeterminate nodules is still unclear) may form a preferential route for assimilates to be symplastically transported within the infected zone. Apparently, symplastic transport between infected cells is negligible (Table 1).

3. METABOLITE UPTAKE AND RELEASE BY ISOLATED PROTOPLASTS

Symplastic metabolite transport in the infected zone should be reflected by the transport capabilities of the plasma membranes of infected and uninfected cells. Initial attempts to isolate plasma membranes from cells in the infected zone were unsuccessful because of the similar properties of the peribacteroid membranes. Therefore, we decided to isolate infected and uninfected protoplasts from the central zone.

Protocols for the preparation of infected protoplasts adopted from the literature yielded non-spherical and osmotically inactive material. We were able to show that conventional isolation procedures produce non-intact protoplasts (Peiter et al., 2003). Consequently, a new isolation and separation method was developed, based on the dissection of the nodule prior to cell-wall digestion. Digestion occurred in a hypertonic medium without shaking the fragile cells. Protoplasts were allowed to be released into a slightly hypotonic medium and infected and uninfected protoplasts were separated by isopycnic density gradient centrifugation. Protoplasts obtained with this method were spherical, osmotically active and excluded propidium iodide. It was shown that the widely accepted viability test of fluorescein diacetate staining is not appropriate for infected cells because bacteroids also stain in ruptured cells (Peiter et al., 2003).

Both infected and uninfected protoplasts actively released protons (Peiter and Schubert, 2003). In both types of protoplasts this process was inhibited by vanadate, a specific inhibitor of the P-type ATPase indicating the involvement of the plasmamembrane H^+-ATPase (Yan et al., 1998, 2002) Interestingly, only proton release by uninfected protoplasts was stimulated by the fungal toxin fusicoccin (FC), which stabilizes binding of 14-3-3-protein to the carboxy terminus and thus stimulates the ion pump (Maudoux et al., 2000). The absence of stimulation of proton pumping by FC in infected protoplasts hints at differences in the plasma membranes of the two cell types (Fig. 4). However in both cell types, secondary active transport of ions and/or metabolites driven by the electro-chemical proton gradient seems feasible.

Clear differences between infected and uninfected protoplasts were also observed for glucose uptake (Peiter and Schubert, 2003). Only uninfected protoplasts showed high-affinity glucose uptake, apparently energized by H^+ co-transport (Fig. 4). This transport was not competitively inhibited by fructose or sucrose. Also, biphasic kinetics were found for sucrose uptake by uninfected cells. This was inhibited by glucose and apparently energized by

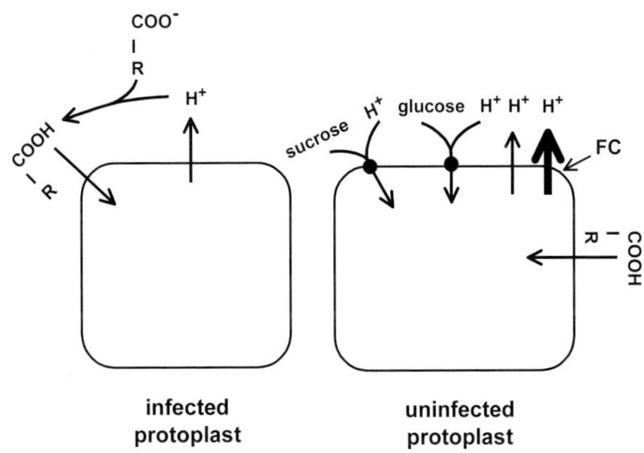

Fig. 4. Plasma-membrane transport-capabilities of infected (right) and uninfected (left) protoplasts isolated from the central tissue of Vicia faba nodules. Both types of protoplasts release protons, but only in uninfected protoplasts this process is stimulated by the fungal toxin fusicoccin (FC). Active uptake of glucose and sucrose driven by the electro-chemical proton gradient only occurs in uninfected protoplasts. Retrieval of organic acids is possible in both types of protoplasts.

H^+ co-transport (Peiter and Schubert, 2003). In contrast to the specific transport systems in uninfected protoplasts, glucose or sucrose uptake against a concentration gradient was not found for infected protoplasts.

4. CONCLUSIONS

Microstructural evidence in the indeterminate faba bean nodule suggests that strong apoplastic barriers in radial and transversal cell walls of nodule and vascular endodermis cells prevent apoplastic flow of metabolites from vascular bundles to the infected zone and vice versa (Figs. 2, 3). Consequently, symplastic routes are responsible for metabolite transport, both for export of amino acids and amides and for the import of sucrose. Direct support for this hypothesis is provided by the high plasmodesmatal frequencies between all cells except for infected cells among themselves (Table 1). Symplastic transport of sucrose into the infected zone implies unselective import also of reduced nitrogen compound recycled from the shoot.

The role of uninfected cells in the infected zone of indeterminate nodules remains obscure. On the one hand, their efficient sugar uptake systems may be responsible for sugar retrieval lost into the apoplast of the central zone. This may be the reason for low apoplastic sucrose concentrations measured in indeterminate soybean nodules (Streeter, 1992). Frequent plasmodesmatal connections to infected cells make uninfected cells predestined to deliver assimilates to infected cells, possibly after degradation of sugars to organic acids which in the undissociated form may be retrieved from the apoplast by infected cells themselves (Fig. 4). Uninfected cells not only function in metabolite transport, but also in collection of amino acids released from infected cells into the apoplast (Peiter et al., 2004).

Unselective symplastic import of reduced nitrogen recycled of surplus reduced N from the host shoot is the prerequisite for the functioning of a feedback regulation system, which coordinates dinitrogen reduction by nitrogenase activity and plant demand for reduced N. Since, however, decreased demand for dinitrogen fixation due to stress-induced growth retardation or mineral N supply will also decrease oxygen consumption by respiration, a tight control of oxygen diffusion into the infected zone and oxygen buffering by leghemoglobin are essential. According to this model primary control of dinitrogen fixation is exerted by a feedback mechanism, whereas control of oxygen diffusion is a secondary control mechanism that prevents oxygen damage of the nitrogenase enzyme. This is in agreement with the conclusion that (despite the significance of the oxygen diffusion barrier for dinitrogen fixation) other factors than oxygen are responsible for the inhibiton of dinitrogen fixation of faba bean during drought (Guerin et al, 1990, Herdina and Silsbury, 1990). Whether or not an inadequate interplay of these two regulation systems is responsible for the premature senescence of nodules in inefficient symbiotic systems such as *Phaseolus vulgaris* (Becher et al., 1997) requires further investigations.

ACKNOWLEDGMENTS

The regulation of biological dinitrogen fixation first roused my interest when we noticed that various environmental stresses decreased nitrogenase activity without changing total N concentrations. Many colleagues contributed with their work to the idea that feedback control of biological dinitrogen fixation may be the decisive factor in determining the absolute amounts of N_2 reduced: Konrad Mengel, Elke Plies-Balzer, Erika Schubert, Rajid Serraj, and Tao Kong. Their findings initiated this project to investigate the structural and functional properties of faba bean nodules to understand the physiological background. I am particularly thankful to the Deutsche Forschungsgemeinschaft (DFG) for financial support and to my co-workers Mohamed Abd-Alla, Edgar Peiter, Feng Yan, and Christina Plachta who made this project a success.

REFERENCES

Abd-Alla, M. H., Koyro, H.-W., Yan, F., Schubert S. and Peiter, E. (2000). Functional structure of the indeterminate *Vicia faba* L. root nodule: Implications for metabolite transport. *J. Plant Physiol.*, 157, 335–343.

Becher, M., Schepl, U. and Schubert, S. (1997). N_2 Fixation during different physiological stages of *Phaseolus vulgaris* OAC Rico and its supernodulating mutant R32BS15: The role of assimilate supply to and export from nodules. *J. Plant Physiol.*, 150, 31–36.

Brown, S. M. and Walsh, K. B. (1996). Anatomy of the legume cortex: Species survey of suberisation and intercellular glycoprotein. *Austr. J. Plant Physiol.*, 23, 211–255.

Gallacher, A. E. and Sprent, J. I. (1978). The effect of different water regimes on growth and nodule development of greenhouse-grown *Vicia faba*. *J. Exp. Bot.*, 29, 413–423.

Giller, K. E. and Wilson, K. J. (Eds.). (1991). Nitrogen Fixation in Tropical Cropping Systems. Oxon, UK: CAB International.

Guerin, V., Trinchant, J.-C. and Regaud, J. (1990). Nitrogen fixation (C_2H_2 reduction) by broad bean (*Vicia faba* L.) nodules and bacteroids under water-restricted conditions. *Plant Physiol.*, 92, 595–601.

Hansen, A. P., Yoneyama, T., Kouchi, H. and Martin, P. (1993). Respiration and nitrogen fixation of hydroponically cultured *Phaseolus vulagris* L. I. Growth, mineral composition and effect of sink removal. *Planta*, 189, 538–545.

Hardarson, G. (1993). Methods for enhancing symbiotic nitrogen fixation. *Plant Soil*, 152, 1–17.

Hartmann, K., Peiter, E., Koch, K., Schubert, S. and Schreiber, L. (2002). Chemical composition and ultrastructure of broad bean (*Vicia faba* L.) nodule endodermis in comparison to the root endodermis. *Planta*, 215, 14–25.

Herdina, J.A. and Silsbury, J. H. (1990). Estimating nitrogenase activity of faba bean (*Vicia faba* L) by acetylene reduction (AR) assay. *Aust. J. Plant Physiol.*, 17, 489–502.

Hunt, S. and Layzell, D. B. (1993). Gas exchange of legume nodules and the regulation of nitrogenase activity. *Annu. Rev. Plant Physiol. Plant Mol. Biol.*, 44, 483–511.
Iannetta, P. P. M., De Lorenzo, C., James, E. K., Fernandez-Pascual, M., Sprent, J. I., Lucas, M. M., Witty, J. F., DeFelipe, M. R. and Minchin, F. R. (1993). Oxygen diffusion in lupin nodules. I. Visualization of diffusion barrier operation. *J. Exp. Bot.*, 44, 1461–1467.
Kennedy, C., Doetsch, N., Meletzus, D., Patriarca, E., Amar, M. and Iaccarino, M. (1994). Ammonium sensing in nitrogen fixing bacteria: Functions of the gln B and gln D gene products. *Plant Soil*, 161, 43–57.
LaRue, T. A. and Patterson, T. G. (1981). How much nitrogen do legume fix ? *Adv. Agron.*, 34, 15–38.
Maudoux, O., Batoko, H., Oecking, C., Gevaert, K., Vandekerckhove, J., Boutry, M. and Morsomme, P. (2000). A plant plasma membrane H^+-ATPase expressed in yeast is activated by phosphorylation at its penultimate residue and binding of 14-3-3 regulatory proteins in the absence of fusicoccin. *J. Biol. Chem.*, 275, 17762–17770.
Merbach, W. and Schilling, G. (1980). Wirksamkeit der symbiontischen N_2-Fixierung der Körnerleguminosen in Abhängigkeit von Rhizobienimpfung, Substrat, N-Düngung und ^{14}C- Saccharoselieferung. *Zentralblatt für Bakteriologie II. Abteilung*, 135, 99–118.
Minchin, F. R., Sheehy, J. E., Minguez, M. I. and Witty, J. F. (1985). Characterization of the resistance to oxygen diffusion in legume nodules. *Ann. Bot.*, 55, 53–60.
Oti-Boateng, C. and Silsbury, J. H. (1993). The effects of exogenous amino acid on acetylene reduction activity of *Vicia faba* L. cv. Fiord. *Ann. Bot.*, 71, 71–74.
Parsons, R., Stanforth, A., Raven, J. A. and Sprent, J. I. (1993). Nodule growth and activity may be regulated by a feedback mechanism involving phloem nitrogen. *Plant, Cell and Environ*, 16, 125–136.
Pate, J. S. (1976). Transport in symbiotic systems fixing nitrogen. In U. Lüttge and M. G. Pitman (Eds.), Encyclopedia of Plant Physiology, New Series, (Vol 2 Transport in Plants, pp. 278–303). Berlin: Springer.
Peiter, E., Imani, J. and Schubert, S. (2003). A novel procedure for gentle isolation and separation of intact infected and uninfected protoplasts from the central tissue of *Vicia faba* L. root nodules. *Plant, Cell and Environ.*, 26, 1117–1126.
Peiter, E. and Schubert, S. (2003). Sugar uptake and proton release by protoplasts from the infected zone of *Vicia faba* L. nodules: Evidence against apoplastic sugar supply of infected cells. *J. Exp. Bot.*, 54, 1691–1700.
Peiter, E., Yan, F. and Schubert, S. (2004). Amino acid export from infected cells of *Vicia faba* root nodules: Evidence for an apoplastic step in the infected zone. *Physiologia Plantarum*, 122, 107–114.
Plies-Balzer, E., Kong, T., Schubert, S. and Mengel, K. (1995). Effect of water stress on plant growth, nitrogenase activity and nitrogen economy of four different cultivars of *Vicia faba* L. *Eur. J. Agr.*, 4, 167–173.
Schubert, S. (1995). Nitrogen assimilation by legumes – processes and ecological limitations. *Fert. Res.*, 42, 99–107.
Schubert, E., Mengel, K. and Schubert, S. (1990). Soil pH and calcium effect on nitrogen fixation and growth of broad bean. *Agron. J.*, 82, 969–972.
Schulze, J. (2003). Source-sink manipulations suggest an N-feedback mechanism for the drop in N_2 fixation during pod-filling in pea and broad bean. *J. Plant Physiol.*, 160, 531–537.
Streeter, J. G. (1992). Analysis of apoplastic solutes in the cortex of soybean nodules. *Physiologia Plantarum*, 84, 584–592.
Streeter, J. G. (1993). Translocation – A key factor limiting the efficiency of nitrogen fixation in legume nodules. *Physiologia Plantarum*, 87, 616–623.
Whitehead, L. F. and Day, D. A. (1997). The peribacteroid membrane. *Physiologia Plantarum*, 100, 30–44.
Witty, J. F. and Minchin, F. R. (1998). Hydrogen measurements provide direct evidence for a variable physical barrier to gas diffusion in legume nodules. *J. Exp. Bot.*, 49, 1015–1020.
Yan, F., Feuerle, R., Schäffer, S., Fortmeier, H. and Schubert, S. (1998). Adaptation of active proton pumping and plasmalemma ATPase activity of corn roots to low root medium pH. *Plant Physiol.*, 117, 311–319.
Yan, F., Zhu, Y., Müller, C., Zörb, C. and Schubert, S. (2002). Adaptation of H^+-pumping and plasma membrane H^+-ATPase activity in proteoid roots of white lupin under phosphate deficiency. *Plant Physiol.*, 129, 50–63.

INDEX